SYNOPTIC EDDIES IN THE OCEAN

Synoptic Eddies in the Ocean

Edited by
V. M. KAMENKOVICH,
M. N. KOSHLYAKOV,

and

A. S. MONIN
P.P. Shirshov Institute of Oceanology
Academy of Sciences of the U.S.S.R.

Translated by V.M. Volosov

With a Preface by the late E.S.W. Simpson

D. Reidel Publishing Company
A MEMBER OF THE KLUWER ACADEMIC PUBLISHERS GROUP
Dordrecht / Boston / Lancaster / Tokyo

Library of Congress Cataloging-in-Publication Data

CIP

Kamenkovich, V. M. (Vladimir Moiseevich)
Synoptic eddies in the ocean.

(Environmental fluid mechanics)
Translation of: Sinopticheskie vikhri v okeane.
Bibliography: p.
Includes index.
1. Ocean mixing. 2. Eddies. I. Koshlîâkov, M. N. (Mikhail Nikolaevich) II. Monin, A. S. (Andreĭ Sergeevich), 1921– . III. Title. IV. Series.
GC299.K3613 1986 551.47′01 85–23249
ISBN 90–277–1925–X (U.S.)

Published by D. Reidel Publishing Company
P.O. Box 17, 3300 AA Dordrecht, Holland

Sold and distributed in the U.S.A and Canada
by Kluwer Academic Publishers,
190 Old Derby Street, Hingham, MA 02043, U.S.A.

In all other countries, sold and distributed
by Kluwer Academic Publishers Group,
P.O. Box 322, 3300 AH Dordrecht, Holland

Originally published in 1982 in Russian by Gidrometeoizdat under the title
СИНОПТИЧЕСКИЕ ВИХРИ В ОКЕАНЕ
This edition is an expanded edition of the Russian original

Table of Contents

Preface by the President of SCOR

Not long ago the activities of SCOR* Working Group 34 led to the publication of the book *Eddies in Marine Science* edited by the Chairman, Professor A. R. Robinson. It was intended to provide an overview of present knowledge on mesoscale eddies of the ocean and their influence in other fields of marine science, and to be of interest and value to a wide range of marine scientists. However, it was recognized that the rapidly expanding knowledge of mesoscale eddies and the development of the underpinning hydrodynamics would mean that a full and complete account of this most important field could not be achieved in one book. Accordingly, SCOR invited Professor A. S. Monin to head the preparation of a new book, the first of its kind, devoted to the dynamics of eddies in the ocean. This book is now presented to the reader.

The first comprehensive survey of several eddies in the ocean by direct measurement was accomplished by the Soviet expedition POLYGON-70 in which six months of continuous current measurements were made at a network of seventeen moorings in the tropical North Atlantic. This experiment revealed the basic parameters of free ocean eddies and indicated their characterization in terms of Rossby wave dynamics with baroclinic instability of the large-scale current as an eddy-generating mechanism. Further successful field studies culminated in the international POLYMODE experiment during 1977–1978 in which year-long measurements of currents have made it possible to interpret open-ocean synoptic eddies as a complicated synthesis of Rossby waves and large-scale quasigeostrophic turbulence, and have yielded particularly rich material for the verification and improvement of the theoretical models of ocean eddies.

The first chapter of the present book is introductory in character and will be very useful to physicists and mathematicians who may wish to familiarize themselves with one of the most important problems of oceanography. The second and third chapters successively present the theory of ocean eddies. A particularly detailed description is given of the most contemporary results, including the theories of baroclinic instability of large-scale currents and large-scale oceanic turbulence, numerical models of eddies, and the theory of Rossby solitary waves. The fourth and fifth chapters contain several results of actual studies of eddies which give an especially vivid illustration of their physical properties. It has appeared more convenient to consider separately eddies of the western boundary currents and

*SCOR: Scientific Committee on Oceanic Research.

those of the open ocean. In the fifth chapter particular emphasis is laid upon the results of the special regional experiments mentioned above. The experimental material in the fourth and fifth chapters is considered both in its own right and in the light of theoretical considerations. Finally, the sixth chapter is devoted to practical applications of eddy science.

The international theoretical and experimental studies presented in this book are devoted to one of the most important problems of contemporary oceanography. These studies provide a good example of work requiring international coordination, which is a responsibility of the Scientific Committee on Oceanic Research. It therefore gives me particular pleasure to accept the invitation from Professor Monin to contribute this Preface.

President, SCOR

E. S. W. SIMPSON

Preface by the Authors to the English Edition

The great interest shown by oceanographers and scientists of many other specialities in synoptic ocean eddies can primarily be accounted for by the simple but very significant fact that among various oceanic phenomena it is mainly eddies that determine the 'oceanic weather' – that is, the instantaneous distributions of current velocities, temperature, salinity, speed of sound, and other oceanographic characteristics. Eddies also seem to play an important part in the formation of the ocean climate, i.e. the average distributions of oceanographic characteristics and their long-period variability. The time scales of synoptic eddies range from weeks to months; their horizontal scales vary from tens of kilometers to the low hundreds of kilometers, and their vertical scales are of the order of a kilometer. The velocities of translatory motion of synoptic eddies are of the order of several kilometers a day, whereas the velocities of water motion in the eddies are much greater than those of mean currents. Observation data demonstrate a great variety of types of ocean eddies. They can be crudely classified as *eddies (rings) of western boundary currents* which have been known for several decades, and *eddies in the open ocean* discovered in the 1960–1970s.

The discovery of synoptic eddies in the open ocean was a great event in post-war oceanography. As far back as the 1930s the existence of strong synoptic inhomogeneities in seas and oceans was anticipated by V. B. Shtokman. In 1935 he conducted a series of long-term current measurements in the Caspian Sea which were continued in the post-war period in the Black Sea (1956) and the North Atlantic (1958). An important step was made by the British oceanographer J. C. Swallow, who discovered strong nonstationary currents at great depths in regions west of Portugal (1958) and near Bermuda (1959–1960). The existence of strong synoptic disturbances was also confirmed by the results of processing the data of current and temperature measurements performed by American oceanographers in the Bermuda region in 1954–1969 and north of the Gulf Stream in 1965–1967.

The first specialized (two-month) experiment intended for studying the spatial structure of synoptic inhomogeneities in the ocean was proposed by V. B. Shtokman and was carried out by the P. P. Shirshov Institute of Oceanology of the USSR Academy of Sciences in 1967 in the Arabian Sea ('Polygon-67'). The results of processing the hydrographic observation data by the dynamic method made it possible to chart synoptic eddies. The second specialized experiment, which made a decisive contribution to the study of the synoptic variability of the ocean, was the Soviet six-month expedition 'Polygon-70' in the tropical zone of the North Atlantic.

The data of direct current measurements in this expedition were, for the first time, used to construct charts of synoptic currents which proved the existence of synoptic eddies in the open ocean and revealed their basic properties. In 1973 American scientists performed an analogous three-month experiment (MODE) in the Sargasso Sea which confirmed the discovery made by Soviet oceanographers. Finally, in 1977–1979 the grandiose Soviet–American experiment POLYMODE was performed. It revealed some new interesting specific properties of the structure and dynamics of ocean eddies.

At present the problem of synoptic eddies has a central role in oceanography and, therefore, an acute need is felt for monographs presenting both the basic experimental results and the modern theoretical concepts of generation and evolution of synoptic eddies in the ocean. This book attempts to fulfill this need.

The material of the book is clear from the table of contents where the authors of different chapters and sections are indicated. The sections were discussed by the authors to give the reader a consistent presentation of the modern state of this branch of knowledge based on a unified approach to the phenomena under study. The final editing of the text in view of this objective was performed by A. S. Monin.

We are indebted to many of our colleagues, mainly in the P. P. Shirshov Institute of Oceanology of the USSR Academy of Sciences, for valuable discussion and help in the preparation of the manuscript. We are particularly grateful to A. L. Berestov, V. M. Kurtepov, G. M. Reznik, A. M. Chernyakova, and Yu. A. Shishkov for their participation in writing some sections of the book. In comparison with the first 1982 Russian edition, the text of the book has been completely revised and has grown almost twice its size in content owing to the inclusion of the latest results obtained in recent years and also to a more detailed presentation of some of the problems.

V. M. KAMENKOVICH
M. N. KOSHLYAKOV
A. S. MONIN

CHAPTER 1

Stratification and Circulation of the Ocean

1. OCEANIC PROCESSES WITH DIFFERENT TEMPORAL AND SPATIAL SCALES

Various physical processes in the ocean (many of which are caused by atmospheric factors) lead to *inhomogeneities* in the distribution of properties or characteristics of sea water. Among such properties and characteristics are the space occupied by sea water in the gravity field (sea level), its phase state (ice cover), the basic thermodynamic characteristics reflecting the state of the water (pressure, temperature, and salinity), derived thermodynamic characteristics (density, electrical conductivity, speed of sound, refractive index, and entropy), the concentration of dissolved gases, bubbles, and organic and mineral suspended matter. Finally, they include characteristics of motion (velocity components and the sea surface level).

Inhomogeneities created by different processes can have different spatial scales, L, ranging from minimum values (for which the inhomogeneities can be preserved for some time despite the smoothing effect of molecular viscosity, heat conduction, and diffusion) of the order of fractions of a millimeter, to maximum values (i.e. the dimensions of the entire ocean) of the order of 10^4 km.

Small-scale inhomogeneities (with scales from fractions of a millimeter to tens of meters and sometimes even to hundreds of meters) are characteristic of (1) quasi-isotropic small-scale turbulence producing vertical mixing (with scales from fractions of a millimeter to a meter or sometimes even to tens of meters); (2) a vertical-layered microstructure (with vertical scales from several millimeters to tens of meters); (3) acoustic waves (whose wavelengths range from a centimeter to hundreds of meters for frequencies from 10^5 to 1 Hz, which are the most important for the ocean); (4) capillary waves (from millimeters to centimeters) and surface gravitational waves (from centimeters to hundreds of meters); (5) internal gravitational waves (from decimeters to kilometers).

Mesoscale inhomogeneities (with scales of hundreds of meters or several kilometers) are characteristic of (5) internal waves, (6) inertial oscillations (whose horizontal coherence scale is of the order of a few kilometers or sometimes even in the low tens of kilometers), and (7) tidal oscillations in shallow waters (whereas in the deep ocean tides are characterized by the scales of the ocean as a whole).

Synoptic inhomogeneities (with scales of tens or low hundreds of kilometers) are characteristic of (8) frontal and free oceanic eddies or Rossby waves having horizontal scales of the order of the Rossby deformation radius:

$$L_R = \frac{\bar{N}}{f} H, \tag{1.1}$$

where $\bar{N}$ is the depth-averaged Brunt–Väisälä frequency, f is the inertial frequency (the Coriolis parameter), and H is the depth of the ocean; the typical L_R value is 50 km.

It should be noted that for some oceanological fields, which characterize near-surface conditions in the ocean and respond immediately to atmospheric factors, e.g. wind waves, drift currents, and sea level, we also observe (9) forced inhomogeneities with scales of barotropic synoptic processes in the atmosphere which are of the order of $L_a = \sqrt{gH/f}$, where g is the acceleration of gravity and H is the effective thickness of the atmosphere – that is, as a rule, one or one and a half orders of magnitude greater than L_R.

In this book, by complete analogy with the atmosphere (see the comparison of atmospheric and oceanic kinetic energy spectra in Figure 1.1.1), we shall use the term *synoptic eddies*, rather than 'mesoscale eddies', for baroclinic quasigeostrophic eddies or Rossby waves in the ocean having horizontal scales of the order of the Rossby deformation radius, which are likely to be formed primarily as a result of the baroclinic instability of large-scale currents and are responsible for most of the kinetic energy spectrum. We shall retain the term *mesoscale eddies* for inhomogeneities with frequencies between the inertial frequency f and the Brunt–Väisälä frequency N and horizontal scales between $L_f = (\varepsilon/f^3)^{1/2}$ and $L_N = (\varepsilon/N^3)^{1/2}$, where ε is the rate of dissipation of kinetic energy. These scales L are of the order of the effective thickness of the ocean or the atmosphere. (Kinetic energy spectra for the atmosphere are minimal in this scale region which separates the synoptic and small-scale energy-containing regions B and D. The mesoscale region of the ocean is probably overlapped by the energy spectrum of the longest period internal waves and inertial oscillations; see spectrum C in Figure 1.1.1, obtained in the IWEX experiment on measuring internal waves in the deep ocean.)

Global inhomogeneities (with scales of thousands or tens of thousand kilometers) are characteristic of processes extending to the oceans as a whole. In particular, they include (10) seasonal variations, (11) the major (quasistationary) oceanic currents, and (12) effects of latitudinal zonality of the climate.

The global and synoptic inhomogeneities of thermo- and hydrodynamic fields describing the states of the ocean can be called large-scale components of the state of the ocean. The general circulation of the ocean can then be defined as the *statistical ensemble of large-scale components of its states*. It should be noted that in this definition the general circulation does not include tidal oscillations (although energetically they can form a notable fraction of the oceanic water motion). Similarly, diurnal oscillations are not included, as a rule, in the notion of the general circulation of the atmosphere (whereas they are contained in the more general notion of climate).

All the enumerated spatial inhomogeneities have definite 'lifetimes' τ, i.e. typical times of the processes generating them. For example, small-scale inhomogeneities are mainly characterized by periods from fractions of a second to tens of minutes. These are (1) small-scale turbulences with periods from 10^{-3} to 10^2 s; (2) vertical

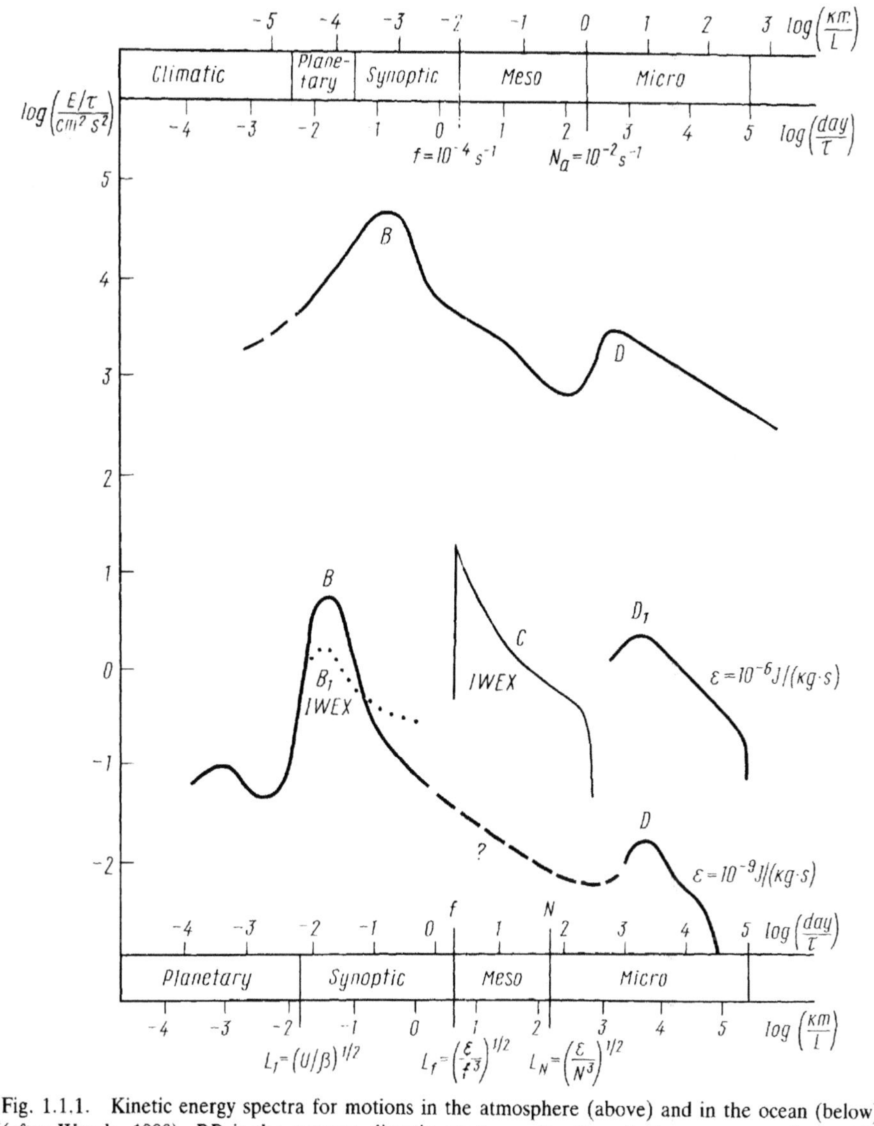

Fig. 1.1.1. Kinetic energy spectra for motions in the atmosphere (above) and in the ocean (below) (after Woods, 1980). *BD* is the average climatic spectrum; B_1, C, and D_1 are spectra obtained in individual expeditions.

microstructures having much longer 'lifetimes' (probably from several minutes to at least tens of hours); (3) acoustic waves with periods from 10^{-5} to 1 s; (4) capillary waves (with periods of the order of 10^{-2} to 10^{-1} s) and surface gravitational waves (primarily with periods from several seconds to a few tens of seconds); and (5) internal waves whose periods range from tens of seconds to the inertial period $2\pi/f$, i.e. at least to many hours.

Mesoscale inhomogeneities have typical periods from hours to a few days. Namely, (6) inertial oscillations have periods around $2\pi/f$, varying from half a day at the poles to a day at latitudes ±30° and increasing further towards the equator, and (7) tidal oscillations have tidal periods of

$$\left(\sum_{i=1}^{6} \frac{n_i}{\tau_i}\right)^{-1}$$

where $n_i = 0, \pm 1, \pm 2, \ldots$; $\tau_1 = 24$ h 50.47 min (a lunar day); $\tau_2 = 27.321582$ days (a tropical month); τ_3 is equal to one year (for $n_1 = n_2 = -n_3 = 1$ this is the diurnal period); and τ_4, τ_5 and τ_6 are longer periods in the Sun–Earth–Moon system. The principal tides have periods equal to half a lunar day, half a solar day, and lunar and solar days. This range of periods also includes thermally induced diurnal fluctuations caused by diurnal variations of insolation.

Synoptic inhomogeneities are characterized by periods from days to months: (8) oceanic eddies or Rossby waves have periods from weeks to months; e.g. according to theory (see Chapter 2, Section 2.1), the typical time scale of a first-mode zonal baroclinic Rossby wave is $\tau = 2(\beta L_R)^{-1}$, where β is the meridional derivative of the Coriolis parameter; for $L_R = 50$ km and $\beta = 2 \times 10^{-8}$ km^{-1} s^{-1} it is $\tau = 2 \times 10^6$ s = 23 days. Further, (9) atmospheric synoptic processes have periods of the order of a few days.

Formation times typical of global inhomogeneities in the ocean are likely to range from years to hundreds of years. For instance, (10) seasonal variations naturally have a 12-month period; (11) the major oceanic currents in the upper ocean are formed (probably as a result of the action of the wind) over periods of several years. The weather feedback in the atmosphere can generate a year-to-year variability of the upper ocean–atmosphere–land system. Further, (12) the vertical stratification of the ocean reflecting the latitudinal zonality on its surface (primarily, the temperature difference in the upper ocean between the equator and polar regions) is probably formed (mainly by slow thermohaline circulations) over periods of the order of hundreds of years. This process can be controlled by the feedback with the states of the atmosphere (which is rapidly adapted to the state of the upper ocean) and the land, and this control can create secular and century-to-century variability of the climate.

The enumerated regions of spatial and temporal scales of various processes in the ocean are shown schematically in Figure 1.1.2. The region of the most probable scales is shaded in the figure. It would be desirable to indicate the distribution of the space–time spectral density of oscillation energy in this region, e.g. of the kinetic energy of oceanic motions. The arrows at the top mark the intervals of spatial scales in which the influx of kinetic energy to the ocean (from the atmosphere and owing to tide-generating forces) occurs (according to Ozmidov, 1965). This is, first, the large-scale region from hundreds to tens of thousands of kilometers where the major oceanic currents and synoptic eddies are generated (the kinetic energy influx per unit mass is probably of the order of $\varepsilon_1 \sim 10^{-9}$ J/kg·s) so that for an ocean depth of the order of 5 km the kinetic energy influx to the ocean across its surface is of the order of $F_1 \sim 5 \times 10^{-3}$ J/m^2·s and the effective horizontal mixing coefficient is of the order of $k_1 \sim 10^4$ m^2/s. Second, this is the mesoscale region from kilometers to tens of kilometers where inertial and tidal oscillations are generated

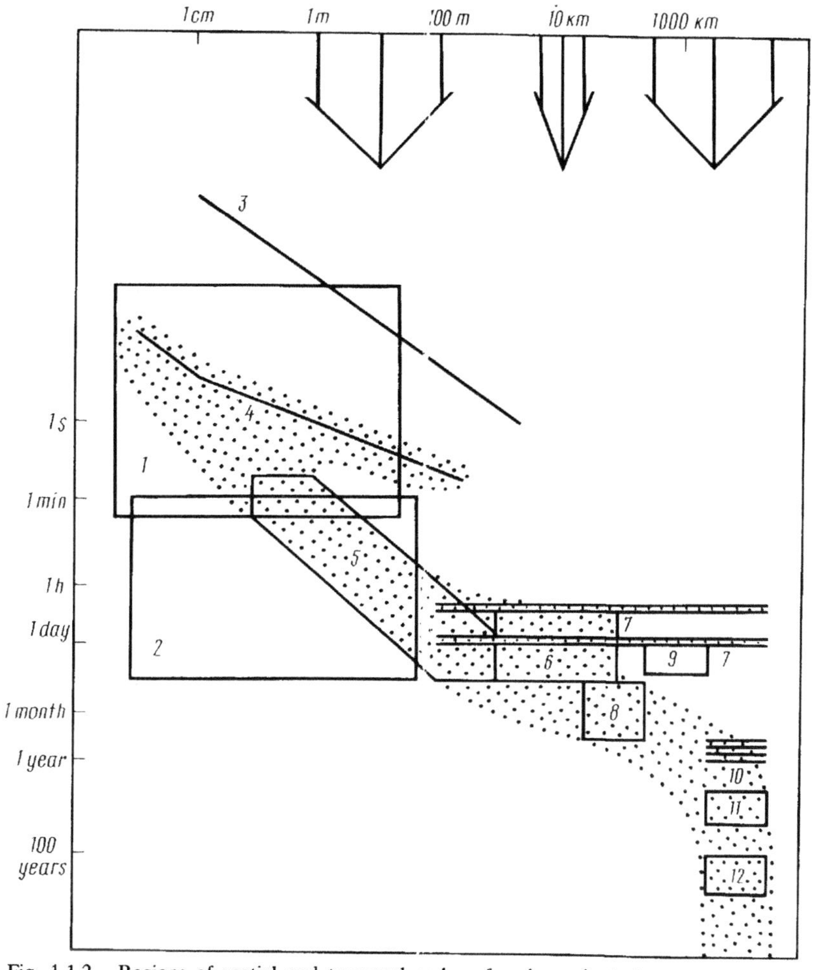

Fig. 1.1.2. Regions of spatial and temporal scales of various physical processes in the ocean. 1: small-scale turbulence; 2: vertical microstructure; 3: acoustic waves; 4: capillary and surface gravitational waves; 5: internal waves; 6: inertial oscillations; 7: tidal oscillations; 8: oceanic eddies and Rossby waves; 9: atmospheric synoptic processes; 10: seasonal variations; 11: major oceanic currents; 12: stratification of the ocean.

(the energy influx rate is $\varepsilon_2 \sim 10^{-7}$ J/kg·s, the energy influx to the ocean is $F_2 \sim 5 \times 10^{-1}$ J/m^2·s, and the horizontal mixing coefficient is $k_2 \sim 10^{-1}$ m^2/s). Third, this is the small-scale region from meters to tens of meters where gravitational waves are generated (the energy influx rate in the upper hundred meter layer of the ocean is $\varepsilon \sim 10^{-5}$ J/kg·s, the energy influx to the ocean is $F_3 \sim 1$ J/m^2·s, and the mixing coefficient is $k_3 \sim 10^{-3}$ m^2/s).

Accordingly, maximum values of the space–time energy spectral density should be expected in the regions of gravitational waves (4 and 5), inertial and tidal oscillations (6 and 7), synoptic eddies (8), and global motions (11 and 12). Integrating the space–time spectrum over all the spatial wavelengths (along the

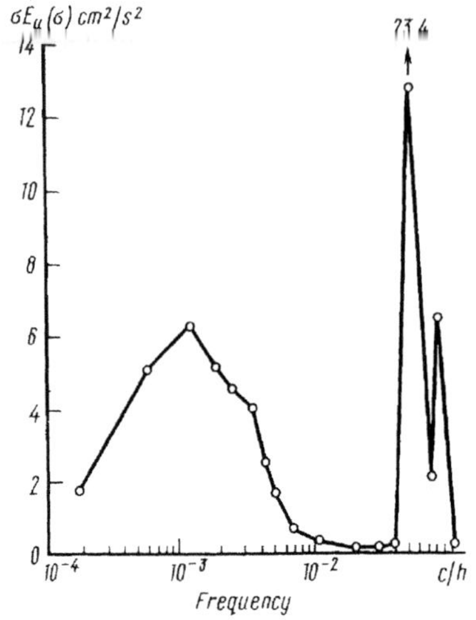

Fig. 1.1.3. Spectrum of oscillations of the zonal component of the current velocity at a depth of 500 m at station 'D' in the West Atlantic according to the data of three-year measurements (after Thompson, 1971).

horizontal lines in Figure 1.1.2), we obtain a time spectrum describing the energy distribution over the oscillation periods or frequencies. A typical example of such a spectrum (namely, $\sigma E_u(\sigma)$ cm^2/s^2, where σ is the frequency and E_u (σ) is the oscillation spectral density of the zonal component u of the current velocity) in the frequency range $10^{-4} < \sigma < 10^{-1}$ c/h (to which the period range 400 days $< \tau <$ 10 h corresponds) is presented in Figure 1.1.3. The figure clearly demonstrates a range of synoptic oscillations with periods from 8 to 200 days with a maximum of the spectrum $\sigma E_u(\sigma)$ near the period of 30 days, an almost complete absence of oscillation energy in the range of periods from 6 days to approximately the inertial period, a very high and narrow inertial maximum, and a semidiurnal tidal maximum four times as small as the former.

2. STRATIFICATION OF THE OCEAN

The 'stratification' of the ocean refers to its density separation into layers in the gravity field, which is possible owing to the compressibility of sea water, i.e. the dependence of the density ϱ on temperature T, salinity S, and pressure p. This dependence is described by the empirical formula

$$\varrho\,(T, S, p) = \varrho_0\,(1 + 10^{-3}\,\sigma_t)\left[1 - \frac{p - p_a}{K^0 + A\,(p - p_a) + B\,(p - p_a)^2}\right]^{-1}, \qquad (2.1)$$

Here σ_t, K^0, A, and B are functions of T and S (representable as low-degree polynomials of T °C and $S^{1/2}$); $\varrho_0 = \varrho(4\,°\mathrm{C}, 0, p_a)$, where p_a = 10.13 × 10^4 N/m^2 is the standard atmospheric pressure, and hence $\varrho_0\,(1 + 10^{-3}\,\sigma_t)$ is the density reduced to atmospheric pressure for constant T and S. It is convenient to measure the density in the units $\sigma_t = 10^3\ (\varrho(T, S, p_a)/\varrho_0 - 1)$. As p or S increases or as T decreases (down to a certain temperature T_1 of maximum density), the density of the water increases. Therefore, when there is cooling or salinization at the sea surface, the surface water sinks. This creates the so-called thermohaline circulation in the ocean and forms its stratification such that, generally, the temperature decreases with increasing depth down to values close to the minimum winter water temperatures in the coldest regions on the sea surface, and the salinity increases with depth.

It should be noted that for $p = p_a$ and $S = 0$ the temperature of maximum density is $T_1 \approx 4$ °C and the freezing temperature is $T_2 \approx 0$ °C, and T_1 and T_2 decrease with increasing p or S, with T_1 decreasing faster than T_2. Therefore, for not very large p (< 270 atm) and S (< 24‰) there is a temperature interval $T_1 < T < T_2$ where the density dependence on temperature is of opposite character. We also note that ice is lighter than water, and therefore it floats on the surface. If water were a normal liquid compressed on freezing, ice would sink and eventually fill vast regions in the ocean.

To estimate the effects of an increase (decrease) in the surface water density when cooling (heating) or salinization (desalinization) takes place, we can make use of the vertical mass flux at the sea surface determined by the formula (Monin, 1970)

$$M = (P - E)\,S - \frac{\alpha}{c}(\mathscr{L}E + Q) + \alpha P\,(T_p - T_w), \tag{2.2}$$

where $M > 0$ when the mass flux is upward (i.e. increases the buoyancy). Here P and E are the precipitation and evaporation rates, $\alpha \approx 2 \times 10^{-4}$ (°C)$^{-1}$ is the thermal expansion coefficient of the water, c is the heat capacity, $\mathscr{L}$ is the latent heat of vaporization, Q is the sum of the radiative and turbulent heat fluxes in the surface air layer (which is positive when the flux Q is upward), and T_p and T_w are the temperatures of the precipitation and the water surface. The first term in (2.2) describes the salinization and desalinization effects; the second and third terms describe the effects of cooling and heating (since $S \approx 0.03$ and $\alpha\mathscr{L}/c \approx 0.12$, for the evaporation the effect of cooling is four times that of salinization). The effects at the water surface of ice freezing and melting and of river run-off are not taken into consideration here. The annual average values of M turn out to be of the order of 10^2 kg/m^2·yr (which corresponds to the generation rate of kinetic energy in the thermohaline circulation per unit mass $gM/\varrho \sim 3 \times 10^{-10}$ J/kg · s.

Thus, the generation of the thermohaline circulation that produces stratification in the ocean is determined by the heat budget of its surface (mainly by $\mathscr{L}E + Q$; according to existing estimates, the contributions to the budget from evaporation, effective radiation, and turbulent heat exchange with the atmosphere are, on average, in the ratio 51:42:7 although the fraction apportioned to evaporation is probably underestimated here) and by the water budget (primarily by $P - E$; according to existing estimates, $P \approx 4.12 \times 10^{17}$ kg/yr and $E \approx 4.53 \times 10^{17}$ kg/yr for the ocean as a whole, and the difference $E - P \approx 0.41 \times 10^{17}$ yr^{-1} is compensated for by river run-off). The annual heat and water budgets of the ocean calculated by Stepanov (1974) are plotted on the charts in Figures 1.2.1 and 1.2.2, and their zonal average values are demonstrated by curves 1 and 4 in Figure 1.2.3. The charts show that the heat and, particularly, water budgets possess latitudinal zonality (which is disturbed in the heat budget by the Gulf Stream and Kuroshio regions). The heat budget is positive (the ocean is heated) in the tropical zone

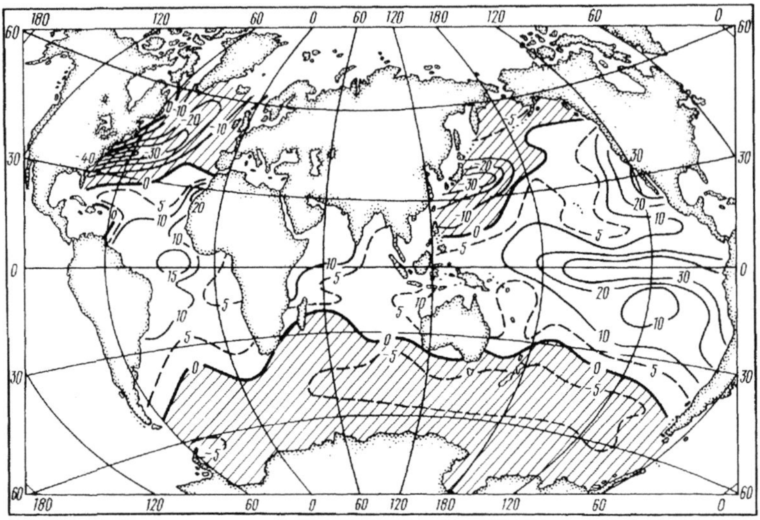

Fig. 1.2.1. Annual heat budget of the ocean in 10^8 J/m^2·yr (after Stepanov, 1974). The shaded parts indicate the regions of negative budget where the ocean is cooled.

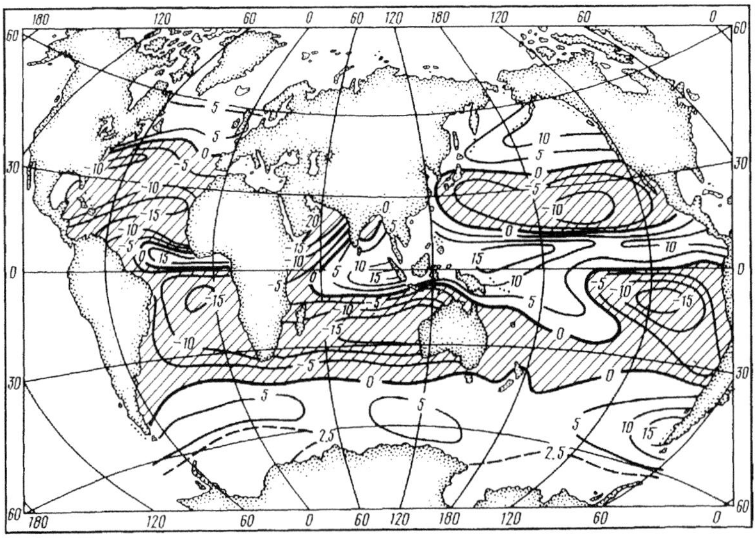

Fig. 1.2.2. Annual water exchange between the ocean and the atmosphere in 10^2 kg/m^2·yr (after Stepanov, 1974). Shaded in the figure are the regions of negative water exchange where the surface waters become more saline.

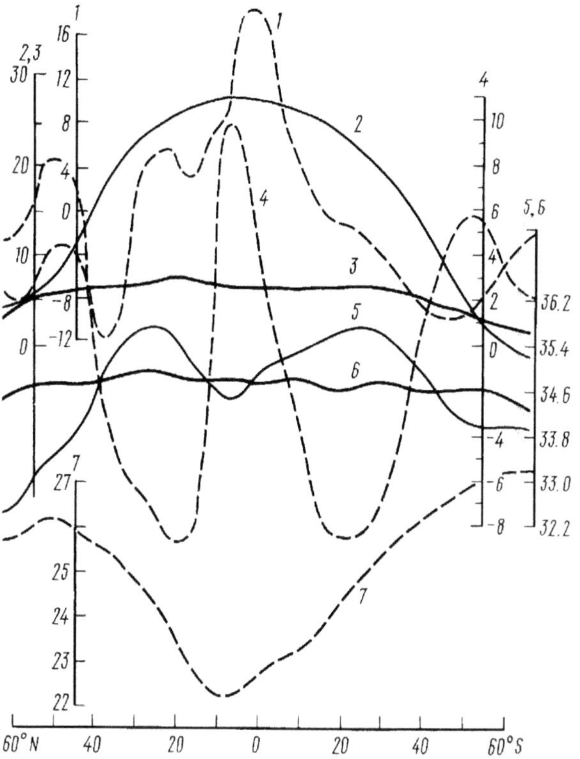

Fig. 1.2.3. Zonal climate of the ocean (after Stepanov, 1974). 1: the annual heat budget, 10^8 J/m^2·yr; 2: the temperature of the surface layer of the ocean, T_w °C; 3: the average temperature over the depth of the ocean, T_{av} °C; 4: the annual water exchange of the ocean with the atmosphere, 10^2 kg/m^2·yr; 5: the salinity at the sea surface, S_0 ‰; 6: the average salinity over the depth of the ocean, S_{av} ‰; 7: the density anomaly σ_t at the sea surface.

between 30°N and 15°S and negative (the ocean is cooled) outside this zone. The maximum positive budget (up to 34–42 × 10^8 J/m^2·yr) is observed in the equatorial zone of the Pacific Ocean, and the maximum negative budget (31–42 × 10^8 J/m^2·yr) is observed in the Gulf Stream and Kuroshio regions. The moisture exchange is positive (precipitation exceeds evaporation) in the equatorial zone between 10°N and 5°S and also in regions north and south of latitudes ±40° and is negative (evaporation exceeds precipitation) in the tropical and subtropical regions. The maximum positive moisture exchange (up to 1.5–2.0 × 10^3 kg/m^2·yr) is observed in the western part of the equatorial zone of the Pacific Ocean, and the maximum negative moisture exchange (1.5 × 10^3 kg/m^2·yr) is observed in the subtropics, particularly in the Atlantic.

The annual average values of vertical mass flux M were calculated by Agafonova *et al.* (1972), and are plotted on the chart in Figure 1.2.4. Positive fluxes are observed in the equatorial zone and also at eastern coasts of the Pacific Ocean, the maxima attaining 150 kg/m^2·yr. Negative fluxes are observed from tropical regions to middle latitudes; they have maxima (up to 200 kg/m^2·yr) in the Gulf Stream and Kuroshio regions (and also possibly in the Antarctic and the Arctic). It should be

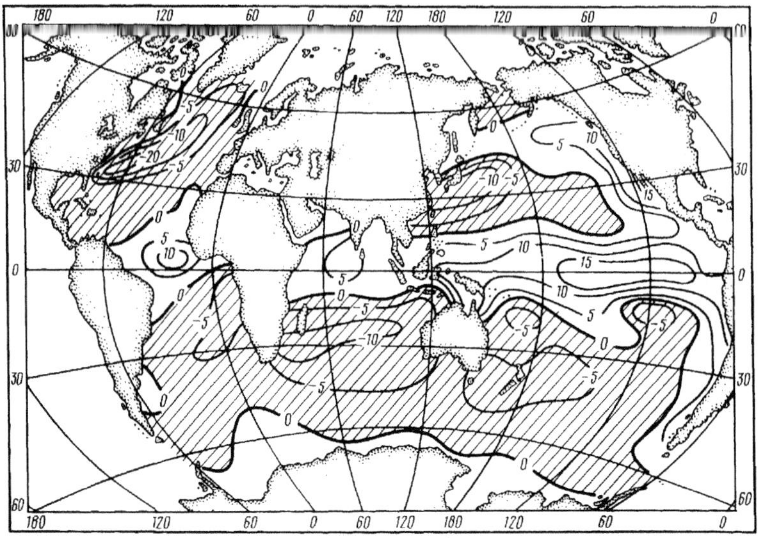

Fig. 1.2.4. Vertical mass flux at the sea surface in 10 kg/m^2·yr (after Agafonova *et al.*, 1972). The shaded parts are the regions of negative flux, i.e. of the sinking of waters.

noted that the charts for $\mathcal{L}E + Q$, $P - E$, and, particularly, M must possess a strong seasonal variability, and the actual regions of generation of the thermohaline circulation (with maximum negative values of M) should be sought in winter charts (however, no seasonal M-charts have yet been prepared).

The annual average temperature field T_w of the surface layer in the ocean is approximately zonal (however, the isotherms slightly converge at the western coasts of the oceans and create higher latitudinal temperature gradients, and diverge at the eastern coasts where cold waters are driven out of high latitudes and the isotherms are bent towards the equator). Therefore the zonal average values shown by curve 2 in Figure 1.2.3 provide a good representation of the field. The average temperature in the upper surface layer of the ocean is equal to 17.82 °C and exceeds the average air temperature at the Earth's surface by 3.6 °C. Hence, according to this characteristic, the ocean is a warmer shell than the atmosphere. (Below it will be indicated that this relates not only to the surfaces but also to the depths of these shells.) For our further aims it will sometimes be advisable to subdivide the ocean into four parts: (1) the Pacific Ocean (52.8% of the mass and 49.8% of the area of the ocean; the average temperature of its surface layer is 19.37 °C); (2) the Atlantic Ocean (24.7% of the mass and 25.9% of the area; the average temperature of its surface layer is 17.58 °C); (3) the Indian Ocean (21.3% of the mass and 20.7% of the area; the average temperature of its surface layer is 17.85 °C); and (4) the Arctic Ocean (1.2% of the mass and 3.6% of the area; the average temperature of its surface layer is about −0.75 °C). Here the seas are also included in the oceans; they account for a total of 3% of the mass and 10% of the area of the ocean.

It should be noted that the average temperature of the surface layer of the ocean in the Northern Hemisphere is approximately 3 °C higher than in the Southern

Hemisphere. The temperature of the surface water layer in the tropical zone (one-third of the ocean area) exceeds 25 °C and attains a maximum of 27.4 °C somewhat north of the equator, and in the middle latitudes it rapidly decreases towards the poles and passes through zero in the zones 60–65°S and 70–75°N. We note that the details of the latitudinal variations of T_w are not in one-to-one correspondence with the local values of heat budget. The minima of the budget at 32–40°N and 42–52°S and, more particularly, its maxima at 42–52°N and >55°S, are not marked in the field T_w. This violation of the relation between T_w and the local heat budget of the sea surface is probably created by warm and cold oceanic currents. We also note that the field T_w undergoes small seasonal oscillations with minimum amplitudes in the equatorial zone somewhat north of the equator (around 1 °C) and maximum amplitudes in the subtropics at 40–45°S (around 9 °C) and 30°S (around 5.5 °C) whereas in polar regions these amplitudes decrease down to 2–3 °C.

The average temperature stratification of the Pacific, Atlantic, and Indian Oceans is presented in Table 1.2.1 (after Galerkin, 1976). The presence of the following factors is typical of this stratification: (1) the upper mixed layer (UML) where the temperature varies little with increasing depth (the UML is approximately 100 m thick in tropical regions, is 10–20 m thick in high latitudes in summer, and is hundreds of meters thick and sometimes even extends to the bottom in winter); (2) the seasonal thermocline, tens of meters thick, where the temperature sharply decreases with increasing depth (by several degrees); (3) the main thermocline, with a lower boundary approximately at a depth of 1500 m, where the temperature decreases smoothly and with deceleration and attains 10.3–11.2 °C at a depth of 300 m, 4.0–4.8 °C at 1000 m, and 2.7–3.5 °C at 1500 m; and (4) the deep layer where the temperature decreases very slowly with depth, reaching 1.0–1.5 °C at the bottom (from 2.5 °C in the north to −0.5 °C in the south in the Atlantic).

The average temperature over the depth of the ocean from the surface to 4000 m (excluding the Arctic Ocean) is T_{av} = 3.8 °C (3.7 °C in the Pacific Ocean, 4.2 °C in the Atlantic Ocean, and 3.8 °C in the Indian Ocean; the Northern Hemisphere is 2° warmer than the Southern Hemisphere). As it is 20.8 °C higher than the mass-averaged atmospheric temperature (−17.0 °C), the ocean as a whole is therefore much warmer than the atmosphere. The zonal average values T_{av} are represented by curve 3 in Figure 1.2.3. They are maximum at latitudes 25–15°N (where they are equal to 7.6–7.3 °C), exceed 6 °C in the zone 40°N–35°S, and decrease towards the poles outside this zone (down to 3.6 °C in latitudes 65–60°N and 2.3 °C in latitudes 65–60°S).

The temperature stratification in some specific regions of the oceans can somewhat differ from the average stratification presented in Table 1.2.1. Stepanov (1974) identified five types of temperature stratification of sea waters (polar, subantarctic, subarctic Atlantic and Pacific, and moderate-tropical) with several subtypes and published a chart of their geographical distribution. The most notable distinction from the average stratification is shown by polar waters in which, under a very thin summer heated layer, there is a layer of extremely cold subsurface water with a warmer layer below it, where the temperature gradually decreases to a depth of 1–2 km; and still deeper, an isothermality (around 0 °C) is observed.

TABLE 1.2.1

Stratification of the oceans

Depth, m	Temperature, T°C			Salinity, S°/oo			Density, σ_t			Speed of sound, m/s		Electrical conductivity, S/m (North Atlantic)
	Pacific Ocean	Atlantic Ocean	Indian Ocean	Pacific Ocean	Atlantic Ocean	Indian Ocean	Pacific Ocean	Atlantic Ocean	Indian Ocean	North Pacific	North Atlantic	
1	2	3	4	5	6	7	8	9	10	11	12	13
0	19.37	17.58	17.85	34.56	35.30	34.68	24.33	25.24	24.46	1524	1515	4.887
25	18.92	17.10	17.61		35.45	34.73	24.39	25.48	24.56	1521	1511	4.810
50	18.06	16.11	16.80	34.68	35.54	34.81	24.74	25.80	24.81	1519	1507	4.732
100	16.13	14.12	14.59	34.79	35.59	34.91	25.30	26.33	25.52	1512	1505	4.510
150	14.31	12.79	12.54	34.82	35.51		25.79	26.62	26.13	1506	1504	4.370
200	12.57	11.68	11.24	34.77	35.39	34.97	26.17	26.79	26.48	1501	1503	4.230
250	11.22	10.86	10.34	34.69	35.31	34.97	26.40	26.89	26.69	1498	1502	4.153
300	10.06	10.11	9.70	34.61	35.21	34.95	26.69	26.97	26.83	1494	1501	4.076
400	8.32	8.88	8.69	34.52	35.08	34.89	26.81	27.10		1490	1500	3.955
500	7.00	7.68	7.93	34.46	34.94	34.84	26.99	27.20	27.08	1487	1496	3.847
600	6.05	6.69	7.26	34.43	34.85	34.80	27.10	27.29	27.16	1486	1494	3.773
800	4.80	5.28	5.96	34.44	34.77	34.74	27.27	27.43	27.31	1484	1491	3.623
1000	3.98	4.41	4.76		34.78	34.69	27.39	27.56	27.45	1483	1488	3.477
1500	2.70	3.48	3.10	34.58	34.88	34.70	27.60	27.75	27.65	1488	1493	3.389
2000	2.08	2.94	2.22	34.64	34.89	34.73	27.70	27.81	27.76	1492	1499	3.301
2500	1.78	2.56	1.77	34.67	34.88	34.74	27.74	27.84	27.80	1499	1506	3.271
3000	1.60	2.27	1.44	34.68	34.87	34.75	27.77	27.86	27.82	1507	1513	3.240
3500	1.47	2.03	1.20	34.69	34.86	34.73	27.79	27.87	27.83	1516	1519	3.220
4000	1.36	1.81	0.99	34.70	34.84	34.72	27.80	27.87	27.83	1524	1525	3.240
Average	3.7	4.2	3.8	34.63	34.87	34.78				1504	1506	

We now pass to the salinity field S. It should be noted that the latitudinal zonality is marked notably less clearly even in the annual average field of surface salinity S_0 in comparison with the field T_w. In particular, the desalinization patches in coastal regions of big river run-off are imposed on it. Nevertheless, some definite regularities are observed in the zonal average values of S_0 represented by curve 5 in Figure 1.2.3; namely, the lowest values of S_0 in the equatorial zone (with the smallest values of 34.43‰ in latitudes 5–10°N) and in polar regions (32.35‰ in latitudes 60–65°N and 33.90‰ in latitudes 65–70°S) and the highest values in the subtropics (35.76‰ at 25–30°N and 35.74‰ at 20–25°S) are in agreement with the corresponding maximum values (precipitation exceeds evaporation) and minimum values (evaporation exceeds precipitation) of water budget of the sea surface. The average salinity of the ocean surface is equal to 34.84‰ (34.56‰ in the Pacific Ocean, 35.30‰ in the Atlantic Ocean, and 34.68‰ in the Indian Ocean). Seasonal variations of the field S_0 are rather weak.

The temperature and salinity of the sea surface are rather closely interrelated statistically. As a rule, cold waters contain less salt ($T_w \sim 2$ °C and $S_0 \sim 33.9$‰ in subarctic waters) and warm waters contain more salt ($T_w \sim 27$ °C and $S_0 \sim 36.4$‰ in equatorial–subtropical Atlantic waters, in the Arabian Sea, and in the subtropical anticyclones of the Pacific Ocean). Exceptions to this rule are the very warm ($T_w \sim 27$ °C) and freshened ($S_0 \sim 34.8$‰) equatorial–tropical waters of the Pacific Ocean and Indian Ocean, the very freshened ($S_0 \sim 33.4$‰) waters of the Bay of Bengal, the east-equatorial zone of the Pacific Ocean, and the water area at the mouths of the Amazon and the Congo and some other African rivers.

The average salinity stratification in the Pacific, Atlantic, and Indian Oceans is presented in Table 1.2.1. The presence of the following factors is typical of this stratification: (1) the upper (quasihomogeneous) mixed layer; (2) a seasonal halocline tens of meters thick where salinity considerably increases with depth; (3) a subsurface high-salinity layer (with maximum salinity along the whole vertical) at depths of 100–250 m; (4) an intermediate low-salinity layer (with minimum salinity along the whole vertical) at depths of 600–1000 m (where hydrostatic stability is due to the effect on water density of the temperature decrease with increasing depth, which is stronger than the effect from the decrease in salinity); (5) the main halocline with depths to 1500–2000 m (where salinity slowly increases with depth); and (6) a deep layer of approximately constant salinity.

The average salinity over the whole depth of the ocean (excluding the Arctic basin) is equal to 34.71‰ (34.63‰, 34.87‰, and 34.78‰ in the Pacific, Atlantic, and Indian Oceans, respectively; salinity is 0.13‰ higher in the Northern Hemisphere than in the Southern Hemisphere). The zonal average values S_{av} are represented by curve 6 in Figure 1.2.3. They vary weakly within the limits 34.34–34.94‰ and generally follow the latitudinal variations of S_0. $S_{av} > S_0$ in the equatorial zone and north and south of latitudes ± 40°, and $S_{av} < S_0$ in the subtropics and outside the equatorial tropical regions. We see that vertical distributions of salinity in different regions of the ocean can deviate in different directions from the average salinity stratification.

Indeed, Stepanov (1974) identified seven types of vertical distributions of salinity (polar, subpolar, moderate-tropical, equatorial-tropical, North Atlantic, Mediterranean, and Indo-Malayan) and several subtypes and published a chart of their geographical distribution. Stratification close to the average is typical only of the equatorial-tropical waters. The surface salinity minimum disappears in moderate-equatorial waters. On the other hand, it is marked very strongly in subpolar and, particularly, polar waters, but as there is no subsurface maximum or intermediate minimum there, the salinity increases everywhere with depth. Conversely, in the North Atlantic waters the salinity monotonously decreases with increasing depth. There is one maximum of S (at a depth of 600 m) in the Indo-Malayan waters and two maxima in the Mediterranean waters (on the surface and at depths of 500–1000 m).

The variety of $S(z)$ profiles is accounted for by the fact that stable density stratification $\varrho(z) = \varrho[T(z), S(z), p(z)]$ can be produced by means of different combinations of $T(z)$ and $S(z)$ profiles. It is convenient to represent these combinations by the so-called T, S curves on plots with coordinates T and S, where different depths z are marked by points. The average T, S curves for the Pacific, Atlantic, and Indian Oceans are demonstrated in Figure 1.2.5, from which, in particular, it is seen that the medium position among the three oceans is occupied by the Indian Ocean to a depth of 200 m, the Atlantic Ocean at depths of 200–600 m, and the Pacific Ocean at greater depths. Stepanov (1974) classified the T, S curves into eight regional types. These are the same types as those for salinity, with the additional separation of tropical and equatorial waters. The greatest departure from the average curves in Figure 1.2.5 is shown by the T, S curves of polar waters

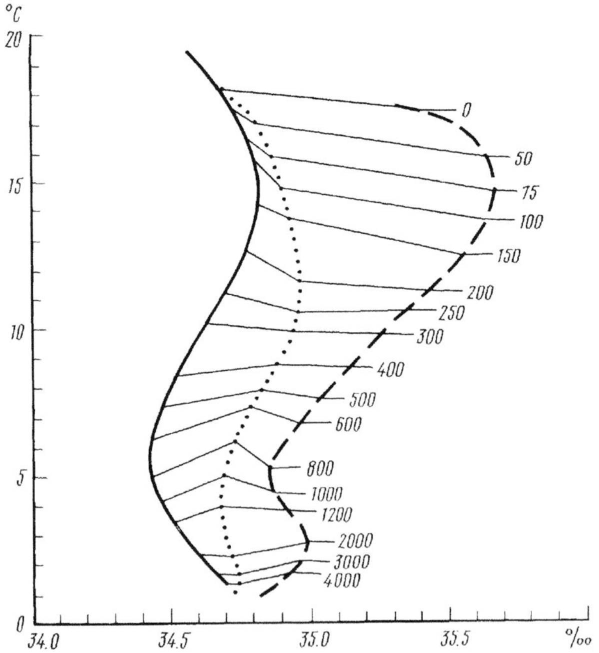

Fig. 1.2.5. Average T, S curves for the Pacific Ocean (solid line), the Atlantic Ocean (dashed line), and Indian Ocean (dotted line).

lying to the left of and below the average curves, the subpolar curves lying on the left, and the North Atlantic and Mediterranean curves lying on the right.

We now pass to the density anomaly field σ_t (reduced to standard atmospheric pressure). First, we note that σ_t and, more particularly, the total density ϱ increase with depth almost everywhere. Hence, the density stratification is almost always hydrostatically stable. Consequently, σ_t is minimum at the sea surface. The average value of σ_t over the whole ocean surface is equal to 24.74 (24.33 in the Pacific Ocean, 25.24 in the Atlantic Ocean, and 24.46 in the Indian Ocean; it is by 1.2 smaller in the Northern Hemisphere than in the Southern Hemisphere). The annual-average zonal values of σ_t on the ocean surface, represented by curve 7 in Figure 1.2.3, have a minimum equal to 22.18 in the zone 10–15°N where, together with high temperature, the desalinization effect of precipitation in the intratropical convergence zone also decreases the water density. The quantity σ_t increases smoothly in the northward direction up to a maximum of 26.19 in the zone 55–50°N, and slightly decreases further towards the pole. Also, σ_t increases towards the south up to a maximum of 27.30 in the zone 60–65°S, after which it seems to decrease slightly. The isopycnic lines at the ocean surface basically repeat the isotherm configuration; they undergo substantial seasonal variations.

The average density stratification of the Pacific, Atlantic, and Indian Oceans in terms of σ_t is presented in Table 1.2.1. The presence of the following factors is typical of this stratification: (1) the upper mixed layer; (2) the seasonal pycnocline where the density sharply increases with depth (at a rate of the order of a unit of σ_t per 10 m or 10^{-6} g/cm^4); (3) the main pycnocline extending to a depth about 1.5 km, where σ_t increases slowly with depth (approximately by 1.5 units of σ_t); and (4) a deep layer where σ_t increases very slowly with depth. As a rule, the major contribution to this stratification is made by temperature effects (the most important exception is the Arctic where the density increase with depth in the upper pycnocline is mainly due to salinity).

Comparing the average stratification of σ_t in, say, the Pacific Ocean (Table 1.2.1), with the zonal values of σ_t at the sea surface (curve 7 in Figure 1.2.3) we derive a crude rule for estimating the latitudes of formation of deep waters – namely, to depths of 100 m (σ_t = 25.30), 200 m (σ_t = 26.17), 300 m (σ_t = 26.69), 500 m (σ_t = 26.99), and 1000 m and more ($\sigma_t \geqslant$ 27.39) the following latitudes correspond, respectively, 35°N and 31°S, 52°N and 41°S, 50°S, 55°S, and 65°S and further to the south.

It is convenient to measure the rate of density increase with depth $\partial\varrho/z$ in units of the Brunt–Väisälä frequency:

$$N = \left\{ \frac{g}{\varrho} \left[\frac{\partial \varrho}{\partial z} - \left(\frac{\partial \varrho}{\partial z} \right)_a \right] \right\}^{1/2}, \tag{2.3}$$

where g is the acceleration due to gravity and $(\partial\varrho/\partial z)_{\mathrm{a}} = g\varrho/c^2 \approx 4.4 \times 10^{-8}$g/cm^4 is the adiabatic correction (where c is the speed of sound).

On average, this frequency usually increases with depth from the sea surface to the seasonal pycnocline where the period $2\pi/N$ is of the order of 10 min (in micropycnoclines separating microstructure layers of the ocean, this period can be several times smaller), and $2\pi/N$ increases tens of times between the seasonal pycnocline and the sea bottom.

The vertical distribution of the speed of sound $c = [(\partial p/\partial \varrho)_{\eta, S}]^{1/2}$ (the subscripts η and S indicate that the derivative $\partial p/\partial \varrho$ is taken for constant entropy η and salinity S) is a characteristic of the thermodynamic stratification of the sea that is very important in hydroacoustics. The speed of sound is a function of T, S, and p, which is described by the Frye–Pugh empirical formula (Frye and Pugh, 1971) for temperature, salinity, and pressure ranges characteristic of the ocean. The formula implies that c increases together with temperature, salinity, and pressure. The effect of a temperature decrease usually prevails in the upper ocean and c decreases with increasing depth, while the effect of a pressure increase is dominant in the lower layers where c increases with depth. As a result, the speed of sound has a minimum at an intermediate depth z_m, and an underwater acoustic waveguide with the axis z_m is formed. As examples, Table 1.2.1 presents the vertical average distributions of the speed of sound in the northern halves of the Pacific and Atlantic Oceans (according to V. P. Kut'ko). For example, on average, the speed of sound at the surface is $c_0 = 1524$ m/s in the northern part of the Pacific Ocean. The axis of the underwater acoustic waveguide is at depth $z_m = 1000$ m, the speed of sound on the axis is $c_m = 1483$ m/c, the waveguide width (i.e. the depth $z_w > z_m$ where the speed of sound attains the same value c_0 as at the sea surface) is $z_w = 4000$ m, and the speed of sound at the bottom, whose average depth in the Pacific Ocean is 4028 m, is somewhat greater than its value at the sea surface. In the North Atlantic $c_0 = 1515$ m/s, $z_m = 1000$ m, $c_m = 1488$ m/s, and $z_w \approx 3130$ m.

The depth z_m of the waveguide axis increases to 2000 m in tropical regions and decreases to 500–200 m in middle latitudes, and the waveguide axis passes still closer to the sea surface in high latitudes. In less deep-water regions where $c_H < c_0$ (c_H is the speed of sound at the bottom), the waveguide extends from the bottom upward to a depth $z'_w < z_m$ at which the speed of sound attains the value c_H. When $z_m = 0$, a subsurface acoustic waveguide is formed. This stratification (with c monotonously increasing to the bottom) is typical of polar regions of the ocean and of cold seasons in subtropical and tropical Mediterranean waters. When $z_m = H$, a bottom acoustic waveguide is formed; this stratification (with c monotonously decreasing to the bottom) is typical of shallow waters in middle latitudes in warm seasons when the ocean is heated from above and, in addition, undergoes salinization at the surface owing to evaporation. Finally, there can exist stratification with two acoustic waveguides when, below the upper waveguide (called thermic), there are waters with higher temperature and salinity.

The electrical conductivity $K(T, S, p)$ of sea water is another important thermodynamic characteristic of the stratification of the sea (whose *in situ* measurements have begun to be widely used in recent years for determining the salinity instead of the earlier chlorinity measurements in water samples taken by bathometers). Like the speed of sound c, electrical conductivity increases together with temperature, salinity, and pressure although, of course, its behavior is qualitatively different from that of c. Instruments for measuring the electrical conductivity are usually calibrated so that they show the relative electrical conductivity $R(T, S, p) = K(T, S, p)/K(15\,°\mathrm{C}, 35‰, p_a)$ (when the temperature scale of 1968 is used, the denominator is equal to 4.2906 S/m, where S=siemens=ohm^{-1}). For the determination of the function $R(T, S, p)$, empirical formulas were constructed (see Background Papers and Supporting Data on the Practical Salinity Scale 1978, UNESCO, 1981; Lewis and Perkin, 1981).

Table 1.2.1 presents the vertical average distribution of electrical conductivity of sea water in the northern half of the Atlantic Ocean (according to S. A. Oleinikov). It shows that, on average, the electrical conductivity monotonously decreases with increasing depth from 4.887 S/m at the sea surface to 3.477 S/m at 1 km and 3.240 S/m at 4 km. We note that, at the sea surface, electrical conductivity has maximum values of 5.6–5.5 S/m in the tropical regions and decreases with increasing latitude down to 3.5–3.0 S/m in the Strait of Labrador; however, the maximum values of electrical conductivity in the deep ocean are shifted to the subtropics. Accordingly, the rate of decrease of electrical conductivity with increasing depth generally decreases from the equator to the pole and becomes very small in subarctic waters. In the north-west part of the subtropic waters, in the Labrador Current, after a subsurface minimum at a depth of about 70 m this rate increases to a depth of 300 m and remains constant in deeper waters.

Vertical distributions of the refractive index n_λ (T, S, p) of sea water for electromagnetic waves with various wavelengths λ are thermodynamic characteristics of the stratification of the ocean which are important for hydro-optics. These quantities are related to the water density ϱ by the Lozentz–Lorenz formula

$$\frac{n_\lambda^2 - 1}{n_\lambda^2 + 2} = R_\lambda \varrho, \tag{2.4}$$

where R_λ is the so-called specific index of refraction which depends weakly on T, S, and p (and increases slightly as λ, T, S, and p increase; e.g. at atmospheric pressure it varies from 0.21193 cm^3/g for $\lambda = 0.4047$ μm, $T = 1$ °C, and $S = 0‰$ to 0.20352 cm^3/g for $\lambda = 0.6438$ μm, $T = 30$ °C, and $S = 35‰$).

According to the Mathäus (1974) empirical formula, the refractive index n_λ decreases with increasing λ and T (e.g. for $S = 35‰$ it varies from 1.35099 for $\lambda = 0.4047$ μm and $T = 0$ °C to 1.33665 for $\lambda = 0.6438$ μm and $T = 30$ °C), and increases with S (e.g. as S increases from 0 to 40‰, for the natrium spectral line D with $\lambda = 0.5893$ μm it varies from 1.33402 to 1.34186 for $T = 0$ °C and from 1.33196 to 1.33914 for $T = 30$ °C). As the pressure increases, the refractive index increases (approximately linearly with the derivative of the order of 1.28/10^{10} Pa^{-1}). Thus, for the average stratification of the ocean, when T decreases and S increases with increasing depth the refractive index monotonously increases with depth.

To conclude this section we describe briefly the stratification of the most important impurities contained in the sea waters (the so-called major nutrients): carbonic-acid gas CO_2 and other carbon compounds, dissolved oxygen O_2, and compounds of silicon Si, nitrogen N, and phosphorus P. They amount to a total of trillions of tons (1 $T_t = 10^{12}$ t). Their average concentrations in the oceans in mg/l are given in Table 1.2.2 (in particular, it is seen from the table that the concentrations of O_2 are maximum in the Atlantic Ocean and minimum in the Pacific Ocean, and vice versa for the concentrations of Si, N, and P).

TABLE 1.2.2

Major Nutrients in the oceans

Ocean	Mass, T_t	Content, T_t					Average concentration, mg/l				
		C	O_2	Si	N_{fixed}	P	C	O_2	Si	N	P
Pacific Ocean	723 699		3.13	1.968	0.3691	0.0579		4.32	2.72	0.51	0.08
Atlantic Ocean	337 699		2.54	0.388	0.1047	0.0200		7.52	1.15	0.31	0.06
Indian Ocean	291 945		1.65	0.555	0.1372	0.0204		5.66	1.90	0.47	0.07
Global Ocean	1 370 323	40	7.48	2.918	0.6123	0.0988	29.19	5.46	2.13	0.45	0.07

Of the 40 T_t of carbon contained in the sea waters, 38.2 T_t relate to dissolved inorganic matter forming the so-called carbonate system comprising free dissolved carbon-acid gas and nondissociated carbonic acid H_2CO_3 (which are almost indistinguishable), bicarbonate ions HCO_3^-, and carbonate ions C_3^{2-} (the remaining 1.8 T_t of carbon relates almost exclusively to the dissolved organic matter; the nondissolved, dead organic particulate matter, detrite, contains only 2.7×10^{-2} T_t of carbon, and the amount of carbon in the living matter is 20 times smaller still; but here we shall not dwell on these organic components).

As result of the chemical equilibrium $CO_2 + H_2O \rightleftharpoons H^+ + HCO_3^-$ and $HCO_3^- \rightleftharpoons H^+ + CO_3^{2-}$, the relationship between the concentrations $[CO_2]$, $[HCO_3^-]$, and $[CO_3^{2-}]$ of the components of the carbonate system (here the square brackets designate the concentrations) is determined primarily by the concentration $[H^+]$ of hydrogen ions. This concentration is usually characterized by the so-called pH value: pH $\approx -\log\, [H^+]$ (equal to 7 in neutral solutions at 25 °C; in sea water, which has a weak alkaline reaction owing to the separation of hydroxyl OH^- in the hydrolysis of bicarbonates and carbonates, $[H^+]$ is smaller than in neutral solutions and the pH varies within the limits 7.5–8.4 and decreases with increasing T and p). The plot representing the dependence of the percentage amounts of $[CO_2]$, $[HCH_3^-]$, and $[CO_3^{2-}]$ on pH at $T = 0$ °C and $p = 1$ atm shows that for $7 \leqslant \text{pH} \leqslant 8.5$ the major part consists of bicarbonate. For pH = 7 this amounts to 80% and almost all the rest relates to CO_2, and for pH = 8.5 it again amounts to 80% and the rest almost entirely relates to CO_3^{2-}. As T increases, these plots are shifted to the right (but $\Sigma\, CO_2 = [CO_2] + [HCO_3] + [CO_3^{2-}]$ decreases), and as S and p increase, they are shifted to the left. On measuring the pH and the total alkalinity Alk (which is determined by hydrochloric acid neutralizing the sea water) we can calculate all the components of the carbonate system.

The annual average values of pH at the sea surface decrease slowly with increasing latitude from 8.25 in tropical and subtropical regions to 8.10–8.05 in polar regions. As the depth increases, the pH generally decreases and the latitudinal maxima are shifted to the subtropics. It is characteristic of tropical and subtropical regions that the pH has a minimum of the order of 7.80–7.85 at 500–1000 m, increases up to 7.90 at a depth of 1500 m (which is not the case for polar regions), and is constant at greater depths. In the Atlantic, particularly in northern latitudes, the pH is greater than in the Pacific or Indian Oceans. The alkalinity in the ocean has values of the order of 2.4 mg-eqv/l and, on average, amounts to 0.0695 of the salinity and 0.125 of the chlorinity. The ratio Alk/Cl at the sea surface increases slowly with latitude from 0.121 in tropical regions to 0.124–0.126 in polar regions; it increases monotonically up to 0.128–0.129 with depth; in the Atlantic, particularly in northern latitudes, it is notably smaller than in the Indian Ocean and much smaller than in the Pacific Ocean. The partial pressure of CO_2 in water at the sea surface increases with latitude from 2.9–3.0 × 10^{-4} atm in the tropical regions to 3.2 × 10^{-4} in the Arctic and 3.6 × 10^{-4} in the Antarctic. Generally it increases with depth and has maxima of the order of 7.6–7.9 × 10^{-4} atm at depths of 500–1000 m and in the bottom waters (and of the order of 5.9 × 10^{-4} atm in the Arctic). It is notably smaller in the Atlantic than in the Indian or Pacific Oceans.

The solubility of oxygen in the sea waters decreases as T, S, and p increase (almost twice as T increases from 0 to 30 °C, by one-quarter as S increases from 0 to 40‰, and at a rate of 0.01 ml/(l × 100 atm) as p increases). The actual concentrations of oxygen are less than its solubility almost everywhere in the ocean (except the upper 50–100 m layer in the vegetative season when photosynthesis takes place), i.e. the waters are undersaturated with oxygen since it is expended on the oxidation of organic and other matters and on the respiration of living organisms (at a rate of 0.15 T_t/yr or 0.11 mg O_2/l·yr).

The concentration of dissolved oxygen in the surface sea waters generally increases with latitude from 4.4–4.6 ml/l in the equatorial zone to 7.0–7.9 in polar regions, particularly in the Antarctic.

In the vertical distributions of O_2 at intermediate depths there is a minimum and sometimes two and even three minima. This minimum lies at depths less than 400 m and is 1–2.5 ml/l in the Atlantic equatorial zone; it is located at the greatest depth in the subtropics (>800 m and 3.5–4 ml/l in the northern subtropics; > 1400 m and 4.2–4.4 ml/l in the southern subtropics); it rises higher than 600–400 m in the Arctic (5.5–6 ml/l); and higher than 600 m in the Antarctic (4.5 ml/l). The minimum is deeper in the Indian Ocean, and the corresponding concentration of O_2 is lower: the minimum lies at higher than 600 m in the Arabian Sea and in the Bay of Bengal, and the concentration is less than 0.5 ml/l. It is at the greatest depth in the southern subtropics at about 40 °S (> 1600 m and around 3.5–3.7 ml/l) and rises above 800–600 m (4–4.5 ml/l) in the Antarctic. The minimum is still deeper and the O_2 concentration is still smaller in the Pacific: < 600–400 m and 0.1–0.5 ml/l in northern tropical regions, the deepest location being in the subtropics (> 1400 m and 0.5 ml/l in the northern subtropics; > 2400 m and 3.4–3.5 ml/l in the southern subtropics) and in polar regions (< 800–600 m and < 0.5 ml/l in the north and > 4 ml/l in the south). The oxygen concentrations are 4.4–5.9 ml/l, 4.1–5.2 ml/l, and 3.5–4.6 ml/l in the bottom waters (at a depth of 5 km) of the Atlantic, Indian, and Pacific Oceans, respectively.

The saturation concentrations of silicon compounds in sea water probably exceed 100 mg/l. (The solubility of amorphous silica SiO_2 for $T \sim 25$ °C, $S \sim 35$‰, and $p = p_a$ is equal to 120–140 mg/l; it decreases twice as T decreases down to 0–5 °C and increases with p at a notable rate of about 2 mg/(l × 100 atm); the solubility of crystalline silica is tens of times lower; for quartz at T = 5–25 °C, S = 35‰, and $p = p_a$ it is equal to 3.2–5.1 mg/l.) Table 1.2.2 shows that, on average, the sea waters are sharply undersaturated with dissolved silicon compounds. In the upper 50–100 m ocean layer silicon is extracted from water by living organisms and is included in frustules of diatoms (where its content is equivalent to 99.3% of the carbon content; we note that these algae form 77% of the entire oceanic phytoplankton), spines of radiolarians, and spicules of siliceous sponges. Below 200 m the silicate skeletons begin to dissolve and the concentration of silicic acid increases with depth (monotonically everywhere except the intermediate water layer of Mediterranean origin in the Atlantic where there is a minimum of silicic acid at depths from 1000–1200 to 1400–1700 m). As a result, more than 95% of Si in sea water is in the form of dissolved meta- and polysilicic acids, about 2–3% is in the form of organogenic amorphous silica, and about 1% is in crystalline form (quartz).

In surface sea waters the concentration of dissolved silicic acid in tropical and subtropical regions (in the latitudinal zone with boundaries about 35°N–50°S and in the North Atlantic, which generally has little SiO_2 owing to its relation to the Arctic where there is almost no silicon anywhere, except the Strait of Labrador) does not exceed 10 μmol Si/l. It increases up to 40 μmol Si/l to the north in the Pacific Ocean and up to 60 μmol Si/l in the Antarctic. At depths of 500–1000 m the latitudinal minimum of SiO_2 in the Pacific and Indian Oceans is shifted to the southern subtropics (25–45°S), and at depths of 2000–3000 m it is shifted to their southern part (40–50°S). The concentrations of SiO_2 remain small at the bottom (30–50 μmol Si/l) in the North Atlantic; they increase up to 120–140 μmol Si/l in the Indian Ocean and up to 130–160 μmol Si/l in the Pacific Ocean, particularly in its northern regions.

There is a great deal of dissolved free molecular nitrogen N_2 in the ocean (e.g. in equilibrium with the air, at $T = 20$ °C and $S = 35$‰, the surface sea waters contain 9.51 ml N_2/l in contrast with 5.17 ml O_2/l; these concentrations decrease as T and S increase). However, it plays no biogenic role and, further, we shall discuss only fixed nitrogen, both organic (of which more than 95% is in dissolved organic matter and less than 5% in suspended matter) and inorganic (nitrate, nitrite, and ammonium, i.e. in the form of NO_3^-, NO_2^-, and NH_4^+ ions).

In the growth of oceanic plankton, nutrients pass into it from sea water in the ratio O : C : N : P, approximately equal to 141 : 41 : 7.2 : 1 in mass and to 276 : 106 : 16 : 1 in the number of atoms. Hence, for the indicated oxygen expenditure of 0.15 T_t/yr on the oxidation of organic matter in the ocean about 7.5×10^9 t of inorganic nitrogen are produced each year. This oxidation yields, in succession, ammonium nitrogen (concentrated mainly in the photosynthesis layer), nitrites (in the seasonal pycnocline), and finally nitrates. (According to the Richards model (1965), plankton organic matter contains 106 $CH_2O \cdot 16\ NH_3 \cdot H_3PO_4$ whose combination with 138 O_2, yields 106 CO_2 + 122 H_2O + 16 HNO_3 + H_3PO_4.) When lacking O_2, further oxidation of organic matter takes place owing to the reduction of nitrates to free nitrogen (denitrification), and when lacking nitrates as well, it goes on, owing to the reduction of sulphates SO_4^{2-}, to free sulphur (SO_4 reduction) with the separation of all the nitrogen in the form of ammonia, NH_3. The resulting distribution of forms of nitrogen over the depth in the Pacific Ocean is demonstrated by Table 1.2.3 (see Ivanenkov, 1979). It is seen from the table that, with the exception

TABLE 1.2.3

Stratification of forms of nitrogen (mg N/l) in the Pacific Ocean

Depth, m	N_{org}	NH_4^+	NO_2	NO_3
0–50	0.140	0.049	0.0014	0.007
50–1000	0.126	0.018	0.0007	0.308
1000–4000	0.025	0.0056	0.00014	0.504

of the photosynthesis layer where N_{org} dominates, the basic form of fixed nitrogen in the ocean are nitrates.

In the surface sea waters the concentrations of nitrates are minimum (< 1 μmol N/l) in tropical and subtropical regions (with the exception of the east-equatorial zone of the Pacific Ocean where there is a local maximum up to 15–20 μmol N/l produced by equatorial upwelling), and they increase up to values > 25 towards the Antarctic and in the northern part of the Pacific Ocean (however, they remain small in the North Atlantic). At a depth of 100 m the equatorial maximum (for the Atlantic in eastern tropical regions) and subtropical minima are marked in all the oceans. At intermediate depths the concentration of nitrates has a maximum at depths of about 800 m in the tropical and northern Atlantic, at depths > 1400 m in its southern subtropics, and < 400 m towards the Antarctic with values up to 25–30. This maximum lies above 800 m and has values > 40 in the northern part of the Indian Ocean; it goes down to a depth of 1600 m and more and slightly decreases in value in the southern subtropics, and becomes planar again towards the Antarctic. In the Pacific Ocean it lies at the greatest depth in the southern subtropics (> 2400 m, 35–40 μmol N/l) and in the northern subtropics (> 1800 m, > 45 μmol N/l), and goes upward in polar waters. The content of nitrates is notably less in Atlantic waters (with the exception of the Antarctic) than in Indian Ocean waters and even less in Pacific waters.

Phosphorus is extremely important for living organisms since it is contained in the main biologic 'fuel', namely ATP and phospholipids forming the base of cell membranes. The phosphorus in sea water is contained in organic matter (> 95% in soluble organic matter and < 5% in suspended organic matter) and in inorganic forms (mainly in salts of orthophosphoric acid H_3PO_4; e.g. for $T = 20$ °C, $S = 34.8$‰, and pH = 8 the content of phosphorous is 41.4% in the neutral salt $MgHPO_4^0$, 28.7% in HPO_4^{2-} ions, 15.0% in $NaHPO_4^-$ and 4.7% in $CaHPO_4^0$; for these T and S the ratio of the ions $H_2PO_4^-$: HPO_4^{2-} : PO_4^{3-} varies from 11.2 : 87.9 : 1.0 for pH = 7 to 0.3 : 75.4 : 24.3 for pH = 8.5). In highly productive regions of the Pacific Ocean $P_{org} \sim 0.5$ and $P_{inorg} \sim 1.0$ μmol/l in the upper 100 m layer; $P_{org} \sim$ 0.4–0.3 and $P_{inorg} \sim$ 2.0–2.5 in the layer 100–500 m; $P_{org} \sim 0.2$ and $P_{inorg} \sim 3.2$ at depths of 500–1000 m; $P_{org} \sim$ 0.1–0.05 and $P_{inorg} \sim$ 3.0–2.8 at depths of 1000–4000 m. In low production regions there is very little P_{org} even in the photosynthesis layer.

In surface waters the phosphate concentrations are minimum (< 0.2 μmol P/l) in tropical and subtropical regions (however, they have a local minimum down to 0.5–1.0 in the east-equatorial zone of the Pacific Ocean) and increase up to 1.5–2.0

towards the Antarctic and the northern regions of the Pacific Ocean and to > 0.5 in the North Atlantic. In deep waters latitudinal subtropical minima are formed. At intermediate depths there is a phosphate maximum (which is similar to the nitrate maximum but is less deep and sharper) with concentrations of the order of 2.0 in the Atlantic, 2.5 in the Indian Ocean, and > 3.0 in the northern half of the Pacific Ocean. The phosphate concentrations have a diffuse minimum at depths about 2000 m and slightly increase towards the bottom.

At present no mathematical models have yet been constructed to explain the vertical distribution of nutrients in the ocean.

3. LARGE-SCALE CURRENTS

Large-scale currents on the sea surface are known from ship drift measurement data, 'bottle mail', and rare measurements with mooring buoy stations. First, these data demonstrate the presence of a quasistationary system of large-scale currents on the sea surface (see Figure 1.3.1 where 55 currents are shown) that are

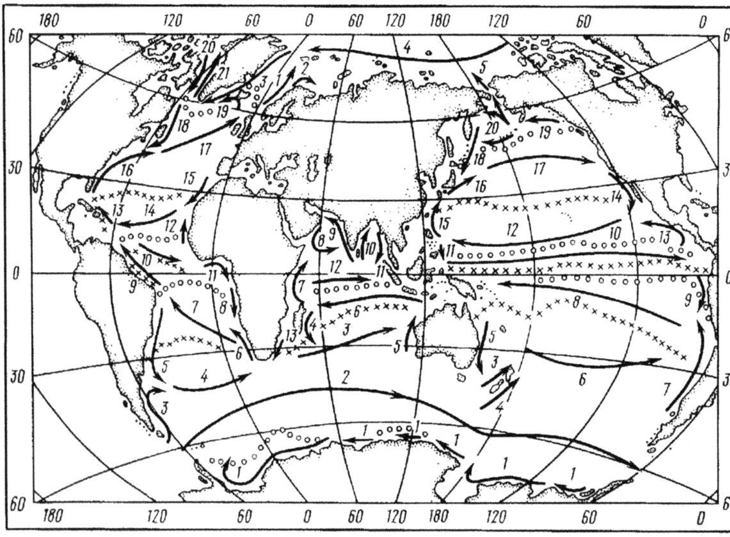

Fig. 1.3.1. Large-scale currents at the surface of the ocean. *The Antarctic* – 1: Antarctic Coastal; 2: Antarctic Circumpolar. *The Pacific Ocean* – 3: West New Zealand; 4: East New Zealand; 5: East Australian; 6: South Pacific; 7: Peru; 8: South Equatorial; 9: El Niño; 10: Equatorial Counter Current; 11: Mindanao; 12: North Equatorial; 13: Mexico; 14: California; 15: Taiwan; 16: Kuroshio; 17: North Pacific; 18: Kurile; 19: Alaska; 20: East Bering Sea. *The Indian Ocean* – 3: South Indian Ocean; 4: Madagascar; 5: West Australian; 6: South Equatorial; 7: Somali; 8: West Arabian; 9: East Arabian; 10: West Bengal; 11: East Bengal; 12: Equatorial Counter Current; 13: Agulhas Stream. *The Atlantic Ocean* – 3: Falkland; 4: South Atlantic; 5: Brazil; 6: Benguela; 7: South Equatorial; 8: Angola; 9: Guiana; 10: Equatorial Counter Current; 11: Guinea; 12: Cape Verde; 13: Antillas; 14: North Equatorial; 15: Canary; 16: Gulf Stream; 17: North Atlantic; 18: Labrador; 19: Irminger; 20: Baffin Bay; 21: West Greenland. *The Arctic* – 1: Norwegian; 2: Nordkapp; 3: East Greenland; 4: West Arctic Current; 5: Pacific. Lines of circles = convergences; lines of crosses = divergences.

permanently present in definite areas although in some places they undergo substantial seasonal and synoptic variations.

Second, these data are in good agreement with the chart of the sea surface dynamic topography (i.e. its heights above the deep level with pressure of 1500 dbar calculated using the hydrostatic equation from hydrographic station data on vertical water-density distributions) whose isolines (dynamic horizontals) coincide approximately with the streamlines of geostrophic currents.

In particular, the axes of the dynamic topography troughs (shown by circles in Figure 1.3.1) are in good correspondence with the divergence lines of surface currents on which the set-down of surface waters occurs and, consequently, the rise (upwelling) of deep waters takes place. Conversely, the axes of the dynamic topography crests (shown by crosses in Figure 1.3.1) correspond to the convergence lines of surface currents on which the set-up of surface waters occurs and consequently their sinking (downwelling) takes place. Figure 1.3.1 shows that the divergence and convergence lines divide the dynamic topography and surface current chart into quasilatitudinal dynamic zones. Namely, from south to north the following divergences and convergences are located in succession: the Antarctic divergence (AD); the Antarctic convergence (AC – also called the southern polar front, SPF) which coincides approximately with the core of the Antarctic Circumpolar Current (ACC); the southern subtropical convergence (SSTC – also called the subantarctic front, SAF); the southern tropical convergence (STC); the northern tropical convergence (NTC; this convergence line is slightly shifted to the north relative to the equatorial line of dynamic symmetry owing to the lack of symmetry in the Northern and Southern Hemispheres); the northern tropical divergence (NTD); the northern subtropical convergence (NSTC), and the subpolar divergence (SPD). The northern polar front is located between NSTC and SPD.

The South and North Equatorial Currents play a very important role in the ocean. They go between SSTC and STD (in the Southern Hemisphere) and NTD and NSTC (in the Northern Hemisphere) with a substantial western component in complete accordance with trades in the atmosphere. For example, their total transport at 150°E is estimated as 130×10^6 m^3/s (see Table 1.3.1).

In the southern and northern halves of the oceans, south of STD and north of NTD, there are huge anticyclonic gyres with axes at SSTC and NSTC, respectively. They go around the corresponding quasipermanent atmospheric subtropical highs (which intensify from winter to summer). In the Northern Hemisphere these are the Azore and Honolulu highs in the Atlantic and Pacific Oceans, and, in the Southern Hemisphere, the St Helena (the Atlantic Ocean), Mauritius (the Indian Ocean), and South Pacific highs. The periods of water circulation in the gyres are of the order of several years. (If the radius of a gyre is taken as 2500 km and the average velocity of the current around its periphery is taken as 10 cm/s, then the period is equal to 5 yr.) The western branches of these gyres form intensive narrow-jet-type boundary currents owing to the so-called β-effect (i.e. the increase of the vertical projection of the angular velocity of the Earth's rotation with latitude): examples include the Gulf Stream in the Azore gyre, the Brazil Current in the St Helena gyre, the Madagascar Current and the Agulhas Current in the Mauritius gyre, the Kuroshio in the Honolulu gyre, and the East Australian

TABLE 1.3.1

Velocity (cm/s) and transport (10^6 m^3/s) of oceanic currents

Current	Point of measurement			Velocities at depths of				Cross	Transport	
				0 m	100 m	250 m	500 m	section	Winter	Summer
North Equatorial	17°N,	120°W;	winter	4	7	3	2	150°E	95	85
	13°N,	150°E;	summer	20	23	14	1	170°E	65	65
	13°N	170°W;	summer	27	26	10	6			
South Equatorial	4°N,	148°W;	summer	192	191	85	65	150°E	35	
	2°N,	160°E;	summer	65	38	20	39	170°W	35	
	5°S,	155°W;	summer	56	56	26	20	100°W	85	
ACC	57°S,	165°W;	winter	7	7	6	4	170°W	75	
	62°S,	97°W;	winter	12	10	8	5	110°W	55	
	62°S,	79°W;	winter	8	7	7	5	60°W		214
Gulf Stream	25°N,	81°W		140	110	60	40	80°W		30
	33°N,	77°W		160	140	80	60	70°W		90
Kuroshio	31°N,	133°E;	summer	31	34	33	23	30°N	35	45
	34°N,	141°E;	summer	46	45	40	27	145°E	45	60
	36°N,	149°E;	summer	63	55	40	25			
North Pacific	38°N,	160°E;	summer	7	5	3	1	180°W	35	25
Equatorial Counter	5°N,	140°E;	summer	17	24	1	5			
	5°N,	170°W;	summer	70	80	2	4			
	8°N,	125°W;	summer	16	10	17	16			
South Pacific	28°S,	177°W;	winter	3	4	4	3	170°W	5	
	30°S,	107°W;	winter	12	17	15	6			
	32°S,	107°W;	winter	8	10	10	9			
East Australian	33°S,	154°E;	winter	51	44	34	19	33°S	20	
Taiwan	21°N,	122°E;	summer	29	34	27	9	20°N	30	30
	22°N,	122°E;	summer	17	20	27	9			
Kurile	46°N,	152°E;	summer	6	7	7	7	45°N	25	15
Alaska	55°N,	152°E;	summer	9	8	7	5	150°W	15	15
California	26°N,	117°W;	summer	6	4	3	1	30°W	12	8
	40°N,	128°W;	summer	8	5	3	2			
Peru	27°S,	73°W;	winter	1	6	5	2	30°S	10	
	35°S,	74°W;	winter	11	8	8	7			

Current in the South Pacific gyre. On the other hand, no intensification of this kind is observed in the boundary currents of the eastern branches of the gyres.

There are special conditions in the northern part of the Indian Ocean where there is no subtropical anticyclone and where sharp seasonal (monsoon) variability of winds in the atmosphere and, consequently, of currents in the ocean is observed. During the winter north-east monsoon (November–March) a relatively weak cyclonic monsoon gyre is formed in the northern part of the Indian Ocean, including the North Equatorial Current (the North-East Monsoon Current), which turns to the south along Somali at the African coast and the East Equatorial Counter Current in the equatorial zone between 3°N and 5–10°S with a maximum in February. During the summer south-west monsoon (May–September) a stronger anticyclonic gyre is formed here which includes the South Equatorial Current turning to the north at the western coast in the form of the intensified Somali Boundary Current and, in the north, the eastward Monsoon Current (merging into the Equatorial Counter Current which is shifted to the north) with a maximum in July.

In the regions north and south of the subtropical convergences there are cyclonic water gyres lying under the corresponding cyclonic wind systems in the atmosphere. In the Southern Ocean this is ACC, the largest current in the ocean (its transport can sometimes exceed 210×10^6 m^3/s). In the North Atlantic and in the northern part of the Pacific Ocean there are cyclonic gyres under the Icelandic Low and the Aleutian Low.

In the American–Asian subbasin of the Arctic basin there is a vast anticyclonic gyre whose period is estimated as 4 yr. Along its Asian periphery from the Bering Strait to the Fram Strait runs the West Arctic (Transarctic) Current, which then turns into the East Greenland Current carrying Arctic waters to the North Atlantic. The reverse transport of Atlantic waters to the Arctic is carried out by the Norwegian Current which then branches into the Nordkapp Current and the Spitsbergen Current (the water budget of the Arctic basin is estimated as 182×10^3 km^3/yr: the inflow through the Fram Strait is 112×10^3, through the Nordkapp–Sörkapp 35×10^3, and through the Bering Strait 30×10^3; the river run-off is 3.8×10^3, and the excess of precipitation over evaporation is 1.0×10^3; the outflows through the Fram Strait and the Straits of Canada are 124×10^3 and 57×10^3, respectively, and the transport of ice is 1.3×10^3 km^3/yr).

Typical values for the velocity and transport of a number of large-scale oceanic currents are given in Table 1.3.1, which shows that typical velocities of the largest surface currents are tens of centimeters per second and typical transport values are of the order of 10^7 m^3/s. According to the estimates obtained by Stepanov *et al.* (1977) with the aid of calculations from the density field on the basis of Sarkisyan's model, the average velocities of surface currents are 19.3 cm/s in the Indian Ocean, 12.3 cm/s in the Pacific Ocean, and 11.6 cm/s in the Atlantic Ocean. Galerkin and Gritsenko (1980) give more detailed results for the Pacific Ocean. The average kinetic energy per unit mass of surface currents is $100\ cm^2/s^2 = 10^{-2}$ J/kg (the root-mean-square velocity is equal to 14 cm/s). Further, 77% of the energy corresponds to zonal motions (64% of this amount corresponds to western motions and 36% to eastern motions) whose root-mean-square velocity is equal to

12.3 cm/s; 23% of the energy corresponds to meridional motions (60.6% of this energy corresponds to northern motions and 39.4% to southern motions) whose root-mean-square velocity is 6.8 cm/s. Among the zonal motions, the strong ($>$ 20 cm/s) western currents carry 40.6% of the energy but occupy only 11% of the ocean area (primarily, these are the equatorial currents), and the strong eastern currents carry 12.7% of the energy and occupy 3.7% of the area. Among the meridional motions, the strong northern currents contain 4.2% of the energy and occupy 0.4% of the area, and the strong southern currents have 13.6% of the energy and 1.2% of the area.

The fact that the major large-scale surface currents are directed along the dominant winds (and their strongest seasonal variability takes place in the regions of the strongest variability of winds, namely in the monsoon regions of the Indian Ocean) shows that basically they are wind-driven. The piling up and removal of water (and, to a certain extent, atmospheric pressure differences, thermohaline expansion and compression of waters, precipitation, and evaporation) produced by these currents create the above-mentioned dynamic topography of the sea surface, i.e. its deviations from the equilibrium geoid level which are of the order of several decimeters. The greatest upward deviations are found in western peripheries of the oceans, particularly in the subtropics, and the greatest downward deviations are found in polar regions. The dynamic height difference of the sea surface between NSTC and SPD in the Atlantic is 170 cm, and in the Pacific it is 120 cm. The differences between the heights of the surface of the ocean and those of its other isobaric surfaces create horizontal pressure differences in its depths generating deep currents.

As was already mentioned, the intensive western boundary currents have a narrow-jet-type character. It often happens that, near a jet current (on its side or below it), a jet counter current is located. The most vivid examples are the narrow ($\pm 2.5°$ lat.) high-salinity jets of east equatorial subsurface counter currents located at depths of 50–300 m below the western surface equatorial currents. These are the Cromwell Current in the Pacific, the Lomonosov Current in the Atlantic, and the Tareev Current in the Indian Ocean (which is clearly marked during the winter monsoon) with core velocities up to 150 cm/s and transport values up to 40×10^6 m^3/s.

Generally, according to calculations from density fields and rather scarce measurement data, the circulation of subsurface and intermediate-depth waters to a depth of 1500 m and a temperature of about 3.5 °C follows the surface circulation in a form weakening with increasing depth (the tropical circulation is almost completely damped and the subtropical gyres are slightly displaced towards the poles). This leads to the propagation of intermediate waters from polar fronts to subtropical and tropical regions (low-salinity waters) and to subpolar regions (high-temperature waters).

According to the existing approximate calculation data, the deep-water circulation (deeper than 1500 m), with the exception of ACC, is not so closely related to the surface circulation and the wind field above the oceans. In the greater part of its area this circulation is directed opposite to the surface circulation (including the deep counter current below the Gulf Stream, the recirculation in the South Atlantic and in the Indian Ocean, the cyclonic circulation in middle northern latitudes of the Pacific, and the anticyclonic circulation still further to the north). Therefore, deep-water circulation is weakest at intermediate depths of 1.5–2 km, and nearer to the bottom it slightly increases and begins to follow the isobaths of the bottom relief. It is likely to be mainly of thermohaline origin.

The Antarctic bottom waters (AABW) in the Southern Ocean move to the west together with ACC (which probably penetrates to the bottom). In the Atlantic they go to the north mainly through the western basins to 40°N where they meet the North Atlantic deep waters (NADW) and the Arctic bottom waters (ABW), AABW and NADW moving in the opposite directions with a boundary at a depth of approximately 4 km. AABW fill all deep basins in the Indian Ocean. The main AABW flow in the Pacific goes along the Kermadek and Tonga Trenches. At 10°S it issues a branch to the east which goes to the south-west part of the northern half of the ocean while the main flow bifurcates in the Northern Hemisphere and reaches approximately the northern tropic moving along the basins. The velocities of these bottom-water flows are 0.1–1 cm/s.

Using the equations of convective diffusion of heat and salt and typical meridional sections of temperature and salinity fields, Stepanov (1969) estimated the absolute value of the meridional velocity averaged over the entire ocean as 2.4 cm/s (for the vertical velocity in the upper ocean he obtained 5–10 × 10^{-5} cm/s and the value for deep layers was an order of magnitude more). The more detailed data obtained by Galerkin and Gritsenko (1980) for the Pacific Ocean are presented in Table 1.3.2. In particular, these data show that the root-mean-square zonal velocities are approximately one and a half times as great as the meridional velocities at all depths in this ocean.

Taking into consideration the data in Table 1.3.2 we can estimate the total velocity of large-scale currents averaged over the entire depth of the ocean as 4.5 cm/s. The corresponding kinetic energy density of these currents is around 1 J/m^3, i.e. 120 times smaller than in the atmosphere (which is quite natural since the ocean receives kinetic energy mainly from the atmosphere and the 'coupling' between them is very weak). For comparison we note (Vulis and Monin, 1975) that the

TABLE 1.3.2

Area-averaged ($\bar{u}$, $\bar{v}$) and root-mean-square (σ_u, σ_v) zonal and meridional velocities (cm/s) at various depths in the Pacific Ocean

Depth, m	$\bar{u}$	$\bar{v}$	σ_u	σ_v
0	-1.64	1.06	13.06	7.01
100	-0.03	-0.25	7.49	4.66
250	-0.42	-0.22	6.00	3.94
500	-0.20	-0.15	5.02	3.60
1000	-0.21	-0.15	4.53	3.31
1500	-0.24	-0.19	4.08	2.96
2000	-0.26	-0.20	3.82	2.73
2500	-0.37	-0.26	3.43	2.38
3000	-0.44	-0.26	3.35	2.35
3500	-0.51	-0.10	3.23	2.13
4000	-0.54	-0.03	3.02	2.30

internal energy density $\rho c T$ in the ocean is much greater than in the atmosphere (1.2×10^9 in comparison with 1.6×10^5 J/m^3) and the potential energy density $\frac{1}{2}\rho g H$ is also much greater (2×10^7 in comparison with 4×10^4 J/m^3) while the available potential energy densities are of the same order (7×10^2 and 5×10^2 J/m^3). However, we emphasize that, besides large-scale motions, substantial contributions to the total kinetic energy in the ocean must be made by synoptic motions, to which this book is primarily devoted, and also by inertial and tidal oscillations (see the energy spectra in Figures 1.1.1 and 1.1.3).

4. SYNOPTIC PROCESSES

There are intensive synoptic-scale motions in the world ocean, namely eddies moving together with the water contained in them, and also longer scale Rossby waves travelling over the water without carrying it along, which develop against the background of large-scale motions. These synoptic processes are in many respects qualitatively analogous to the well-known and thoroughly studied synoptic processes in the atmosphere although there are substantial quantitative distinctions between them. A comparison with atmospheric processes, and elucidation of the existing analogies and distinctions, may facilitate the study of synoptic processes in the ocean and the elaboration of methods of forecasting them, which is now becoming one of the urgent problems of ocean hydrodynamics. To this end we begin with a brief description of atmospheric large-scale currents and synoptic motions forming the general atmospheric circulation.

The primary source of the atmosphere's general circulation is the influx of solar heat. This influx has a purely zonal daily-average distribution on the outer boundary of the atmosphere (as a consequence, the zonal components are dominant in large-scale currents of the atmosphere's general circulation). Solar radiation is partly absorbed in the atmosphere but a substantial fraction reaches the Earth's surface where it is absorbed and reradiated in the form of longwave radiation which is then partly absorbed by lower atmospheric layers (a weak greenhouse effect). As a result, the atmosphere is heated primarily from beneath (and not very strongly so that the troposphere stratification is moderately stable). This heating retains a chiefly zonal character, and the equatorial zone is found to be heated more than polar regions (the annual insolation at the equator is 2.4 times that at the poles). The heated air expands and therefore its masses rise so that the pressure at a fixed height is greater in the equatorial atmosphere than in polar regions. In this way the zonal available potential energy $\bar{P}$ of the atmosphere is formed.

The zonal pressure difference creates an air outflow from the equator to the poles at upper levels, which obviously compensates for the air inflow from middle latitudes to the equator at lower levels (trades). The air flow from the equator to the poles at upper levels is turned to the east by the Coriolis force, which creates the west-to-east transport in the upper troposphere, i.e. cyclonic circumpolar currents. Below we shall explain the fact that relatively narrow currents are formed in these circumpolar currents (their width between the points where the velocity decreases down to half the maximum value is of the order of 300–400 km and their thickness is 1–2 km). These are the so-called subtropical jet currents at latitudes, on average, about ±35° and at a height of about 12 km (with pressure about 200 mbar) having maximum velocities of the order of 60 m/s or more.

Jet currents have been found to be baroclinically unstable (their energy is transferred to disturbances at an average rate equal to $\overline{\varrho'\mathbf{v}'} \cdot \nabla\,(g\bar{z} + c_v\bar{T}) < 0$). Small initial disturbances appearing in these currents increase and become Rossby waves with large latitudinal amplitudes (of the order of 400 km) and zonal wave numbers $k = 4, \ldots, 8$ and particularly $k = 5, 6$ (to which wavelengths of the order of 4000 km correspond). The Rossby waves travel to the east more slowly than the air in the main current (relative to which they propagate to the west with phase velocities of the order of 10 m/s).

Cyclonic and anticyclonic eddies formed in the troughs and crests of Rossby waves are in chessboard arrangement. Between their quadruples there appear saddle regions, or high-altitude deformation fields, along whose compression axes high-altitude frontal zones are formed. The collision of warm and cold air masses transformed (i.e. heated or cooled) by the underlying surface in the lower troposphere and at the Earth's surface below these frontal zones results in the formation of atmospheric fronts which, thus, are consequences of the existence of high-altitude eddies. Atmospheric fronts in the lower atmosphere are often also baroclinically unstable (particularly below the mouths of high-altitude frontal zones). Wave disturbances appearing in these fronts increase and generate less extensive near-ground eddies, i.e. higher order baroclinic modes (since the fronts are particularly sharp at the Earth's surface), or, in other words, low-level cyclones and anticyclones (sometimes the effects of currents flowing over the irregularities of the Earth's surface contribute to the formation of eddies). In this way the available potential energy P' of synoptic processes is generated (and their kinetic energy K' is generated by P') at the expense of the available potential and kinetic energies $\bar{P}$ and $\bar{K}$ of large-scale currents.

Low-level cyclonic eddies, on passing the immature cyclone stage with a warm sector, are sometimes occluded and penetrate the upper troposphere, i.e. are 'barotropized'. The stationary anticyclones concluding cyclonic series are also 'barotropized'. It is likely that the statistics of these 'barotropized' synoptic eddies is such that, owing to the work $\overline{\varrho \mathbf{v}'_\alpha \mathbf{v}'_\beta}\,\nabla^\alpha\,\bar{\mathbf{v}}^\beta$ of the corresponding Reynolds stresses, their ensemble transforms part of its kinetic energy K' into the kinetic energy $\bar{K}$ of jet currents (which are barotropically stable). Thus, the ensemble acts on these currents as negative viscosity, and this accounts for their existence. (The angular momentum is imparted to a jet current from tropical regions by wave disturbances since the axes of their troughs and crests are inclined from these regions towards the current, and therefore the angular momentum before the troughs (where it is greater) is transferred along the meridian to the current, while the angular momentum behind the troughs (where it is smaller) is carried away from the current.)

The major distinction between the oceanic circulation and the general atmospheric circulation is that the origin of large-scale oceanic currents differs from that of atmospheric large-scale currents. The main large-scale currents of the atmosphere's general circulation – i.e. the equatorial circulations and the west-to-east transport in middle latitudes are produced by heating from below (which decreases from the equator to the poles) and the Earth's rotation (the Coriolis force). However, the basic source of the main large-scale currents in the ocean is probably the wind (the wind stress at the sea surface) in combination with the influence of the coasts (which are absent in the atmosphere) and the Earth's rotation (or, more precisely, the variation of the Coriolis parameter with latitude owing to the sphericity of the Earth). West-intensified jet currents in the ocean, of the type of the Gulf Stream and the Kuroshio and also ACC, can be regarded as analogs of atmospheric subtropical jet currents. They are ten times as narrow and tens of times as slow as the atmospheric jet currents. For example, the width of the Gulf Stream east of the Blake Plateau between the points where the velocity decreases down to half its maximum value is 30–40 km, and the maximum velocity attains 1.5–2 m/s.

It is very important from the dynamical viewpoint that the average stratification of the ocean characterized by the depth-averaged Brunt–Väisälä frequency $\bar{N}$ is considerably less stable than that of the atmosphere even when only the troposphere is taken into consideration and the stratosphere, which possesses extremely stable stratification but contains only a very small fraction of the mass, is ignored. The troposphere stratification is determined by the heating of the stratosphere (which creates stability) owing to the absorption of ultraviolet solar radiation by ozone and the heating of the lower troposphere and the cooling of the upper troposphere (which produce instability) owing to the greenhouse effect. This results in the formation of a temperature difference between the Earth's surface, say $T_0 = +15$ °C, and the stratosphere, say $T_1 = -50$ °C at a height of 10 km. The corresponding average vertical temperature gradient is equal to $\gamma = 6.5$ °C/km and is smaller than the adiabatic gradient $\gamma_a \approx 10$ °C/km, i.e. the average stratification is found to be stable. Accordingly, the Brunt–Väisälä frequency is equal to $\bar{N} = [gT_0^{-1}(\gamma_a - \gamma)]^{1/2} \sim 10^{-2}$ s^{-1}. We should say that the ocean is cooled from above in contrast to the troposphere which is heated from below.

The heating from above can directly affect only a very thin upper layer of the ocean tens of meters thick since it is realized either in a contact way or as a result of strong absorption of light penetrating the water. The heated layer cannot mix on its own with lower waters since it is hydrostatically stable, and the forced mixing produced by the breaking surface waves does not penetrate deeply. The mixing in the upper layer of the ocean creates the seasonal pycnocline at its lower boundary with values of $\partial\varrho/\partial z \sim 10^{-6}$ g/cm^4, i.e. one and a half orders of magnitude greater than in the strong and rare temperature inversions in the near-ground atmospheric layer where $\partial T/\partial z \sim 1$ °C/m and $\partial\varrho/\partial z \sim 4 \times 10^{-8}$ g/cm^4 (and, more particularly, than in weak inversions at the upper boundary of the unstable atmospheric boundary layer which is often marked by rows of stratocumuli formed on internal waves). The seasonal pycnocline in the ocean is stratified considerably more stably in terms of the Brunt–Väisälä frequency ($N \sim 3 \times 10^{-2}$ s^{-1}) than, on average, the troposphere ($\bar{N} \sim 10^{-2}$ s^{-1}) but less stably than strong atmospheric inversions ($N \sim 1.7 \times 10^{-1}$ s^{-1}). Since the upper mixed layer is spread everywhere in the ocean and strong inversions in the atmosphere are rare, what has been said accounts for the fact that internal waves are much more widely developed in the ocean than in the atmosphere. However, the seasonal pycnocline is thin and therefore its stratification, i.e. the effect of the ocean heating from above, influences the average stratification of the ocean insignificantly.

Conversely, in the regions where the ocean is cooled from above, the cooled waters sink – i.e. the mixing which weakens the stratification stability is analogous to the action of the troposphere heating from below (although the mixing is limited since the cooling temperature cannot be lower than the freezing temperature of water). At the same time, there is no analog of the troposphere heating from above which creates its stability, i.e. there is no cooling from below in the ocean. As a result, the average stratification in the ocean, however stable, is much closer to the adiabatic stratification than in the atmosphere: on average, $\bar{N} \sim 10^{-3}$ s^{-1} for the ocean.

Knowing the average stratification $\bar{N}$, we can determine the typical horizontal scale of synoptic processes – say, of first-mode baroclinic Rossby waves – by calculating it as the Rossby deformation radius $L_R = (\bar{N}/f)H$ by formula (1.1) (see Chapter 2, Section 2.1). We obtain $L_R = 2000$ km for Rossby waves in an atmospheric jet current taking the average value $\bar{N} = 2 \times 10^{-2}$ s^{-1} for the neighborhood of the current in the upper troposphere and the lower stratosphere and putting $f = 10^{-4}$ s^{-1} and $H = 10$ km. (If the dimension of an eddy is meant, then the wavelength is twice as large as L_R and is equal to 4000 km.) For the ocean we take $\bar{N} = 10^{-3}$ s^{-1} and $H = 5$ km and obtain $L_R = 50$ km. Further, the typical time scale $\tau = 2(\beta L_R)^{-1}$ of the first-mode zonal baroclinic Rossby wave for $\beta = 2 \times 10^{-8}$ km^{-1} s^{-1} is $\tau = 5 \times 10^4$ s = 14 h for the atmosphere and $\tau = 2 \times 10^6$ s = 556 h = 23 days for the ocean. The typical phase velocity $c = L_R/\tau = \beta L_R^2/2$ of oceanic eddies is 2.5 cm/s. Thus, owing to the fact that the average stratification (effective thickness) of the ocean is an order of magnitude smaller than (twice as small as) that of the atmosphere, synoptic eddies in the ocean are tens of times as small in their dimensions and hundreds of times as slow (and long-lived) as those in the atmosphere.

Meanders in oceanic jet currents and ring-shaped frontal synoptic eddies (called *rings*), which are formed when instability sets in and meanders are cut off, are analogs of Rossby waves in atmospheric jet currents. For instance, in the Gulf Stream, which flows from west to east and separates the cold northern and warm southern waters, the cut-off eddies north of the stream have anticyclonic sign and contain warm cores with southern water (hampering downwelling, which is characteristic of anticyclones) whereas the cut-off eddies south of the stream have cyclonic sign and contain cold cores with northern water (hampering upwelling, which is characteristic of cyclones). It seems that rings extend only to the upper half of the ocean. They move together with the water contained in them and in this sense they are eddies and not waves. However, their motion is characterized by a systematic displacement to the west, which is typical of Rossby waves.

Meanders in jet currents and rings do not produce analogs of atmospheric high-altitude deformation fields and high-altitude frontal zones. Hence, they do not induce frontogenesis at the underlying surface (at the bottom), particularly because there is almost no transformation of water masses under the influence of the bottom; and near-bottom fronts, say, between Arctic and Antarctic water masses and bottom waters of some other origin, have not yet been observed. Thus, there seem to be no oceanic analogs of atmospheric low-level frontal cyclones and anticyclones. However, their role is probably played by synoptic eddies in the open ocean which can be formed owing to the baroclinic instability of even weak oceanic currents. (The instability is likely to be strengthened because of the presence of regional and local thermohaline fronts which are rather widely spread in the ocean; then there can appear analogs of cyclones with a warm sector and even occlusions. It seems that occlusions were observed in meanders of the winter regional front in the Sargasso Sea.) Apparently, these *free* synoptic eddies in the ocean are mainly concentrated in its upper half (in contrast to atmospheric low-level cyclones and anticyclones).

On the other hand, such factors as eddy formation processes and the generation of the so-called topographic Rossby waves, when currents flow over the bottom irregularities, must play a substantially more important role in the ocean than in the atmosphere. The contribution of the bottom relief is determined by the ratio h/H, where h is the root-mean-square height of the irregularities of the bottom relief. It becomes comparable with the β-effect for $h/H \sim L/a$, where L is the wavelength and a is the radius of the Earth. The critical value $h = HL/a$ in the ocean is tens of times as small as that in the atmosphere. Topographic eddies and Rossby waves must be most clearly marked in bottom layers of the ocean and, particularly, behind obstacles (however, so far there have been no direct observations of this kind).

Finally, there is another possible mechanism of generating synoptic eddies and Rossby waves which has no analogs in the atmosphere – namely, their direct generation by atmospheric action on the sea surface, i.e. by inhomogeneities with corresponding scales in the fields of the wind stress curl, and atmospheric pressure at the surface. Synoptic eddies and waves generated in this way must be most clearly marked in the uppermost layers of the ocean.

Thus, synoptic motions in the ocean can be classified according to their generation mechanisms into frontal motions, i.e. meanders in jet currents and rings (I), free motions or eddies in the open ocean generated by internal baroclinic instability (II), topographic effects (III), and atmospheric factors (IV). They are divided into cyclonic (C) and anticyclonic (A) motions depending on the sense of their rotation. Further, they can be classified into eddies moving together with the water contained in them (E) and waves travelling over the water (W), and, in the latter case, solitary waves or solitons (S) can be separated out. Finally, it may be useful to distinguish between these formations according to their ages.

Rossby waves are encountered in the atmosphere only in the form of isolated and rather rapidly damped wave trains. By contrast, in the ocean the variety of generation mechanisms and the extent to which they are spread, the weak damping, and the additional possibility of reflection from the coasts (which is one more distinction between the ocean and the atmosphere) allow Rossby waves to form considerably more developed wave fields whose complete diagnosis requires a rather complicated space–time analysis.

By analogy with low-level synoptic eddies in the atmosphere, eddies in the open ocean (II, III, and IV) seem to have a tendency towards 'barotropization'. Again by analogy with the atmosphere, their ensemble probably has a negative viscosity effect on oceanic jet currents. This means that it imparts a fraction of its kinetic energy to them owing to the work $\varrho \overline{\mathbf{v}'_\alpha \mathbf{v}'_\beta} \nabla^\alpha \overline{\mathbf{v}}^\beta > 0$ of the corresponding Reynolds stresses (and thus carries out the spectral transfer of kinetic energy from small scales to large scales, which is characteristic of quasi-two-dimensional turbulence). This probably accounts for the narrow-jet character of a number of large-scale currents in the ocean.

CHAPTER 2

Theory of Rossby Waves

1. THE QUASIGEOSTROPHIC APPROXIMATION

1.1. Continuous Stratification

We shall now consider adiabatic motions without external forces. Then in the quasistatic and Boussinesq approximations the equations of motion, mass conservation, and entropy evolution have the form (see Kamenkovich and Monin, 1978, §§2 and 5):

$$\frac{\mathrm{d}u}{\mathrm{d}t} - \frac{uv}{a} \tan \varphi - fv = - \frac{1}{\varrho_0} \frac{\partial p}{a \cos \varphi \, \partial \lambda}; \tag{1.1}$$

$$\frac{\mathrm{d}v}{\mathrm{d}t} + \frac{u^2}{a} \tan \varphi + fu = - \frac{1}{\varrho_0} \frac{\partial p}{a \, \partial \varphi}; \tag{1.2}$$

$$\frac{\partial p}{\partial z} = - g\varrho; \tag{1.3}$$

$$\frac{\partial u}{a \cos \varphi \, \partial \lambda} + \frac{1}{\cos \varphi} \frac{\partial}{a \, \partial \varphi}(v \cos \varphi) + \frac{\partial w}{\partial z} = 0; \tag{1.4}$$

$$\frac{\mathrm{d}\varrho}{\mathrm{d}t} = \frac{1}{c^2} \frac{\mathrm{d}p}{\mathrm{d}t}, \tag{1.5}$$

where

$$\frac{\mathrm{d}}{\mathrm{d}t} = \frac{\partial}{\partial t} + \frac{u}{a \cos \varphi} \frac{\partial}{\partial \lambda} + \frac{v}{a} \frac{\partial}{\partial \varphi} + w \frac{\partial}{\partial z}; \qquad f = 2\omega_e \sin \varphi. \tag{1.6}$$

Here λ, φ, and z are spherical coordinates; λ is longitude $(-\pi < \lambda \leqslant \pi)$; φ is latitude $(-\pi/2 \leqslant \varphi \leqslant \pi/2)$; $z = r - a$, r is the distance from the center of the Earth; a is the Earth's radius; t is time; u, v, and w are velocity components along the axes λ, φ, and z; ϱ is the density; p is the pressure; c is the speed of sound; g is the acceleration due to gravity; ω_e is the Earth's angular velocity; and ϱ_0 is the mean density.

We are interested in synoptic motions (Section 4 of Chapter 1) with a time scale T of the order of several days or more, horizontal scales L of the order of 100 km, and vertical scales H of the order of the mean depth of the ocean (4 km) for which the ratio H/L is small. Therefore, the equation of motion along the vertical (1.3) is written in the quasistatic approximation. Accordingly, in Equations (1.1) and (1.2) the terms uw/r, $2\omega_e w \cos\varphi$, and vw/r are not included, and r is replaced by a ($H/a \ll 1$) in all the expressions. In the mass conservation equation the term $d\varrho/dt$ is discarded; its inclusion in the equation is necessary only when acoustic phenomena are described. The factor $1/\varrho$ on the right-hand sides of (1.1) and (1.2) is replaced by $1/\varrho_0$ (the Boussinesq approximation).

The speed of sound c varies very slightly in the ocean; therefore, we replace c in (1.5) by an average (equilibrium) value $c(z)$. Then the system (1.1)–(1.5), with unknowns u, v, w, p, and ϱ, becomes closed.

The typical length scales H and L and the horizontal velocity scale U are assumed to be given. We exclude the near-equator region from our consideration and introduce the Kibel' number[1] $Ki = U/f_0L$, where $f_0 = 2\omega_e \sin\varphi_0$ (the ratio of the typical value of nonlinear terms in (1.1) and (1.2) to the typical value of the Coriolis force) and the parameter $\varepsilon_T = 1/(Tf_0)$ (the ratio of the typical value of local acceleration in (1.1) and (1.2) to the typical value of the Coriolis force). For synoptic motions we have $U \lesssim 10$ cm/s and, therefore,

$$Ki = \frac{U}{f_0L} \ll 1; \qquad \varepsilon_T = \frac{1}{Tf_0} \ll 1; \qquad \varepsilon_L = \frac{L}{a} \ll 1. \tag{1.7}$$

The density and pressure distributions, ϱ and p, in the ocean can be represented in the form

$$\varrho = \varrho_s(z) + \varrho'; \quad p = p_a + \int_z^0 g\varrho_s(z)\,dz + p', \tag{1.8}$$

where p_a is the atmospheric pressure, ϱ_s is the average (equilibrium) density distribution for the region under consideration, and p' and ϱ' are pressure and density disturbances.

We now state the boundary conditions along the vertical. Usually the departure ζ of the free surface of the ocean from the equilibrium surface $r = a$ ($z = 0$) is small and, therefore, the dynamic and kinematic conditions can be set for $z = 0$. Using (1.8) it can easily be shown (see Kamenkovich, 1973, Ch. 4, §7) that

$$w = \frac{\partial\zeta}{\partial t} + \frac{u}{a\cos\varphi}\frac{\partial\zeta}{\partial\lambda} + \frac{v}{a}\frac{\partial\zeta}{\partial\varphi}; \qquad p' = g\varrho_0\zeta \quad \text{for} \quad z = 0. \tag{1.9}$$

At the bottom of the ocean

$$w = \frac{u}{a\cos\varphi}\frac{\partial b}{\partial\lambda} + \frac{v}{a}\frac{\partial b}{\partial\varphi} \quad \text{for} \quad z = -H + b(\lambda, \varphi), \tag{1.10}$$

where H is the mean depth and b is the height of bottom irregularities.

[1] In many papers the dimensionless ration $U/f_0 L$ is called the Rossby number. However, it was first introduced in the paper by Kibel' (1940) devoted to the analysis of the quasigeostrophic approximation (also see Phillips, 1963).

We consider a region in the ocean in the vicinity of a point with coordinates λ_0, φ_0 and determine the scales W, Z, $\mathcal{P}$ and R for the vertical velocity, the departure of the free surface from its rest state, and the pressure and density disturbances, respectively. By virtue of the smallness of parameters ε_T and Ki, the motion is nearly geostrophic. It follows that $\mathcal{P} = \varrho_0 f_0 UL$, and, by virtue of (1.3) and the second condition (1.9), we have $R = \gamma^2\ Ki\ \varrho_0$ and $Z = \gamma^2\ Ki\ H$, where $\gamma^2 = f_0^2 L^2/gH$. The scale W is found from the entropy evolution equation (1.5) (the density disturbances are connected with vertical motion). Using (1.8), introducing the Brunt–Väisälä frequency

$$N^2 = -\frac{g}{\varrho_0}\frac{\mathrm{d}\varrho_s}{\mathrm{d}z} - \frac{g^2}{c^2}, \tag{1.11}$$

and equating, for instance, the orders of the terms $\partial\varrho'/\partial t$ and $(\varrho_0 N^2/g)\, w$, we obtain

$$W = \left(\frac{L}{L_\mathrm{R}}\right)^2 \varepsilon_T \frac{H}{L} U, \tag{1.12a}$$

where L_R is the Rossby deformation radius (see Section 1 of Chapter 1).

Expanding sin φ, cos φ, and tan φ into Taylor series in the neighborhood of φ_0 and introducing the local Cartesian coordinates x and y

$$x = a \cos \varphi_0 (\lambda - \lambda_0); \qquad y = a(\varphi - \varphi_0) \tag{1.13}$$

to replace λ and φ, we write Equations (1.1)–(1.5) and conditions (1.9) and (1.10) in dimensionless form:

$$\varepsilon_T \frac{\partial u}{\partial t} + Ki\left(\frac{u}{1 + O(\varepsilon_L)}\frac{\partial u}{\partial x} + v\frac{\partial u}{\partial y}\right) + \left(\frac{L}{L_\mathrm{R}}\right)^2 \varepsilon_T\, Ki\, w \frac{\partial u}{\partial z} -$$

$$- Ki\ \varepsilon_L\ uv \tan \varphi_0 [1 + O(\varepsilon_L)] - [1 + \varepsilon_L y \cot \varphi_0 + O(\varepsilon_L^2)]\, v$$

$$= -\frac{1}{1 - \varepsilon_L y \tan \varphi_0 + O(\varepsilon_L^2)}\frac{\partial p'}{\partial x}; \tag{1.14}$$

$$\varepsilon_T \frac{\partial v}{\partial t} + Ki\left(\frac{u}{1 + O(\varepsilon_L)}\frac{\partial v}{\partial x} + v\frac{\partial v}{\partial y}\right) + \left(\frac{L}{L_\mathrm{R}}\right)^2 \varepsilon_T\, Ki\, w \frac{\partial v}{\partial z} +$$

$$+ Ki\ \varepsilon_L u^2 \tan \varphi_0 [(1 + O(\varepsilon_L)]$$

$$+ [1 + \varepsilon_L y \cot \varphi_0 + O(\varepsilon_L^2)]\, u = -\frac{\partial p'}{\partial y}; \tag{1.15}$$

$$\frac{\partial p'}{\partial z} = -\varrho'; \tag{1.16}$$

$$\frac{1}{1-\varepsilon_L y \tan\varphi_0 + O(\varepsilon_L^2)}\frac{\partial u}{\partial x} + \frac{\partial v}{\partial y} - \varepsilon_L v \tan\varphi_0 [1 + O(\varepsilon_L)] + $$

$$+ \left(\frac{L}{L_R}\right)^2 \varepsilon_T \frac{\partial w}{\partial z} = 0; \tag{1.17}$$

$$\varepsilon_T \frac{\partial \varrho'}{\partial t} + Ki\left(\frac{u}{1 + O(\varepsilon_L)}\frac{\partial \varrho'}{\partial x} + v\frac{\partial \varrho'}{\partial y}\right) + \left(\frac{L}{L_R}\right)^2 \varepsilon_T Ki\, w \frac{\partial \varrho'}{\partial z} - \varepsilon_T N^2 w$$

$$= \frac{gH}{c^2}\left[\varepsilon_T \frac{\partial p'}{\partial t} + Ki\left(\frac{u}{1 + O(\varepsilon_L)}\frac{\partial p'}{\partial x} + v\frac{\partial p'}{\partial y}\right) + \left(\frac{L}{L_R}\right)^2 \varepsilon_T Ki\, w\frac{\partial p'}{\partial z}\right]; \tag{1.18}$$

$$\left(\frac{L}{L_R}\right)^2 \varepsilon_T w = \gamma^2\left\{\varepsilon_T \frac{\partial \zeta}{\partial t} + Ki\left[\frac{u}{1 + O(\varepsilon_L)}\frac{\partial \zeta}{\partial x} + v\frac{\partial \zeta}{\partial y}\right]\right\},$$

$$p' = \zeta \quad \text{for} \quad z = 0; \tag{1.19}$$

$$\left(\frac{L}{L_R}\right)^2 \varepsilon_T w = \varepsilon_b\left(\frac{u}{1 + O(\varepsilon_L)}\frac{\partial b}{\partial x} + v\frac{\partial b}{\partial y}\right) \quad \text{for} \quad z = -1 + \frac{b_0}{H} b, \tag{1.20}$$

where $\varepsilon_b = b_0 L/HL_b$, b_0 is the typical vertical scale of bottom irregularities, and L_b is the typical horizontal scale of these irregularities. For the sake of simplicity, the dimensionless variables x, y, z, t, u, v, w, p', ϱ', N, b, and ζ are denoted by the same letters as before. For our further aims we introduce $\beta = (2\,\omega_e/a)\cos\varphi_0$.

We now assume that the parameters Ki, ε_T, ε_L, γ, and gH/c^2 are of the same order of magnitude and ε_b and b_0/H are of that same order or smaller. This assumption usually holds for the typical values of the determining parameters of the ocean. For instance, if $H = 4$ km, $g = 9.8$ m/s^2, $a = 6.37 \times 10^3$ km, $c = 1.5$ km/s, $U = 5$ cm/s, $L = 50$ km, $f_0 = 7.29 \times 10^{-5}$ s^{-1} ($\varphi_0 = 30°$), $L = L_b$, $T = 1/\beta L$ (T is the typical time scale for Rossby waves equal to $1/(1.98 \times 10^{-11} \times 5 \times 10^4) = 10^6$s $= 11.6$ days; see Section 2 of Chapter 2), and $b_0 = 40$ m, we have $\varepsilon_L = 0.79 \times 10^{-2}$; $Ki = 1.4 \times 10^{-2}$; $\varepsilon_T = 1.4 \times 10^{-2}$; $\gamma = 1.8 \times 10^{-2}$; $gH/c^2 = 1.7 \times 10^{-2}$; $b_0/H = 1.0 \times 10^{-2}$. It should be noted that in some special cases the small parameters ε_L, Ki, ε_T, and γ are not necessarily of the same order; however, we shall not consider these cases here.

We seek the solution of the problem in the form of the following series:

$$(u, v, w, p', \varrho', \zeta) = (u_0, v_0, w_0, p'_0, \varrho'_0, \zeta_0) + \\ + \mu(u_1, v_1, w_1, p'_1, \varrho'_1, \zeta_1) + \ldots, \tag{1.21}$$

where μ is a small parameter of the same order as the parameters ε_l, Ki, ε_T, γ, and gH/c^2.

Substituting the series (1.21) into Equations (1.14)–(1.18) and into boundary conditions (1.19) and (1.20), and assuming that $L \sim L_R$, we equate to zero the coefficients in the zeroth power of the small parameter (the zeroth approximation), the first power (the first approximation), etc. We write down the zeroth and first approximations for Equations (1.14), (1.15), and (1.17), the zeroth approximation for (1.16), and the first approximation for (1.18), (1.19), and (1.20):

$$-v_0 = -\frac{\partial p'_0}{\partial x}; \tag{1.22$_0$}$$

$$\varepsilon_T \frac{\partial u_0}{\partial t} + Ki\left(u_0 \frac{\partial u_0}{\partial x} + v_0 \frac{\partial u_0}{\partial y}\right) - \varepsilon_L y \cot \varphi_0 v_0 - \mu v_1$$

$$= -\mu \frac{\partial p'_1}{\partial x} - \varepsilon_L y \tan \varphi_0 \frac{\partial p'_0}{\partial x}; \tag{1.22$_1$}$$

$$u_0 = -\frac{\partial p'_0}{\partial y}; \tag{1.23$_0$}$$

$$\varepsilon_T \frac{\partial v_0}{\partial t} + Ki\left(u_0 \frac{\partial v_0}{\partial x} + v_0 \frac{\partial v_0}{\partial y}\right) + \varepsilon_L y \cot \varphi_0 u_0 +$$

$$+ \mu u_1 = -\mu \frac{\partial p'_1}{\partial y}; \tag{1.23$_1$}$$

$$\frac{\partial p'_0}{\partial z} = -\varrho'_0; \tag{1.24$_0$}$$

$$\frac{\partial u_0}{\partial x} + \frac{\partial v_0}{\partial y} = 0; \tag{1.25$_0$}$$

$$\varepsilon_L y \tan \varphi_0 \frac{\partial u_0}{\partial x} + \mu\left(\frac{\partial u_1}{\partial x} + \frac{\partial v_1}{\partial y}\right) - \varepsilon_L \tan \varphi_0 v_0 +$$

$$+\left(\frac{L}{L_R}\right)^2 \varepsilon_T \frac{\partial w_0}{\partial z} = 0; \tag{1.25$_1$}$$

$$\varepsilon_T \frac{\partial \varrho'_0}{\partial t} + Ki\left(u_0 \frac{\partial \varrho'_0}{\partial x} + v_0 \frac{\partial \varrho'_0}{\partial y}\right) - \varepsilon_T N^2 w_0 = 0; \tag{1.26$_1$}$$

$$w_0 = 0; \quad p'_0 = \zeta_0 \quad \text{for} \quad z = 0; \tag{1.27$_1$}$$

$$\left(\frac{L}{L_R}\right)^2 \varepsilon_T w_0 = \varepsilon_b \left(u_0 \frac{\partial b}{\partial x} + v_0 \frac{\partial b}{\partial y}\right) \quad \text{for} \quad z = -1. \tag{1.28$_1$}$$

It should be noted that Equation (1.25_0) is a simple consequence of (1.22_0) and (1.23_0) and, therefore, at first glance the system of zero-order equations seems to be 'linked' with the system of first-order equations. However, we shall show that it is possible to derive a closed system of equations and boundary conditions for u_0, v_0, w_0, p'_0, ϱ'_0, and ζ_0.

Eliminating the pressure p'_1 from (1.22_1) and (1.23_1) by means of cross differentiation with respect to x and y and expressing $\partial u_1/\partial x + \partial v_1/\partial y$ with the aid of Equation (1.25_1) we obtain, taking into account (1.22_0), (1.23_0), and (1.25_0), the vorticity equation

$$\varepsilon_1 \frac{\partial}{\partial t}\left(\frac{\partial v_0}{\partial x} - \frac{\partial u_0}{\partial y}\right) + Ki\left(u_0 \frac{\partial}{\partial x} + v_0 \frac{\partial}{\partial y}\right)\left(\frac{\partial v_0}{\partial x} - \frac{\partial u_0}{\partial y}\right) +$$

$$+ \varepsilon_L \cot \varphi_0 v_0 - \left(\frac{L}{L_R}\right)^2 \varepsilon_T \frac{\partial w_0}{\partial z} = 0. \tag{1.29}$$

Equations (1.22_0), (1.23_0), (1.24_0), (1.26_1), and (1.29) and boundary conditions (1.27_1) and (1.28_1), involve only the unknown functions u_0, v_0, w_0, p'_0; ϱ'_0, and ζ_0.

We now derive some additional relations. Consider a closed region V with vertical walls (Figure 2.1.1). We denote the components of the horizontal velocity in the coordinate system x, y, by $\hat{u}$ and $\hat{v}$; then according to (1.13),

$$\hat{u} = \frac{dx}{dt} = a \cos \varphi_0 \frac{d\lambda}{dt} = \frac{\cos \varphi_0}{\cos \varphi}\, u,$$

$$\hat{v} = \frac{dy}{dt} = \frac{a\,d\varphi}{dt} = v.$$

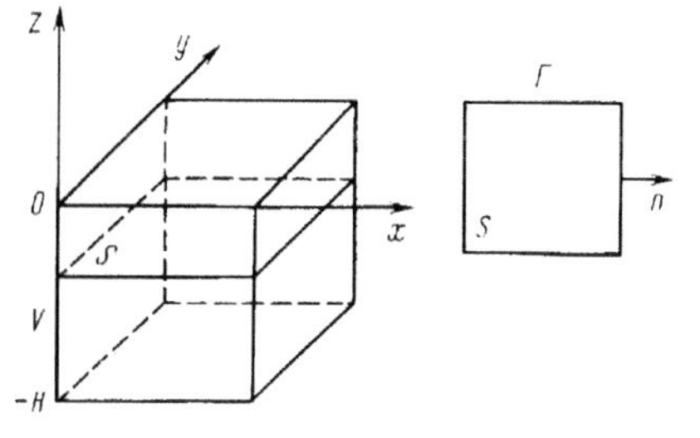

Fig. 2.1.1. Region of integration of the equations under study. Shown on the right are the section S and the boundary Γ.

Expanding into series of the form of (1.21) we obtain

$$\hat{u}_0 = u_0, \ \hat{v}_0 = v_0; \ \mu \hat{u}_1 = \mu u_1 + \varepsilon_L y \tan \varphi_0 u_0, \ \hat{v}_1 = v_1. \tag{1.30}$$

Integrating (1.25_1) over the section S of the region V by a plane z = const, and taking into account that in the coordinate system x, y the normal components of the velocities

$$(\hat{u}_0, \hat{v}_0), (\hat{u}_1, \hat{v}_1) \ldots .$$

must vanish at the boundary Γ, we conclude, by virtue of (1.22_0), (1.23_0), and (1.30), that

$$\iint_S \frac{\partial w_0}{\partial z} dx\, dy = 0 \quad \text{for} \quad -1 \leqslant z \leqslant 0. \tag{1.31}$$

Currents in a zonal channel are important. Since v_0 and v_1 vanish on the walls of the channel, assuming that the functions u_0, p'_0, and p'_1 are bounded for $x \to \pm\infty$ we find from (1.22_1) that

$$\lim_{X \to \infty} \frac{1}{2X} \int_{-X}^{X} \frac{\partial u_0}{\partial t} dx = 0 \tag{1.32}$$

on the walls of the channel.

Thus the solution of the problem in the zeroth approximation depends on the following five parameters:

$$\frac{\varepsilon_T}{\varepsilon_L \cot \varphi_0} = \frac{1}{T\beta L};\ \frac{\text{Ki}}{\varepsilon_L \cot \varphi_0} = \frac{U}{\beta L^2};\ \frac{Ki}{\varepsilon_T} = \frac{UT}{L};\ \frac{L}{L_R};\ \frac{\varepsilon_b}{\varepsilon_T}.$$

We remind the reader that $\beta, f_0, U, L, H, \bar{N}$ (the average value of N), b_0, and L_b are the determining parameters of the problem under study. If $1/(\beta L)$ (the time scale of Rossby waves) is taken as T, the number of parameters reduces to three: $U/(\beta L^2)$, L/L_R, and $\varepsilon_b/\varepsilon_T$; in the case of an ocean of constant depth, the number of parameters reduces to two: $U/(\beta L^2)$ and L/L_R.

Confining ourselves to the solution of the problem in the zeroth approximation and returning to the dimensional variables we obtain the system of equations of the quasigeostrophic approximation:

$$u = -\frac{1}{f_0\varrho_0}\frac{\partial p'}{\partial y};\quad v = \frac{1}{f_0\varrho_0}\frac{\partial p'}{\partial x};\quad \varrho' = -\frac{1}{g}\frac{\partial p'}{\partial z}; \tag{1.33}$$

$$\left(\frac{\partial}{\partial t} + u\frac{\partial}{\partial x} + v\frac{\partial}{\partial y}\right)\left(\frac{1}{f_0\varrho_0}\Delta p' + \beta y\right) - f_0\frac{\partial w}{\partial z} = 0; \tag{1.34}$$

$$\left(\frac{\partial}{\partial t} + u\frac{\partial}{\partial x} + v\frac{\partial}{\partial y}\right)\varrho' - \frac{\varrho_0 N^2}{g} w = 0 \tag{1.35}$$

with boundary conditions

$$w = 0;\quad p' = g\varrho_0\zeta \text{ for } z = 0; \tag{1.36}$$

$$w = u\frac{\partial b}{\partial x} + v\frac{\partial b}{\partial y} \quad \textit{for } z = -H; \tag{1.37}$$

$$v_n\,|_\Gamma = 0,\quad \int_\Gamma \frac{\partial^2 p'}{\partial n\, \partial t} d\Gamma = 0 \text{ for } -H \leqslant z \leqslant 0, \tag{1.38a}$$

where Δ is the horizontal Laplace operator.

The second condition (1.38a) follows from condition (1.31) and Equation (1.34). Here we do not dwell on setting boundary conditions for an open region when fluid particles can flow across the lateral boundary (see Kamenkovich *et al.*, 1981).

In the case of motion in a zonal channel $0 \leqslant y \leqslant \mathscr{L}$, conditions (1.38a) are replaced by

$$v = 0, \ \lim_{X \to \infty} \frac{1}{2X} \int_{X}^{X} \frac{\partial u}{\partial t} \, dx = 0 \ \text{ for } \ y = 0, y = \quad \text{, and all } z. \tag{1.38b}$$

Eliminating w from Equations (1.34) and (1.35) we obtain the transfer equation for the quasigeostrophic potential vorticity ω_*:

$$\left(\frac{\partial}{\partial t} + u \frac{\partial}{\partial x} + v \frac{\partial}{\partial y} \right) \omega_* = 0;$$

$$\omega_* = \frac{1}{f_0 \varrho_0} \Delta p' + \beta y + \frac{\partial}{\partial z} \left(\frac{f_0}{\varrho_0 N^2} \frac{\partial p'}{\partial z} \right). \tag{1.39a}$$

This equation follows (for the indicated approximations) from the general transfer equation for the potential vorticity (see Section 5 of Chapter 3 and also Kamenkovich and Reznik, 1978, §1).

We determine the displacement ζ_ϱ of the isopycnic surface lying at a depth $z = z_0$ in the undisturbed state. We have

$$\varrho_s (z_0 + \zeta_\varrho) + \varrho' (x, y, z_0 + \zeta_\varrho, t) = \varrho_s (z_0),$$

whence, using the smallness of ζ_ϱ/H, we find

$$\zeta_\varrho = - \frac{\varrho'}{d\varrho_s/dz.} \tag{1.40}$$

The above derivation holds for baroclinic motions when the density inhomogeneity plays a significant role. However, we are also interested in motions of a stratified fluid for which the density inhomogeneity is of minor importance. These motions will be called *barotropic*. Formula (1.12a) is inapplicable to barotropic motions. Assuming that the pressure p' does not depend on z we see that, by virtue of the geostrophic relations, u and v do not depend on z either. It readily follows that w depends linearly on z; assuming that the displacement of the free surface is insignificant, by virtue of (1.9) and (1.10), we obtain

$$W = \varepsilon_b \frac{H}{L} U. \tag{1.12b}$$

As before, equating the orders of the terms $\partial\varrho'/\partial t$ and $(\varrho_0 N^2/g)w$ in Equation (1.5) we find the scale R for the density disturbances. The condition that the motion is barotropic obviously has the form

$$\frac{\partial p'/\partial z}{\mathcal{P}/H} \sim \frac{gR}{\mathcal{P}/H} \ll 1.$$

Substituting the expressions of the scales R and $\mathcal{P}$ into this relation we find (Kamenkovich and Reznik, 1978, §1; Reznik, 1976):

$$\left(\frac{L_R}{L}\right)^2 \frac{\varepsilon b}{\varepsilon_T} \ll 1. \tag{1.41}$$

For barotropic quasigeostrophic motions the vorticity equation (1.34) and boundary conditions (1.36) and (1.37) are valid; their derivation is analogous to the above derivation for baroclinic motions.

Since

$$w = -\frac{z}{H}\left(u\frac{\partial b}{\partial x} + v\frac{\partial b}{\partial y}\right),$$

the transfer equation for the quasigeostrophic potential vorticity $\omega_*^{(b)}$ of barotropic motions has the form

$$\left(\frac{}{t} + u\frac{\partial}{\partial x} + v\frac{\partial}{\partial y}\right)\omega_*^{(b)} = 0;$$

$$\omega_*^{(b)} = \frac{1}{f_0\varrho_0}\Delta p' + \beta y + f_0\frac{b}{H}. \tag{1.39b}$$

To this equation we must add the first two equations (1.33) and the corresponding boundary conditions on the basin walls.

It should be noted that Equation (1.39b) holds on condition that the parameters ε_T, Ki, ε_L, b_0/H, ε_b, γ, gH/c^2, and $(L_R/L)^2(\varepsilon_b/\varepsilon_T)$ are small (the scale L is not necessarily of the same order as L_R). Under the same assumptions Equation (1.39b) describes quasigeostrophic motions in a homogeneous ocean as well.

The quasigeostrophic approximation was considered in many works (e.g. see Kamenkovich and Reznik, 1978, §1; Monin, 1973, §7; Charney and Flierl, 1981, §18.4; Pedlosky, 1979, Ch. 6, and the literature indicated in these works). The necessity of setting an additional boundary condition of type (1.31) was proved by Buchwald (1973) and Larichev (1974). Condition (1.32) was also considered in a number of works (e.g. see Pedlosky, 1979, Ch. 7).

1.2. Multilayer Model

To simplify the above equations various approximations of the vertical structure of the ocean are often used. In this connection we consider the so-called n-layer fluid model consisting of n nonmissible homogeneous layers with thicknesses h_k in the undisturbed state (Figure 2.1.2). It is assumed that the densities ϱ_k of the layers are chosen so that $\varrho_1 = \varrho_s(-h_1/2)$, $\varrho_2 = \varrho_s(-h_1 - h_2/2)$, etc. We obviously have $\varrho_1 < \varrho_2 < \varrho_3 < \cdots < \varrho_n$.

The basic equations for the multilayer fluid in the quasigeostrophic approximation can easily be derived directly from Equations (1.33)–(1.35), assuming that $\varrho_s(z)$ is a step function and putting $N^2 = -g/\varrho_0(\mathrm{d}\varrho_s/\mathrm{d}z)$ (e.g. see Flierl, 1978). However, it is advisable to present an alternative derivation based on the analysis of the original equations (1.1)–(1.4).

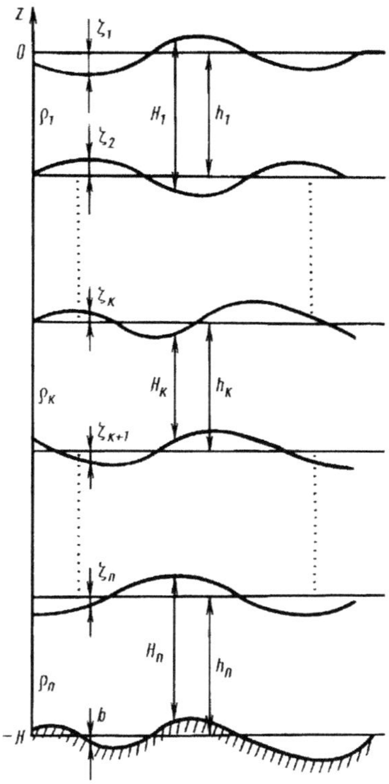

Fig. 2.1.2. Schematic representation of the n-layer fluid.

We supply the velocity components and the pressure in the kth layer with subscript k and introduce the thickness $H_k = h_k + \zeta_k - \zeta_{k+1}$ $(k = \overline{1, n})$ of the layer in the disturbed state, assuming that, $\zeta_{n+1} = b$. Equations (1.1)–(1.4) hold within the limit of each of the layers. The equilibrium pressure in the undisturbed state is denoted $p_k^{(0)}$. Integrating (1.3) with respect to z and taking into account the continuity of the pressure we find

$$p_k^{(0)} = p_k^{(00)} - g\,\varrho_k\, z, \quad p_k^{(00)} = p_a + g \sum_{i=1}^{k-1} (\varrho_i - \varrho_k)\, h_i; \quad k = \overline{1, n}.$$

We now introduce the pressure disturbance p'_k. It is obvious that within the limits of the kth layer p'_k does not depend on z. Writing the condition of continuity of the total pressure $p_k = p_k^{(0)} + p'_k$ at the interface:

$$z = -\sum_{i=1}^{k-1} h_i + \zeta_k,$$

we find

$$p'_{k-1} - g\varrho_{k-1}\zeta_k = p'_k - g\varrho_k\zeta_k, \quad k = \overline{2,n}. \tag{1.42a}$$

By virtue of the second condition (1.9),

$$p'_1 = g\varrho_1\zeta_1. \tag{1.42b}$$

Since p'_k does not depend on z, the geostrophic relations imply that u_k and v_k do not depend on z and hence, according to (1.4), $\partial w_k/\partial z$ does not depend on z either. Naturally, this assertion holds only in the zeroth approximation. Since the layers are nonmissible, the following relations hold at the interface:

$$w_{k-1}^{+} = \frac{\partial \zeta_k}{\partial t} + u_{k-1}\frac{\partial \zeta_k}{a\cos\varphi\,\partial\lambda} + v_{k-1}\frac{\partial \zeta_k}{a\,\partial\varphi}, \quad k = \overline{2,n}; \tag{1.43a}$$

$$w_k^{-} = \frac{\partial \zeta_k}{\partial t} + u_k\frac{\partial \zeta_k}{a\cos\varphi\,\partial\lambda} + v_k\frac{\partial \zeta_k}{a\,\partial\varphi}, \quad k = \overline{1,n}; \tag{1.43b}$$

where '±' means above (or below) the interface.

Integrating Equations (1.4) over the thickness H_k of each of the layers we find

$$\frac{\partial H_k}{\partial t} + \operatorname{div}(H_k\,\mathbf{v}_k) = 0, \tag{1.44}$$

where by 'div' is meant the divergence operator on a sphere of radius a.

Discarding the small terms $w(\partial u/\partial z)$ and $w(\partial v/\partial z)$ (which, generally, are nonzero) in Equations (1.1) and (1.2) we obtain the following relations for the kth layer:

$$\frac{\partial u_k}{\partial t} + \frac{u_k}{a\cos\varphi}\frac{\partial u_k}{\partial\lambda} + \frac{v_k}{a}\frac{\partial u_k}{\partial\varphi} - \frac{u_k v_k}{a}\tan\varphi - fv_k = -\frac{1}{\varrho_0}\frac{\partial p_k'}{a\cos\varphi\,\partial\lambda}; \tag{1.45}$$

$$\frac{\partial v_k}{\partial t} + \frac{u_k}{a\cos\varphi}\frac{\partial v_k}{\partial\lambda} + \frac{v_k}{a}\frac{\partial v_k}{\partial\varphi} + \frac{u_k^2}{a}\tan\varphi + fu_k = -\frac{1}{\varrho_0}\frac{\partial p_k'}{a\,\partial\varphi}. \tag{1.46}$$

Eliminating the pressure disturbance p_k' from (1.45)–(1.46) we derive the vorticity equation

$$\left(\frac{\partial}{\partial t} + \frac{u_k}{a\cos\varphi}\frac{\partial}{\partial\lambda} + \frac{v_k}{a}\frac{\partial}{\partial\varphi}\right)(\omega_k + f) + (\omega_k + f)\operatorname{div}\mathbf{v}_k = 0, \quad k = \overline{1,n}, \tag{1.47}$$

where

$$\omega_k = \frac{\partial v_k}{a\cos\varphi\,\partial\lambda} - \frac{1}{\cos\varphi}\frac{\partial}{a\,\partial\varphi}(u_k\cos\varphi), \quad k = \overline{1,n}. \tag{1.48}$$

Eliminating div $\mathbf{v}_k$ from Equations (1.44) and (1.47) we have

$$\left(\frac{\partial}{\partial t} + \frac{u_k}{a\cos\varphi}\frac{\partial}{\partial\lambda} + \frac{v_k}{a}\frac{\partial}{\partial\varphi}\right)\frac{\omega_k + f}{H_k} = 0, \quad k = \overline{1,n}. \tag{1.49}$$

We now estimate the amplitudes of the interface displacements ζ_k ($k = \overline{1,n}$). By virtue of (1.42) and the geostrophic relations, the typical scales Z_k for the displacements ζ_k are equal to

$$Z_1 = \frac{Lf_0U}{g}; \quad Z_k = \frac{Lf_0\varrho_0U}{g\delta\varrho_k}; \quad \delta\varrho_k = \varrho_k - \varrho_{k-1}; \quad k = \overline{2,n} \tag{1.50}$$

Let us now assume that the layer thicknesses h_k are such that for the ocean motions under study

$$\left\| \left(\frac{H_k - h_k}{h_k} \right) \right\| \ll 1, \ k = \overline{1, n}. \tag{1.51}$$

We have $\delta \varrho_k \simeq 10^{-3} \varrho_0$, and for the values of the determining parameters indicated above, condition (1.51) implies that the layer thicknesses h_k ($k = \overline{1, n}$) must be greater than or of the order of 100 m.

By virtue of condition (1.51) and the estimates $\omega_k/f_0 \sim Ki \ll 1$ and $\beta y/f_0 \sim \varepsilon_L \ll 1$, we find

$$\frac{\omega_k + f}{H_k} \simeq \frac{f_0}{h_k}\left(1 + \frac{\omega_k + \beta y}{f_0} - \frac{\zeta_k - \zeta_{k+1}}{h_k}\right), \quad k = \overline{1, n}. \tag{1.52}$$

Further, by virtue of conditions (1.7), passing to the local Cartesian coordinates x, y (see (1.13)) we obtain

$$u_k = -\frac{1}{f_0 \varrho_0} \frac{\partial p_k'}{\partial y}; \quad v_k = \frac{1}{f_0 \varrho_0} \frac{\partial p_k'}{\partial x};$$

$$\omega_k = \frac{1}{f_0 \varrho_0} \Delta\, p_k', \quad k = \overline{1, n}. \tag{1.53}$$

Using (1.52), (1.53), and (1.41) and taking into account that $Z_1 \ll Z_2$, we derive from (1.49) the transfer equation for the quasigeostrophic potential vorticity $\omega_*^{(k)}$:

$$D_k \omega_*^{(k)} = 0, \quad k = \overline{1, n}, \tag{1.54}$$

where

$$\omega_*^{(k)} = \frac{1}{f_0 \varrho_0} \Delta p_k' + \beta y + \frac{f_0}{g h_k} \frac{p_{k-1}' - p_k'}{\delta \varrho_k} +$$

$$+ \frac{f_0}{g h_k} \frac{p_{k+1}' - p_k'}{\delta \varrho_{k+1}} + f_0 \frac{b}{h_n} \delta_{kn}; \tag{1.55}$$

$$D_k = \frac{\partial}{\partial t} + \frac{1}{f_0 \varrho_0} J(p_k', \ ); \quad J(p_k',) = \frac{\partial p_k'}{\partial x} \frac{\partial}{\partial y} - \frac{\partial p_k'}{\partial y} \frac{\partial}{\partial x}. \tag{1.56}$$

Here δ_{kn} is the Kronecker delta and, formally, $\delta \varrho_1 = \infty$ and $\delta \varrho_{n+1} = \infty$.

These equations are analogous to Equation (1.39a) describing the transfer of ω_* in the case of a continuous stratification.

We note that since the unknown functions p_k' ($k = \overline{1, n}$) depend only on x, y, and t, the system of equations (1.54)–(1.56) is simpler than the system (1.33)–(1.35). Formulas (1.43) and (1.42a) imply

$$w_{k-1}^+ = w_k^- = \frac{\partial \zeta_k}{\partial t} + \frac{1}{f_0 \varrho_0} J(p_{k-1}', \zeta_k), \quad k = \overline{2, n}, \tag{1.57}$$

i.e. in the quasigeostrophic approximation the vertical velocity is continuous at the interfaces.

The normal component of the velocity must vanish at the rigid boundaries of the region V for each of the layers. It can easily be seen that the analog of condition (1.31) is written

$$\iint_S \frac{\partial H_k}{\partial t}\, dx\, dy = 0, \quad k = \overline{1, n}, \tag{1.58a}$$

or, using the definition of H_k and the vorticity equation (1.54),

$$\int_\Gamma \frac{\partial^2 p'_k}{\partial n\, \partial t}\, d\Gamma = 0, \quad k = \overline{1, n}. \tag{1.58b}$$

Condition (1.58b) is analogous to the second condition (1.38a). The second condition (1.38b) for the multilayer fluid obviously has the form

$$\lim_{X \to \infty} \frac{1}{2X} \int_{-X}^{X} \frac{\partial u_k}{\partial t}\, dx = 0 \text{ for } y = 0;\, y = \mathscr{L},\, k = \overline{1, n}. \tag{1.59}$$

1.3. Multilevel Model

Let us now consider a finite-difference approximation of Equations (1.33)–(1.35) along the vertical. We split the fluid into n layers, generally, with different thicknesses h_k. Following Charney and Phillips (1953) we introduce levels with integral and fractional indices (Figure 2.1.3) assuming that the pressure disturbance p'_k and the horizontal velocities u_k, v_k are defined at the levels z_k and the density disturbance $\varrho'_{k+1/2}$ and the vertical velocity $w_{k+1/2}$ are defined at the levels $z_{k+1/2}$. We write the first two of Equations (1.33) and Equation (1.34) at the levels z_k ($k = \overline{1, n}$) and the third of Equations (1.33) and Equation (1.35) at the levels $z_{k+1/2}$ ($k = \overline{1, n}$). Using the finite-difference approximations

$$\left(\frac{\partial w}{\partial z}\right)_k = \frac{w_{k-1/2} - w_{k+1/2}}{h_k}; \quad \left(\frac{\partial p'}{\partial z}\right)_{k-1/2}$$

$$= 2\left(\frac{p'_{k-1} - p'_k}{h_{k-1} + h_k}\right), \tag{1.60a}$$

$$D_{k-1/2} = \left(\frac{\partial}{\partial t} + u\frac{\partial}{\partial x} + v\frac{\partial}{\partial y}\right)_{k-1/2}$$

$$= \left(\frac{h_k}{h_{k-1} + h_k}\right) D_{k-1} + \left(\frac{h_{k-1}}{h_{k-1} + h_k}\right) D_k, \tag{1.60b}$$

$$D_k = \frac{\partial}{\partial t} + u_k \frac{\partial}{\partial x} + v_k \frac{\partial}{\partial y} = \frac{\partial}{\partial t} + \frac{1}{f_0 \varrho_0} J(p'_k,\ ) \tag{1.60c}$$

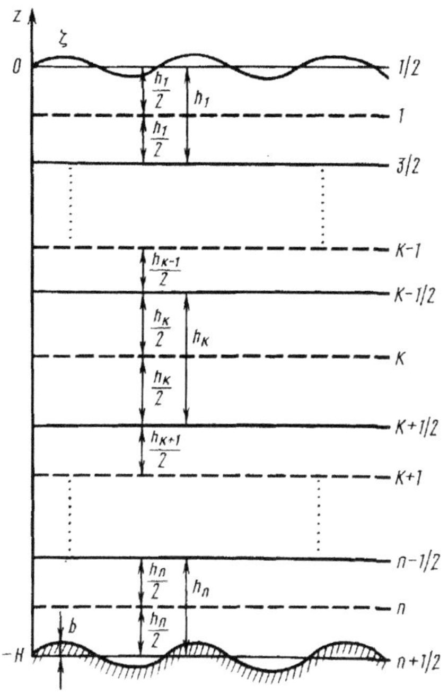

Fig. 2.1.3. To the derivation of the equations for the n-level model.

and writing the first boundary condition (1.36) at the level $z_{1/2}$ and boundary condition (1.37) at the level $z_{n+1/2}$ we obtain

$$w_{1/2} = 0, \qquad (1.61_{1/2})$$

$$\vdots$$

$$D_{k-1/2}\left[-\frac{1}{g}\left(\frac{\partial p'}{\partial z}\right)_{k-1/2}\right] - \frac{\varrho_0 N_{k-1/2}^2}{g} w_{k-1/2} = 0, \qquad (1.61_{k-1/2})$$

$$D_k\left(\frac{1}{f_0 \varrho_0}\Delta p'_k + \beta y\right) - f_0\left(\frac{w_{k-1/2} - w_{k+1/2}}{h_k}\right) = 0, \qquad (1.61_k)$$

$$D_{k+1/2}\left[-\frac{1}{g}\left(\frac{\partial p'}{\partial z}\right)_{k+1/2}\right] - \frac{\varrho_0 N_{k+1/2}^2}{g} w_{k+1/2} = 0, \qquad (1.61_{k+1/2})$$

$$w_{n+1/2} = \frac{1}{f_0 \varrho_0} J(p'_n, b). \qquad (1.61_{n+1/2})$$

For unequal h_k and $b \neq 0$ the order of this approximation is $O(h_k)$, for equal h_k and $b = 0$ the order of the approximation is $O(h_k^2)$.

To transform the system of equations (1.61) we need the following identities that can easily be proved. Proceeding from definitions (1.60c) and the properties of the Jacobian J, we obtain

$$D_k\,(p'_{k-1} - p'_k) = D_{k-1}\,(p'_{k-1} - p'_k);\ k = \overline{2,\ n}. \tag{1.62a}$$

By virtue of definition (1.60b) and identity (1.62a), we find

$$D_{k-1/2}\,(p'_{k-1} - p'_k) = D_k\,(p'_{k-1} - p'_k);\ k = \overline{2,\ n}; \tag{1.62b}$$

$$D_{k+1/2}\,(p'_k - p'_{k+1}) = D_k\,(p'_k - p'_{k+1});\ k = \overline{1,\ n-1}. \tag{1.62c}$$

We eliminate the vertical velocities $w_{k-1/2}$ and $w_{k+1/2}$ from the vorticity equation (1.61_k) $(k = \overline{1,\ n})$ using Equations $(1.61_{k-1/2})$ and $(1.61_{k+1/2})$ and boundary conditions $(1.61_{1/2})$ and $(1.61_{n+1/2})$. Taking into account identities (1.62b) and (1.62c) we obtain the transfer equations for the quasigeostrophic potential vorticity:

$$D_k \left\{ \frac{1}{f_0\,\varrho_0} \Delta\, p'_k + \beta y + \right.$$

$$+ \frac{f_0}{g h_k} \left[\frac{2g\,(p'_{k-1} - p'_k)}{\varrho_0\, N^2_{k-1/2}\,(h_{k-1} + h_k)} \right] +$$

$$+ \frac{f_0}{g h_k} \left[\frac{2g\,(p'_{k+1} - p'_k)}{\varrho_0\, N^2_{k+1/2}\,(h_k + h_{k+1})} \right] +$$

$$\left. + \ \delta_{kn}\, f_0\, \frac{b}{h_n} \right\} = 0;\ k = \overline{1,\ n}; \tag{1.63}$$

where, formally, $N^2_{1/2} = \infty$ and $N^2_{n+1/2} = \infty$.

By virtue of the second formula (1.60a) and formulas (1.62a, b), formula $(1.61_{k-1/2})$ implies

$$w_{k-1/2} = \left[\frac{2}{\varrho_0\, N^2_{k-1/2}\,(h_{k-1} + h_k)} \right] D_{k-1}\,(p'_k - p'_{k-1});\ k = \overline{2,\ n}; \tag{1.64}$$

and the third formula (1.33) and the second formula (1.60a) imply

$$\varrho'_{k-1/2} = \left[\frac{2}{g(h_{k-1} + h_k)} \right] (p'_k - p'_{k-1});\ k = \overline{2,\ n}. \tag{1.65}$$

Instead of the density disturbance we can consider the displacement of the isopycnic surface. According to formula (1.40), the approximation of $d\varrho_s/dz$ by a finite-difference ratio of the type of the second formula (1.60a) results in

$$\zeta_{k-1/2} = \frac{\varrho'_{k-1/2}}{2\delta\varrho_k}(h_{k-1} + h_k);\ k = \overline{2,\ n}. \tag{1.66}$$

Let us now compare the n-layer and n-level models whose corresponding layer thicknesses h_k are equal. Since in the n-layer model it was assumed that the fluid was incompressible, we put $c = \infty$ in formula (1.11). Then

$$N^2 = -\frac{g}{\varrho_0}\frac{d\varrho_s}{dz}.$$

Using a finite-difference approximation of the same type as the second formula (1.60a), we write

$$N^2_{k-1/2} = \frac{2g}{\varrho_0}\frac{\varrho_k - \varrho_{k-1}}{h_{k-1} + h_k}. \tag{1.67}$$

Taking formula (1.67) into account, we see that the transfer equations for the quasigeostrophic potential vorticity (1.54) and (1.63) in the n-layer and n-level models coincide indentically, and if $\zeta_{k-1/2}$ and $w_{k-1/2}$ are expressed in terms of $p'_k - p'_{k-1}$ and $D_{k-1}\,\zeta_{k-1/2}$ respectively with the aid of formulas (1.64), (1.65), and (1.66), then the resulting formulas coincide with the corresponding formulas (1.57) and (1.42a) for w_k and ζ_k in the n-layer model.

Table 2.1.1 sums up the results of the comparison of the two models.

How must the number of layers and their thicknesses be chosen? The number of layers and their thicknesses in the multilevel model are chosen so that the required accuracy of the solution of the problem is attained, as is usually done in the solution of any problem by the finite-difference method. The number of layers and their thicknesses in the multilayer model can be chosen from the condition of approximation of the profile $\varrho_s(z)$ or from the condition of accuracy in the reproduction of various characteristics of the solution of the problem (see Flierl, 1978). However, it should be borne in mind that, by virtue of constraint (1.51), the layers must not be too thin. It is useful to note that the multilayer model describes the motion of a real fluid and therefore, generally speaking, it can be reproduced in a laboratory.

TABLE 2.1.1
Model characteristics which coincide when the number of the layers and the corresponding layer thicknesses are equal for $N^2 = -(g\varrho_0)\,(d\varrho_s/dz)$

n-Layer model	n-Level model
The horizontal velocities u_k, v_k and the pressure disturbance p'_k independent of z within the limits of the layer	The horizontal velocities u_k, v_k and the pressure disturbance p'_k defined at the middle of the layer
The vertical velocity w_k at the interface between the $(k-1)$th and kth layers (w_k depends linearly on z within the limits of the layer)	The vertical velocity $w_{k-1/2}$ defined at the interface between the $(k-1)$th and kth layers
The displacement ζ_k of the interface between the $(k-1)$th and kth layers	The displacement $\zeta_{k-1/2}$ of the isopycnic surface at the interface between the $(k-1)$th and kth layers

1.4. Energy Equation

We begin with the case of a continuously stratified fluid. We multiply Equation (1.34) by $(1/f_0)p'$ and Equation (1.35) by $(-g^2/\varrho_0 N^2)\,\varrho'$ and add the results. Using the identity

$$p'\Delta\frac{\partial p'}{\partial t} = \operatorname{div}\left[p'\nabla\frac{\partial p'}{\partial t}\right] - \frac{1}{2}\left[\frac{\partial(\nabla p')}{\partial t}\right] \tag{1.68}$$

and Equations (1.33) we derive the energy equation

$$\frac{\partial}{\partial t}\left\{\varrho_0\left[\frac{u^2+v^2}{2}\right] + \frac{g^2}{\varrho_0 N^2}\frac{\varrho'^2}{2}\right\} = \operatorname{div}\left\{\frac{p'}{\varrho_0 f_0^2}\nabla\frac{\partial p'}{\partial t} + \left[\frac{1}{\varrho_0 f_0^2}p'\Delta p' + \left(\frac{\beta y}{f_0}\right)p' - \frac{g^2}{\varrho_0 N^2}\frac{\varrho'^2}{2}\right]\mathbf{v}_h\right\} - \frac{\partial(p'w)}{\partial z}, \tag{1.69}$$

where ∇ is understood as a plane operator and $\mathbf{v}_h$ is the horizontal velocity vector.

We integrate Equation (1.69) over a closed region V (Figure 2.1.1). From the geostrophic relations and the first of conditions (1.38a) we have

$$p'|_\Gamma = p'_\Gamma(z, t). \tag{1.70}$$

Using (1.70), the first of conditions (1.36), conditions (1.37) and (1.38a), and the geostrophic relations we derive the energy equation for a closed region in the integral form

$$\int_V\left[\varrho_0\left(\frac{u^2+v^2}{2}\right) + \frac{g^2}{\varrho_0 N^2}\frac{\varrho'^2}{2}\right]\mathrm{d}V = \text{const.} \tag{1.71}$$

In the case of barotropic motions, formula (1.71) does not involve the second summand. The quantity

$$P = \int_V \frac{g^2}{\varrho_0 N^2}\frac{\varrho'^2}{2}\,\mathrm{d}V$$

denotes the total available potential energy of quasigeostrophic motions.

Let us now pass to the case of a multilayer fluid. We multiply the kth equation (1.54) by $h_k(1/f_0)p'_k$ and perform the summation of the resulting relations with respect to k from 1 to n. Taking into account relations (1.68) we obtain, after some simple transformations, the energy equation for a multilayer fluid:

$$\frac{\partial}{\partial t}\sum_{k=1}^{n}\left\{h_k\varrho_0\left(\frac{u_k^2+v_k^2}{2}\right) + \frac{1}{g\,\delta\varrho_{k+1}}\left[\frac{p'_{k+1}-p'_k}{2}\right]^2\right\} =$$

$$= \operatorname{div}\left\{\sum_{k=1}^{n} h_k\frac{p'_k}{\varrho_0 f_0^2}\nabla\frac{\partial p'_k}{\partial t} + \sum_{k=1}^{n}\left[h_k\frac{p'_k}{\varrho_0 f_0^2}\Delta p'_k + h_k\frac{\beta y}{f_0}p'_k + \right.\right.$$

$$+ \frac{p'_k}{g\,\delta\varrho_k}(p'_{k-1} - p'_k) + \frac{p'_k}{g\,\delta\varrho_{k+1}}(p'_{k+1} - p'_k)\Bigg]\mathbf{v}_{k(h)} + b\mathbf{v}_{n(h)}\Bigg\}, \tag{1.72}$$

where $\delta\varrho_l = \infty$, $\delta\varrho_{n+1} = \infty$. Repeating the derivation of Equation (1.71) we find the energy equation for a closed region in the integral form

$$\sum_{k=1}^{n}\iint_S \left\{ h_k\varrho_0\left(\frac{u_k^2 + v_k^2}{2}\right) + \frac{1}{g\,\delta\varrho_{k+1}}\left[\frac{(p'_{k+1} - p'_k)^2}{2}\right]\right\} \mathrm{d}x\,\mathrm{d}y = \text{const}, \tag{1.73}$$

where $\delta\varrho_{n+1} = \infty$.

Taking into account (1.42a) we can rewrite the expression for the total available potential energy in the form

$$P = \sum_{k=1}^{n}\iint_S g\,\delta\varrho_{k+1}\left(\frac{\zeta_{k+1}^2}{2}\right)\mathrm{d}x\,\mathrm{d}y. \tag{1.74}$$

It is also easy to prove the conservation of the quasigeostrophic potential vorticity ω_* and enstrophy $\Omega_* = \frac{1}{2}\,\omega_*^2$. In the case of a closed region V and continuous stratification we have

$$\frac{\partial}{\partial t}\iint_S \omega_*\,\mathrm{d}x\,\mathrm{d}y = 0, \qquad \frac{\partial}{\partial t}\iint_S \Omega_*\,\mathrm{d}x\,\mathrm{d}y = 0, \; -H \leqslant z \leqslant 0, \tag{1.75}$$

and for the multilayer model we find

$$\frac{\partial}{\partial t}\iint_S \omega_*^{(k)}\,\mathrm{d}x\,\mathrm{d}y = 0, \qquad \frac{\partial}{\partial t}\iint_S \Omega_*^{(k)}\,\mathrm{d}x\,\mathrm{d}y = 0, \quad k = \overline{1, n}. \tag{1.76}$$

It is obvious that formulas (1.72)–(1.74) and (1.76) hold for the multilevel model (with the replacement of ζ_{k+1} by $\zeta_{k+1/2}$). In other words, this finite-difference approximation along the vertical does not violate the laws of conservation of the energy and the quasigeostrophic potential vorticity and enstrophy.

1.5. Inclusion of Ekman Boundary Layers

In the foregoing sections we excluded external forces and friction from our consideration. We shall now study the effect of vertical turbulent friction, ignoring the influence of horizontal turbulent friction. It is well known that vertical turbulent friction is significant only within the limits of the Ekman surface and bottom layers which are thin relative to the scale H. Let us first consider the bottom layer. We have

$$-f_0 v_b = A_z\,\frac{\partial^2 u_b}{\partial z^2}, \qquad f_0 u_b = A_z\,\frac{\partial^2 v_b}{\partial z^2}, \tag{1.77}$$

$$\frac{\partial u_b}{\partial x} + \frac{\partial v_b}{\partial y} + \frac{\partial w_b}{\partial z} = 0 \tag{1.78}$$

with boundary conditions

$$u_b = -u_g, \qquad v_b = -v_g, \qquad w_b = -w_g \quad \text{for} \quad z = -H + b, \tag{1.79}$$

$$u_b, v_b, w_b \to 0 \quad \text{for} \quad z \to \infty \tag{1.80}$$

where A_z is the coefficient of vertical turbulent friction, the subscript g indicates the velocity of the geostrophic motion, and the subscript b indicates the corrections to these velocities which are substantially nonzero only within the limits of the boundary layer (condition (1.80)). Using the third condition (1.80), we conclude from (1.78) that

$$w_b(-H + b) = \int_{-H+b}^{\infty} \left(\frac{\partial u_b}{\partial x} + \frac{\partial v_b}{\partial y} \right) dz \tag{1.81}$$

Solving Equations (1.77) for the first two conditions (1.79) and (1.80) and taking into account the third condition (1.79) and the geostrophic relations (1.33), we derive for $b_0/H \ll 1$ the condition

$$w_g = \sqrt{A_z/2f_0}\, \frac{1}{f_0\varrho_0} \Delta p' + u_g \frac{\partial b}{\partial x} + v_g \frac{\partial b}{\partial y} \quad \text{for } z = -H. \tag{1.82}$$

Similarly, considering the Ekman boundary layer at the sea surface we obtain (see also Section 3.1 in Chapter 3)

$$w_g = \frac{1}{f_0\,\varrho_0} \operatorname{rot}_z \boldsymbol{\tau} \quad \text{for} \quad z = 0, \tag{1.83}$$

where $\tau = (\tau_x, \tau_y)$ is the wind stress at the atmosphere–ocean interface.

Conditions (1.82) and (1.83) must be used instead of condition (1.37) and the first of conditions (1.36), respectively, when Equations (1.33)–(1.35) are solved. The corresponding changes that must be introduced into the equations of the multilayer or multilevel model are obvious.

In conclusion, we write the basic equations for a two-layer fluid, which is the simplest model for studying baroclinic effects:

$$D_1 \left[\frac{1}{f_0\,\varrho_0} \Delta p_1' + \beta y + \frac{\Lambda_1}{f_0\,\varrho_0} (p_2' - p_1') \right] = \frac{1}{\varrho_0\, h_1} \operatorname{rot}_z \boldsymbol{\tau}, \tag{1.84}$$

$$D_2 \left[\frac{1}{f_0\,\varrho_0} \Delta p_2' + \beta y + \frac{\Lambda_2}{f_0\,\varrho_0} (p_1' - p_2') + f_0 \frac{b}{h_2} \right] = -\frac{1}{\varrho_0\, h_2} \sqrt{\frac{A_z}{2f_0}}\; p_2', \tag{1.85}$$

where

$$\Lambda_1 = \frac{f_0^2}{g'\, h_1}, \qquad \Lambda_2 = \frac{f_0^2}{g'\, h_2}, \qquad g' = g \frac{\delta\varrho}{\varrho_0}, \delta\varrho = \varrho_2 - \varrho_1. \tag{1.86}$$

2. ROSSBY WAVES

2.1. Linear Waves; Constant Depth

Assume that $\delta = Ki/(\varepsilon_L \cot \varphi_0) = U/(\beta L^2) \ll 1$. Then the nonlinear terms in the vorticity equation (1.29) can be neglected and we obtain (in dimensionless form)

$$\varepsilon_T \frac{\partial}{\partial t}\left(\frac{\partial v_0}{\partial x} - \frac{\partial u_0}{\partial y}\right) + \varepsilon_L \cot \varphi_0 v_0 - \left(\frac{L}{L_R}\right)^2 \varepsilon_T \frac{\partial w_0}{\partial z} = 0. \tag{2.1}$$

Assuming that all the terms in Equation (2.1) are of the same order we find that $\varepsilon_L \cot \varphi_0 = \varepsilon_T$ and therefore the typical time scale is $T = 1/(\beta L)$. Then $Ki/\varepsilon_T = \delta \ll 1$ and the nonlinear terms in Equation (1.26_1) can also be neglected:

$$\varepsilon_T \frac{\partial \varrho'_0}{\partial t} - \varepsilon_T N^2 w_0 = 0. \tag{2.2}$$

We remind the reader that Equation (2.2) is also written in dimensionless form. Eliminating the vertical velocity w_0 from (2.1) and (2.2) and using (1.22_0), (1.23_0), and (1.24_0) we derive the linearized transfer equation for the quasigeostrophic potential vorticity. In dimensionless form we have

$$\frac{\partial}{\partial t}\left[\Delta p' + \frac{\partial}{\partial z}\left(\frac{f_0^2}{N^2}\frac{\partial p'}{\partial z}\right)\right] + \beta \frac{\partial p'}{\partial x} = 0. \tag{2.3}$$

Using (2.2), boundary conditions (1.27_1) and (1.28_1), and relations (1.22_0), (1.23_0), and (1.24_0) we obtain in dimensional form

$$\frac{\partial^2 p'}{\partial z\, \partial t} = 0 \quad \text{for} \quad z = 0; \tag{2.4}$$

$$\frac{\partial^2 p'}{\partial z\, \partial t} = -\frac{N^2}{f_0}\left(\frac{\partial b}{\partial y}\frac{\partial p'}{\partial x} - \frac{\partial b}{\partial x}\frac{\partial p'}{\partial y}\right) \quad \text{for} \quad z = -H. \tag{2.5}$$

We begin with the case of an unbounded ocean of constant depth. The solution of problem (2.3)–(2.5) is sought in the form

$$p' = \operatorname{Re}\{p(z)\exp[i(k_x x + k_y y - \sigma t)]\}. \tag{2.6}$$

Substituting (2.6) into (2.3)–(2.5) we have

$$\frac{\mathrm{d}}{\mathrm{d}z}\left(\frac{f_0^2}{N^2}\frac{\mathrm{d}p}{\mathrm{d}z}\right) + \lambda p = 0; \ -\lambda = \frac{k_x \beta}{\sigma} + k^2; \ k^2 = k_x^2 + k_y^2; \tag{2.7}$$

$$\frac{\mathrm{d}p}{\mathrm{d}z} = 0 \quad \text{for} \quad z = 0; \qquad \frac{\mathrm{d}p}{\mathrm{d}z} = 0 \quad \text{for} \quad z = -H. \tag{2.8}$$

Assume that $N^2 = \text{const}$. Then for $\lambda < 0$, problem (2.7), (2.8) possesses only a trivial solution and for $\lambda \geqslant 0$ there is a sequence of eigenvalues $\lambda_m = (f_0^2/N^2)$

$(m\pi/H)^2$ ($m = 0, 1, \ldots$) and the corresponding sequence of eigenfunctions $p^{(m)} = \cos(m\pi/H)(z + H) = (-1)^m \cos(m\pi/H)z$. Remembering the definition of λ we write

$$\sigma_m = -\frac{k_x\beta}{k^2 + \lambda_m}, \quad m = 0, 1, \ldots, \tag{2.9}$$

where

$$\lambda_m = \frac{(m\pi)^2}{L_R^2}; \quad L_R = H\frac{N}{f_0}, \quad m = 0, 1, \ldots \tag{2.10}$$

Formula (2.9) implies that the zonal component of the phase velocity of the wave is always directed towards the west.

Formulas (2.9) are called dispersion relations. For $m = 0$ we obtain a barotropic Rossby wave on which the stratification of the ocean exerts no effect; $p^{(0)} = 1$ for this wave. For $m > 0$ we obtain baroclinic Rossby waves; the amplitude $p^{(m)} = (-1)^m \cos(m\pi/H)z$ has exactly m zeros inside the interval $(-H, 0)$.

In the case of an arbitrary dependence of N on z there appear no new qualitative features (e.g. see Kamenkovich, 1973, Ch. 3). As in the case $N =$ const, problem (2.7), (2.8) has eigenvalues $\lambda_0 = 0$, $\lambda_m > 0$ ($m = 1, 2, \ldots$); the eigenfunction $p^{(m)}$ has exactly m zeros inside the interval $(-H, 0)$ and, naturally, formulas (2.9) do not change. Figure 2.2.1 demonstrates the results of a specific calculation; Figure 2.2.2 shows the plots of the corresponding dispersion relations.

We rewrite the linearized equations (1.84), (1.85) of a two-layer fluid for $b = 0$ and $\tau = 0$ and without bottom friction:

$$\frac{\partial}{\partial t}[\Delta p_1' + \Lambda_1(p_2' - p_1')] + \beta\frac{\partial p_1'}{\partial x} = 0; \tag{2.11}$$

$$\frac{\partial}{\partial t}[\Delta p_2' + \Lambda_2(p_1' - p_2')] + \beta\frac{\partial p_2'}{\partial x} = 0.$$

The solution of Equations (2.11) is sought in the form (cf. (2.6))

$$(p_1', p_2') = \text{Re}\,\{(p_1, p_2)\exp[i(k_x x + k_y y - \sigma t)]\}.$$

Substituting this expression into (2.11) we find

$$[\sigma(k^2 + \Lambda_1) + k_x\beta]\,p_1 - \sigma\Lambda_1 p_2 = 0; \tag{2.12}$$

$$-\sigma\Lambda_2\,p_1 + [\sigma(k^2 + \Lambda_2) + k_x\beta]\,p_2 = 0.$$

This system possesses a nontrivial solution only if its determinant vanishes. Solving the resulting quadratic equation with respect to σ we obtain

$$\sigma_0 = -\frac{k_x\beta}{k^2}; \tag{2.13a}$$

$$\sigma_1 = -\frac{k_x\beta}{k^2 + \Lambda_1 + \Lambda_2}. \tag{2.13b}$$

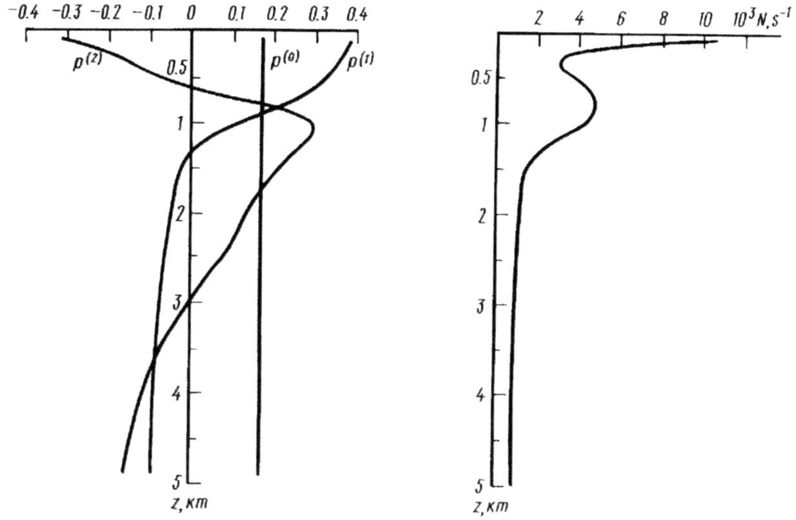

Fig. 2.2.1. Average frequency $N = N(z)$ according to the data of the POLYMODE expedition and the calculated barotropic ($p^{(0)}$) and first two baroclinic ($p^{(1)}$) and $p^{(2)}$) modes (eigenfunctions of problem (2.7), (2.8)). The corresponding eigenvalues are $\lambda_0 = 0$, $\lambda_1 = 5.23 \times 10^{-4}$ km^{-2}, and $\lambda_2 = 3.00 \times 10^{-3}$ km^{-2}.

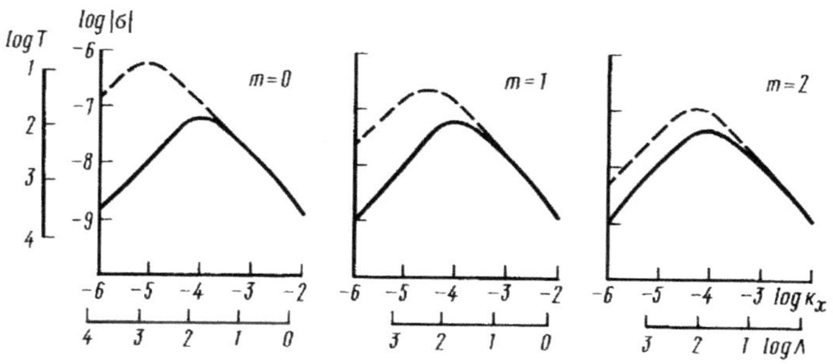

Fig. 2.2.2. Dispersion relations ($|\sigma|$ as a function of k_x) for the barotropic ($m = 0$) and first two baroclinic ($m = 1, 2$) modes calculated by formula (2.9) for $\lambda_0 = 0$, $\lambda_1 = 5.23 \times 10^{-4}$ km^{-2}, and $\lambda_2 = 3.00 \times 10^{-3}$ km^{-2}. Dashed line: $k_y = 10^{-5}$ m^{-1}; solid line: $k_y = 10^{-4}$ m^{-1}. Units of measurement: σ s^{-1}; k_x m^{-1}; the period $T = 2\pi/\sigma$ days; the wavelength $\Lambda = 2\pi/k_x$ km; $\beta = (2\,\omega_e/a)\cos\varphi$, $\varphi = 29°$; $\omega_e = 7.29 \times 10^{-5}$ s^{-1}; $a = 6.37 \times 10^3$ km.

The first of Equations (2.12) yields

$$p_1^{(0)} = p_2^{(0)} \text{ for } \sigma = \sigma_0 \tag{2.14a}$$

and, by virtue of (1.86),

$$h_1 p_1^{(1)} = -h_2 p_2^{(1)} \text{ for } \sigma = \sigma_1. \tag{2.14b}$$

Thus we have obtained a barotropic wave with dispersion relation (2.13a) and vertical structure $p^{(0)} = (1, 1)$ and a first-mode baroclinic wave with dispersion relation (2.13b) and vertical structure $p^{(1)} = (1, -h_1/h_2)$. Qualitatively, the structures of these two modes are analogous to those of the barotropic and first baroclinic modes in the case of continuous stratification (Figure 2.2.1).

In the case of an n-layer fluid the linearized transfer equations (1.54) for the quasigeostrophic potential vorticity for $b = 0$ have the form ($\delta\varrho_1 = \infty$; $\delta\varrho_{n+1} = \infty$)

$$\frac{\partial}{\partial t}\left(\Delta p'_k + \frac{\varrho_0 f_0^2}{g h_k}\,\frac{p'_{k-1} - p'_k}{\delta\varrho_k} + \frac{\varrho_0 f_0^2}{g h_k}\,\frac{p'_{k+1} - p'_k}{\delta\varrho_{k+1}}\right) + {}$$

$$+ \beta\frac{\partial p'_k}{\partial x} = 0,\ k = \overline{1, n}. \tag{2.15}$$

The solution of system (2.15) is sought in the form

$$(p'_1, \ldots, p'_n) = \mathrm{Re}\,\{(p_1, \ldots, p_n)\exp\,[i(k_x x + k_y y - \sigma t)]\} \tag{2.16}$$

Introducing the three-diagonal matrix A:

$$\begin{array}{ccccc}
-\dfrac{\varrho_0 f_0^2}{g h_1}\dfrac{1}{\delta\varrho_2} & \dfrac{\varrho_0 f_0^2}{g h_1}\dfrac{1}{\delta\varrho_2} & 0 & \cdots & 0 \\
\dfrac{\varrho_0 f_0^2}{g h_2}\dfrac{1}{\delta\varrho_2} & -\dfrac{\varrho_0 f_0^2}{g h_2}\left(\dfrac{1}{\delta\varrho_2} + \dfrac{1}{\delta\varrho_3}\right) & \dfrac{\varrho_0 f_0^2}{g h_2}\dfrac{1}{\delta\varrho_3} & \cdots & 0 \\
\vdots & \vdots & \vdots & & \vdots \\
0 & 0 & 0 & & -\dfrac{\varrho_0 f_0^2}{g h_n}\dfrac{1}{\delta\varrho_n}
\end{array} \tag{2.17}$$

and the n-dimensional vector $\mathbf{p} = (p_1, \ldots, p_n)$ we obtain

$$A\mathbf{p} + \lambda\mathbf{p} = 0;\quad -\lambda = \frac{k_x\beta}{\sigma} + k^2. \tag{2.18}$$

We see that nontrivial solutions $(p_1, \ldots, p_n)$ of system (2.18) are eigenvectors of the matrix A. Since the sum of the elements of each of the rows of the matrix A is equal to zero, it is obvious that $\lambda_0 = 0$ is an eigenvalue of A and $(1, \ldots, 1)$ is the corresponding eigenvector; this is a barotropic Rossby wave. It also follows that $\det A = 0$ (the homogeneous system $A\mathbf{p} = 0$ possesses a nontrivial solution). Further, it can readily be verified that for a baroclinic wave ($\lambda \neq 0$) we have

$$\sum_{k=1}^{n} h_k p_k = 0. \tag{2.19}$$

Formula (2.19) generalizes formula (2.14b); as can easily be seen, the analogous formula $\int_{-H}^{0} p^{(i)}\,dz = 0$ holds for the eigenfunctions of problem (2.7), (2.8) for an eigenvalue $\lambda_i \neq 0$ (baroclinic waves).

It is easy to show that all the eigenvalues λ_i ($i = \overline{0, n-1}$) are real, distinct, and $\lambda_i > 0$ ($i = \overline{1, n-1}$), and the eigenvectors $\mathbf{p}^{(i)}$ of matrix A are orthogonal with weight h_k (see Kamenkovich *et al.*, 1981):

$$\frac{1}{H}\sum_{k=1}^{n} p_k^{(i)} p_k^{(j)} h_k = \delta_{ij},\quad i, j = \overline{0, n-1}. \tag{2.20}$$

Putting $i = 0$ in (2.20) we obtain formula (2.19).

The second formula (2.18) implies the dispersion relations (2.9) for Rossby waves. There are exactly n such relations for an n-layer fluid. Formulas (2.13), (2.14) for a two-layer fluid are easily obtained from the general formulas for an n-layer fluid.

We estimate the Rossby deformation radius from POLYMODE data. According to Figure 2.2.1, $\bar{N} = (1/H) \int_H^0 N(z)\mathrm{d}z = 1.86 \times 10^{-3}\ \mathrm{s}^{-1}$, whence $L_\mathrm{R} = H\bar{N}/f_0 =$ 132 km at latitude 29°. It is also advisable to note that $\lambda_1^{1/2} L_\mathrm{R} = 3.02$ where λ_1 is an eigenvalue of the problem (2.7), (2.8) (in the case of a constant N this product is equal to π).

For a two-layer (or two-level) fluid we define L_R by the following formula (see formula (1.67) according to which $N_{3/2} = (2g'/H)^{1/2}$):

$$L_\mathrm{R} = \frac{\sqrt{g'H}}{f_0}, \qquad g' = \frac{g(\varrho_2 - \varrho_1)}{\varrho_0}. \tag{2.21}$$

Using (2.13b) we can calculate $(\Lambda_1 + \Lambda_2)^{1/2} L_\mathrm{R} = (H^2/h_1 h_2)^{1/2}$; for example, for $h_1 = 1000$ m and $h_2 = 4000$ m this product is equal to 2.5.

The large-scale expeditions ('Polygon-70', MODE, and POLYMODE) which have been carried out recently to study synoptic eddies in the ocean prove that Rossby waves actually exist in the ocean. It has been shown that the relationship between the spatial and temporal scales of the observed eddies (determined from both synoptic charts and an analysis of maxima of space–time spectra) are in agreement with the dispersion relations for barotropic and first baroclinic Rossby waves (see Sections 1 and 2 of Chapter 5).

When we solve Cauchy's problem for Equation (2.3) with boundary conditions (2.4) and (2.5) for a boundless ocean for $b = 0$, the initial disturbance splits into noninteracting barotropic and baroclinic Rossby waves which propagate in accordance with the corresponding dispersion relations (2.9).

We introduce the group velocity $\mathbf{c}_g(\mathbf{k})$, $\mathbf{k} = (k_x, k_y)$:

$$\mathbf{c}_g = \left(\frac{\partial \sigma}{\partial k_x}, \frac{\partial \sigma}{\partial k_y} \right) = \frac{\beta}{(k^2 + \lambda_m)^2} \quad (k_x^2 - k_y^2 - \lambda_m,\ 2k_x k_y) \tag{2.22}$$

with which a Rossby wave packet propagates; such a packet is a disturbance consisting of plane waves with wavenumbers (k_x, k_y) lying within a very narrow interval. We note that, in contrast to the zonal component σ/k_x of the phase velocity, the zonal component of the group velocity can be both positive and negative. This means that, although the crests of an individual plane wave always travel in the westward direction, a wave packet can propagate both to the west $(k_x^2 < k_y^2 + \lambda_m)$ and to the east $(k_x^2 > k_y^2 + \lambda_m)$.

The notion of group velocity makes it possible to set the problem of reflection of a plane Rossby wave from a coast, say $x = 0$, since it allows us to distinguish between incident and reflected waves, which cannot be done using the phase velocity alone. For $x > 0$, let there be an incident wave

$$p_i' = \cos(k_{ix} x + k_y y - \sigma t).$$

The sum of the reflected wave p_r' and the incident wave must satisfy the impermeability condition on the wall $x = 0$ ($\partial p_i'/\partial y + \partial p_r'/\partial y = 0$ at $x = 0$) and therefore these waves must have the same wavenumber k_y and the same frequency σ. Hence,

$$p_r' = -\cos(k_{rx}x + k_y y - \sigma t)$$

For given k_y and σ, dispersion relation (2.9) is a quadratic equation in k_x, and $k_{ix}k_{rx} = k_y^2 + \lambda_m$. It is obvious that $k_{ix}^2 < k_y^2 + \lambda_m$ ($c_{gi} < 0$) and $k_{rx}^2 > k_y^2 + \lambda_m$ ($c_{gr} > 0$). Therefore,

$$k_{rx} = -\frac{\beta}{2\sigma} + \sqrt{\frac{\beta^2}{4\sigma^2} - k_y^2 - \lambda_m}.$$

We note that for group velocities the law of equality of the angle of incidence to the angle of reflection holds. It is important that the modulus of the wave vector of the wave reflected from the western coast is greater than that of the wave vector of the incident wave, i.e. its wavelength is smaller. It is obvious that in the reflection from the eastern coast the opposite phenomenon takes place and the wavelength of the reflected wave exceeds that of the incident wave. Hence, large-scale disturbances produced, for instance, by wind stress and propagating towards the western coast are reflected in the form of disturbances with a smaller spatial scale. This leads to the formation of a narrow region of intensive currents at the western coast of the ocean (i.e. the formation of a nonstationary boundary layer) (Pedlosky, 1965; Lighthill, 1969; Il'in, 1972).

The motion of a fluid in bounded basins can be represented as a superposition of Rossby eigenmodes (e.g. see Kamenkovich and Reznik, 1978, §2.3; Pedlosky, 1979, §3.25; Novitskiĭ, 1977). As an example, we present simple formulas for Rossby eigenfrequencies and eigenmodes in the case of barotropic motions in a rectangular basin ($0 \leqslant x \leqslant L_x$, $0 \leqslant y \leqslant L_y$) of constant depth:

$$\sigma_{mn}^2 = \frac{\beta^2}{4}\left[\left(\frac{m\pi}{L_x}\right)^2 + \left(\frac{n\pi}{L_y}\right)^2\right]; \; m, n = 1, 2, \ldots \tag{2.23a}$$

$$p'_{mn} = \sin\left(\frac{m\pi x}{L_x}\right)\sin\left(\frac{n\pi y}{L_y}\right)\cos\left(\frac{\beta}{2\sigma_{mn}}x + \sigma_{mn}t\right);$$

$$m, n = 1, 2, \ldots \tag{2.23b}$$

Since we consider geostrophic motions, the isolines p'_{mn} = const determine streamlines. We see that in each of the modes the motion is split by fixed and moving node lines into individual eddies. We also note that in lower eigenmodes there can be eddies whose scale along the x-axis is much smaller than the dimension L_x of the basin, which is due to the specific character of the reflection of Rossby waves from the western coast (formula (2.23b) can easily be written in the form of a superposition of incident and reflected Rossby waves). Thus, generally speaking, a large-scale wind field can excite motions with a substantially smaller scale *inside* a closed basin.

2.2 Linear Waves; Variable Depth

We begin the study of the effect of the bottom relief with the case when N = const, $\partial b/\partial x = b_x$ = const, and $\partial b/\partial y = b_y$ = const (Rhines, 1970). The solution of problem (2.3)–(2.5) is sought in the form of (2.6).

We have

$$\frac{d^2p}{dz^2} - \frac{N^2}{f_0^2}\left(\frac{k_x\beta}{\sigma} + k^2\right) p = 0; \tag{2.24a}$$

$$\frac{dp}{dz} = 0 \quad \text{for} \quad z = 0; \qquad \frac{dp}{dz} = \frac{N^2}{f_0\sigma} \times$$

$$\times\ (k_xb_y - k_yb_x)p \quad \text{for} \quad z = -H. \tag{2.24b}$$

It is advisable to introduce a parameter γ characterizing the relative contribution of the bottom relief in comparison with the β-effect:

$$\gamma = \frac{f_0}{H\beta}\left(b_y - \frac{k_y}{k_x}b_x\right). \tag{2.25}$$

First, let $(k_x\beta/\sigma) + k^2 > 0$. We introduce $L_R = HN/f_0$ and denote $\mu = (N/f_0)(k_x\beta/\sigma + k^2)^{1/2}$. By virtue of (2.24a) and the first condition (2.24b), we have $p^{(0)} = \cosh \mu z$ and, according to the second condition (2.24b),

$$\tanh \mu H = -\gamma\left[\mu H - \frac{(kL_R)^2}{\mu H}\right]. \tag{2.26}$$

Figure 2.2.3 implies that Equation (2.26) always has a single root μ_0. Remembering the definition of μ we obtain a dispersion relation for this wave. Hence

$$\sigma_0 = -\frac{k_x\beta}{k^2 - (\mu_0H)^2L_R^{-2}}, \quad p^{(0)} = \cosh \mu_0 z. \tag{2.27}$$

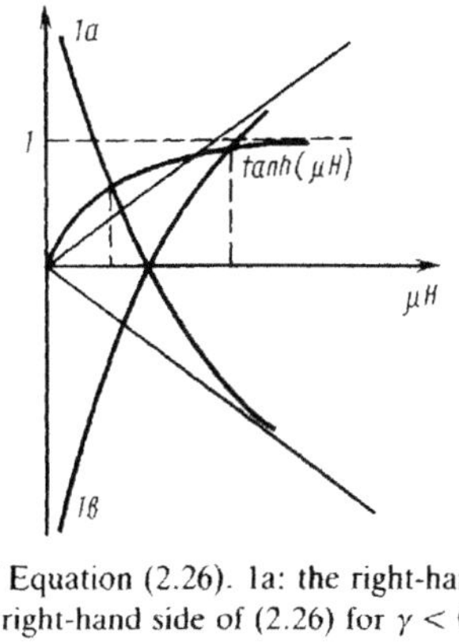

Fig. 2.2.3 Graphical solution of Equation (2.26). 1a: the right-hand side of (2.26) for $\gamma > 0$. 1b: the right-hand side of (2.26) for $\gamma < 0$.

Now let $(k_x\beta/\sigma) + k^2 < 0$. Introducing the notation $\lambda = (N/f_0)(-(k_x\beta/\sigma) - k^2)^{1/2}$ and repeating the derivation of Equation (2.20) we find an equation determining λ:

$$\tan \lambda H = -\gamma\left[\lambda H + \frac{(kL_R)^2}{\lambda H}\right]. \tag{2.28}$$

Figure 2.2.4 shows that Equation (2.28) has infinitely many roots $\lambda_1, \lambda_2, \ldots$. If λ_m is a root of Equation (2.28), then

$$\sigma_m = -\frac{k_x\beta}{k^2 + (\lambda_m H)^2 L_R^{-2}}, \quad p^{(m)} = \cos \lambda_m z, \quad m = 1, 2, \ldots \tag{2.29}$$

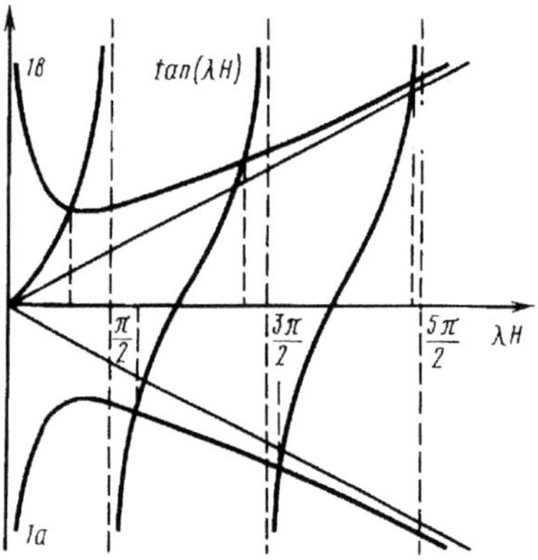

Fig. 2.2.4. Graphical solution of Equation (2.28). 1a: the right-hand side of (2.28) for $\gamma > 0$. 1b: the right-hand side of (2.28) for $\gamma < 0$.

The obtained wave solutions are called topographic Rossby waves (accordingly, the first, second, etc., modes). The quantities $\mu_0^{-1}, \lambda_1^{-1}, \lambda_2^{-1}, \ldots$ denote the corresponding vertical scales. It is advisable to discuss these waves separately for the cases $\gamma > 0$ and $\gamma < 0$.

(a) $\gamma > 0$. We note that the zeroth mode has no zeros in the closed interval $[-H, 0]$, the first mode has one zero in this closed interval, the second mode has two zeros, etc.; according to Figure 2.2.4, $\pi/2 < \lambda_1 H < \pi$, $3\pi/2 < \lambda_2 H < 2\pi, \ldots$. Further, remembering the parameters introduced in Section 1, we write $(k_x \sim k_y)$

$$|\gamma| \sim \frac{b_0 a}{L_b H} = \frac{\varepsilon_b}{\varepsilon_L}. \tag{2.30}$$

We begin with the case of $\gamma \ll 1$ and finite kL_R. Then, to within $O(\gamma)$; relations (2.26) and (2.27) imply

$$(\mu_0 H)^2 = \gamma (kL_R)^2, \quad \sigma_0 = -\frac{k_x\beta}{k^2}, \quad p^{(0)} = 1, \tag{2.31}$$

and (2.28) and (2.29) imply

$$\lambda_m H = m\pi, \qquad \sigma_m = -\frac{k_x\beta}{k^2 + (m\pi)^2 L_R^{-2}},$$

$$p^{(m)} = \cos\frac{m\pi}{H}z, \; m = 1, 2, \ldots \tag{2.32}$$

Hence, for $\gamma \to 0$ the zeroth topographic mode transforms into the barotropic mode for an ocean with $b = 0$, the first topographic mode transforms into the first baroclinic mode for an ocean with $b = 0$, etc.

Now we assume that $\gamma \sim 1$ and consider the limiting cases of long ($kL_R \ll 1$) and short ($kL_R \gg 1$) waves. For $kL_R \ll 1$ Equations (2.26) and (2.27) imply the following relations for the zeroth mode:

$$(\mu_0 H)^2 = (kL_R)^2 \frac{\gamma}{1+\gamma};$$

$$\sigma_0 = -\frac{1}{k^2}\left[k_x\left(\beta + \frac{f_0}{H}b_y\right) - \frac{f_0}{H}k_y b_x\right], \qquad p^{(0)} = 1. \tag{2.33}$$

It is interesting to note that in the dispersion relation the bottom relief effect adds 'linearly' to the β-effect. Thus for $kL_R \ll 1$ the stratification of the fluid exerts no effect on the zeroth topographic mode; the motion corresponding to this mode has a barotropic character. We note that the specific cases we have considered confirm the general criterion (1.41) for the existence of a barotropic motion. Indeed, taking $T = 1/(\beta L) \sim a/(f_0 L)$ we obtain $\varepsilon_L \sim \varepsilon_T$; since $k \sim 1/L$, taking into account (2.30) we can rewrite the criterion (1.41) in the form $(kL_R)^2\gamma \ll 1$. It is clear that in the cases $\gamma \ll 1$, $kL_R \sim 1$, and $\gamma \sim 1$, $kL_R \ll 1$ under consideration the criterion (1.41) holds. It is important to emphasize that when (1.41) holds, baroclinic motions are also possible.

For the first, second, etc., topographic modes Equation (2.28) reduces to $\tan(\lambda H) = -\gamma\lambda H$ for $kL_R \ll 1$ and finite γ, and the first summand in the denominator of formula (2.29) can be discarded, i.e. these waves are found to be dispersionless.

For short waves ($kL_R \gg 1$) and finite γ barotropic motions are impossible; by virtue of (2.26)–(2.29), we have

$$\mu_0 H = kL_R, \qquad \sigma_0 = -N\frac{k_x b_y - k_y b_x}{k},$$

$$p^{(0)} = \exp\left[-kL_R\left(1 + \frac{z}{H}\right)\right]; \tag{2.34}$$

$$\lambda_m H = \pi(m - \tfrac{1}{2}), \qquad \sigma_m = -\frac{k_x\beta}{k^2},$$

$$p^{(m)} = \cos\pi(m - \tfrac{1}{2})\frac{z}{H}, \quad m = 1, 2, \ldots \tag{2.35}$$

Thus for the short-wave zeroth topographic mode the β-effect is insignificant and $p^{(0)}$ is nonzero only near the bottom. For the first, second, etc., short-wave topographic modes the bottom relief exerts only an indirect effect on $p^{(m)}$ (here $\lambda_m H$ is equal to $(m - 1/2)\pi$ and not to $m\pi$ as in the case $b = 0$) and does not affect σ_m. We note that, by virtue of (2.35), we have $u = v = 0$ for $z = -H$ but $w \neq 0$. This fact does not mean that boundary condition (2.24b) is violated at the bottom; this condition holds to within $O[(kL_R)^2]$ because, according to (2.2), (1.24_0), (1.22_0), and (2.35), we have $W = (kL_R)^{-2}\, \varepsilon_L (H/L) U$.

(b) $\gamma < 0$. According to Figure 2.2.4, the zeroth and first modes have no zeros in the closed interval $[-H, 0]$, the second mode has one zero, the third mode has two zeros, etc. ($0 < \lambda_1 H < \pi/2$, $\pi/2 < \lambda_2 H < 3\pi/2$, etc.). We consider the same limiting cases as those for $\gamma > 0$.

For $|\gamma| \ll 1$ and finite kL_R, relations (2.26) and (2.27) imply that, to within $O(\gamma)$,

$$\mu_0 H = -\frac{1}{\gamma}; \qquad \sigma_0 = \frac{N^2}{\beta} k_x \left(b_y - \frac{k_y}{k_x} b_x \right)^2;$$

$$p^{(0)} = \exp\left[\frac{1}{\gamma} \left(1 + \frac{z}{H} \right) \right], \tag{2.36}$$

and (2.28) and (2.29) imply

$$\lambda_1 H = \gamma (kL_R)^2; \qquad \sigma_1 = -\frac{k_x \beta}{k^2}; \quad p^{(1)} = 1; \tag{2.37a}$$

$$\lambda_{m+1} H = m\pi; \qquad \sigma_{m+1} = -\frac{k_x \beta}{k^2 + (m\pi)^2 L_R^{-2}};$$

$$p^{(m+1)} = \cos\left(\frac{m\pi}{H} z \right), \quad m = 1, 2, \ldots \tag{2.37b}$$

Hence for $\gamma \to 0$ the zeroth topographic mode $p^{(0)}$ is concentrated at the bottom and $\sigma_0 \to 0$, the first topographic mode transforms into a barotropic mode for an ocean with $b = 0$, the second mode transforms into a first baroclinic mode for an ocean with $b = 0$, etc. (cf. (2.32)).

For $kL_R \ll 1$ and finite γ we have

$$(\mu_0 H)^2 = (kL_R)^2\, \gamma (1 + \gamma)^{-1}; \; \sigma_0 = -k_x k^{-2} \beta\, (1 + \gamma);$$
$$p^{(0)} = 1 \text{ for } \gamma < -1; \tag{2.38a}$$

$$\tanh(\mu_0 H) = -\gamma \mu_0 H; \; \sigma_0 = \beta L_R^2 (\mu_0 H)^{-2} k_x;$$
$$p^{(0)} = \cosh(\mu_0 z) \text{ for } \gamma > -1; \tag{2.38b}$$

$$\tan(\lambda_1 H) = -\gamma \lambda_1 H; \; \sigma_1 = -\beta L_R^2 (\lambda_1 H)^{-2} k_x;$$
$$p^{(1)} = \cos(\lambda_1 z) \text{ for } \gamma < -1; \tag{2.38c}$$

$$(\lambda_1 H)^2 = -(kL_R)^2\, \gamma (1 + \gamma)^{-1}; \; \sigma_1 = -k_x k^{-2} \beta\, (1 + \gamma);$$
$$p^{(1)} = 1 \text{ for } \gamma > -1; \tag{2.38d}$$

$$(\mu_0 H)^2 = (\lambda_1 H)^2 = kL_R\sqrt{3};\ \sigma_0 = k_x k^{-1} \cdot 3^{-1/2}\beta L_R;$$
$$\sigma_1 = -k_x k^{-1} \cdot 3^{-1/2}\beta L_R \text{ for } \gamma = -1. \tag{2.38e}$$

For the second, third, etc., topographic modes, finite γ, and $kL_R \ll 1$ we have $\tan(\lambda H) = -\gamma\lambda H$ and the first summand in the denominator of formula (2.29) can be discarded (dispersionless waves).

Formulas (2.38a, d) describe a barotropic wave; by virtue of expression (2.25) for γ, the corresponding formulas for σ_0 and σ_1 coincide with the second formula (2.33) describing the zeroth mode for $\gamma > 0$.

For $kL_R \gg 1$ and finite γ the case $\gamma < 0$ is described by the same formulas as the case $\gamma > 0$; see formulas (2.34) and (2.35) and their discussion.

Figure 2.2.3 implies that $(\mu_0 H)^2 < (kL_R)^2$ for $\gamma > 0$ and $(\mu_0 H)^2 > (kL_R)^2$ for $\gamma < 0$ – i.e., according to (2.27), for $\gamma < 0$ the zeroth mode moves to the east ($\sigma_0/k_x > 0$); the effect of the bottom relief on this mode 'supresses' the β-effect. For $\gamma > 0$ we have $\sigma_0/k_x < 0$ and the zeroth mode moves to the west. Generally, the above formulas clearly indicate that for all the modes the bottom relief can both strengthen and weaken the β-effect; however, by virtue of (2.29), we have $\sigma_m/k_x < 0$ for $m = 1,2,\ldots$ and all γ. The plots of the dispersion relations for $k_x = k_y$ and the zeroth and first topographic modes and also the structure of the modes are presented in Figure 2.2.5.

For the discussion of the above problem for a two-layer ocean see Rhines (1970) and for an arbitrary $N(z)$ profile see the review by Charney and Flierl (1981; §18.5.1). The extension of the formulas obtained for barotropic motions to the case when $\partial b/\partial x$ and $\partial b/\partial y$ vary in such a way that $L \ll L_b$ can easily be carried out by the short-wave approximation method (e.g. see Kamenkovich and Reznik, 1978, §2.2). The existence of topographic waves in a real ocean was demonstrated in a number of papers (e.g. see Thompson, 1971, 1977; Thompson and Luyten, 1976; Rhines, 1977).

Now we discuss a peculiar effect of bottom relief. It turns out that a small-scale bottom relief of a cylindrical form (i.e. not varying along a certain direction) can support the propagation of a wave with a typical horizontal scale L substantially exceeding the relief scale L_b (Rhines and Bretherton, 1973). Let $L_b/L \sim b_0/H \ll 1$ and $\varepsilon_b = b_0 L/HL_b \sim 1$. First, we consider a homogeneous ocean. Note that for quasigeostrophic motions in a homogeneous ocean Equation (1.39b) holds for $\varepsilon_b \sim 1$ as well provided that the parameters Ki, ε_T, and ε_L are small (conditions (1.7)). Relation (1.39b) implies ($b = b(y)$)

$$\frac{\partial}{\partial t}\Delta p' + \left[\beta + \frac{f_0}{H}\frac{\mathrm{d}b}{\mathrm{d}y}\right]\frac{\partial p'}{\partial x} = 0. \tag{2.39}$$

We seek a bounded wave solution of Equation (2.39):

$$p' = P(y)\exp\left[i(k_x x - \sigma t)\right].$$

Substituting this expression into (2.39) we find

$$\frac{\mathrm{d}^2 P}{\mathrm{d}y^2} - \left[k_x^2 + \frac{k_x}{\sigma}\left(\beta + \frac{f_0}{H}\frac{\mathrm{d}b}{\mathrm{d}y}\right)\right]P = 0. \tag{2.40}$$

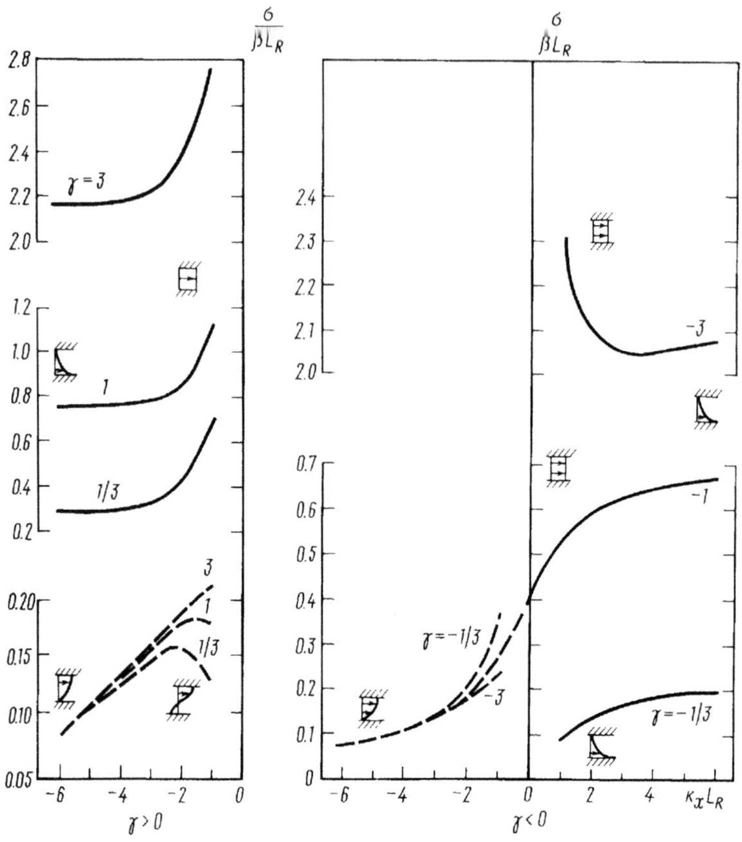

Fig. 2.2.5. Dispersion relations (σ as a function of k_x) for the zeroth (solid line) and first (dashed line) topographic modes for different values of γ. The schemes demonstrate the vertical structures of the modes; $k_x = k_y$, $\gamma = (f_0/\beta H)\,(b_y - b_x)$, $N = \text{const}$, $L_R = H(N/f_0)$. To obtain the curves lying in the half-plane $\sigma < 0$ the values of (k_x, σ) must be replaced by $(-k_x, -\sigma)$. The figure is taken from Rhines (1970) and has been reconstructed with a different scale.

The solution of this equation is sought in the form

$$P = P_0\left(\frac{y}{L}\right) + \frac{L_b}{L}\,P_1\left(\frac{y}{L}, \frac{y}{L_b}\right) + \cdots \tag{2.41}$$

where all the functions P_j are bounded. The corresponding analysis (Rhines and Bretherton, 1973) revealed the existence of a large-scale motion $P_0 = \exp(ik_y y)$ with a dispersion relation of the form

$$\sigma^2 + \frac{\beta k_x}{k^2}\,\sigma - f_0^2\,\frac{\overline{b^2}}{H^2}\,\frac{k_x^2}{k^2} = 0 \tag{2.42}$$

where

$$\overline{b^2} = \lim_{Y\to\infty}\frac{1}{2Y}\int_{-Y}^{Y} b^2\,dy.$$

We note that, as follows from (2.42), for $\beta = 0$ the wave motion does not disappear; for $\overline{b^2} = 0$ it transforms into a barotropic Rossby wave in an ocean of constant depth.

In the case of a stratified fluid there also exists a large-scale barotropic wave over a small-scale bottom relief of cylindrical form (see Volosov and Zhdanov, 1980a, b, 1982, 1983). Let $L_R \sim L_b$ and let $L_b/L \sim b_0/H \ll 1$ as in the case of a homogeneous fluid; naturally, conditions (1.7) are fulfilled. Then, although $\varepsilon_b \sim 1$, condition (1.41) for the existence of barotropic motion in a stratified fluid holds (e.g. $\varepsilon_T \sim L_R/L \ll 1$). In the case of the simplest bottom relief $b(y) = \sqrt{2}\, b_0 \sin(y/L_b)$ and $N = \text{const}$ we have

$$\sigma^2 + \frac{\beta k_x}{k^2}\sigma - f_0^2 \frac{b_0^2}{H^2}\left(\frac{L_R}{L_b}\coth\frac{L_R}{L_b}\right)\frac{k_x^2}{k^2} = 0. \tag{2.43}$$

This dispersion relation differs from (2.42) in the correcting factor (L_R/L_b) coth (L_R/L_b); for $L_R = L_b$ this coefficient is equal to 1.31. For a variable $N(z)$ the correcting factor is close to unity.

Expansion (2.41) is also used in the case of a stratified fluid; here $P_1, \ldots$ depend not only on y/L and y/L_b but also on z/H. Since $L_b/L\, \partial P_1/\partial y \sim \partial P_0/\partial y$, the velocities of small-scale baroclinic and large-scale barotropic motions are of the same order. Therefore, although stratification has very little effect on the dispersion relation, it leads, as the calculation shows, to the intensification of near-bottom currents.

Of interest is the effect of isolated irregularities of the bottom relief on the propagation of Rossby waves. The presence of such irregularities leads, on the one hand, to the scattering of Rossby waves and, on the other, to the appearance of trapped wave modes whose energy is concentrated in a narrow region in the vicinity of the irregularities of the bottom relief (e.g. see Kamenkovich and Reznik, 1978, §2; Charney and Flierl, 1981, §18.5).

We also mention the paper by Odulo and Pelinovskiĭ (1978) where consideration is given to the damping of Rossby waves caused by random irregularities of the bottom relief.

2.3 Nonlinear Waves; Weak Interaction

Even when the nonlinearity parameter $\delta = U/(\beta L^2)$ is small, nonlinear phenomena can exert a substantial effect on the propagation of waves. For simplicity, we consider barotropic waves in a homogeneous ocean of constant depth and pass to dimensionless variables by the formulas

$$(x, y) = L(x_1, y_1); \quad p' = f_0 \varrho_0 U L \psi; \quad t = (\beta L)^{-1} t_1.$$

According to Equation (1.39b), omitting the subscripts in x_1, y_1, and t_1, we write

$$\frac{\partial}{\partial t}\Delta\psi + \frac{\partial \psi}{\partial x} = -\delta J(\psi, \Delta\psi); \quad \delta = \frac{U}{\beta L^2} \ll 1. \tag{2.44}$$

Let, for instance, the function ψ be representable for $t = 0$ as a sum of three harmonics:

$$\psi(x, y, 0) = \psi_{\mathbf{k}_1} \cos(\mathbf{k}_1\mathbf{r}) + \psi_{\mathbf{k}_2} \cos(\mathbf{k}_2\mathbf{r}) + \psi_{\mathbf{k}_3} \cos(\mathbf{k}_3\mathbf{r}), \tag{2.45}$$

where $\mathbf{r} = (x, y)$ and $\mathbf{k}_m = (k_{mx}, k_{my})$ $(m = 1, 2, 3)$.

We construct a solution of problem (2.44), (2.45) bounded in the plane x, y and for large times t in the form of the following series:

$$\psi = \psi_0(x, y, t; X, Y, T) + \delta\psi_1(x, y, t; X, Y, T) + \cdots \tag{2.46}$$

Here the 'slow' variables $X = \delta x$, $Y = \delta y$, and $T = \delta t$ have been introduced. Substituting (2.46) into (2.44) we obtain

$$\frac{\partial}{\partial t} \Delta\psi_0 + \frac{\partial \psi_0}{\partial x} = 0; \tag{2.47$_0$}$$

$$\frac{\partial}{\partial t} \Delta\psi_1 + \frac{\partial \psi_1}{\partial x} = -J(\psi_0, \Delta\psi_0) - \frac{\partial}{\partial T} \Delta\psi_0 -$$

$$-2\frac{\partial}{\partial t}\left(\frac{\partial^2 \psi_0}{\partial x\, \partial X} + \frac{\partial^2 \psi_0}{\partial y\, \partial Y}\right) - \frac{\partial \psi_0}{\partial X}, \tag{2.47$_1$}$$

where the operators J and Δ involve differentiation with respect to x and y.

Introducing the notation

$$\theta_m = \mathbf{k}_m\mathbf{r} - \sigma(\mathbf{k}_m)t; \ \sigma(\mathbf{k}_m) = -\frac{k_{mx}}{k_m^2}, \ m = 1, 2, 3, \tag{2.48}$$

we find the function ψ_0 in the form of a sum of three wave packets:

$$\psi_0 = \psi_{\mathbf{k}_1}(X, Y, T) \cos\theta_1 + \psi_{\mathbf{k}_2}(X, Y, T) \cos\theta_2 +$$

$$+ \psi_{\mathbf{k}_3}(X, Y, T) \cos\theta_3, \tag{2.49}$$

where the dependence of the amplitudes $\psi_{\mathbf{k}_1}$, $\psi_{\mathbf{k}_2}$, and $\psi_{\mathbf{k}_3}$, on X, Y, and T is arbitrary. Assume that the given wave vectors $\mathbf{k}_1$, $\mathbf{k}_2$, and $\mathbf{k}_3$, are such that the resonant conditions hold:

$$\mathbf{k}_1 + \mathbf{k}_2 = \mathbf{k}_3; \tag{2.50a}$$

$$\sigma(\mathbf{k}_1) + \sigma(\mathbf{k}_2) = \sigma(\mathbf{k}_1 + \mathbf{k}_2). \tag{2.50b}$$

The existence of such resonant triads $\mathbf{k}_1$, $\mathbf{k}_2$, $\mathbf{k}_3$ is proved in the paper by Longuet-Higgins and Gill (1967).

We now consider Equation (2.47$_1$). Substituting expression (2.49) for ψ_0 and taking into account that, according to (2.50), $\theta_1 + \theta_2 = \theta_3$ we rewrite Equation (2.47$_1$) in the form

$$\frac{\partial}{\partial t} \Delta\psi_1 + \frac{\partial \psi_1}{\partial x} = D_{\mathbf{k}_1\mathbf{k}_2}\psi_{\mathbf{k}_1}\psi_{\mathbf{k}_2} \cos(\theta_1 - \theta_2) +$$

$$+ \left[-D_{\mathbf{k}_1\mathbf{k}_2}\psi_{\mathbf{k}_1}\psi_{\mathbf{k}_2} + k_3^2\left(\frac{\partial \psi_{\mathbf{k}_3}}{\partial T} + \mathbf{c}_g(\mathbf{k}_3)\,\nabla\psi_{\mathbf{k}_3}\right)\right]\cos\theta_3 +$$

$$+\left[D_{\mathbf{k}_1\mathbf{k}_3}\psi_{\mathbf{k}_1}\psi_{\mathbf{k}_3}+k_2^2\left(\frac{\partial\psi_{\mathbf{k}_2}}{\partial T}+\mathbf{c}_g(\mathbf{k}_2)\,\nabla\psi_{\mathbf{k}_2}\right)\right]\cos\theta_2-$$

$$-D_{\mathbf{k}_1\mathbf{k}_3}\psi_{\mathbf{k}_1}\psi_{\mathbf{k}_3}\cos(\theta_1+\theta_3)+\left[D_{\mathbf{k}_2\mathbf{k}_3}\psi_{\mathbf{k}_2}\psi_{\mathbf{k}_3}+\right.$$

$$\left.+k_1^2\left(\frac{\partial\psi_{\mathbf{k}_1}}{\partial T}+\mathbf{c}_g(\mathbf{k}_1)\,\nabla\psi_{\mathbf{k}_1}\right)\right]\cos\theta_1-D_{\mathbf{k}_2\mathbf{k}_3}\psi_{\mathbf{k}_2}\psi_{\mathbf{k}_3}\cos(\theta_2+\theta_3),\quad(2.51)$$

where $\mathbf{c}_g(\mathbf{k})$ is the group velocity of Rossby waves (see (2.22)); $\nabla=(\partial/\partial X,\partial/\partial Y)$; and the quantities

$$D_{\mathbf{k}_m\mathbf{k}_n}=\frac{1}{2}(k_{nx}k_{my}-k_{mx}k_{ny})(k_m^2-k_n^2),\quad m,n=1,2,3,\qquad(2.52)$$

introduced here are called interaction coefficients.

We eliminate the resonant terms on the right-hand side of Equation (2.51) by equating to zero the coefficients in $\cos\theta_1$, $\cos\theta_2$, and $\cos\theta_3$. In this way we derive equations for the amplitudes $\psi_{\mathbf{k}_1}$, $\psi_{\mathbf{k}_2}$, and $\psi_{\mathbf{k}_3}$ of the wave packets forming the field ψ_0:

$$\frac{\partial\psi_{\mathbf{k}_1}}{\partial T}+\mathbf{c}_g(\mathbf{k}_1)\,\nabla\psi_{\mathbf{k}_1}=-\frac{D_{\mathbf{k}_2\mathbf{k}_3}}{k_1^2}\psi_{\mathbf{k}_2}\psi_{\mathbf{k}_3};$$

$$\frac{\partial\psi_{\mathbf{k}_2}}{\partial T}+\mathbf{c}_g(\mathbf{k}_2)\,\nabla\psi_{\mathbf{k}_2}=-\frac{D_{\mathbf{k}_1\mathbf{k}_3}}{k_2^2}\psi_{\mathbf{k}_1}\psi_{\mathbf{k}_3};\qquad(2.53)$$

$$\frac{\partial\psi_{\mathbf{k}_3}}{\partial T}+\mathbf{c}_g(\mathbf{k}_3)\,\nabla\psi_{\mathbf{k}_3}=\frac{D_{\mathbf{k}_1\mathbf{k}_2}}{k_3^2}\psi_{\mathbf{k}_1}\psi_{\mathbf{k}_2}.$$

Now the function ψ_1 can be represented as a sum of the corresponding wave packets the equations of whose amplitude are obtained by eliminating the resonant terms from the right-hand sides of the equations for ψ_2, etc.

We note that, owing to the dependence of the coefficients ψ_0, ψ_1, . . . of expansion (2.46) on the 'slow' variables X, Y, and T, we have the possibility of eliminating the resonant terms from the right-hand sides of the equations for ψ_1, . . . If otherwise, even the function ψ_1 would increase linearly with time and expansion (2.46) would no longer hold for times of the order of $1/\delta$ (the terms ψ_0 and ψ_1 would become comparable in their orders of magnitude).

If the resonant conditions (2.50) are not fulfilled, the right-hand sides of Equations (2.53) must be equated to zero; in this case the amplitudes of the wave packets move in space with the group velocity $\mathbf{c}_g$ without changing their form. In the general case of interacting wave packets the system of equations (2.53) is extremely complicated. However, if the amplitudes $\psi_{\mathbf{k}_1}$, $\psi_{\mathbf{k}_2}$, and $\psi_{\mathbf{k}_3}$ do not depend on X and Y at $t=0$, they do not depend on the coordinates X and Y for any T, and, by virtue of (2.53), we obtain

$$\frac{d\psi_{\mathbf{k}_1}}{dT} = \frac{k_3^2 - k_2^2}{2k_1^2} b\psi_{\mathbf{k}_2}\psi_{\mathbf{k}_3};$$

$$\frac{d\psi_{\mathbf{k}_2}}{dT} = \frac{k_1^2 - k_3^2}{2k_2^2} b\psi_{\mathbf{k}_1}\psi_{\mathbf{k}_3}; \tag{2.54}$$

$$\frac{d\psi_{\mathbf{k}_3}}{dT} = \frac{k_2^2 - k_1^2}{2k_3^2} b\psi_{\mathbf{k}_1}\psi_{\mathbf{k}_2},$$

where $b = k_{3x}k_{2y} - k_{2x}k_{3y} = k_{1x}k_{3y} - k_{3x}k_{1y} = k_{1x}k_{2y} - k_{2x}k_{1y}$; the equality of these expressions simply follows from the fact that each of them is equal to the area of the parallelogram constructed on vectors $\mathbf{k}_1$ and $\mathbf{k}_2$ (positive if $\mathbf{k}_1$, $\mathbf{k}_2$ is a right-handed pair of vectors and negative if $\mathbf{k}_1$, $\mathbf{k}_2$ is a left-handed pair).

The system of ordinary differential equations (2.54) was studied sufficiently thoroughly (e.g. see Longuet-Higgins and Gill, 1967). This system has two simple integrals of energy and enstrophy and its solution is expressed in terms of elliptic functions. The analysis of the solutions shows that there occurs energy exchange between various modes, which leads to finite variations of amplitudes. However, it should be borne in mind that since these variations are caused by weak nonlinearity ($\delta \ll 1$), they are very slow (relative to the scale of t).

The above asymptotic scheme admits of setting the initial field (2.45) in the form of a sum of an arbitrary number of wave packets with wave vectors $\mathbf{k}_j (j = \overline{1, n})$. It can easily be verified that in this case the resonant conditions have the form

$$\sigma(\mathbf{k}_i) \pm \sigma(\mathbf{k}_j) = \sigma(\mathbf{k}_i \pm \mathbf{k}_j), \quad i, j = \overline{1, n}. \tag{2.55}$$

(For the construction of the solution in this case see, for instance, Kamenkovich and Reznik, 1978, §5.1. We also mention the works devoted to interaction of Rossby waves on a sphere (e.g. see Galin, 1974).)

In conclusion we note that if the approximate expression for the kinematic boundary condition at the sea surface (the first condition (1.9)) in the form of the first condition (1.36) is abandoned and the boundary condition is written in full as $w = \partial\zeta/\partial t$ for $z = 0$ (in the geostrophic approximation we have $u(\partial\zeta/\partial x) + v(\partial\zeta/\partial y) = 0$), then it can easily be shown that the transfer equation for the quasigeostrophic potential vorticity for barotropic motions (1.39b) has the form

$$\left(\frac{\partial}{\partial t} + u\frac{\partial}{\partial x} + v\frac{\partial}{\partial y}\right)\omega_*^{(b)} = 0, \tag{2.56}$$

$$\omega_*^{(b)} = \frac{1}{f_0\varrho_0}\Delta p' + \beta y - \frac{f_0^2}{gH}\frac{p'}{f_0\varrho_0} + f_0\frac{b}{H}.$$

In this case the dispersion relation for the barotropic Rossby wave has the form (cf. (2.9) and (2.10))

$$\sigma = -\frac{k_x\beta}{k^2 + \dfrac{f_0^2}{gH}} \tag{2.57}$$

The length scale

$$L_R = \frac{\sqrt{gH}}{f_0} \tag{2.58}$$

is usually called the Rossby deformation radius for barotropic motions. For H = 5 km and latitude equal to 30° this scale is equal to 3.04×10^3 km; therefore, for motions with scales of the order of 100 km it is allowable to apply the rigid lid approximation, i.e. the third term in expression (2.56) for $\omega_*^{(b)}$ can be ignored. In this case it turns out that the two-dimensional divergence for barotropic Rossby waves in an ocean of constant depth is equal to zero (nondivergent waves). However, to study qualitatively the effect of the two-dimensional divergence on Rossby waves this term can be taken into account in the expression for $\omega_*^{(b)}$ (for the discussion of boundary conditions at the sea surface also see Section 2.5 of this chapter and Section 3 of Chapter 3).

We choose L_R as the scale L, introduce the stream function, and write Equation (2.56) in the following dimensionless form (cf. (2.44)):

$$\frac{\partial}{\partial t}(\Delta\psi - \psi) + \frac{\partial\psi}{\partial x} = -\,\delta J(\psi, \Delta\psi); \; \delta \ll 1. \tag{2.59}$$

The results of the above analysis can be extended without any changes to the case of interaction of divergent waves satisfying Equation (2.59); here one must only replace the dispersion relation $\sigma(\mathbf{k}) = -k_x k^{-2}$ by $\sigma(\mathbf{k}) = -k_x(k^2 + 1)^{-1}$. In Section 3 we study weak interaction of divergent barotropic Rossby waves with random amplitudes forming a continuous spectrum.

2.4 Nonlinear Stationary Waves

Synoptic motions in the ocean develop on the background of the large-scale oceanic circulation. The ratio of the typical horizontal scale L of synoptic eddies to the typical scale L_c of large-scale motions is small ($L_c \sim 1000$ km; $L \sim 50$ km) and the typical velocity U_c of large-scale currents is of the order of the typical velocity U_{ph} of motion of synoptic eddies. Since the typical particle velocity U in a synoptic eddy is substantially greater than the typical velocity $U_{\text{ph}} = \beta L^2$ of motion of the eddy itself, the nonlinearity parameter satisfies the condition $\delta = U/\beta L^2 > 1$ and nonlinear effects must play a significant role in the eddy dynamics. Therefore, we present the basic aspects of the theory of short nonlinear stationary Rossby waves propagating on the background of a slowly varying mean current (see Larichev and Reznik, 1976a, c).

Again, for simplicity, we consider barotropic motions in an ocean of constant depth. We study solutions propagating with a constant velocity U_{ph} along the x-axis without changing their form:

$$p' = p'(x - U_{\text{ph}}\, t, y). \tag{2.60}$$

Substituting expression (2.60) into Equation (1.39b) we obtain, after some simple transformations, the equation

$$\frac{1}{f_0\varrho_0}\Delta p' + \beta y = F\left(\frac{1}{f_0\varrho_0}p' + U_{\mathrm{ph}}y\right), \tag{2.61}$$

where F is an arbitrary function, $p' = p'(\xi, y)$, $\xi = x - U_{\mathrm{ph}}t$, $\Delta = \partial^2/\partial\xi^2 + \partial^2/\partial y^2$.

Within the framework of expression (2.60) the pressure p' can be represented in the form of a sum of a *given* pressure $\bar{p}'(y)$ generating a purely zonal current and a stationary wave p'_w. It is advisable to write it in the form:

$$\frac{p'}{f_0\varrho_0} = U_c L_c \bar{\psi}\left(\frac{y}{L_c}\right) + UL\psi_w\left(\frac{\xi}{L}, \frac{y}{L}\right). \tag{2.62}$$

We introduce the dimensionless coordinates $\xi_1 = \xi/L$ and $y_1 = y/L$. Then after substitution of (2.62) into (2.61) we have

$$\frac{\varepsilon^2 d^2\bar{\psi}}{dY^2} + \alpha\Delta\psi_w + Y = F(\bar{\psi} + \alpha\psi_w + c_{\mathrm{ph}}Y) \tag{2.63}$$

where

$$\varepsilon = \frac{L}{L_c}, \quad \alpha = \frac{U}{U_c}\varepsilon, \quad Y = \varepsilon y_1, \quad U_c = \beta L^2, \quad c_{\mathrm{ph}} = \frac{U_{\mathrm{ph}}}{\beta L^2}. \tag{2.64}$$

Here c_{ph} is the dimensionless phase velocity of the order of unity; the typical velocity of the large-scale motion is equated to βL^2 and the unknown function F is defined accordingly.

Assume that the function F does not depend on the amplitude of the wave motion. Then for $\alpha \to 0$ we obtain

$$\frac{\varepsilon^2 d^2\bar{\psi}}{dY^2} + Y = F(\bar{\psi} + c_{\mathrm{ph}}Y). \tag{2.65}$$

Given the function $\bar{\psi}(Y)$, we can determine the function $F(\tau)$ using (2.65).

Thus we seek a wave solution ψ_w of Equation (2.63) where the function F is determined by (2.65). By hypothesis, $L \ll L_c$ and hence $\varepsilon \ll 1$; therefore, the short-wave approximation can be applied (e.g. see Kogelman and Keller, 1973). We represent the functions ψ_w and F as:

$$\alpha\psi_w = \varphi_0(\theta, X, Y) + \varepsilon\varphi_1(\theta, X, Y) + \cdots \tag{2.66}$$

$$F = F_0 + \varepsilon F_1 + \cdots \tag{2.67}$$

where $X = \varepsilon\xi_1$, $\theta = \Phi(X, Y)/\varepsilon$, $\Phi = O(1)$, and all φ_n are assumed to be 2π-periodic functions of the variable θ. From the physical viewpoint, expansion (2.66) means that we seek a solution of the type of a plane wave with slowly varying wave vector $(\partial\Phi/\partial X, \partial\Phi/\partial Y)$ and energy. (This is a generalization of the short-wave expansion for linear problems where the functions φ_n depend on θ, as $\exp(i\theta)$.)

Substituting (2.66) and (2.67) into (2.63) and (2.65) and equating to zero the coefficients in the various powers of ε we find

$$(\nabla\Phi)^2\varphi''_0 - F_0(\bar{\psi} + c_{\text{ph}}Y + \varphi_0) + Y = 0, \tag{2.68$_0$}$$

$$(\nabla\Phi)^2\varphi''_1 - \frac{\partial F_0}{\partial\varphi_0}\varphi_1 = -2\frac{\partial\Phi}{\partial X}\frac{\partial\varphi'_0}{\partial X} - 2\frac{\partial\Phi}{\partial Y}\frac{\partial\varphi'_0}{\partial Y} - $$

$$-\varphi'_0\left(\frac{\partial^2\Phi}{\partial X^2} + \frac{\partial^2\Phi}{\partial Y^2}\right), \tag{2.68$_1$}$$

$$F_0(\bar{\psi} + c_{\text{ph}}Y) - Y = 0, \tag{2.69$_0$}$$

where the differentiation with respect to θ is symbolized by the prime.

Equation (2.68$_0$) is an ordinary differential equation for the function $\varphi_0(\theta)$. It is known that for a nonconstant periodic solution of this equation to exist it is required that the function

$$Q(\varphi_0) = -\int F_0(\bar{\psi} + c_{\text{ph}}Y + \varphi_0)\,\mathrm{d}\varphi_0 + Y\varphi_0 \tag{2.70}$$

should attain minimum. By virtue of (2.69$_0$), we have

$$Q'(0) = -F_0(\bar{\psi} + c_{\text{ph}}Y) + Y = 0, \tag{2.71a}$$

$$Q''(0) = -\left.\frac{\partial F_0}{\partial\varphi_0}\right|_0 = -\frac{1}{\bar{\psi}' + c_{\text{ph}}} > 0. \tag{2.71b}$$

The inequality (2.71b) implies that stationary waves must move westward relative to the given zonal current $u(Y) = -\bar{\psi}'(Y)$.

It can readily be shown that if inequality (2.71b) is fulfilled, the solution of Equation (2.68$_0$) and the expression for its period 2π are given by

$$\theta - \theta_0 = \pm\sqrt{\frac{(\nabla\Phi)^2}{2}}\int_{r_1}^{\varphi_0}[E - Q(\tau)]^{-1/2}\,\mathrm{d}\tau, \tag{2.72}$$

$$2\pi = \sqrt{2(\nabla\Phi)^2}\int_{r_1}^{r_2}[E - Q(\tau)]^{-1/2}\,\mathrm{d}\tau, \tag{2.73}$$

where θ_0 is the reference phase and $E = E(X, Y)$ is an arbitrary function; r_1 and r_2 are the minimum and maximum roots of the equation $E - Q(\tau) = 0$. The function F_0 is found with the aid of (2.69$_0$) by inverting the function $\bar{\psi}(Y) + c_{\text{ph}}Y$ (according to (2.71b), this function is monotonous).

Consider Equation (2.68$_1$). Differentiating (2.68$_0$) with respect to θ we find that φ'_0 satisfies the homogeneous equation corresponding to (2.68$_1$). Therefore, multiplying (2.68$_1$) by φ'_0 we obtain, after some simple transformations,

$$(\nabla\Phi)^2\frac{\partial}{\partial\theta}(\varphi'_1\varphi'_0 - \varphi_1\varphi''_0) = -\frac{\partial}{\partial X}\left(\frac{\partial\Phi}{\partial X}\varphi'^2_0\right) - \frac{\partial}{\partial Y}\left(\frac{\partial\Phi}{\partial Y}\varphi'^2_0\right).$$

Integrating this equation with respect to θ from 0 to 2π and using the 2π-periodicity of $\varphi_0(\theta)$ and $\varphi_1(\theta)$ we find

$$\frac{\partial}{\partial X}\left(\frac{\partial \Phi}{\partial X}\int_0^{2\pi} \varphi_0'^2 \, \mathrm{d}\theta\right) + \frac{\partial}{\partial Y}\left(\frac{\partial \Phi}{\partial Y}\int_0^{2\pi} \varphi_0'^2 \, \mathrm{d}\theta\right) = 0. \tag{2.74}$$

Equations (2.73) and (2.74) are used to determine the functions $\Phi = \Phi(X, Y)$ and $E = E(X, Y)$; together with (2.72) they yield the complete solution of the problem under study in the first approximation. Equation (2.73) serves as a dispersion relation for the waves under consideration.

Larichev and Reznik (1976a, c) analyzed the above formulas for both the weakly nonlinear case ($\alpha \ll 1$) and the strongly nonlinear case ($\alpha \sim 1$). Summing up their conclusions for the weakly nonlinear case we first note that, owing to the nonlinearity, the dispersion relation involves dependence on the wave amplitude. This is the general property of all nonlinear waves (Karpman, 1973, Ch. 3, §12). Further, the structure of a nonlinear wave substantially differs from that of a linear wave since in the expression for the former wave there appear multiple harmonics and terms independent of θ which cause a change of the given zonal current. This effect means that, generally speaking, synoptic eddies can take part in the formation of a large-scale circulation. Finally, we note that nonlinear effects manifest themselves only when the given zonal current has a meridional shear $(\mathrm{d}\bar{u}/\mathrm{d}Y) \neq 0)$.

2.5. Nonlinear Topographic Waves

Nonlinear effects in the problem of large-scale topographic waves over a cylindric bottom relief with a smaller scale were studied in the papers by Volosov (1976a, b, c, 1982) and Volosov and Zhdanov (1980a, b, 1982, 1983). Here we confine ourselves only to the case of motion of a homogeneous fluid. We represent the solution of the problem in the form of the sum of a large-scale wave $f_0 \varrho_0 U L \psi_l$ and a small-scale component $f_0 \varrho_0 U L_b \psi_s$:

$$\frac{p'}{f_0 \varrho_0} = UL\left[\psi_l\left(\frac{x}{L}, \frac{y}{L}, \frac{t}{T}\right) + \right.$$
$$\left. + \frac{b_0}{H}\psi_s\left(\frac{y}{L_b}, \frac{x}{L}, \frac{y}{L}, \frac{t}{T}\right)\right]. \tag{2.75}$$

The equation for $\psi_l(\theta)$ has the form

$$\frac{\mathrm{d}^2\psi_l}{\mathrm{d}\theta^2} - \frac{\beta k_x}{k^2\sigma}\psi_l + \frac{k_x}{k^2\sigma} M\left(\frac{k_x}{\sigma}\psi_l\right) = 0 \tag{2.76}$$

where $\theta = k_x x + k_y y - \sigma t$, $M(0) = 0$, and

$$\frac{dM(\tau)}{d\tau} = \frac{1}{H^2} \langle b(\tilde{y})b(\tilde{y} + \tau) \rangle$$

$$= \frac{1}{H^2} \lim_{A \to \infty} \frac{1}{2A} \int_{-A}^{A} b(\tilde{y})b(\tilde{y} + \tau)\, d\tilde{y} = B(\tau). \tag{2.77}$$

If we put $M[(k_x/\sigma)\ \psi_l] \simeq (k_x/\sigma)\ \psi_l B(0)$ in (2.76), we obtain the equation of the linear problem (see (2.42)). By formal analogy with the theory of stationary random functions, the expression $B(\tau)$ can be called the correlation function of the relative elevation of the bottom relief.

3. WEAK TURBULENCE ON THE β-PLANE

3.1. Kinetic Equation for Divergent Barotropic Rossby Waves

In Section 2.3 we considered the weakly nonlinear interaction of three Rossby wave packets satisfying the resonance conditions (2.50a, b). During this interaction there occurs an energy exchange between various modes (which is slow in comparison with the typical times of variation of the phases of the modes). This results in periodic variation of the amplitudes of the modes. The process of interaction of a very large number of Rossby waves (an infinite number of waves in the case of a continuous spectrum) when each of the modes can be in resonance (i.e. satisfy conditions (2.55)) with a multitude of other modes is much more complicated. In this case the variation of the amplitudes of the modes is no longer periodic but is much more complex (essentially, chaotic). Therefore, when studying these wave systems the wave field is assumed to be random (and, for simplicity, homogeneous) and emphasis is laid not upon the behavior of the amplitudes of individual modes but upon statistical characteristics of the field and, primarily, the energy spectral density $F(\mathbf{k}, \tau)$ defined as

$$\tfrac{1}{2}\,(k^2 + 1) \langle \tilde{\psi}(\mathbf{k},t\ \tilde{\psi}(\mathbf{k}', t\,) \rangle = F(\mathbf{k},\tau)\ \delta\,(\mathbf{k} + \mathbf{k}'),$$

where $\mathbf{k}$ is the wave vector, $\tau = \delta^2 t$ is slow time, and $\tilde{\psi}$ $(\mathbf{k}, t)$ is the Fourier transform of the stream function ψ (x, y, t). The function F satisfies the so-called kinetic equation (Longuet-Higgins and Gill, 1967; Kenyon, 1964):

$$\frac{\partial F}{\partial \tau} = 8\pi \int D_{\mathbf{k}_1\mathbf{k}_2}\,(D_{\mathbf{k}_1\mathbf{k}_2}F_1F_2 + D_{\mathbf{k}\mathbf{k}_1}FF_1 + D_{\mathbf{k}\mathbf{k}_2}FF_2)\mathcal{K}^{-1} \times$$

$$\times\ \delta(\mathbf{k}_{012})\ \delta(\sigma_{012})\ d\mathbf{k}_1\ d\mathbf{k}_2. \tag{3.1}$$

(For the method of derivation of (3.1) and the corresponding bibliography see, for instance, the paper by Reznik, 1984b.) In (3.1), $\delta(s)$ is the delta function, $D_{\mathbf{k}_1\mathbf{k}_2}$ is the interaction coefficient:

$$D_{\mathbf{k}_1\mathbf{k}_2} = \tfrac{1}{2}(k_{1x}k_{2y} - k_{2x}k_{1y})\ (k_2^2 - k_1^2), \tag{3.2}$$

$\sigma(\mathbf{k}) = -k_y(k^2+1)^{-1}$ is the Rossby wave frequency, $\sigma_m = \sigma(k_m)$, $F_m = F(\mathbf{k}_m, \tau)$; $m = 1, 2$; $\mathbf{k}_0 \equiv \mathbf{k}$, $V = (k^2+1)(k_1^2+1)(k_2^2+1)$, $\mathbf{k}_{012} = \mathbf{k} + \mathbf{k}_1 + \mathbf{k}_2$, and $\sigma_{012} = \sigma + \sigma_1 + \sigma_2$. Equation (3.1) is written in dimensionless form, the dimensionless Rossby deformation radius being equal to 1 (see Equation (2.59).

It can easily be verified that for $\mathbf{k}$, $\mathbf{k}_1$, and $\mathbf{k}_2$ lying on a resonance surface (i.e. satisfying the conditions $\mathbf{k}_{012} = 0$ and $\sigma_{012} = 0$) the interaction coefficients satisfy the relations (see Sazontov, 1981):

$$(pk^2+q)D_{\mathbf{k}_1\mathbf{k}_2} + (pk_2^2+q)D_{\mathbf{k}\mathbf{k}_1} + (pk_1^2+q)D_{\mathbf{k}\mathbf{k}_2} = 0, \tag{3.3a}$$

$$(k_y/\sigma)D_{\mathbf{k}_1\mathbf{k}_2} + (k_{1y}/\sigma_1)D_{\mathbf{k}\mathbf{k}_2} + (k_{2y}/\sigma_2)D_{\mathbf{k}\mathbf{k}_1} = 0, \tag{3.3b}$$

where p and q are arbitrary constants. Using (3.3a, b) one can show (see Reznik, 1984a; Sazontov, 1981) that the function F must satisfy the conditions of conservation of energy E, enstrophy Ω_*, and the meridional component L of the total wave momentum, written as

$$E = \int F\,d\mathbf{k} = \text{const}, \tag{3.4a}$$

$$\Omega_* = \int (k^2+1)F\,d\mathbf{k} = \text{const}, \tag{3.4b}$$

$$L = \int k_y\sigma^{-1}F\,d\mathbf{k} = \text{const}. \tag{3.4c}$$

It is important to note that conditions (3.4a, b) also remain valid when the nonlinearity is not weak (i.e. the parameter δ is not small). This follows (e.g. see Reznik 1984b) from the equation

$$\frac{\partial F}{\partial t} = 2\delta^2 \int D_{\mathbf{k}_1\mathbf{k}_2}(\Gamma_{012} + \Gamma^*_{012})\,\delta(\mathbf{k}_{012})\,d\mathbf{k}_1\,d\mathbf{k}_2$$

which is implied by the potential vorticity equation for arbitrary nonlinearity (here $\Gamma_{012} = \Gamma(\mathbf{k}, \mathbf{k}_1, \mathbf{k}_2)$ is the spectral cumulant of the third order defined as

$$\langle \tilde{\psi}(\mathbf{k})\,\tilde{\psi}(\mathbf{k}_1)\,\tilde{\psi}(\mathbf{k}_2)\rangle = \Gamma(\mathbf{k}, \mathbf{k}_1, \mathbf{k}_2)\,\delta\,(\mathbf{k} + \mathbf{k}_1 + \mathbf{k}_2)$$

and relation (3.3a) which holds not only on the resonance surface but also for any $\mathbf{k}$, $\mathbf{k}_1$, and $\mathbf{k}_2$ such that $\mathbf{k}_{012} = 0$. By virtue of the condition of conservation of enstrophy (3.4b), we have

$$F = o(k^{-4})\ (k \to \infty), \tag{3.5}$$

i.e. the spectrum F rapidly decreases for sufficiently large k.

We now discuss the limits of applicability of Equation (3.1). According to Reznik (1984b), the decisive role in the derivation of the kinetic equation is played by the behavior of the group velocities of the interacting waves in the vicinity of a resonance surface S_R determined by the relations $\mathbf{k}_{012} = 0$ and $\sigma_{012} = 0$. If the modulus of the difference between the group velocities on the resonance surface S_R which is equal to

$$\Delta = |\nabla\sigma_1 - \nabla\sigma_2|_{\mathbf{k}_1,\,\mathbf{k}_2 \in S_R}$$

is uniformly bounded below for all S_R, then the kinetic equation holds for all $\mathbf{k}$. The specific property of Rossby waves is that their frequencies and group velocities decrease with increasing moduli of wave vectors. Therefore, the boundedness of Δ from below is violated when the length of the wave vectors $\mathbf{k}_1$ and $\mathbf{k}_2$, which are in resonance with the wave vector $\mathbf{k}$, is very large. First, this is the case for $k_x \to 0$, i.e. for strongly nonlocal interaction of very short waves with a wave having a wavelength of the order of 1 and propagating approximately along the meridian; then the maximum values of k_1 and k_2 are of the order of k_x^{-1}, i.e. the difference Δ can be made arbitrarily small. Second, this takes place in the interaction between three very short waves when $k, k_1, k \to \infty$; in this case $\Delta \sim k^{-2} \to 0$.

The estimates obtained from the system of equations for the cumulants F and Γ_{012} (this system can, for instance, be found in the paper by Reznik, 1984b) in the quasi-Gaussian approximation (i.e. under the assumption that the fourth-order cumulant is equal to zero) show that, by virtue of the fast damping of F for large k (see formula (3.5)), the interactions of the first type mainly reduce to an energy exchange between short waves with wave vectors $\mathbf{k}_1$ and $\mathbf{k}_2$ and exert almost no effect on the wave with $k_x \to 0$ which serves as a 'catalyzer' for the interaction between the waves with $\mathbf{k}_1$ and $\mathbf{k}_2$ (Longuet-Higgins and Gill, 1967). Therefore, one can assume that the kinetic equation remains valid in the neighborhood of the axis k_y although $\sigma(\mathbf{k}) \to 0$ for $k_x \to 0$. It is important to stress that the applicability of the kinetic equation is violated not because $\sigma(\mathbf{k}) = 0$ for some $\mathbf{k}$ but because there exist resonantly interacting waves in which the difference between the *group velocities* is arbitrarily small (cf. Holloway and Hendershott, 1977). For example, although the frequencies of waves with wavenumbers k_x close or equal to zero are almost equal to zero, the group velocities of their 'resonant partners' are substantially different (provided that the interactions with very short waves are not taken into account). Accordingly, the kinetic equation (3.1) remains valid for these $\mathbf{k}$.

Considering the interactions between very short waves with $\mathbf{k}, \mathbf{k}_1, \mathbf{k}_2 \gtrsim \delta^{-2}$, the frequencies of all waves are small ($\sim\delta^2$) in this region and, therefore, here the interactions between wave vectors,[2] lying in a boundary layer of thickness δ^2 in the vicinity of any resonance surface (see Reznik, 1984b) for which $\sigma_{012} \sim \delta^2$, do not differ from arbitrary nonresonant interactions for which σ_{012} is also of the order of δ^2. It follows that the approximation of weak nonlinearity and, accordingly, the kinetic equation (3.1) are inapplicable in the region of very short waves with $k \sim \delta^{-2}$.

However, it is important that, first, very short waves mainly exchange their energy between themselves and, second, by virtue of (3.5), they make a very small contribution to the total energy of the system. Therefore, we can assume that no significant error is made if the effect of very short waves is ignored when studying the nonlinear evolution of waves with moderate wavelengths ($\ll \delta^{-2}$). This can be done in two ways. In the first place, we can simply assume that $F(\mathbf{k})$ is equal to zero outside a finite region $\Omega(\mathbf{k})$ which must be sufficiently large for the estimate (3.5) to hold for $\mathbf{k}$ not belonging to $\Omega(\mathbf{k})$. This 'truncation' is in fact equivalent to 'switching on' infinite viscosity outside $\Omega(\mathbf{k})$, which instantaneously 'eliminates' all

[2] For brevity, we sometimes speak about interactions between wave vectors meaning the interactions between the waves having these wave vectors.

the modes with wave vectors not contained in $\Omega(\mathbf{k})$. Accordingly, the energy and enstrophy of such a system are not conserved and must slowly decrease with increasing time (the greater the region Ω, the lower the rate of decrease). In the second place, we can neglect *all* the nonlinear interactions involving wave vectors lying outside $\Omega(\mathbf{k})$ (including, for instance, those interactions where only one of the interacting wave vectors does not belong to $\Omega(\mathbf{k})$). This is equivalent to replacing the kinetic equation (3.1) by

$$\frac{\partial F}{\partial \tau} = 8\pi \int_{\Omega_{12}} D_{\mathbf{k}_1\mathbf{k}_2} (D_{\mathbf{k}_1\mathbf{k}_2} F_1 F_2 + D_{\mathbf{k}\mathbf{k}_1} FF_1 + D_{\mathbf{k}\mathbf{k}_2} FF_2) \mathcal{K}^{-1} \times$$

$$\times\, \delta(\mathbf{k}_{012})\, \delta(\sigma_{012})\, d\mathbf{k}_1\, d\mathbf{k}_2 \tag{3.1'}$$

where the integration no longer extends to the infinite region $\{-\infty \leqslant k_{1x}, k_{1y} \leqslant \infty, -\infty \leqslant k_{2x}, k_{2y} \leqslant -\infty\}$ but is carried out over the finite region $\Omega_{12} = \Omega(\mathbf{k}_1) \times \Omega(\mathbf{k}_2)$.

There is an important distinction between the first and second types of truncation, namely Equation (3.1′) possesses conserved quantities for arbitrary $\Omega(\mathbf{k})$ (see Reznik, 1984a):

$$E = \int_{\Omega} F\, d\mathbf{k} = \text{const}, \tag{3.4'a}$$

$$K = \int_{\Omega} (k^2 + 1) F\, d\mathbf{k} = \text{const}, \tag{3.4'b}$$

$$L = \int_{\Omega} k_y \sigma^{-1} F\, d\mathbf{k} = \text{const}. \tag{3.4'c}$$

They are completely analogous to the integrals (3.4) of the original system. If the region $\Omega(\mathbf{k})$ is sufficiently large, then, by virtue of what has been said, the difference between the values of $F(\mathbf{k})$ calculated by the above two methods must be small for $\mathbf{k}$ lying sufficiently far from the boundary of the region $\Omega(\mathbf{k})$ (which is confirmed by the corresponding numerical experiments).

In what follows the numerical scheme based on the first truncation method is called scheme A and the one based on the second method is called scheme B.

3.2. Entropy and Statistical Equilibrium States (Reznik, 1984a)

The kinetic equation (3.1′) admits of the derivation of the entropy theorem analogous to the H-theorem for the Boltzmann equation. To this end, the following relations (which can easily be verified; see Kenyon, 1964) are used:

$$\frac{D_{\mathbf{k}_1\mathbf{k}_2}}{\sigma} = \frac{D_{\mathbf{k}\mathbf{k}_1}}{\sigma_2} = \frac{D_{\mathbf{k}\mathbf{k}_2}}{\sigma_1} = A. \tag{3.6}$$

Relations (3.6) hold on a resonance surface, i.e. for $\mathbf{k}_{012} = 0$ and $\sigma_{012} = 0$. We divide (3.1′) by F and integrate with respect to $\mathbf{k}$ over the region Ω; after some simple transformations, taking into account (3.6), we find

$$\int_{\Omega} \frac{1}{F}\frac{\partial F}{\partial \tau}\,\mathbf{dk} = 8\pi \int_{\Omega_{012}} \frac{A^2 F F_1 F_2}{\mathcal{X}} \frac{\sigma}{F}\left(\frac{\sigma}{F} + \frac{\sigma_1}{F_1} + \frac{\sigma_2}{F_2}\right) \times$$

$$\times\ \delta(\mathbf{k}_{012})\ \delta(\sigma_{012})\ \mathbf{dk}_{012}, \tag{3.7}$$

where

$$\Omega_{012} = \Omega(\mathbf{k}) \times \Omega(\mathbf{k}_1) \times \Omega(\mathbf{k}_2) \text{ and } \mathbf{dk}_{012} = \mathbf{dk}\ \mathbf{dk}_1\ \mathbf{dk}_2.$$

Performing the change of variables $\mathbf{k} \leftrightarrow \mathbf{k}_1, \mathbf{k} \leftrightarrow \mathbf{k}_2$ in (3.7) and adding together (3.7) and the analogous equations obtained by the permutations, we arrive at the relation

$$\frac{\mathrm{d}}{\mathrm{d}\tau}\int_{\Omega} \ln F\,\mathbf{dk} = \frac{8\pi}{3}\int_{\Omega_{012}} \frac{A^2 F F_1 F_2}{\mathcal{X}}\left(\frac{\sigma}{F} + \frac{\sigma_1}{F_1} + \frac{\sigma_2}{F_2}\right)^2 \times$$

$$\times\ \delta(\mathbf{k}_{012})\ \delta(\sigma_{012})\ \mathbf{dk}_{012} \geqslant 0. \tag{3.8}$$

Without loss of generality, we can assume that the integrals in (3.8) are convergent since, by virtue of Equation (3.1′), even if $F(\mathbf{k}, \tau) = 0$ for some $\mathbf{k}$ and $\tau = 0$, the function F increases with increasing time and the corresponding singularity in the integrals disappears. Thus the quantity $H = \int_{\Omega} \ln F\,\mathbf{dk}$ can naturally be interpreted as entropy, and Equation (3.8) indicates the irreversibility of the evolution of the spectrum of weakly nonlinear Rossby waves.

Formula (3.8) readily implies that the maximum of the entropy and, hence, the statistical equilibrium are attained for F satisfying the functional equation

$$\frac{\sigma}{F} + \frac{\sigma_1}{F_1} + \frac{\sigma_2}{F_2} = 0 \tag{3.9a}$$

under the conditions

$$\mathbf{k} + \mathbf{k}_1 + \mathbf{k}_2 = 0,\ \sigma + \sigma_1 + \sigma_2 = 0. \tag{3.9b}$$

It can easily be checked that system (3.9a, b) possesses a solution of the form

$$\frac{\sigma}{F} = \tilde{b}\sigma + \tilde{a}_1 k_x + \tilde{a}_2 k_y \tag{3.10}$$

where $\tilde{a}_1$, $\tilde{a}_2$ and $\tilde{b}$ are arbitrary constants. Since the spectrum must be positive for any k_x and k_y, by virtue of (3.10) we have $\tilde{a}_2 = 0$, and finally

$$F = \frac{1}{a + bk^2}, \tag{3.11}$$

where a and b are arbitrary constants such that $F > 0$. Spectrum (3.11) is nothing other than the well-known equilibrium spectrum of two-dimensional turbulence (see below, Section 5 of Chapter 3), which is a consequence of the existence of two conserved quantities: energy and enstrophy. In the case under consideration, the energy and enstrophy are also conserved, and therefore the existence of a solution in the form of (3.11) is quite natural.

The equilibrium spectrum (3.11) possesses the following two important properties. First, it is stable relative to small disturbances (Reznik, 1984b) and, second, which is less obvious, it is the unique solution of the system of functional equations (3.9) in the class of differentiable functions. (The proof of this important fact is presented in the paper by Reznik (1984a) and in the appendix to this section (see Section 3.8).) Therefore, it can seem that since in the absence of sources and sinks of energy the system tends to an equilibrium state, the spectrum $F(\mathbf{k}, \tau)$, irrespective of any arbitrary initial form, becomes more and more isotropic when approaching the spectrum (3.11).

However, the results of numerical experiments presented below in Section 3.4 indicate quite the opposite. Namely, they show that when a system tends to equilibrium, this is always accompanied by a strong concentration of energy in the vicinity of the axis k_y, i.e. by the appearance of an intensive zonal component of motion. This phenomenon of *anisotropization* of a spectrum takes place even when the spectrum of the initial state differs very slightly from the equilibrium spectrum (3.11). Therefore, it can be assumed that the spectra characterizing a real statistical equilibrium are substantially anisotropic. This assumption does not contradict the uniqueness of solution (3.11) of system (3.9) since anisotropic equilibrium spectra turn out to be *generalized* functions.

3.3. Generalized Spectra

By generalized spectra (see Reznik and Soomere, 1983b), we mean those solutions of Equation (3.1′) which are generalized functions. Consider a stationary spectrum

$$F = \Phi(k_y)\,\delta(k_x), \tag{3.12}$$

where $\Phi(k_y)$ is an arbitrary continuous, positive, and even function and $\delta(k_x)$ is the delta function. Spectrum (3.12) obviously corresponds to a rectilinear stationary zonal flow. Substituting (3.12) into (3.1′) and using the identity $x\delta(x) = 0$ one can readily see that spectrum (3.12) identically satisfies the kinetic equation (3.1′). This fact is quite clear from the physical viewpoint since an arbitrary zonal flow satisfies the original potential vorticity equation; it relates to the property that the interaction coefficient $D_{\mathbf{k}_1\mathbf{k}_2}$ vanishes for collinear wave vectors $\mathbf{k}_1$ and $\mathbf{k}_2$.

Let us now consider the solutions of the kinetic equation having the form

$$F(\mathbf{k}, \tau) = \Phi(k_y)\,\delta(k_x) + \hat{F}(\mathbf{k}, \tau), \tag{3.13}$$

where $\hat{F}$ is an ordinary (not generalized) function. Substituting (3.13) into (3.1′), we obtain, after some simple but rather cumbersome calculations, an equation for $\hat{F}(\mathbf{k}, \tau)$:

$$\frac{\partial \hat{F}}{\partial \tau} = \frac{8\pi|k_x\, k_y|}{1 + 4k_y^2}\,(k^2 - 4k_y^2)^2\,\Phi(2k_y)(\bar{\hat{F}} - \hat{F}) + S(\hat{F}, \hat{F}) \tag{3.14}$$

where $\bar{\hat{F}} = \hat{F}(-k_x, k_y)$ and

$$S(R, P) = 8\pi \int_{\Omega_{12}} D_{\mathbf{k}_1\mathbf{k}_2}\,(D_{\mathbf{k}_1\mathbf{k}_2} R_1 P_2 + D_{\mathbf{k}\mathbf{k}_1} R P_1 + D_{\mathbf{k}\mathbf{k}_2} R P_2)\, \Upsilon^{-1} \times \\ \times \delta(\mathbf{k}_{012})\,\delta(\sigma_{012})\,\mathrm{d}\mathbf{k}_1\,\mathrm{d}\mathbf{k}_2. \tag{3.15}$$

Equation (3.14) describes the nonlinear evolution of a Rossby wave spectrum in the presence of a zonal flow.

It can easily be shown that, by virtue of (3.14), any spectrum of a zonal flow in the form of (3.12) is stable with respect to arbitrary disturbances. We choose a spectrum in the form of (3.12); then the disturbance $\hat{F}(\mathbf{k}, \tau)$ of the spectrum satisfies Equation (3.14). We integrate (3.14) over the whole plane k_x, k_y and after some simple transformations we obtain

$$\int \hat{F}\, \mathbf{dk} = \text{const.} \tag{3.16}$$

By definition, spectrum (3.13) is nonnegative. Therefore $\hat{F}(\mathbf{k}, \tau) \geqslant 0$ for all k_x, k_y (except, possibly, on the line $k_x = 0$). It follows that in the case under consideration the total disturbance energy $\int \hat{F}\, \mathbf{dk}$ can be regarded as a natural measure of stability[3] and therefore, by virtue of (3.16), an arbitrary spectrum corresponding to a zonal flow is stable relative to arbitrary disturbances. The physical meaning of this result is quite clear since weakly nonlinear interactions do not change an exactly zonal flow during the typical time of variation of the amplitudes of the interacting waves (Longuet-Higgins and Gill, 1967).

Relation (3.14) also implies that in the case of a symmetric function $\hat{F}$ when $\bar{\hat{F}} = \hat{F}$ the equation for $\hat{F}$ exactly coincides with Equation (3.1′), i.e. a spectrum equal to a sum of an arbitrary symmetric spectrum satisfying (3.1′) and a spectrum of an arbitrary zonal flow (in the form of (3.12)) is a solution of the kinetic equation. This property does not hold for an asymmetric spectrum $\hat{F}$ – i.e., by virtue of (3.14), the zonal flow which itself remains invariable serves as a 'catalyzer' (Longuet-Higgins and Gill, 1967) for the interaction between the waves with wave vectors $(k_x, k_y)$and $(-k_x, k_y)$ when the amplitudes of these waves (i.e. $\hat{F}(k_x, k_y)$ and $\hat{F}(-k_x, k_y)$) are not equal. However, it can easily be understood that this interaction creates a tendency towards the *symmetrization* of the spectrum $\hat{F}$. To prove this we retain only the interaction with the zonal flow in (3.14), i.e. neglect the term $S(\hat{F}, \hat{F})$ describing the nonlinear self-action of the spectrum $\hat{F}$:

$$\frac{\partial \hat{F}}{\partial \tau} = M\Phi(2k_y)(\bar{\hat{F}} - \hat{F}), \tag{3.17a}$$

where

$$M = 8\pi\, |k_x k_y|\, \frac{(k^2 - 4k_y^2)^2}{1 + 4k_y^2} > 0.$$

An analogous equation is written for $\bar{\hat{F}}$:

$$\frac{\partial \bar{\hat{F}}}{\partial \tau} = M\Phi(2k_y)(\hat{F} - \bar{\hat{F}}). \tag{3.17b}$$

The subtraction of Equations (3.17a, b) yields

$$\frac{\partial(\hat{F} - \bar{\hat{F}})}{\partial \tau} = -\,2M\Phi(2k_y)(\hat{F} - \bar{\hat{F}}), \tag{3.17c}$$

[3] We note that if the disturbed spectrum is not a delta spectrum as in (3.13), then the disturbances are not necessarily positive definite and relation (3.16) does not imply stability.

whence it readily follows that $\hat{F} - \bar{\hat{F}} \to 0$ with increasing time τ, i.e. the spectrum $\hat{F}$ is being symmetrized and the rate of symmetrization is directly proportional to the amplitude $\Phi(2k_y)$ of the zonal flow.

Of particular interest is a spectrum equal to a sum of an equilibrium spectrum in the form of (3.11) and a spectrum of a zonal flow in the form of (3.12):

$$F = \Phi(k_y)\,\delta(k_x) + \frac{1}{a + bk^2}\,; \tag{3.18}$$

it possesses a number of important properties. First, since spectrum (3.11) is symmetric, spectrum (3.18) satisfies the kinetic equation (3.1′). Second, like (3.11), it is stable with respect to small disturbances (for the proof see Reznik and Soomere, 1983b). Finally (which is most important), spectrum (3.18) corresponds to a possible statistical equilibrium state of the system under consideration. To prove this property we derive the entropy equation for Equation (3.14). The derivation does not differ substantially from that of Equation (3.8), namely Equation (3.14) is divided by $\hat{F}$ and integrated over the region Ω, after which the resulting equation is subjected to the same permutations as in the derivation of (3.8). As a result, we find

$$\frac{d}{d\tau}\int \ln \hat{F} d\mathbf{k} = \int_{\Omega} M\phi(2k_y)\left(\frac{\bar{\hat{F}}}{\hat{F}} - 1\right) d\mathbf{k} + $$
$$+ \frac{8\pi}{3}\int_{\Omega_{012}} \frac{D_{\mathbf{k}_1\mathbf{k}_2}\hat{F}\hat{F}_1\hat{F}_2}{\sigma^2 \mathcal{K}}\left(\frac{\sigma}{\hat{F}} + \frac{\sigma_1}{\hat{F}_1} + \frac{\sigma_2}{\hat{F}_2}\right)^2 \times$$
$$\times\ \delta(\mathbf{k}_{012})\,\delta(\sigma_{012})\,d\mathbf{k}_{012}. \tag{3.19}$$

A similar equation for $\bar{\hat{F}}$ is obtained from (3.19) with the aid of a simple replacement of $\hat{F}$ by $\bar{\hat{F}}$; adding together this equation and (3.19) we obtain

$$\frac{d}{d\tau}\int_{\Omega} \ln \hat{F}\, d\mathbf{k} = \frac{1}{2}\int_{\Omega} M\phi(2k_y)\left(\frac{\bar{\hat{F}}}{\hat{F}} + \frac{\hat{F}}{\bar{\hat{F}}} - 2\right) d\mathbf{k} + $$
$$+ \frac{8\pi}{3}\int_{\Omega_{012}} \frac{D_{\mathbf{k}_1\mathbf{k}_2}\hat{F}\hat{F}_1\hat{F}_2}{\sigma^2 \mathcal{K}}\left(\frac{\sigma}{\hat{F}} + \frac{\sigma_1}{\hat{F}_1} + \frac{\sigma_2}{\hat{F}_2}\right)^2 \times$$
$$\times\ \delta(\mathbf{k}_{012})\,\delta(\sigma_{012})\,d\mathbf{k}_{012} \geqslant 0. \tag{3.20}$$

Relation (3.20) is nothing other than the desired law of an increase of entropy for a wave system that includes a zonal flow.

It is easy to see that the right-hand side of (3.20) vanishes only for

$$\hat{F} = \frac{1}{a + bk^2} \tag{3.21}$$

(see above). This leads to the important conclusion that spectrum (3.18), along with spectrum (3.11), *also* corresponds to a statistical equilibrium state of the system described by Equation (3.1′). However, there is an important distinction between spectra (3.18) and (3.11), namely in contrast to (3.11), spectrum (3.18), which is a sum of a spectrum of a zonal flow in the form of (3.12) and an isotropic spectrum in the form of (3.11), is not itself isotropic. Thus the intensive cascade of energy across the spectrum towards wave vectors of predominantly meridional direction observed in numerical experiments (see Section 4) does not contradict the tendency of the system to a statistical equilibrium and, furthermore, can be produced by this very tendency to an equilibrium state with spectrum (3.18). There arises the question of how an ordinary (not generalized) spectrum can tend to spectrum (3.18) in the process of nonlinear evolution. However, here it should be taken into consideration that, generally, the equilibrium states (3.11) and (3.18) are unattainable (e.g. this follows from the fact that, by virtue of (3.1′), the spectrum on the axis k_y does not vary with time) and their form determines only the *direction* of the evolution of the system. Further, if we assume that the system does in fact tend to an equilibrium state with spectrum (3.18), then the energy is not only accumulated near the axis k_y but must also be uniformly distributed in a region lying sufficiently far from that axis since spectrum (3.18) is isotropic in this region. In other words, along with the anisotropization of the spectrum, its *isotropization* outside a neighborhood of the axis k_y must take place. The possibility of realizing the tendencies described can be verified by a numerical study of Cauchy's problem for Equation (3.1′).

3.4. Numerical Experiments

The majority of the numerical experiments on the kinetic equation for Rossby waves have been carried out quite recently, and the results are presented in papers by Reznik and Kozlov (1981) and Reznik and Soomere (1983a, 1984a, b) (see also Kenyon, 1967). Here we briefly describe the essence of the numerical scheme used in the calculations. The collision integral (the right-hand side of (3.1′)) is reduced to a one-fold integral with finite limits of integration (Reznik and Kozlov, 1981) and is replaced by a quadrature formula; similarly, the derivative $\partial F/\partial \tau$ is also represented in finite-difference form. The functions F and $\partial F/\partial \tau$ are calculated at the grid points $\mathbf{k}_{ij} = (k_{xi}, k_{yj})$ of a rectangular grid covering a bounded region Ω in the plane k_x, k_y. The calculation of the values of F at the points (k_x, k_y) not coinciding with the grid points (which is necessary for calculating the collision integral in (3.1′)) is carried out by interpolation from the values at the grid points nearest to (k_x, k_y).

According to the two different techniques of truncation, the problem of interaction with wave vectors lying outside the region Ω is solved in two ways. In the first version, the function $F(\mathbf{k}, \tau)$ is simply assumed to be equal to zero outside Ω (scheme A). From the physical viewpoint this is equivalent to the introduction of infinite dissipation outside Ω (see Section 1). In the second version, all the triple interactions involving wave vectors lying outside Ω are completely neglected (scheme B), i.e. we consider a discrete analog of Equation (3.1′) for which the integrals (3.4′) are conserved.

The major difficulty in the elaboration of the scheme lies in the adequate choice of the quadrature formula and the methods of interpolation and finite-difference representation of the time derivative, which determines the efficiency of the operation of the scheme. A detailed discussion of these questions and the properties of the above-mentioned scheme can be found in the papers by Reznik and Soomere (1983a) and Soomere (1983).

The calculation results show that for a sufficiently large region $\Omega(\mathbf{k})$ the distribution of energy where the energy level is sufficiently high depends weakly on the way the truncation is performed and, therefore, in some cases we do not specify which of the two schemes is used.

The correctness of the operation of the schemes is checked by considering the behavior of the conserved quantities (3.4′). For example, when scheme A is employed, the system involves weak viscosity and therefore the quantities (3.4′a, b) must weakly decrease with increasing time, the enstrophy decreasing faster than the energy since (see Rhines, 1975) the enstrophy is transferred to large wavenumbers whereas the energy is transferred to small wavenumbers, and the effect of the viscosity on the enstrophy is stronger than its effect on the energy. In calculations by scheme B, these quantities must be approximately conserved. All these specific properties are reproduced rather well in real calculations (for more detail see Reznik and Soomere, 1984a), which indicates the correctness of the operation of the scheme. The calculations were mainly performed for the grid points of a rectangular 97 × 114 grid covering the square $\{-4 \leqslant k_x \leqslant 4, -4 \leqslant k_y \leqslant 4\}$ which are contained in the circular region $\Omega = \{\mathbf{k} \mid k \leqslant 4\}$ inscribed in it. To conclude the discussion of the numerical scheme we note that the elaborate calculation algorithm can be used to investigate not only Rossby waves but also arbitrary wave systems involving triple interactions (e.g. capillary waves).

First we consider the following three symmetric initial states[4]:

$$F_{01} = k\,\mathrm{e}^{-k^2}, \quad F_{02} = k\,\mathrm{e}^{-k^2}\cos^4\varphi, \quad F_{03} = k\,\mathrm{e}^{-k^2}\sin^4\varphi. \tag{3.22}$$

The initial spectra (3.22) substantially differ in the sense that the first is isotropic whereas the other two are sharply anisotropic, the functions F_{02} and F_{03} being 'concentrated' near the axes k_x and k_y, respectively. It can be shown (see Reznik and Kozlov, 1981) that a spectrum which is symmetric at a certain time remains symmetric at all subsequent times. Further, it suffices to calculate a symmetric spectrum only for one quadrant in the plane k_x, k_y.

We start our discussion with the case of the isotropic initial spectrum F_{01} and its time derivative $\partial F/\partial\tau$ (Figure 2.3.1). The dots in the figure show the lines $\partial F/\partial\tau = 0$. At the beginning this line is unique and divides the whole region into two parts, namely region I on the right of the line, where $\partial F/\partial\tau < 0$ (the region of outflow of energy), and region II on the left, where $\partial F/\partial\tau > 0$ (the region of inflow of energy). In the course of time this line deviates to the left, 'comes close' to the axis k_y, and

[4] Relations (3.22) and the analogous relations for F_{04} and F_{05} (see below) hold to within constant factors chosen so that the energies of the initial states are equal to unity.

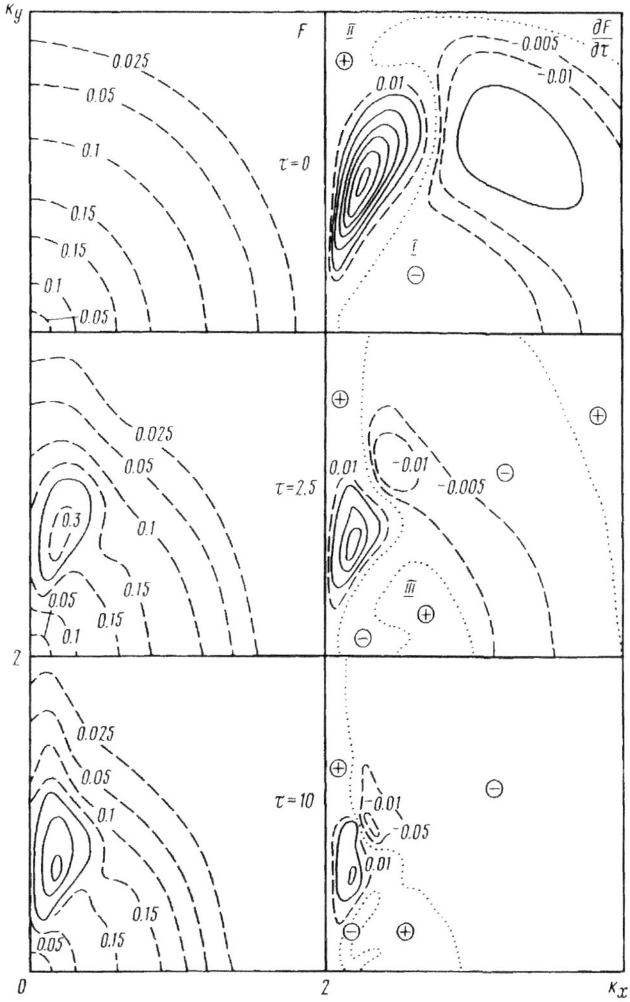

Fig. 2.3.1. Isolines of the functions F (on the left) and $\partial F/\partial \tau$ (on the right) for different times (indicated in the figure) for the initial spectrum $F_{01} = k \exp(-k^2)$. The solid isolines of F and $\partial F/\partial \tau$ are drawn at contour intervals of 0.2 beginning with 0.2 and at intervals of ± 0.02 beginning with ± 0.02, respectively.

divides the whole wavenumber region into a narrow (and permanently contracting) inflow region concentrated in the neighborhood of the axis k_y and a broad outflow region occupying all the remaining range of wavenumbers. The inflow region contains a region of *intensive* energy inflow whose center is at $k \approx 0.6$, and gradually approaches the axis k_y. The maximum values of $\partial F/\partial \tau$ in the inflow region are substantially greater than (approximately ten times as great as) the values of $\partial F/\partial \tau$ with the maximum modulus in the outflow region. Besides, after $T = 0.7$ one more inflow region (III) is formed; however, first, this region is also located near the axis k_y, and, second, the values of $\partial F/\partial \tau$ in III are very small (much smaller than in the zone of intensive inflow). It is easy to understand that this behavior of the distribution of $\partial F/\partial \tau$ must result in concentration of a considerable part of the

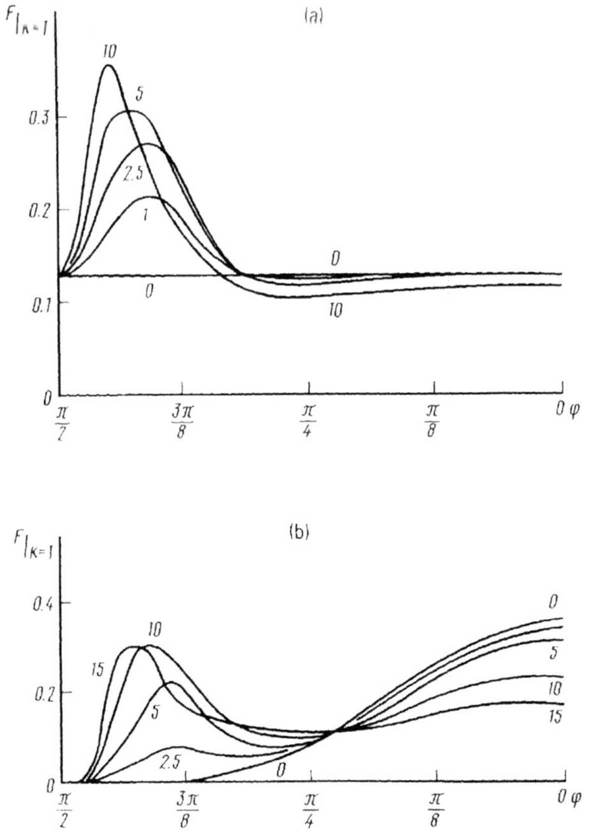

Fig. 2.3.2. Behavior of the energy spectrum at $k = 1$ for the initial states (a) F_{01} and (b) F_{02}. The figures at the curves indicate time.

energy in a comparatively small region near the axis k_y (i.e. in the formation of a 'peak'), which is confirmed by the evolution of the spectrum F. The consideration of the time varying dependence of the spectrum F on the angle φ for k = const (Figures 2.3.2(a), 2.3.3(b), and 2.3.4(b)) leads to the same conclusion. It is seen that the initial isotropy of the spectrum is strongly and still more strongly violated with increasing time, and in the neighborhood of $\varphi = \pi/2$ (i.e. near the axis k_y) a peak is formed which increases with time. However, it is important to note that the isotropy of the spectrum is preserved in a region lying farther from the axis k_y, and, although the energy somewhat decreases here, it does not tend to zero. The behavior of the distribution of the spectrum in Figure 2.3.2 for $k = 1$ and for other values, k = const, in the case of scheme B is approximately the same (see Figures 2.3.3.(b) and 2.3.4(b)), and in the case of scheme A for large values, k = const, the role of the viscosity becomes significant (Figures 2.3.2(a) and 2.3.4(a)) and the isotropy is violated.

In case 2 (Figure 2.3.5) the evolution of the spectrum basically repeats case 1. A clearly marked region of inflow of energy is again formed with increasing time; it is concentrated near the axis k_y (see $T = 15$) and contains a zone of intensive energy

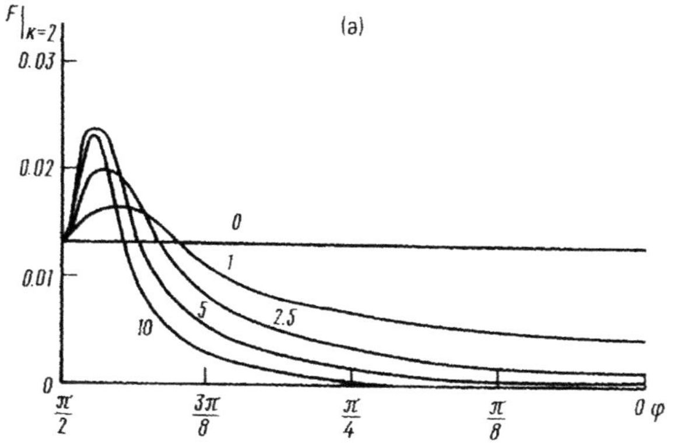

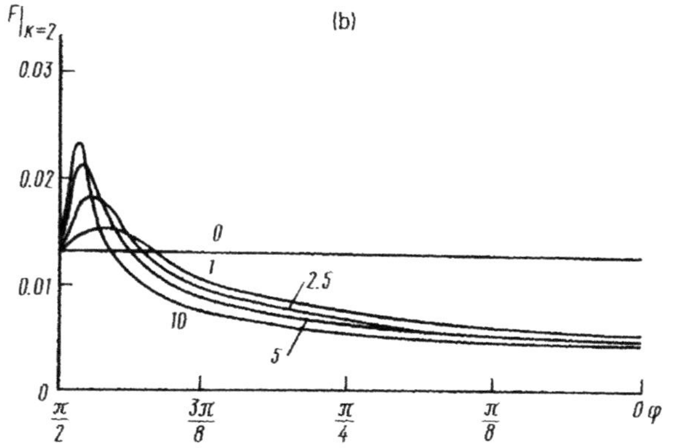

Fig. 2.3.3. Behavior of the spectrum corresponding to the initial state F_{01} for $k = 2$ depending on the calculation scheme: (a) in the presence of dissipation (scheme A) and (b) in the inviscid case (scheme B). The figures at the curves indicate time.

inflow in the neighborhood of $k \approx 1$. The corresponding process of concentration of energy in a localized region in the vicinity of the axis k_y is clearly seen in the patterns of isolines of F. In the course of time the angular distribution of F at $k = 1$ (see Figure 2.3.2(b)) also becomes similar to the corresponding distribution in case 1 (cf. Figures 2.3.2(a) and 2.3.2(b) for $T = 10$ and $T = 15$, respectively). Namely, there is an increasing peak near the axis k_y (near $\varphi = \pi/2$) and an approximate isotropy in the region lying farther from that axis; here the isotropization process for $\varphi \lesssim 3\pi/8$ is marked still clearer than in the foregoing case.

Up to some definite time the evolution of the spectrum in case 3 (Figure 2.3.6) substantially differs from what was described above. The spectrum F_{03} is strongly 'concentrated' near the axis k_y, i. e. the initial motion is nearly zonal, which leads to the formation of two inflow regions. As before, one of them is concentrated near the axis k_y and contains a localized region of comparatively intensive inflow, and the other is concentrated near the axis k_x. Thus, during the initial period there is a rather intensive inflow of energy to wave vectors that have zonal direction, i.e. to the meridional component of motion. However, the intensity of the inflow in the

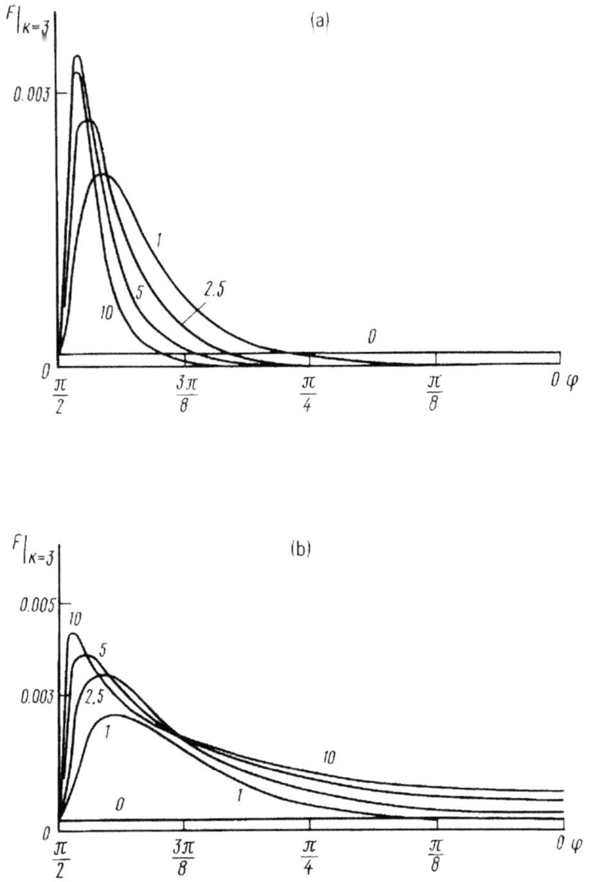

Fig. 2.3.4. Behavior of the spectrum corresponding to the initial state F_{01} at $k = 3$ depending on the calculation scheme: (a) in the presence of dissipation (scheme A) and (b) in the inviscid case (scheme B). The figures at the curves indicate time.

vicinity of the axis k_x rapidly decreases with increasing time, and the region itself is contracted. By the time $T = 2.5$ the distribution of $\partial F/\partial \tau$ qualitatively coincides with the corresponding distribution in case 1 (cf. Figure 2.3.1, $T = 10$; and Figure 2.3.6, $T = 2.5$). The structure of the resulting distribution of F (in case 3 the calculation was carried out up to $T = 2.5$) is also analogous to that of the corresponding distributions in cases 1 and 2. Namely, first, the initial concentration of a part of the energy near the axis k_y increases still more with time and, second, in the region lying far from the axis k_y the spectrum tends to an isotropic state. The latter property is particularly clearly seen from the behavior of F at $k = 1$, demonstrated by Figure 2.3.7(a): in the region $\pi/4 \leqslant \varphi' \leqslant 3\pi/8$ (the boundaries are indicated approximately) the energy level decreases and for $\varphi < \pi/4$ (i.e. where the initial energy is very small) it increases and tends to become the same for all $\varphi < \varphi' \approx 3\pi/8$. This levelling is still more clearly seen in Figure 2.3.7(b) constructed from the results of calculations with the same initial condition F_{03} in the rigid lid

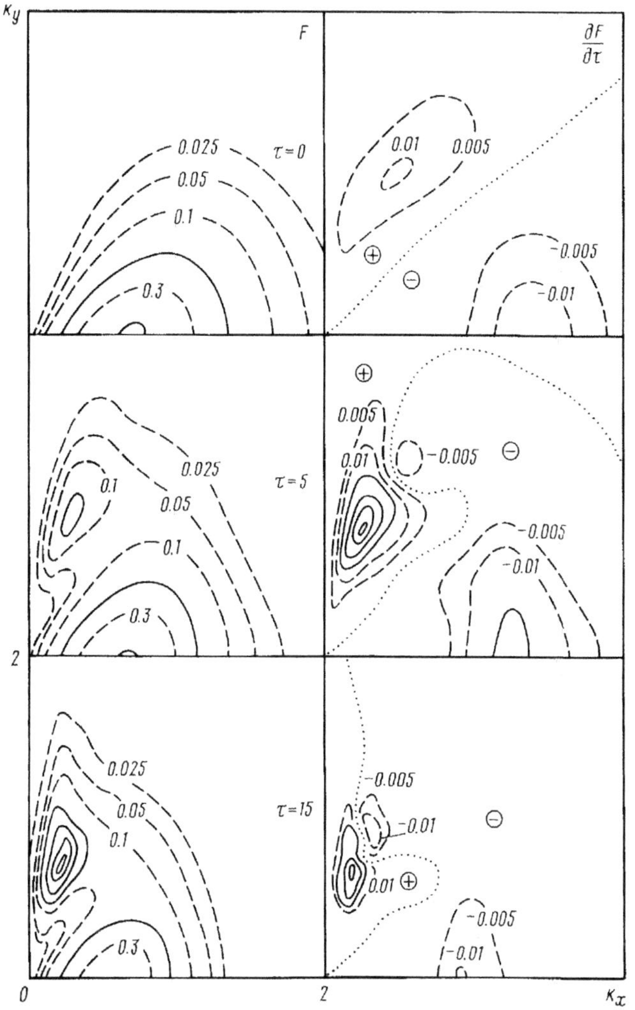

Fig. 2.3.5. Isolines of the functions F (on the left) and $\partial F/\partial\tau$ (on the right) for different times (indicated in the figure) in the case of the initial spectrum $F_{02} = k\, e^{-k^2} \cos^4\varphi$. For the solid isolines see the caption to Figure 2.3.1.

approximation which does not change the structure of the energy exchange qualitatively but accelerates it notably (Reznik and Kozlov, 1981).

Effects analogous to those described above were also observed in calculations with other initial conditions. Of interest is the example of calculation for an initial state which is close to the equilibrium spectrum (3.11) and is determined by

$$F_{04} = \frac{100}{1 + 10k^2} + k\, e^{-k^2}(\sin^4\varphi + \cos^4\varphi). \tag{3.23}$$

Figure 2.3.8 demonstrates the evolution of the deviation from the state (3.11). It is significant that near the axis k_y the concentration of energy takes place again;

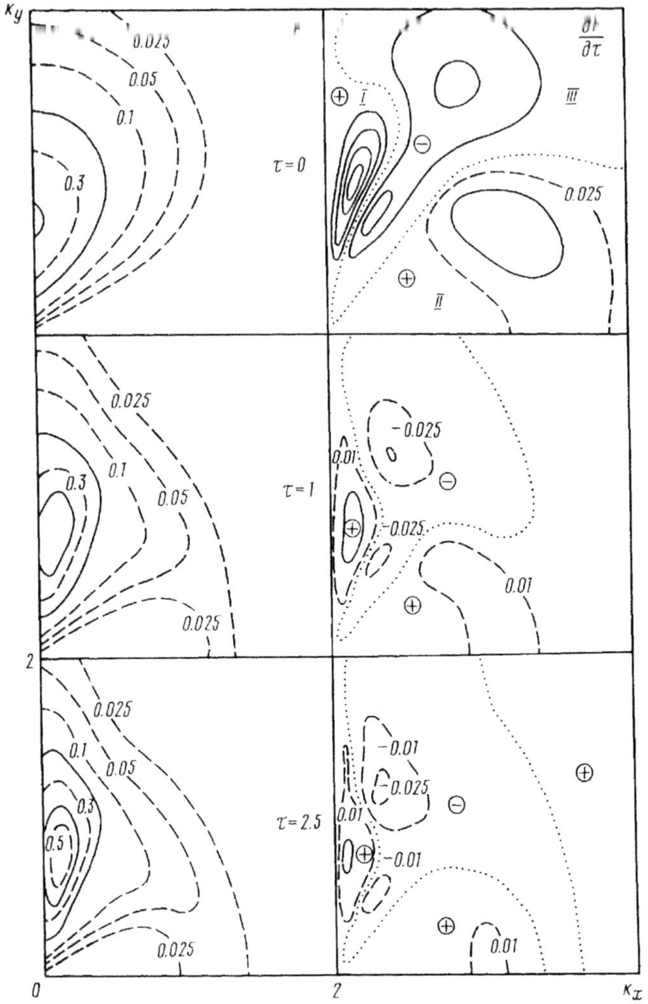

Fig. 2.3.6. Isolines of the functions F (on the left) and $\partial F/\partial\tau$ (on the right) for different times (indicated in the figure) in the case of the initial spectrum $F_{03} = k\,e^{-k^2}\sin^4\varphi$. The solid isolines of F and $\partial F/\partial\tau$ are drawn at contour intervals of 0.2 beginning with 0.2 and at intervals of ± 0.05 beginning with ± 0.05, respectively.

however, in this case the total energy of the deviation does not increase (i.e. the equilibrium 'background' (3.11) is not destroyed) and the resultant zonal component is small. The structure of the total spectrum is obviously the same as that of the spectra considered earlier – namely, a part of the energy is concentrated near the axis k_y and the energy is distributed isotropically in the remaining part of the region. Thus we can assert that the above-mentioned specific properties of the nonlinear evolution do not relate to the fact that spectra (3.22) are 'far' from the equilibrium spectrum (3.11).

Hence, our investigation convincingly indicates the existence of two tendencies.

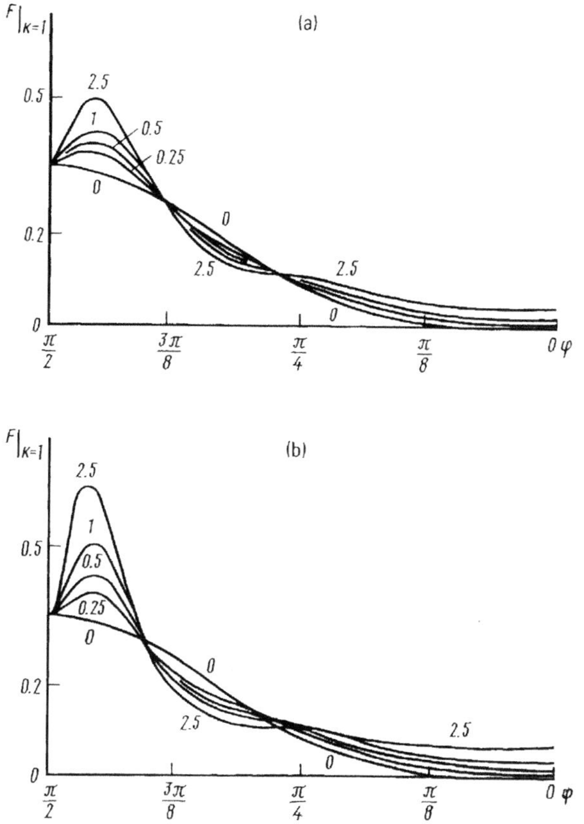

Fig. 2.3.7. Behavior of the spectrum at $k = 1$ for the initial spectrum F_{03}: (a) in the case of an ocean with a free surface and (b) in the rigid lid approximation. The figures at the curves indicate time.

In the first place, a part of the energy is concentrated in the neighborhood of the axis k_y; it forms a clearly marked peak in the spectrum, and thus strengthens the zonal component of motion. In the second place, the spectrum tends to become isotropic in a region lying sufficiently far from the axis k_y (cf. the end of Section 2.3). The question of how these tendencies relate to the fact that the system approaches statistical equilibrium with spectrum (3.18) will be discussed in Section 2.5; and in the remaining part of this section we shall consider the symmetrization effect.

We note that the investigation of spectra of the form of (3.22) in the papers by Kenyon (1967) and Reznik and Kozlov (1981) based on the analysis of the instantaneous distributions of $\partial F/\partial\tau$ corresponding to these spectra (see Figures 2.3.1, 2.3.5, and 2.3.6 for $\tau = 0$) enabled us to draw a conclusion only about the possible strengthening of the zonal component of motion in the cases of F_{01}, F_{02}, and F_{03}. In the last of these cases it was also observed that the energy was predominantly transferred to wave vectors having zonal direction.

Now we consider the effect of symmetrization described for the first time by

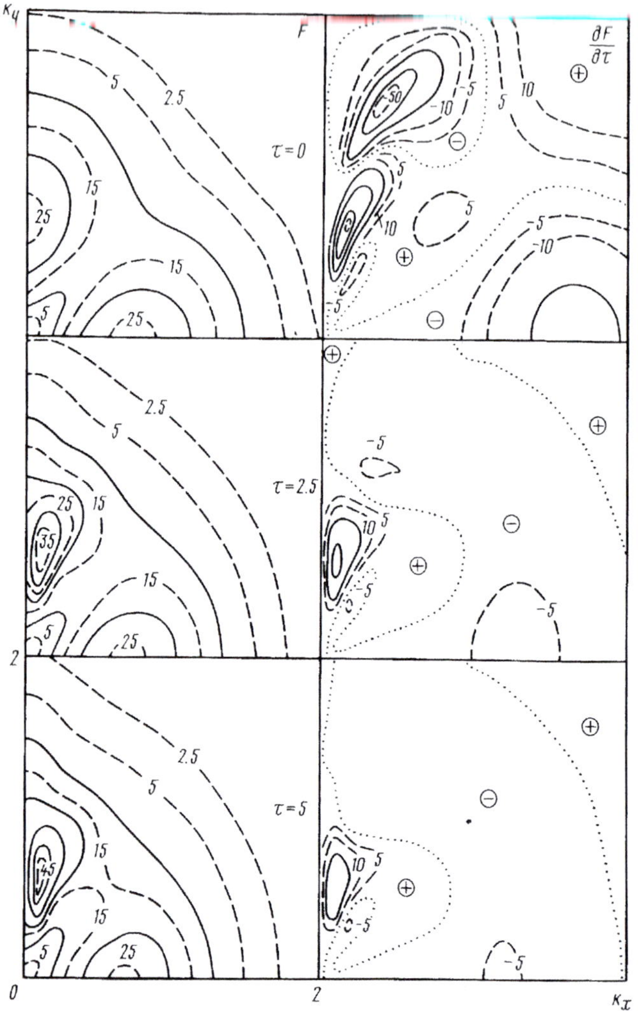

Fig. 2.3.8. Isolines of the deviation of the function F from the equilibrium spectrum $100/(1 + 10k^2)$ (on the left) and the isolines of $\partial F/\partial \tau$ (on the right) for different times (indicated in the figure) for the initial spectrum $F_{04} = 100/(1 + 10k^2) + k\ e^{-k^2}\ (\sin^4\varphi + \cos^4\varphi)$. The isolines of F and $\partial F/\partial \tau$ are expressed in units of 10^{-3} and 10^{-4}, respectively. The solid isolines of F and $\partial F/\partial \tau$ are drawn at contour intervals of 10 beginning with 20 and at intervals of $\pm$ 20 beginning with 20, respectively.

Reznik and Kozlov (1981). It consists in the following. As already mentioned, if the difference $\Delta F = F(k_x, k_y) - F(-k_x, k_y)$ is equal to zero at a certain time (i.e. the spectrum is symmetric), then this property also holds for all subsequent times (see Reznik and Kozlov, 1981). However, if $\Delta F \neq 0$ at $\tau = \tau_0$ (an asymmetric spectrum), then $\partial F/\partial \tau|_{\tau = \tau_0}$ is such that, on average, the modulus of ΔF decreases with increasing time (at least at times τ immediately after τ_0), i.e. the spectrum becomes more symmetric. This result turned out to be valid for various initial distributions $F(k_x, k_y, \tau_0)$ (see Reznik and Kozlov, 1981; Soomere, 1983). However, it was not clear to what extent this tendency to symmetrization is stable. The matter is that, generally, the total symmetrization of a spectrum which is initially asymmetric is impossible owing to the conservation of the meridional component L of the total wave momentum. It can readily be shown that this quantity can be rewritten in the form

$$L = 2 \int_0^4 \int_0^4 \frac{k_y}{k_x} \; (k^2 + 1) \; [F(k_x, k_y) - F(-k_x, k_y)] \, dk_x \, dk_y, \qquad (3.24)$$

whence it follows that for $L \neq 0$ the spectrum must be asymmetric all the time.

The symmetrization effect was investigated for the example of the asymmetric initial spectrum

$$F_{05} = k \exp \left(- k^2 + \frac{k}{2} \sin 2\varphi \right) \cos^2(\varphi - k). \qquad (3.25)$$

Figure 2.3.9(a) gives an indication of the evolution of spectrum (3.25). It is seen that the substantially asymmetric initial spectrum tends rather rapidly to a symmetric state with increasing time. Accordingly, the structure of the derivative $\partial F/\partial \tau$ (see Figure 2.3.9(b)) is such that energy is continuously transferred from the right quadrant, where a large amount of energy is concentrated, to the left quadrant where there is less energy. The symmetrization is also indicated by the monotonic decrease of the integral measure of asymmetry

$$As = \int_\Omega (F - \bar{F})^2 \, d\mathbf{k} \qquad (3.26)$$

with increasing time (see Figure 2.3.10). It is important to note that there appear spectral peaks in the neighborhood of the axis k_y on both its sides. We believe that the strengthening of the zonal component of the flow and the symmetrization are interrelated for the following reasons. In Section 3, when considering a spectrum equal to a sum of a delta spectrum corresponding to a zonal flow and an arbitrary asymmetric spectrum, we came to the conclusion that the presence of the zonal flow produced a tendency towards symmetrization of the spectrum. It can be shown that, in the case of a bounded continuous function F, this tendency exists for $F(0, k_y) \neq 0$. To this end we consider the contribution to $\partial F(k_x, k_y)/\partial \tau$ coming from the interaction of the wave vector $\mathbf{k} = (k_x, k_y)$ with the wave vectors $\mathbf{k}_1 = (0, -2k_y)$ and $\mathbf{k}_2 = (-k_x, k_y)$. It can easily be checked that this triad satisfies resonance conditions. Since $D_{\mathbf{k}\mathbf{k}_2} = -D_{\mathbf{k}_1\mathbf{k}_2}$ and $D_{\mathbf{k}\mathbf{k}_1} = 0$ for these $\mathbf{k}$, $\mathbf{k}_1$, and $\mathbf{K}_2$, the corresponding contribution to the integrand in (3.1′) is proportional (with a positive proportionality factor) to the quantity $D^2_{\mathbf{k}_1\mathbf{k}_2} F(0, 2k_y)[F(-k_x, k_y) - F(k_x, k_y)]$. It is obvious that for $F(k_x, k_y) > F(-k_x, k_y)$ this interaction decreases $\partial F(k_x, k_y)/\partial \tau$ (and, hence, decreases $F(k_x, k_y)$); for $F(k_x, k_y) < F(-k_x, k_y)$ it increases these quantities and for $F(k_x, k_y) = F(-k_x, k_y)$ it does not change them in comparison with the values that would be obtained if such interactions were ignored. In other

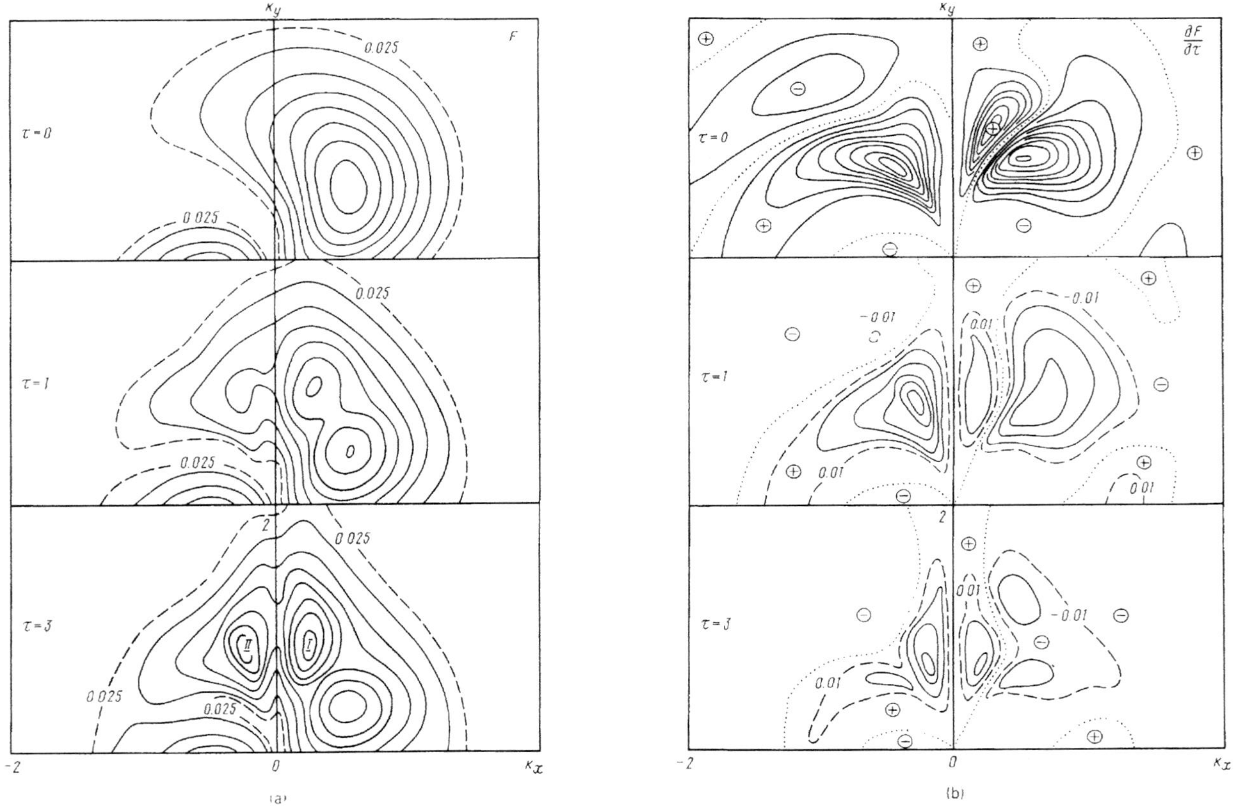

Fig. 2.3.9. Isolines of the functions (a) F and (b) $\partial F/\partial\tau$ for different times (indicated in the figure) for the initial spectrum $F_{05} = k \exp[-k^2 + (k/2)\sin 2\varphi)] \cos^2 (\varphi - k)$. The solid isolines of F and $\partial F/\partial\tau$ are drawn at contour intervals of 0.05 beginning with 0.05 and at intervals of ± 0.02 beginning with ± 0.02, respectively.

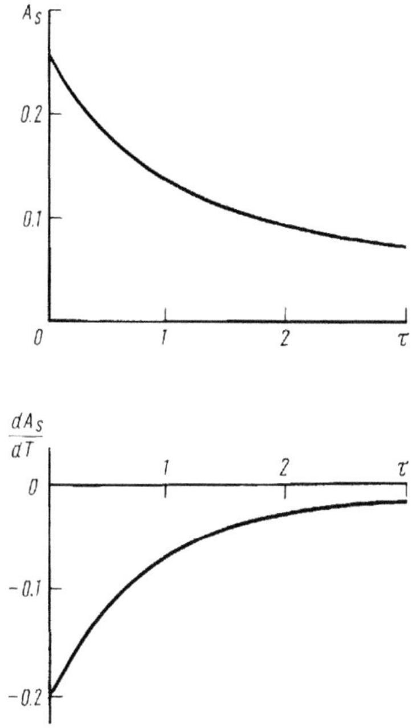

Fig. 2.3.10. Behavior of the integral characteristics As and $\mathrm{d}As/\mathrm{d}\tau$ for the initial spectrum F_{05}.

words, the interaction of the wave vectors $\mathbf{k} = (k_x, k_y)$ and $\mathbf{k}_2 = (-k_x, k_y)$ with the wave vector $\mathbf{k}_1 = (0, -2k_y)$ does in fact tend to 'level' the values of $F(k_x, k_y)$ and $F(-k_x, k_y)$, and the rate of levelling increases together with the increasing 'amplitude' of the zonal component $F(0, 2k_y)$. It is easy to verify that the same tendency is also produced by wave vectors not lying exactly on the axis k_y but having almost meridional directions (on condition that the difference $F(k_x, k_y) - F(-k_x, k_y)$ is not very small). It follows that the strengthening of the zonal component of motion must be accompanied by symmetrization of the spectrum, which is confirmed by Figures 2.3.9(a) and (b).

As previously mentioned, by virtue of the conservation of the meridional component L of the wave momentum, complete symmetrization of the spectrum F is impossible. At the same time, strengthening the zonal component of motion can increase the degree of final symmetrization of the spectrum. To prove this fact we consider formula (3.4′c). The term $\int_0^4 \int_0^4 k_y k_x^{-1}(k^2 + 1)F(k_x, k_y)\, dk_x\, dk_y$ is equal to the momentum contained in the first quadrant of the plane k_x, k_y and the term $-\int_0^4 \int_0^4 k_y k_x^{-1}(k^2 + 1)F(-k_x, k_y)\, dk_x\, dk_y$ is equal to the momentum in the second quadrant. Let a peak be formed in the first quadrant near the axis k_x (region I in Figure 2.3.9(a)). Since the denominator of the integrand in (3.4′c) is equal to k_x, the major fraction of the meridional momentum contained in the first quadrant is concentrated in the neighborhood of this peak. As the peak increases and approaches the axis k_y, the total magnitude of the meridional momentum contained in the first quadrant increases. To provide the conservation of L a peak approximately symmetric with respect to the peak in the first quadrant must also be formed in the second quadrant in the vicinity of the axis k_y (region II in Figure 2.3.9(a)). Here the values of F in regions I and II can only slightly differ since, owing to the closeness of regions I and II to the axis k_y, even a small difference makes a large contribution to the integral in (3.4′c). The closer the peaks are to the axis k_y, the smaller can be the difference between the corresponding values of F. As to the regions lying far from the axis k_y, their contribution to the momentum is small, and therefore here almost complete symmetrization is also possible. Thus, the condition of conservation of L does not contradict the tendency towards symmetrization of a spectrum if, along with this tendency, strengthening of the zonal component of motion takes place.

Of interest also is the behavior of various integral characteristics such as, for instance, the average modulus of the wavenumber k defined as (Rhines, 1975)

$$\langle k \rangle = \frac{\int_{\Omega} kF \, \mathrm{d}\mathbf{k}}{\int_{\Omega} F \, \mathrm{d}\mathbf{k}} . \tag{3.27}$$

Figure 2.3.11 shows that in the cases under consideration $\langle k \rangle$ decreases monotonically with increasing time, which is a consequence of the existence of the energy and enstrophy integrals (3.4′a, b). (For more detail, see Section 3.5.)

The integral rates of the modular (R_k) and azimuthal (R_φ) transfer of energy (Reznik and Kozlov, 1981) are equal to

$$R_k = \int_0^4 \left| \int_0^{2\pi} k \frac{\partial F}{\partial \tau} \, \mathrm{d}\varphi \right| \mathrm{d}k, \tag{3.28a}$$

$$R_\varphi = \int_0^{2\pi} \left| \int_0^4 k \frac{\partial F}{\partial \tau} \, \mathrm{d}k \right| \mathrm{d}\varphi. \tag{3.28b}$$

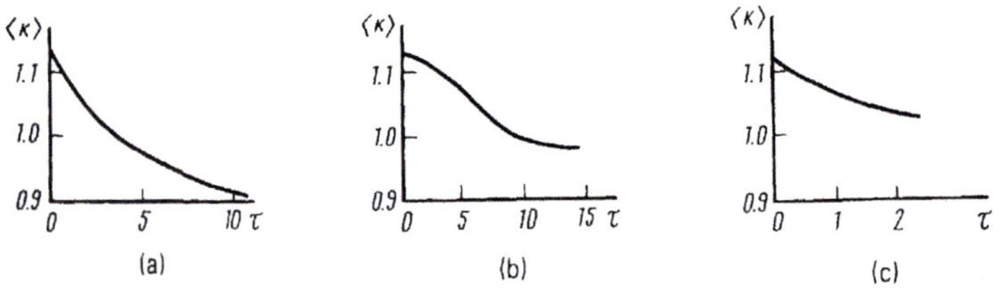

Fig. 2.3.11. Behavior of the average modulus of a wavenumber $\langle k \rangle$ for the cases (a) F_{01}, (b) F_{02} and (c) F_{03}.

It is seen from Figure 2.3.12 that, at the beginning, the rate R_φ of the azimuthal transfer of energy exceeds the rate R_k of the modular transfer. However, in the course of time these rates tend to become equal and in cases 1 and 3 the quantity R_k even becomes greater than R_φ. As to the ratio $R_s = S_k/J$, where

$$S_k = \frac{1}{\langle k \rangle} \frac{\mathrm{d}\langle k \rangle}{\mathrm{d}\tau}, \tag{3.29a}$$

$$J = \frac{\int_\Omega \left| \frac{\partial F}{\partial \tau} \right| \mathbf{dk}}{\int_\Omega F\, \mathbf{dk},} \tag{3.29b}$$

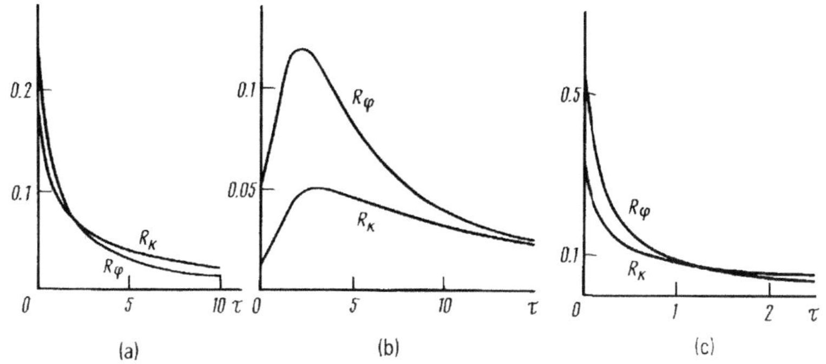

Fig. 2.3.12. Behavior of the rates of the modular (R_k) and azimuthal (R_φ) energy transfer for the initial states (a) F_{01}, (b) F_{02}, and (c) F_{03}.

introduced by Rhines (1975), it is very small at the beginning (~ 0.1), increases with time, and attains the value $R_s \approx 0.25$ in all the three cases (see Figure 2.3.13). This difference between the values of the ratios R_k/R_φ and S_k/J probably relates to the fact that the rate R_k takes into account all the interactions among which there exist both interactions transferring the energy to larger wavenumbers, and interactions transferring energy to smaller wavenumbers, whereas the quantity S_k describes only the resultant tendency to the decrease of the average modulus of the wave vector.

Some of the above-mentioned calculations were repeated in the rigid lid approximation. As was already mentioned above as well as in the paper by Reznik and Kozlov (1981), the introduction of this approximation does not change the results qualitatively and only increases the energy exchange rate.

3.5. Tendency Towards Statistical Equilibrium[5]

The integral intensity J of interactions (see formula (3.2.9b)) is an important characteristic. In a stationary state we have $J = 0$, and therefore a decrease of J with increasing time is a necessary condition for the system to tend towards a stationary state (in particular, towards a state of statistical equilibrium).

In all the cases under consideration the quantity J monotonically decreases, beginning with a certain time (e.g. in cases 1 and 3 the decrease begins with $\tau = 0$ and in case 2 it begins with $\tau \approx 2.5$; see Figure 2.3.14). The behavior of J depends weakly (at least during the calculation times) on the spatial resolution and the presence (scheme A) or the absence (scheme B) of weak dissipation. In cases 1 and

[5] All the calculations described in this section were performed using the nondissipative scheme B.

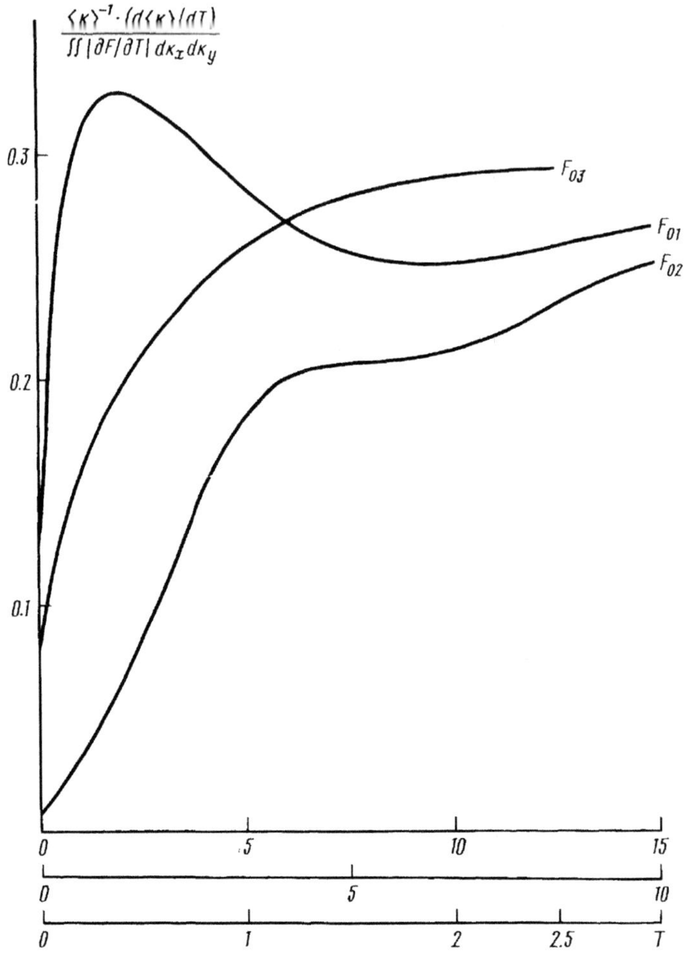

Fig. 2.3.13. Behavior of the ratio $R_s = (\langle k\rangle^{-1} (\mathrm{d}\langle k\rangle/\mathrm{d}\tau))/(\int |\partial F/\partial\tau| \,\mathrm{d}\mathbf{k})$ for the initial states (3.22).

3 the decrease was rather fast; for example, J decreased ten times in comparison with its initial value by the times $T \approx 10$ and $T \approx 2.5$, respectively, after which the calculation was stopped. In case 2 the calculation was carried out until $T = 15$; by this time the intensity of interaction had decreased approximately four times in comparison with its maximum value at $T \approx 2.5$.

Hence, judging from the behavior of J, we can assume that owing to nonlinear interactions the system tends towards a stationary state. However, to show that this is a state of statistical equilibrium (and not an intermediate unstable stationary state) it is necessary to investigate the behavior of the entropy $H = \int_\Omega \ln F \,\mathrm{d}\mathbf{k}$ and, in particular, its derivative $\mathrm{d}H/\mathrm{d}\tau = \int_\Omega (1/F)(\partial F/\partial\tau)\,\mathrm{d}\mathbf{k}$ since the fact that $\mathrm{d}H/\mathrm{d}\tau$ is equal to zero is a necessary and sufficient condition for statistical equilibrium. Figure 2.3.15 demonstrates the dependence of $\mathrm{d}H/\mathrm{d}\tau$ on τ for the initial spectra (3.22). For $\tau = 0$ the values of $\mathrm{d}H/\mathrm{d}\tau$ are very large (approximately equal to 5500

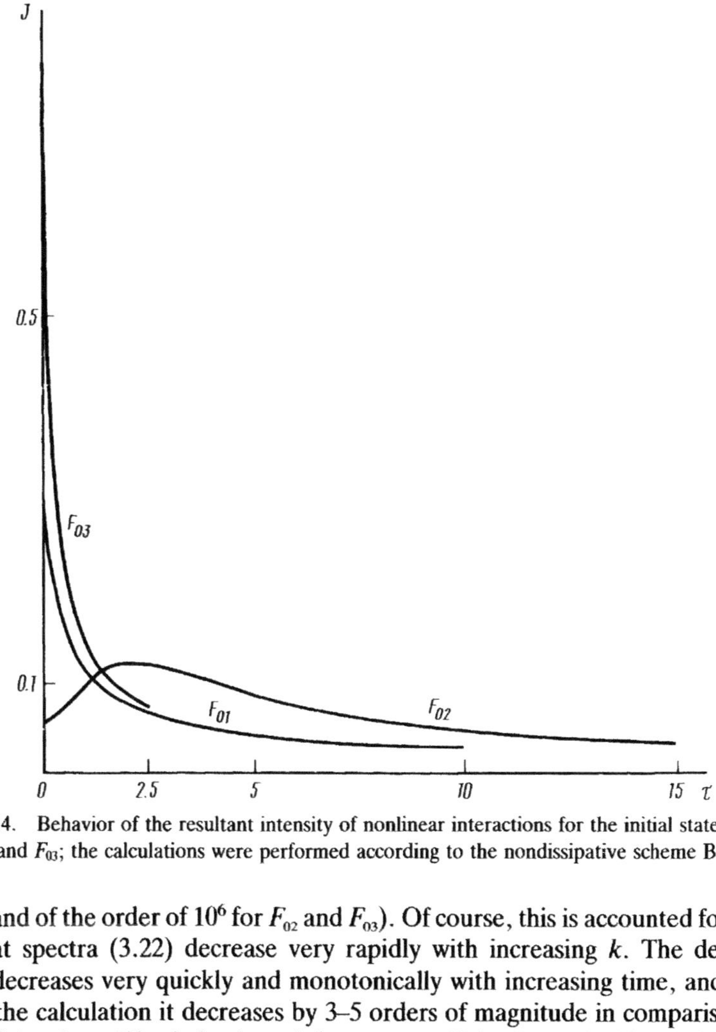

Fig. 2.3.14. Behavior of the resultant intensity of nonlinear interactions for the initial states F_{01}, F_{02}, and F_{03}; the calculations were performed according to the nondissipative scheme B.

for F_{01} and of the order of 10^6 for F_{02} and F_{03}). Of course, this is accounted for by the fact that spectra (3.22) decrease very rapidly with increasing k. The derivative $\mathrm{d}H/\mathrm{d}\tau$ decreases very quickly and monotonically with increasing time, and by the end of the calculation it decreases by 3–5 orders of magnitude in comparison with the initial values. The behavior of the entropy H is shown in Figure 2.3.16. In accordance with the behavior of $\mathrm{d}H/\mathrm{d}\tau$, the entropy H increases very rapidly at the beginning of the calculation and becomes almost constant at the end. This behavior of H and $\mathrm{d}H/\mathrm{d}\tau$ allows us to conclude that the system does in fact tend towards statistical equilibrium in the process of nonlinear evolution, and now the problem is to determine the specific form of the equilibrium spectrum.

As already mentioned, nonlinear interactions result, first, in the concentration of energy near the axis k_y and, second, in an isotropic energy distribution in the region lying sufficiently far from this axis. It follows that the most probable equilibrium spectrum is of the form of (3.18). The direct verification of the fact that the spectrum tends to (3.18) encounters difficulties since it is *a priori* unknown how the energy and the enstrophy are distributed between the almost zonal component of the flow and the motion corresponding to the isotropic part of the spectrum (i.e. the coefficients a and b in (3.18) are unknown). On the other hand, the equilibrium

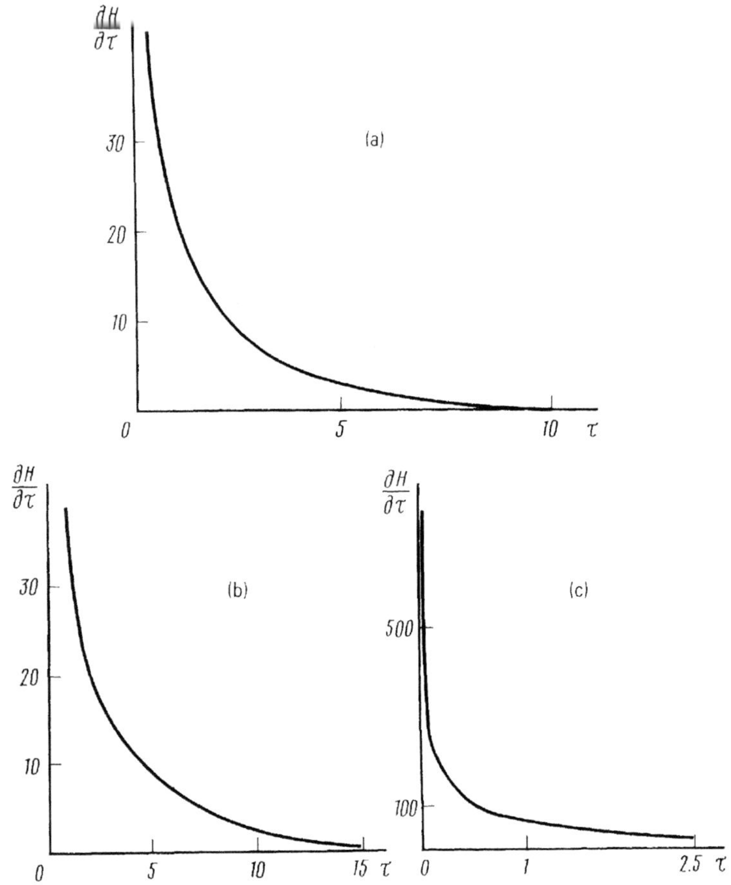

Fig. 2.3.15. Behavior of the derivative of the entropy $dH/d\tau$ for the initial spectra (a) F_{01}, (b) F_{02}, and (c) F_{03}; the calculation was performed according to scheme B.

state can be approached at a different rate in different regions of the plane k_x, k_y. Therefore, it seems natural to try to introduce a measure of deviation of the spectrum under consideration from (3.18) which is applicable to any subregion of Ω and is independent of the parameters a and b. The following expression can serve as a characteristic of this kind:

$$D = \int_{\bar{\Omega}} \int_{\bar{\Omega}_1} \int_{\bar{\Omega}_2} \left[(k_2^2 - k_1^2)F_1F_2 + (k_1^2 - k^2)FF_1 + (k^2 - k_2^2)FF_2\right]^2 d\mathbf{k}_{012}, \quad (3.30)$$

where $\bar{\Omega}(\mathbf{k})$ is a subregion of the calculation region. It can readily be verified that the integral D vanishes only when $F = (a + bk^2)^{-1}$ and therefore it is a suitable measure of the deviation from (3.18) in the region $k_x \neq 0$.

Figure 2.3.17 demonstrates the behavior of D calculated for the whole calculation region for the initial states (3.22). It is seen that in all the cases D first decreases and then begins to increase after some time. This increase seems to relate

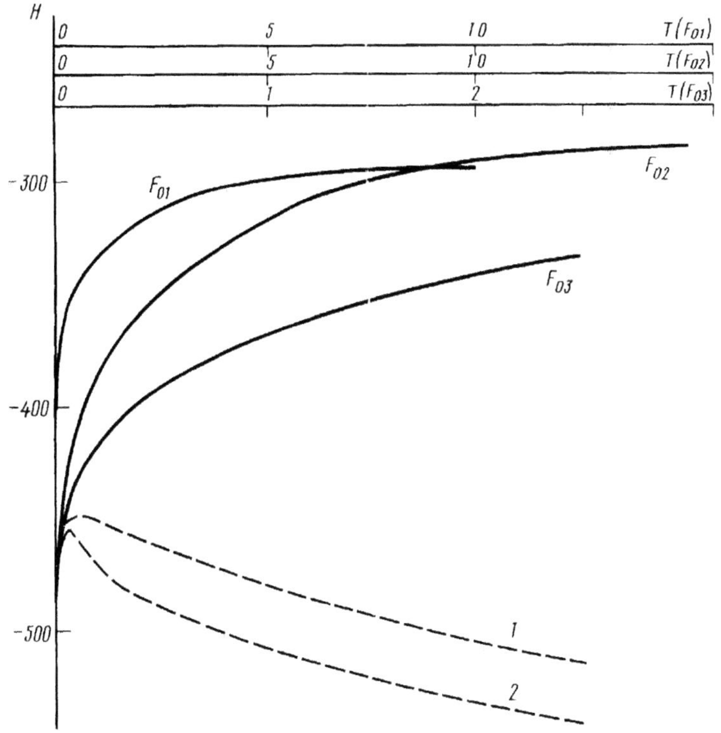

Fig. 2.3.16. Behavior of the entropy $H = \int_\Omega \ln F\, dk$. The solid lines correspond to the initial states (3.22) and are calculated according to scheme B. The dashed lines 1 and 2 correspond to the initial state F_{03} and are calculated according to scheme A in an ocean with a free surface and in the rigid lid approximation, respectively.

to the transfer of energy to the nearly zonal flow; this follows from a comparison with the behaviour of the characteristic D calculated for the region $|k_x| \geqslant 0.5$ (Figure 2.3.18). The quantity D denoted by $D|_{|k_x| \geqslant 0.5}$ in Figure 2.3.18 rapidly decreases with increasing time (approximately ten times during the calculation time), and this decrease does not relate to a decrease of energy in the region $|k_x| \geqslant 0.5$ since here the energy either decreases much slower than D (in the cases of F_{01} and F_{02}) or even increases (in the case of F_{03}). However, we note that the characteristic D calculated for the region $k \geqslant 2$, $|k_x| \geqslant 0.5$ decreases not monotonically with increasing time nor as fast as $D|_{|k_x| \geqslant 0.5}$, and in the case of F_{03} it even slightly increases (see Figure 2.3.19). This probably relates to the fact that the modes with wave vectors **k** lying near the boundary of the region Ω take part in a 'smaller' number of interactions than the modes with **k** lying 'inside' Ω since the majority of interactions of near-boundary modes correspond to wave vectors lying outside Ω, and we ignore these interactions. This slows down the nonlinear evolution of near-boundary modes and, as a result, leads to a decrease in the rate of approach to the equilibrium state as k increases.

Here there is another subtle feature relating to the fact that the derivative of the entropy of an ordinary (not generalized) spectrum tends to zero only when the

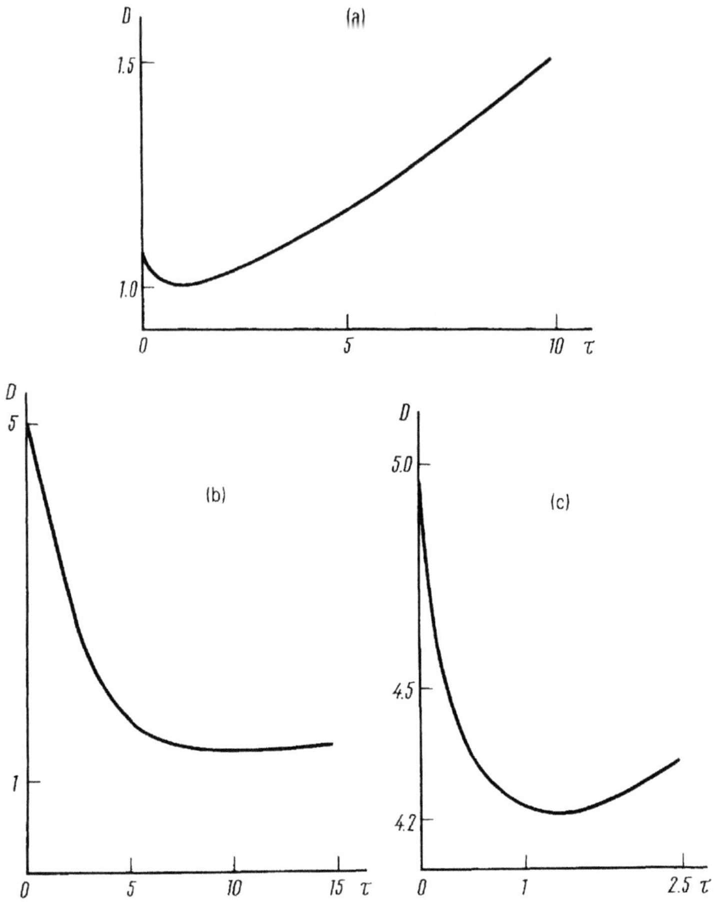

Fig. 2.3.17. Behavior of the characteristic D of the deviation of a spectrum from the equilibrium state calculated for the whole calculation region for the initial spectra (a) F_{01}, (b) F_{02}, and (c) F_{03}.

spectrum tends to (3.11) (see Section 2). As to the case under consideration, here the spectrum tends to (3.18), and nevertheless $dH/d\tau$ tends to zero. The matter is that (for more details see Reznik and Soomere, 1984b) although the energy of the nearly zonal flow increases, its entropy decreases and tends to zero as the energy is being concentrated near the axis k_y. Thus the nearest neighborhood of the axis k_y plays an important role in the *energy* exchange, but the evolution of the *entropy* is completely determined by the behavior of the spectrum in the region of 'isotropization' of the spectrum lying outside this neighborhood.

3.6. Conclusions

Thus, to a certain degree, the numerical results presented above allow us to assert that, irrespective of the initial form of a spectrum, in the absence of sources and sinks of energy the spectrum tends towards an equilibrium spectrum of the form of (3.18), which is a sum of a spectrum of a zonal flow of the form of (3.12) and an

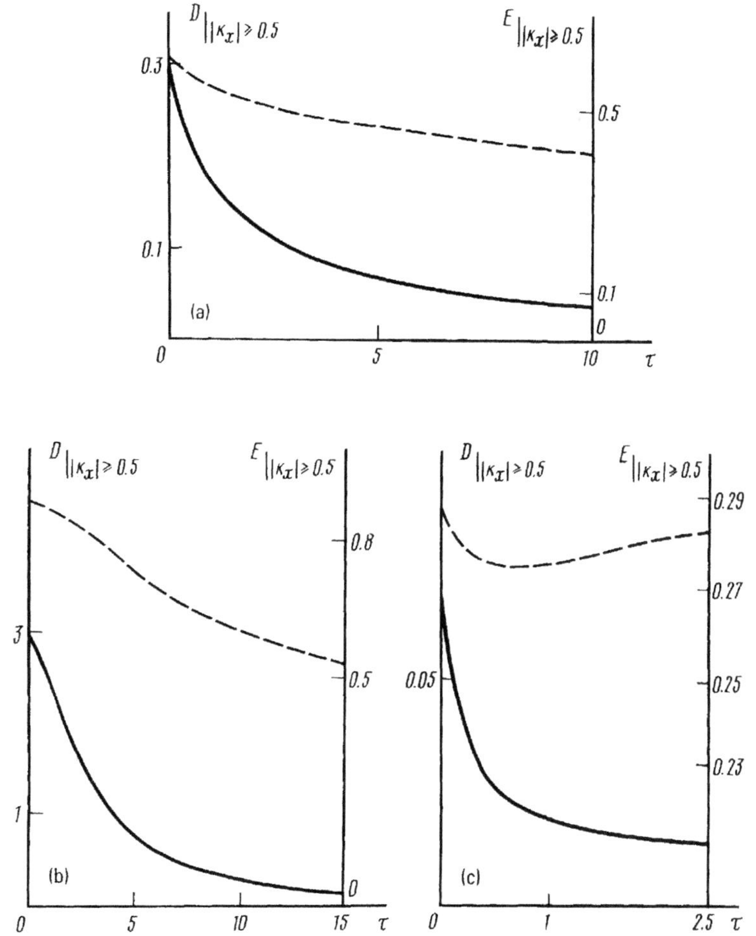

Fig. 2.3.18. Behavior of the characteristic D of the deviation of a spectrum from the equilibrium state (3.11) (solid lines) and energy E (dashed lines) calculated for the region $|k_x| \geqslant 0.5$ for the initial spectra (a) F_{01}, (b) F_{02}, and (c) F_{03}.

equilibrium spectrum of two-dimensional turbulence of the form of (3.11). However, the relative fraction of the energy of the zonal flow depends on the initial conditions, and this problem requires further investigation.

The phenomenon of strengthening the zonal component of motion owing to nonlinear interactions, which was described above, is in fact analogous to the well-known negative viscosity phenomenon which consists in a transfer of energy from turbulent fluctuations to the mean state. Indeed, the transfer of energy to wave vectors of nearly meridional directions implies the existence of a cascade of energy from waves with comparatively large frequencies to low-frequency waves, i.e. from components of motion 'rapidly varying' with time to 'slowly varying' components. We stress that this phenomenon is produced only by the β-effect and nonlinearity and, in our opinion, it is of major importance since it indicates the possibility of a transfer of energy from synoptic eddies to mean motions. This

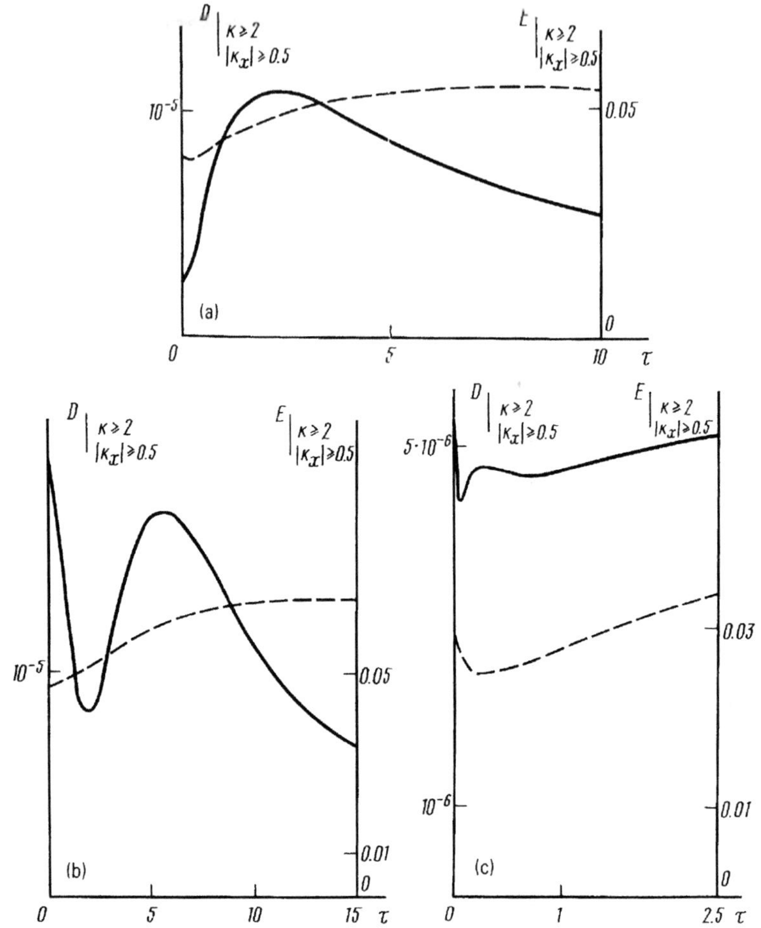

Fig. 2.3.19. Behavior of the characteristic D of the deviation of a spectrum from the equilibrium state (3.11) (solid lines) and energy E (dashed lines) calculated for the region $|k_x| \geq 0.5$, $k \geq 2$ for the initial spectra (a) F_{01}, (b) F_{02}, and (c) F_{03}.

mechanism can also be directly involved in the formation and maintenance of zonal currents in the atmosphere of the Earth and other planets (Williams, 1978) and also ring-shaped flows in plasma (e.g. see Hasegawa *et al.*, 1979).

At the same time it should be stressed once again that the motion does not 'degenerate' into a nearly zonal flow (as, for instance, is asserted in the paper by Rhines, 1975); instead, beginning at a certain time, energy 'saturates' the zonal component and any increase of this component is stopped; the rest of the energy is distributed isotropically over the wave vectors. As a result, a kind of equilibrium occurs between the 'slowly varying' nearly zonal component and the 'rapidly varying' component corresponding to the isotropic part of the spectrum.

In the case of asymmetric initial conditions when $F(k_x, k_y) \neq F(-k_x, k_y)$ (e.g. see (3.25)) it seems, at first glance, that the tendency towards the symmetric equilibrium spectrum (3.18) is hindered by the conservation of the zonal component L of the wave momentum which vanishes for $F = \bar{F}$ and is nonzero for $F \neq \bar{F}$ (see formula (3.24)). However, as calculations show, the strengthening of the concentration of energy in the vicinity of the axis k_y (which is approximately symmetric about this axis) makes symmetrization possible since the major part of the zonal momentum L also turns out to be concentrated in the neighborhood of the axis k_y, and even a small difference $F(k_x, k_y) - F(-k_x, k_y)$ suffices to provide the conservation of L. It is important that, as the nearly symmetric concentration of energy in the vicinity of the axis k_y strengthens, the difference $F(k_x, k_y) - F(-k_x, k_y)$ required for the conservation of L decreases, and the spectrum tends towards a symmetric form.

The phenomenon of symmetrization of a spectrum implies that, on average, the cascade of energy to nearly meridional wave vectors must be accompanied by a 'levelling' of the amplitudes of Rossby waves with wave vectors symmetric about a parallel. It would be of interest to estimate the efficiency of this levelling over spectra determined from experimental evidence.

3.7. Effect of Dissipation. Equilibrium Spectra in the Presence of Sources and Sinks of Energy

As previously mentioned, almost all the calculations were performed simultaneously according to scheme A and scheme B. This makes it possible to estimate the effect of dissipation on the nonlinear evolution of the spectrum. First, it should be noted that dissipation hardly produces any effect (during the calculation times) on the distribution of energy F and its derivative $\partial F/\partial\tau$ for $k \sim 1$, i.e. in those regions where these quantities are substantially nonzero. The dissipation becomes significant for large k since these modes transfer their energy to wave vectors lying outside the region Ω more intensively than the modes with $k \sim 1$.

However, it is interesting that although dissipation itself is 'isotropic' (we remind the reader that in scheme A all the components with $k > k_0 = 4$ dissipate infinitely fast), its effect is sharply 'anisotropic'. The comparison of distributions of $F(\mathbf{k}, \tau)$ at $k = \text{const}$ calculated by scheme A (Figures 2.3.3(a), 2.3.4(a), 2.3.20(a), and 2.3.21(a)) with the analogous distributions calculated by scheme B (Figures 2.3.3(b), 2.3.4(b), 2.3.20(b), 2.3.21(b)) shows that dissipation leads to the damping of sufficiently short-wave components, the direction of whose propagation is sufficiently distinct from the meridional direction. As to the short-wave components propagating primarily along the meridian, their intensity remains almost unchanged and in some cases even somewhat increases in comparison with the inviscid case. Hence, dissipation is an additional factor strengthening the concentration of energy near the axis k_y.

What has been said is also confirmed by the configurations of the regions of inflow of energy (where $\partial F/\partial\tau > 0$; marked by the sign ⊕) and the regions of outflow of energy (where $\partial F/\partial\tau > 0$; marked by the sign ⊖) demonstrated in Figures 2.3.22(a), (b), and (c). We see that in the presence of dissipation (the left columns) the whole field $\partial F/\partial\tau$ splits into two subregions, namely, a narrow region of inflow of energy concentrated near the axis k_y and a broader region of wave vectors whose energy decreases with increasing time. By contrast, if there is no

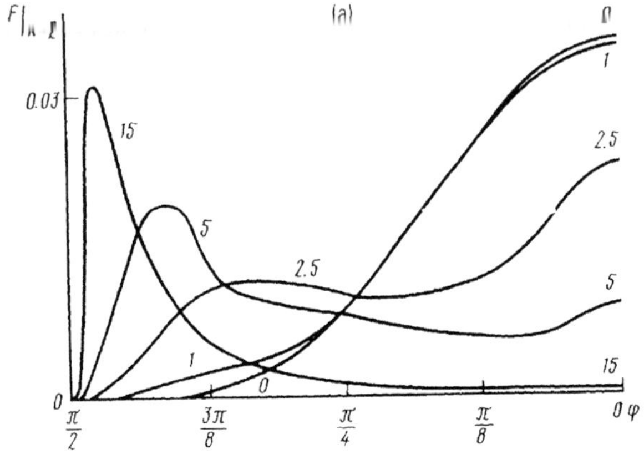

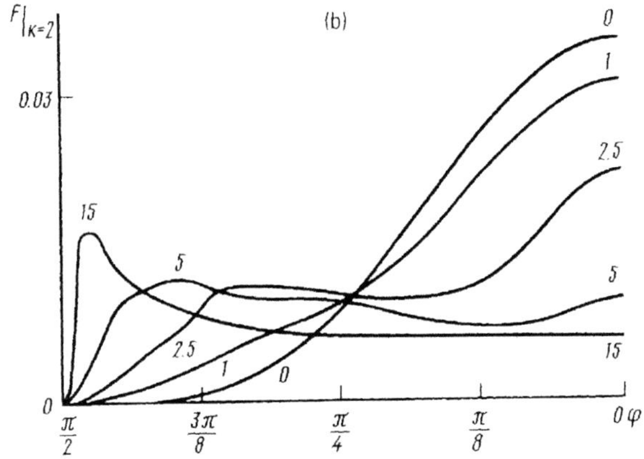

Fig. 2.3.20. Behavior of the spectrum corresponding to the initial state F_{02} at $k = 2$ depending on the calculation scheme: (a) in the presence of dissipation (scheme A) and (b) in the inviscid case (scheme B). The figures at the curves indicate time.

dissipation, then the energy is transferred not only to nearly meridional wave vectors but also to the region lying far from the axis k_y, which leads to the 'isotropization' of the spectrum outside the neighborhood of this axis. Thus, we can assume that in the process of viscous degeneration of weak turbulence the motion is first transformed relatively quickly into a nearly zonal flow and then is damped relatively slowly under the action of dissipation.

It is also interesting to note that dissipation, which does not exert any substantial effect on F and $\partial F/\partial \tau$ where these quantities are sufficiently large, has a very strong influence on the entropy values (see Figure 2.3.16). We see that in the presence of dissipation the entropy does not increase with time but decreases and its modulus exceeds more than one and a half times the modulus of the 'inviscid' entropy at the end of the calculation. This considerable distinction is due to the fact that the major contribution to the entropy is made by those regions where F is close to zero (i.e. $|\ln F|$ is large) and the dissipation 'broadens' these regions. In other words, the broadening of the region of small values of F implies that dissipation reduces the

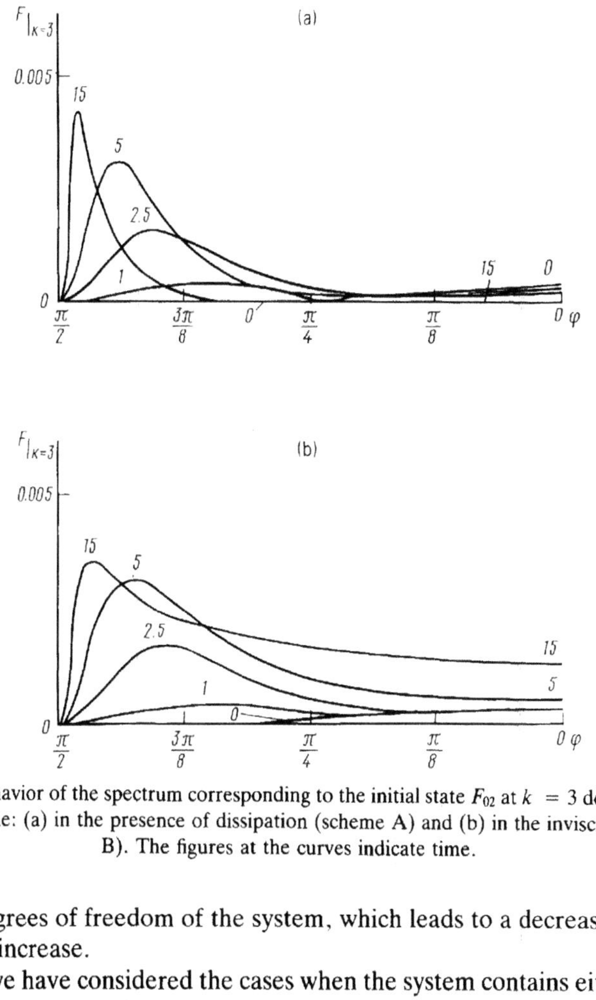

Fig. 2.3.21. Behavior of the spectrum corresponding to the initial state F_{02} at $k = 3$ depending on the calculation scheme: (a) in the presence of dissipation (scheme A) and (b) in the inviscid case (scheme B). The figures at the curves indicate time.

number of degrees of freedom of the system, which leads to a decrease of entropy instead of an increase.

Until now we have considered the cases when the system contains either no sinks and no sources of energy (scheme B) or only sinks (scheme A). Of course, of great interest is the evolution of a system of Rossby waves in the presence of sources and sinks of energy. This problem has not been studied thoroughly yet; the basic results in this field are presented in the papers by Pelinovskiĭ (1978) and Sazontov (1981) who considered the possible form of stationary spectra produced by a stationary source and a stationary sink spaced in the spectrum. The work by Pelinovskiĭ (1978) is in fact based not on the kinetic equation (3.1) but on the assumption that the interaction has a local character and that the fluxes of energy, enstrophy, and momentum depend quadratically on energy (anisotropic effects are not taken into account). The spectra of Rossby waves in the inertial range have the following form: accordingly, spectra of the form $F(k) \sim k^{-5/2}$, $F(k) \sim k^{-3}$, and $F(k) \sim k^{-7/2}$ correspond to constant fluxes of energy, enstrophy, and momentum. A more thorough investigation was carried out by Sazontov (1981) who showed that there exist stationary spectra localized near the axis k_y and approximately satisfying the kinetic equation (3.1). These spectra can, in principle, be generated by a nearly

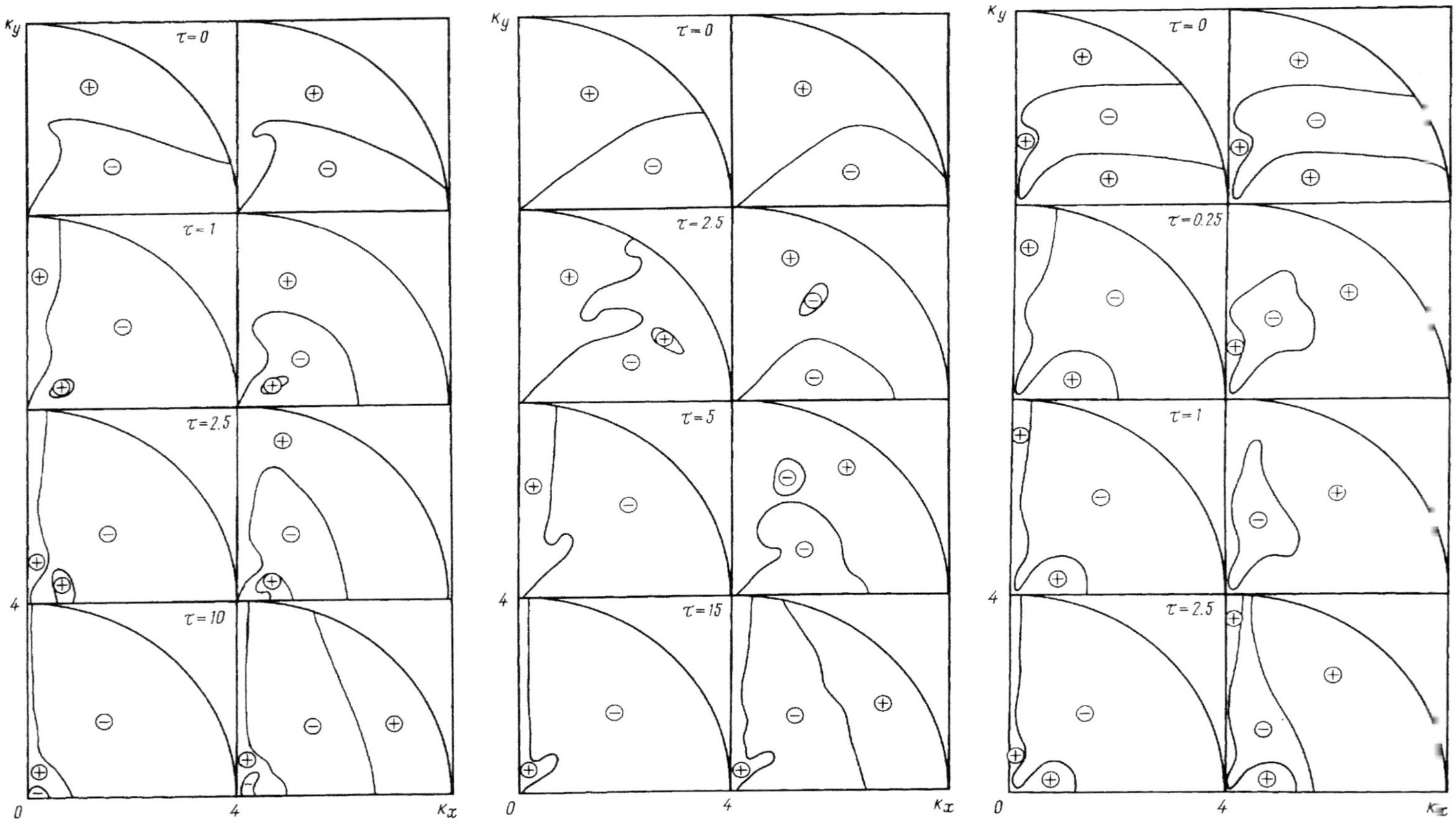

Fig. 2.3.22. Regions of inflow and outflow of energy (marked by the signs '+' and '−' respectively) for the initial spectra (a) F_{01}, (b) F_{02}, and (c) F_{03}. The left and right of each diagram shows, respectively, the calculation results obtained according to scheme A or B.

zonal external action. In this case spectra of the form $F(\mathbf{k}) \sim k_x^{-3/2}k_y^{-2}$, $F(\mathbf{k}) \sim k_x^{-3/2}k_y^{-3}$, and $F(\mathbf{k}) \sim k_x^{-1}k_y^{-7/2}$ correspond to constant fluxes of energy, enstrophy, and the meridional component of momentum.

3.8. Appendix: The Proof of the Uniqueness of the Solution (3.11) of System (3.9a, b)

We rewrite system (3.9a, b) in the form

$$Q(\mathbf{k}_1) + Q(\mathbf{k}_2) = Q(\mathbf{k}_1 + \mathbf{k}_2), \tag{3.31a}$$

$$\sigma(\mathbf{k}_1) + \sigma(\mathbf{k}_2) = \sigma(\mathbf{k}_1 + \mathbf{k}_2), \tag{3.31b}$$

where $Q(\mathbf{k}) = \sigma/F$. Consider an arbitrary wave vector $\mathbf{k}_1$ and a vector $\mathbf{k}_2 = -\mathbf{k}_1 + \mathbf{d}$; $\mathbf{d} = (d_x, d_y)$, $|\mathbf{d}| \to 0$. Substituting $\mathbf{k}_1$ and $\mathbf{k}_2$ into (3.31a, b) and using the relations $\sigma(\mathbf{0}) = Q(\mathbf{0}) = 0$, $\sigma(-\mathbf{k}) = -\sigma(\mathbf{k})$, and $Q(-\mathbf{k}) = -Q(\mathbf{k})$ we find that to within small quantities the vector $\mathbf{d}$ satisfies the equations

$$\begin{aligned}(\sigma_{k_x} - \sigma^0_{k_x})d_x + (\sigma_{k_y} - \sigma^0_{k_y})d_y = 0,\\ (Q_{k_x} - Q^0_{k_x})d_x + (Q_{k_y} - Q^0_{k_y})d_y = 0,\end{aligned} \tag{3.32}$$

where the subscripts k_x and k_y in σ and Q symbolize differentiation with respect to the corresponding variables; $\sigma_{k_x}^0$, $\sigma_{k_y}^0$, $Q^0_{k_x}$, and $Q^0_{k_y}$ are the values of the derivatives at $\mathbf{k} = 0$. After some simple transformations the condition for solvability of system (3.32) with respect to d_x and d_y is written in the form

$$J(\sigma - \sigma^0_{k_x} k_x - \sigma^0_{k_y}, Q - Q^0_{k_x} k_x - Q^0_{k_y} k_y) = 0,$$

$$J(a, b) = \frac{\partial a}{\partial k_x}\frac{\partial b}{\partial k_y} - \frac{\partial a}{\partial k_y}\frac{\partial b}{\partial k_x}, \tag{3.33a}$$

and must hold for any $\mathbf{k}$. It follows that

$$Q - Q^0_{k_x} k_x - Q^0_{k_y} k_y) = F(\sigma - \sigma^0_{k_x} k_x - \sigma^0_{k_y} k_y), \tag{3.33b}$$

where $F(z)$ is an arbitrary differentiable function.

Further, Equations (3.31a, b) are satisfied for arbitrary $\mathbf{k}_1 = (k_x, k_y)$ and $\mathbf{k}_2 = (-k_x, k_y)$. Let $\mathbf{k}_2 = (-k_x, k_y)$ and $\mathbf{k}_1 = (k_x, k_y) + \mathbf{d}$, $|\mathbf{d}| \to 0$. Substituting these values of $\mathbf{k}_1$ and $\mathbf{k}_2$ into (3.31a, b) and using the relations $\sigma_{k_y}(0, k_y) = Q_{k_y}(0, k_y) = 0$ we find, by analogy with the above,

$$\begin{aligned}[\sigma_{k_x} - \sigma_{k_x}(0, 2k_y)]d_x + \sigma_{k_y}d_y = 0,\\ [Q_{k_x} - Q_{k_x}(0, 2k_y)]d_x + Q_{k_y}d_y = 0.\end{aligned} \tag{3.34}$$

The condition for solvability of system (3.34) is written

$$[\sigma_{k_x} - \sigma_{k_x}(0, 2k_y)]\, Q_{k_y} - [Q_{k_x} - Q_{k_x}(0, 2k_y)]\, \sigma_{k_y} = 0. \tag{3.35}$$

Substituting (3.33) into (3.35) and taking into account that $\sigma'_{k_x} = 1$ we obtain, after some simple transformations,

$$F'(\sigma - k_x) = \text{const} \tag{3.36}$$

Hence, by virtue of (3.33) and (3.36),

$$Q = a\sigma + bk_x, \tag{3.37}$$

where a and b are arbitrary constants. Now (3.11) can easily be derived from (3.37).

4. ROSSBY SOLITONS

Solitons are known to be of great importance for analyzing wave phenomena in hydrodynamics, optics, acoustics, radiophysics, etc. There can be no solitons in linear physical systems; their formation is due to nonlinearity and dispersion. From the physical viewpoint the existence of a soliton (a localized disturbance propagating with a constant velocity without changing its form) is accounted for by the fact that the nonlinear increase of the wave slope can be balanced by its spreading owing to dispersion.

The interest in Rossby solitons in recent time is due to the discovery and intensive study of various localized disturbances in the ocean and the atmosphere (synoptic eddies in the open ocean, Gulf Stream eddies, and the Great Red Spot in the Jovian atmosphere). Attempts to interpret synoptic eddies as Rossby solitons seem to be quite natural. However, at present both the level of the development of the theory and the amount of observation data are insufficient to assume that this interpretation is fully justified. Therefore, in Section 5 of Chapter 3 another approach will be presented. In this connection we mention the work by Korotaev (1980) where an attempt was made to interpret synoptic motions in the ocean as a system of moving intensive eddies of nonsoliton type which radiate Rossby waves.

To avoid confusion we note that by 'Rossby solitons' we mean disturbances which are localized, at least with respect to one of the disturbance coordinates, and move with a constant velocity without changing their form. Unfortunately, at present we do not know much about the behavior of such disturbances when they collide with one another and, therefore, we state here the definition of solitons without introducing the requirement that they should retain their form and velocity after collisions. (For the general definition of a soliton, see Scott *et al.*, 1973.) Rossby solitons are considered in the reviews by Berestov and Monin (1980) and Berestov (1981b); for the role of localized solutions in geophysics see Flierl *et al.* (1983).

4.1. Asympotic Solutions; Weak Nonlinearity and Dispersion

We consider barotropic motions and represent the pressure disturbance in the form (cf. (2.62)):

$$\frac{p'}{f_0\varrho_0} = \bar{\psi}(y) + \psi(\xi, y), \quad \xi = x - ct, \tag{4.1}$$

where $\bar{\psi}(y)$ describes a given zonal current and ψ is the unknown function. Substituting (4.1) into the first two of Equations (1.33) and Equation (1.39b), and putting $b = b(y)$ we obtain

$$J\left(\psi + cy + \bar{\psi}(y), \Delta\psi + \frac{\mathrm{d}^2\bar{\psi}}{\mathrm{d}y^2} + \beta y + \frac{f_0 b}{H}\right) = 0, \tag{4.2a}$$

where

$$\Delta = \frac{\partial^2}{\partial\xi^2} + \frac{\partial^2}{\partial y^2}, \qquad J(a, b) = \frac{\partial a}{\partial\xi}\frac{\partial b}{\partial y} - \frac{\partial a}{\partial y}\frac{\partial b}{\partial\xi}. \tag{4.2b}$$

We investigate a class of solutions of Equations (4.2) assuming that their typical scale L_x along the x-axis is much greater than the typical scale L_y along the y-axis (an elongated soliton). We introduce the following dimensionless characteristics:

$$\xi = L_x\,\xi_1, \quad y = L_y y_1, \quad \bar{\psi} = U_c L_y\,\bar{\psi}_1, \quad \psi = UL_y\,\psi_1,$$

$$c = U_c c_1, \quad \frac{\mathrm{d}b}{\mathrm{d}y} = \frac{b_0}{L_b}\frac{\mathrm{d}b_1}{\mathrm{d}y_1},$$

where U is the typical wave motion velocity. Dropping, for simplicity, the subscript 1 in the dimensionless variables and putting $U_c = \beta L_y^2$ (U_c is the typical velocity of the zonal current) we rewrite (4.2a) in the form

$$\delta\varepsilon^2 J\left(\psi, \frac{\partial^2\psi}{\partial\xi^2}\right) + \delta J\left(\psi, \frac{\partial^2\psi}{\partial y^2}\right) + \frac{\partial\psi}{\partial\xi}\left(1 + \frac{\mathrm{d}^3\bar{\psi}}{\mathrm{d}y^3} + \frac{\varepsilon_b}{\varepsilon_L}\frac{\mathrm{d}b}{\mathrm{d}y}\right) -$$

$$- \varepsilon^2\left(c + \frac{\mathrm{d}\bar{\psi}}{\mathrm{d}y}\right)\frac{\partial^3\psi}{\partial\xi^3} - \left(c + \frac{\mathrm{d}\bar{\psi}}{\mathrm{d}y}\right)\frac{\partial^3\psi}{\partial\xi\,\partial y^2} = 0, \tag{4.3}$$

where $\varepsilon = L_y/L_x \ll 1$, $\delta = U/\beta L_y^2$, $\varepsilon_b = b_0 L_y/HL_b$, $\varepsilon_L = L_y/a$ (the parameters δ, ε_b, and ε_L were introduced in Section 1 of the present chapter).

The parameter δ characterizes the degree of nonlinearity of the wave motion (the ratio of the typical particle velocity U to the typical phase velocity βL_y^2 of Rossby waves). One can easily write an expression for the phase velocity of barotropic Rossby waves imposed on a zonal current U_c: $\sigma/k_x = U_c - \beta k^{-2} \simeq U_c - \beta k_y^{-2}(1 - k_x^2/k_y^2)$ where (k_x, k_y) are the wavenumbers along (x, y). We see that the parameter ε^2 characterizes the degree of dispersion of a Rossby wave in its propagation along the axis of the channel. Since we seek soliton solutions (for which the nonlinearity and the dispersion are of the same order), we put $\delta = \varepsilon^2$ in (4.3). Assuming that $\varepsilon_b \sim \varepsilon_L$ we seek the solution of Equation (4.3) in the form

$$\psi = \psi_0 + \delta\psi_1 + \cdots, \quad c = c_0 + \delta c_1 + \cdots \tag{4.4}$$

Substituting (4.4) into (4.3) and equating to zero the coefficients in different powers of ε we find

$$G\frac{\partial\psi_0}{\partial\xi}+\frac{\partial^3\psi_0}{\partial\xi\,\partial y^2}=0, \tag{4.5$_0$}$$

$$G\frac{\partial\psi_1}{\partial\xi}+\frac{\partial^3\psi_1}{\partial\xi\,\partial y^2}+c_1\left(c_0+\frac{d\bar\psi}{dy}\right)^{-1}\frac{\partial^3\psi}{\partial\xi\,\partial y^2}+$$

$$+\frac{\partial^3\psi_0}{\partial\xi^3}-\left(c_0+\frac{d\bar\psi}{dy}\right)^{-1}J\left(\psi_0,\frac{\partial^2\psi_0}{\partial y^2}\right)=0,\ \ldots \tag{4.5$_1$}$$

where

$$G(y)=-\left(1+\frac{d^3\bar\psi}{dy^3}+\frac{\varepsilon_b}{\varepsilon_L}\frac{db}{dy}\right)\left(c_0+\frac{d\bar\psi}{dy}\right)^{-1}. \tag{4.6}$$

We assume that the functions ψ_0, ψ_1, . . . vanish at the walls of the channel ($y=0$ and $y=L$). We represent ψ_0 in the form $\psi_0=\Psi(\xi)\Phi(y)$ where $\Phi(y)$ is the solution of the following Sturm–Liouville problem:

$$\frac{d^2\Phi}{dy^2}+G(y)\Phi=0,\quad \Phi(0)=\Phi(L)=0. \tag{4.7}$$

Equation (4.5$_0$) is obviously satisfied for this choice of the function Φ. Substitute $\psi_0=\Psi(\xi)\Phi(y)$ into the right-hand side of Equation (4.5$_1$), multiply this equation by Φ, and integrate it across the channel. Performing integration by parts and taking into account (4.7) we find, after some simple transformations,

$$\frac{d\Psi}{d\xi}+\frac{a_1}{c_1}\Psi\frac{d\Psi}{d\xi}+\frac{a_2}{c_1}\frac{d^3\Psi}{d\xi^3}=0, \tag{4.8}$$

where

$$a_1=-\int_0^L\Phi^3G'\left(c_0+\frac{d\bar\psi}{dy}\right)^{-1}dy\left[\int_0^L\Phi^2G\left(c_0+\frac{d\bar\psi}{dy}\right)dy\right]^{-1},$$

$$a_2=-\int_0^L\Phi^2\,dy\left[\int_0^L\Phi^2G\left(c_0+\frac{d\bar\psi}{dy}\right)^{-1}dy\right]^{-1}. \tag{4.9}$$

Equation (4.8) is the stationary Korteweg–de Vries equation. It is known to possess a soliton-type solution

$$\Psi(\xi)=\operatorname{sgn}(a_1a_2)A\cosh^{-2}(B\xi), \tag{4.10}$$

where

$$c_1=-\operatorname{sgn}(a_1a_2)\frac{a_1A}{3},\qquad B^2=\frac{1}{12}\left|\frac{a_1}{a_2}\right|A. \tag{4.11}$$

The velocity of soliton motion c_0 is an eigenvalue of the problem (4.7). On finding c_0 and the corresponding eigenfunction Φ we then determine a_1 and a_2 by formulas (4.9). Further, regarding A (the modulus of the soliton amplitude) as a free parameter, we obtain the correction c_1 to the soliton velocity and the 'width' B of the soliton. We note that for $G(y)$ = const (e.g. $\mathrm{d}\bar{\psi}/\mathrm{d}y$ = const and $\mathrm{d}b/\mathrm{d}y$ = const) we have $a_1 = 0$ and the soliton solution does not exist. Thus, for elongated soliton solutions ($L_x \gg L_y$) in a zonal channel to exist it is necessary that there be a velocity shear in the zonal current or that the depth of the channel be variable across its axis and the variation be nonlinear.

The first step in studying Rossby solitons was made by Long (1964) who studied Equation (4.2) for a channel with a linear function $\mathrm{d}\bar{\psi}/\mathrm{d}y$. Similar investigations were carried out by Larsen (1965) and Benney (1966). Later the results of these works were developed by Maxworthy and Redecopp (1976), Redecopp (1977), Redecopp and Weidman (1978), Maxworthy *et al.* (1978), and Weidman and Maxworthy (1978) for the case of a stratified fluid with a mean current $\mathrm{d}\bar{\psi}/\mathrm{d}y$ and by Rizzoli and Hendershott (1980) and Rizzoli (1980) for the case of topographic barotropic Rossby waves without mean current. In the last two works the stability of the constructed Rossby solitons was proved (analytically for small disturbances and numerically for finite disturbances). We note the papers by Volosov (1976a, b, and c) who investigated topographic Rossby solitons over a rough bottom (linear waves of this type were discussed in Subsection 2 of Section 6) and also the paper by Kobaladze *et al.* (1981).

4.2. Asymptotic Solutions; Strong Nonlinearity and Dispersion

The waves considered in Section 4.1 had sharply different scales L_x and L_y ($L_x \gg L_y$) and therefore they had weak dispersion. However, according to observational data on synoptic eddies in the open ocean, Gulf Stream rings, etc., the scales L_x and L_y are of the same order. Such waves possess strong dispersion; however, since, as observations show, the nonlinearity parameter $\delta = U/\beta L^2$ for synoptic eddies is of the order of or greater than unity, the nonlinearity can generally balance the dispersion. This is the type of soliton that will be studied in this subsection.

We consider barotropic motions again and pass to the dimensionless variables $\psi_1 = \psi/UL$, $x_1 = x/L$, $y_1 = y/L$, and $\mathrm{d}b_1/\mathrm{d}y_1 = (b_0/L_b)^{-1}(\mathrm{d}b/\mathrm{d}y)$. Omitting the subscript 1 in the dimensionless variables we write

$$J(\psi, \Delta\psi) + \mu \frac{\partial}{\partial\xi}\left[\alpha(y)\Delta\psi + \gamma(y)\psi\right] = 0, \tag{4.12}$$

where

$$\mu = \frac{c}{U}, \qquad \alpha(y) = -c^{-1}\frac{\mathrm{d}\bar{\psi}}{\mathrm{d}y} - 1,$$

$$\gamma(y) = \frac{\beta L^2}{c} + c^{-1}\frac{\mathrm{d}\bar{\psi}}{\mathrm{d}y} + \varepsilon_b \frac{f_0 L}{c}\frac{\mathrm{d}b}{\mathrm{d}y}. \tag{4.13}$$

The velocity of propagation of the wave disturbance c is of the order of βL^2. Therefore, $\mu \sim \delta^{-1}$. We consider strongly nonlinear waves for which $\mu \ll 1$.

Further, in accordance with observations, we assume that $d\bar{\psi}/dy \sim c \sim \beta L^2$ and $\varepsilon_b(f_0L/c) \sim (\varepsilon_b/\varepsilon_L)(\beta L^2/c) \sim 1$ and seek a solution of Equation (4.12) of the form

$$\psi = \psi_0(x, y) + \mu\psi_1(x, y) + \cdots \tag{4.14}$$

Assume that

$$\alpha = \alpha_0 + \mu\alpha_1(y) + \cdots, \quad \gamma = \gamma_0 + \mu\gamma_1(y) + \cdots \tag{4.15}$$

and that the first terms α_0 and γ_0 of these expansions do not depend on y (e.g. this condition holds for a weak shear of the zonal current velocity and a nearly constant bottom slope). Substituting (4.14) and (4.15) into (4.12) we obtain

$$J(\psi_0, \Delta\psi_0) = 0 \tag{4.16$_0$}$$

$$J(\psi_1, \Delta\psi_0) + J(\psi_0, \Delta\psi_1) + \frac{\partial}{\partial\xi}(\alpha_0\Delta\psi_0 + \gamma_0\psi_0) = 0. \tag{4.16$_1$}$$

Now we study nearly axisymmetric wave solutions for which the first term of the series (4.14) depends only on $r = \sqrt{\xi^2 + y^2}$. It is obvious that any function of the form $\psi_0 = \psi_0(r)$ satisfies (4.16$_0$); we emphasize that the original equation (4.12) does not possess axisymmetric solutions.

Taking into account that $\psi_0 = \psi_0(r)$ and passing to the polar coordinates $\xi = r\cos\theta$, $y = r\sin\theta$ we find that Equation (4.16$_1$) can be integrated once with respect to θ. After some simple transformations we obtain

$$\Delta\psi_1 + \gamma_0 r\sin\theta = \left(\frac{d\Delta\psi_0}{dr} \Big/ \frac{d\psi_0}{dr}\right)(\psi_1 - \alpha_0 r\sin\theta) + P_1(r), \tag{4.17}$$

where $P_1(r)$ is an arbitrary function.

We seek ψ_1 in the form $\psi_1 = \Phi(r)\sin\theta + \eta(r)$. Then for $\Phi(r)$ and $\eta(r)$ we derive the equations

$$\frac{1}{r}\frac{d}{dr}\left(r\frac{d\Phi}{dr}\right) - \left[\frac{1}{r^2} + \left(\frac{d\Delta\psi_0}{dr} \Big/ \frac{d\psi_0}{dr}\right)\right]\Phi = - r\left[\left(\frac{d\Delta\psi_0}{dr} \Big/ \frac{d\psi_0}{dr}\right)\alpha_0 + \gamma_0\right], \tag{4.18a}$$

$$\frac{1}{r}\frac{d}{dr}\left(r\frac{d\eta}{dr}\right) - \left(\frac{d\Delta\psi_0}{dr} \Big/ \frac{d\psi_0}{dr}\right)\eta = P_1(r). \tag{4.18b}$$

First we consider Equation (4.18a) for Φ. We require that this equation should have a smooth solution decreasing faster than any power of r for large r (the condition for the localization of the disturbance). This requirement imposes certain conditions on the function $\psi_0(r)$ and the parameters γ_0 and α_0 (for the derivation and the discussion see Larichev and Reznik, 1976d):

(1) $\alpha_0\gamma_0 < 0$;
(2) $\psi_0 \sim r^{-1/2}\exp[-(-\alpha_0\gamma_0)^{1/2}r]$ for $r \to \infty$;
(3) $\int_0^\infty \psi_0' r^2\, dr = 0$ or $\int_0^\infty \psi_0 r\, dr = 0$;

(4) $\psi_0(0) = 0$;
(5) $\psi_0'''(r_k)\,[\psi_0''(r_k)]^{-1} + r_k^{-1}$, where r_k is a zero of the function $\psi_0'(r)$.

We briefly discuss these conditions. Condition (1) determines the possible values of the phase velocity c. For example, if $d\bar{\psi}/dy = 0$ and $b = 0$, then, according to condition (1), we have $c > 0$, i.e. the disturbance under study can move only to the east. Condition (2) is the condition that the wave disturbance should be localized in space. To interpret condition (3) consider the total angular momentum about the z-axis:

$$\varrho_0 \iint_S (yu - xv)\, dS = -\varrho_0 \iint_S \left(y\frac{\partial\psi}{\partial y} + x\frac{\partial\psi}{\partial x} \right) dS = 2\varrho_0 \iint_S \psi\, dS.$$

According to condition (3), the total angular momentum must be equal to zero in the first approximation. Condition (4) guarantees the continuity of the velocity at the soliton center and condition (5) means that the distribution F of the relative vorticity $\Delta\psi_0$ over the streamlines determined by the function $\psi_0(r)$ is smooth ($\Delta\psi_0 = F(\psi_0)$ and $dF/d\psi_0 = (\Delta\psi_0)'/\psi_0' = \{\psi_0'''(r_k)\,[\psi_0''(r_k)]^{-1} + r_k^{-1}\}\,(r - r_k)^{-1} + \ldots$ in the neighborhood of a zero of the function $\psi_0'(r)$).

One can readily choose a sufficiently smooth function $\psi_0 = \psi_0(r)$ (satisfying conditions (1)–(4) (see Figure 2.4.1); however, we note that these conditions do not determine the function $\psi_0 = \psi_0(r)$ uniquely. After $\psi_0(r)$ has been set, the function Φ is uniquely determined by (4.18a). By virtue of the arbitrariness of $P_1(r)$, the function $\eta(r)$ also remains arbitrary and it can probably be determined from the analysis of the second approximation.

The above axisymmetric barotropic soliton solution was constructed by Larichev and Reznik (1976d); analogous solutions for baroclinic motions were considered by Flierl (1979); see also the paper by Kizner (1984a).

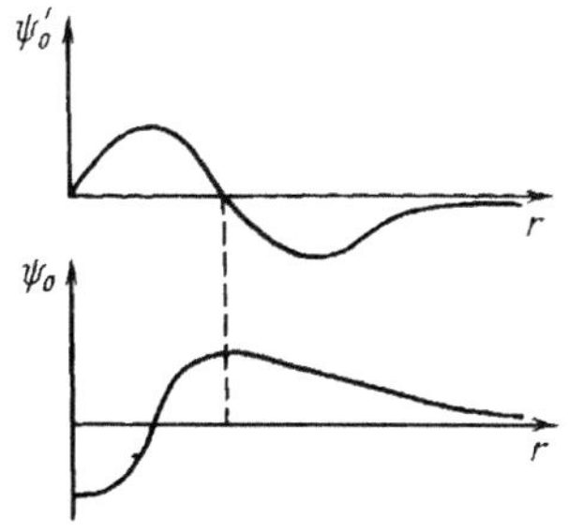

Fig. 2.4.1. Schematic form of the functions $\psi_0'(r)$ and $\psi_0(r)$ satisfying conditions 1–4 (with a change of the sign of circulation).

4.3. Exact Solutions; a Barotropic Ocean

The rest of this section is devoted to a certain class of exact soliton-type solutions of the transfer equation for the quasigeostrophic potential vorticity. We begin with the simplest case of a barotropic ocean of constant depth ($b = 0$). The pressure disturbance is represented in the form

$$p' = f_0 \rho_0 \psi(\xi, y), \quad \xi = x - ct, \tag{4.19}$$

where c is the soliton propagation velocity.

Substituting (4.19) into the first two of Equations (1.33) and Equation (1.39b) we find

$$J(\psi + cy,\ \Delta\psi + \beta y) = 0, \tag{4.20a}$$

where

$$\Delta = \frac{\partial^2}{\partial \xi^2} + \frac{\partial^2}{\partial y^2}\ , \quad J(a, b) = \frac{\partial a}{\partial \xi}\ \frac{\partial b}{\partial y} - \frac{\partial a}{\partial y}\ \frac{\partial b}{\partial \xi}\ . \tag{4.20b}$$

We take a circle of radius a in the plane ξ, y and consider the following equations:

$$\Delta\psi + \beta y = -\,k^2(\psi + cy) \ \text{ for } r < a; \tag{4.21a}$$

$$\Delta\psi + \beta y = p^2(\psi + cy) \ \text{ for } r > a; \tag{4.21b}$$

where k, p, and a are arbitrary constants. It is obvious that any function ψ satisfying Equations (4.21) also satisfies Equation (4.20) everywhere on the plane ξ, y except, possibly, at the circle $r = a$.

We require that the function ψ tends to zero exponentially for $r \to \infty$ (the condition for the localization of the disturbance). Then (4.21b) implies that

$$p^2 = \frac{\beta}{c}\ . \tag{4.22}$$

We introduce the polar coordinates $x = r \cos\,\theta$, $y = r \sin\,\theta$ and consider the following solutions of Equation (4.21):

$$\psi = \left[AJ_1(kr) - \frac{\beta + ck^2}{k^2}\ r \right] \sin\theta \quad \text{for } r < a, \tag{4.23a}$$

$$\psi = BK_1(pr) \sin\theta \text{ for } r > a, \tag{4.23b}$$

where A and B are arbitrary constants; J_1 is a Bessel function of the first kind; and K_1 is the Macdonald function.

Now we require that the function ψ itself and its derivatives $\partial\psi/\partial r$ and $\partial^2\psi/\partial r^2$ be continuous at $r = a$ (the condition of continuity for pressure, velocity, and vorticity). By virtue of (4.21), these requirements are equivalent to the following relations:

$$\psi + cy\,\big|_{r\,=\,a+0} = 0; \ \ \psi + cy\,\big|_{r\,=\,a-0} = 0;$$
$$\left.\frac{\partial\psi}{\partial r}\right|_{r\,=\,a+0} = \left.\frac{\partial\psi}{\partial r}\right|_{r\,=\,a-0}\ . \tag{4.24}$$

From the first two conditions (4.24) we find that

$$A = \frac{\beta a}{k^2 J_1(ka)}\ ; \qquad B = -\,\frac{\beta a}{p^2 K_1(pa)}\ , \tag{4.25}$$

and the third condition (4.24) implies

$$\frac{J_1'(ka)}{J_1(ka)} = \frac{1}{ka}\left(1 + \frac{k^2}{p^2}\right) - \frac{k}{p}\frac{K_1'(pa)}{K_1(pa)} .$$

Using the well-known formulas $zJ_1'(z) = J_1(z) - zJ_2(z)$ and $zK_1'(z) = K_1(z) - zK_2(z)$ and also formula (4.22) we rewrite the above relation as

$$-\frac{J_2(ka)}{kaJ_1(ka)} = \frac{K_2(q)}{qK_1(q)} ; \; q = a\sqrt{\frac{\beta}{c}} . \tag{4.26}$$

Equation (1.26) determines k as a function of a and c. It can be shown that for a given q Equation (4.26) has a countable set of positive roots k. In what follows we consider only the solution with the smallest value of the root ka. (For a discussion of solutions with other roots, see Berestov (1981a) and Berestov and Monin (1980).) The values of this root are given in Table 2.4.1 according to which Figure 2.4.2 is constructed.

TABLE 2.4.1
(After Flierl *et al.*, 1980)

q	ka	q	ka	q	ka
0	3.8317	1	3.9226	3	4.2121
0.01	3.8317	1.5	3.9974	10	4.6985
0.03	3.8318	2	4.0732	30	4.9715
0.1	3.8330	2.5	4.1454	100	5.0848
0.3	3.8426	2.75	4.1795	∞	5.1356

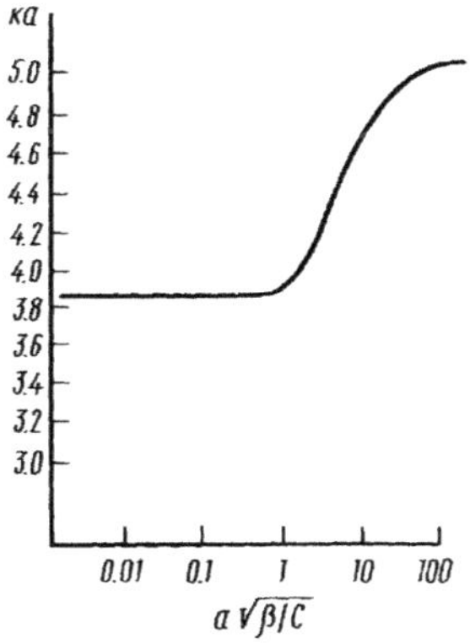

Fig. 2.4.2. The smallest root ka of Equation (4.26) as a function of the parameter $q = a(\beta/c)^{1/2}$. The plot is constructed from Table 2.4.1.

Thus, setting a and c we find k and then determine the constants A and B by formulas (4.25), taking into account (4.22). Figure 2.4.3 demonstrates streamlines (ψ = const) of such a solution consisting of a pair of eddies rotating in opposite directions. Introducing the typical size l and intensity M of the soliton as the

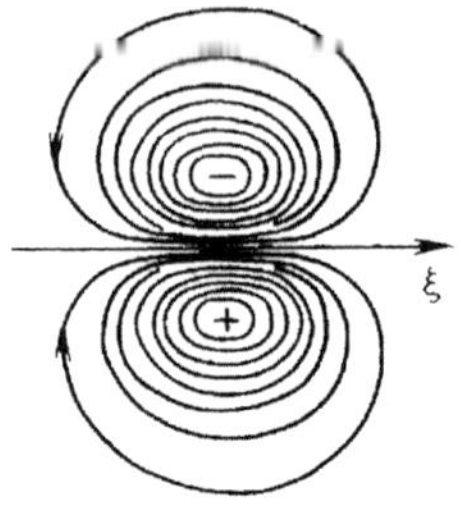

Fig. 2.4.3. Isolines of the stream function ψ (the streamlines) at a fixed time in the case of the Larichev–Reznik barotropic soliton (4.23), (4.25), (4.26).

distance between the eddy centers and the modulus of the pressure disturbance at the center of an eddy we combine formulas (4.25) and (4.26) to obtain a dispersion relation for c, l and M. Calculations show (Flierl *et al.*, 1980; Fig. 6) that for a given M the velocity c decreases with increasing l (the interaction between the two eddies decreases) and for a fixed l the velocity c increases with increasing M (the interaction between the two eddies increases). Ultimately we arrive at the following simple result: the stronger the interaction between the two eddies forming the soliton, the greater the velocity c of its propagation.

Expression (4.23b) implies that the asymptotics of the above soliton-type solution of Equation (4.20) for $r \to \infty$ is $\psi \sim (1/\sqrt{r}) \exp[-(\beta c)^{1/2} r]$. The pressure, velocity, and vorticity of this solution are continuous on the circle $r = a$, whereas the vorticity gradient suffers discontinuity on this line. Nevertheless, by virtue of the finiteness of the discontinuity and conditions (4.20), the left-hand side of Equation (4.20) is continuous throughout the plane ξ, y.

According to (4.22), the above Rossby soliton can propagate only to the east ($c > 0$), in contrast to periodic Rossby waves propagating only in the westward direction. If the rigid lid approximation we have used (the first condition (1.36)) is rejected, it can easily be shown that c must lie outside the interval $(-g\beta H/f_0^2, 0)$; this is the interval in which the possible x-components of the phase velocity of periodic waves lie (the dispersion relation for barotropic Rossby waves without the rigid lid approximation is written $\sigma = -k_x\beta/[k^2 + (f_0^2/gH)]$; see (2.57)). From a physical viewpoint, this is due to the fact that if the velocity of propagation of the soliton lay inside the interval of the possible phase velocities of periodic waves, the propagating soliton would generate undamped periodic waves and soliton-type solutions of the original equation would not exist.

It is advisable to present an additional explanation why the β-effect makes the soliton propagate to the east (see Flierl *et al.*, 1980). Figure 2.4.4 presents a schematic section along the x-axis across a high-pressure region. The particles move clockwise around this region and the planetary vorticity transferred by them to the eastern periphery of the eddy (region E) is greater than the value of the vorticity in the section, whereas the planetary vorticity transferred to the western periphery (region W) is smaller (in comparison with the value of the planetary vorticity in this section). This excess of vorticity can be balanced only by its local variation as the soliton propagates eastward (the nonlinear advection of relative vorticity on the periphery of the soliton is negligibly small).

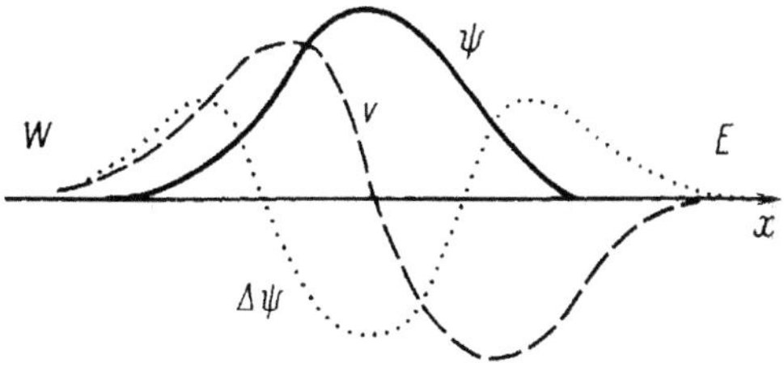

Fig. 2.4.4. Schematic section along the x-axis passing through the center of the high-pressure region (after Flierl *et al.*, 1980). Shown in the figure are the profiles of the function ψ, the meridional velocity v, and the vorticity $\Delta\psi$.

The above solution of Equation (4.20) was discovered by Larichev and Reznik (1976b). An analogous solution in the form of an immovable disturbance ($c = 0$) was constructed by Stern (1975); however, the vorticity in Stern's solution has a discontinuity at $r = a$. These solutions can be regarded as a generalization (the inclusion of the β-effect) of the well-known vortical solution satisfying the equations $\Delta\psi = -k^2(\psi + cy)$ and $\Delta\psi = 0$ for $r < a$ and $r > a$, respectively (Lamb, 1932, §165). These equations are obtained from Equations (4.2) for $\beta = 0$ and $p = 0$; repeating the above derivation we find

$$\psi = \left[\frac{2c}{kJ_1'(ka)} J_1(kr) - cr \right] \sin\theta \quad \text{for } r < a; \tag{4.27a}$$

$$\psi = -\frac{a^2 c}{r} \sin\theta \quad \text{for } r > a; \tag{4.27b}$$

$$J_1(ka) = 0. \tag{4.27c}$$

It can easily be verified that the function ψ and its derivatives $\partial\psi/\partial r$ and $\partial^2\psi/\partial r^2$ are continuous at $r = a$. We note that in the absence of the β-effect (1) the solution damped for large r has the asymptotics $1/r$ and (2) this solution can propagate with an arbitrary velocity c. The latter property is probably related to the fact that in the linear case the vorticity equation $J(\Delta\psi, \psi + cy) = 0$ has no wave solutions at all.

Stern's solution is expressed by

$$\psi = \left[\frac{\beta a}{k^2 J_1(ka)} J_1(kr) - \frac{\beta r}{k^2} \right] \sin\theta \quad \text{for } r < a; \tag{4.28}$$

$$\psi \equiv 0 \quad \text{for } r > a; \tag{4.29}$$

$$J_2(ka) = 0. \tag{4.30}$$

Differentiating the identity $zJ'(z) = J_1(z) - zJ_2(z)$, and taking into account (4.30), we find that $J_1''(ka) = -J_2'(ka)$. Since $J_2'(ka) \neq 0$, the derivative $\partial^2\psi/\partial r^2$ cannot vanish at $r = a$, i.e. the vorticity in Stern's solution does in fact have a discontinuity at $r = a$.

Flierl *et al.* (1980) suggested a generalization of the Larichev–Reznik solution by adding a function to solution (4.23) which depends on r solely; however, as in

Stern's solution, the vorticity in this solution suffers discontinuity at $r = a$. Berestov (1979) suggested a generalization in the form of a 'three-layer' soliton:

$$\psi = AI_1\,(\sqrt{\beta/c}\; r) \sin\theta \text{ for } r < a; \tag{4.31a}$$

$$\psi = \left[BJ_1\,(kr) + E\,\frac{\beta}{k^3}\,N_1\,(kr) - \frac{\beta + ck^2}{k^2}\,r \right] \times$$

$$\times \sin\theta \text{ for } a < r < b; \tag{4.31b}$$

$$\psi = DK_1\,(\sqrt{\beta/c}\; r) \sin\theta \text{ for } r > b, \tag{4.31c}$$

where I_1 is a modified Bessel function and N_1 is a Bessel function of the second kind. Here we do not write down the system of equations for the constants A, B, E, and D implied by the continuity of ψ, $\partial\psi/\partial r$, and $\partial^2\psi/\partial r^2$ and the expressions for the other parameters of the soliton. The solution (4.31) also admits of the addition of a function dependent only on r (Berestov, 1981a).

The stability of the indicated solutions is an important aspect of the problem. McWilliams *et al.* (1981) investigated the stability of solutions (4.23), (4.25), and (4.26) by means of numerical analysis of the problem for the transfer equation for the quasigeostrophic potential vorticity of barotropic motions (see Equation (1.39b) for $b = 0$; in the numerical experiments a small viscosity was introduced in this equation). Setting the initial data in the form of solutions (4.23), (4.25), (4.26) plus a disturbance with various horizontal scales and amplitudes, the authors of this paper carried out a series of numerical experiments which demonstrated the stability of solution (4.23), (4.25), (4.26) relative to small-amplitude disturbances. They also found the critical disturbance amplitude above which the soliton is destroyed; this amplitude decreases with increasing spatial scale of the disturbance.

Makino, *et al.* (1981), Larichev and Reznik (1982, 1983), and McWilliams and Zabusky (1982) studied numerically different types of interaction of barotropic solitons (4.23), (4.25), (4.26) (the displacement of the free surface was taken into account and therefore the solitons could propagate both to the east and to the west). The interaction turned out to be rather complicated, and here we briefly indicate only the most interesting cases. The character of a head-on collision of solitons strongly depends on the distance δy separating their antisymmetry axes. In the central collision ($\delta y = 0$) solitons usually 'survived' after the interaction. Figure 2.4.5 shows that the eddies of the less intensive soliton making rotations about the eddy centers of the more intensive soliton come closer together, after which the original solitons, propagating in opposite directions, are formed again (however, there are also examples of central collisions of solitons differing in their intensities when the interaction resulted in the destruction of one of the solitons). The soliton interaction for small values of δy is probably analogous to the above. For large values of δy a triple of interrelated eddies is formed after the interaction. A further increase of δy leads to the formation of two eddy pairs which differ from the eddy pairs forming the original solitons and are in complicated motions in the plane x, y. For very large δy the solitons are not likely to 'react' to each other.

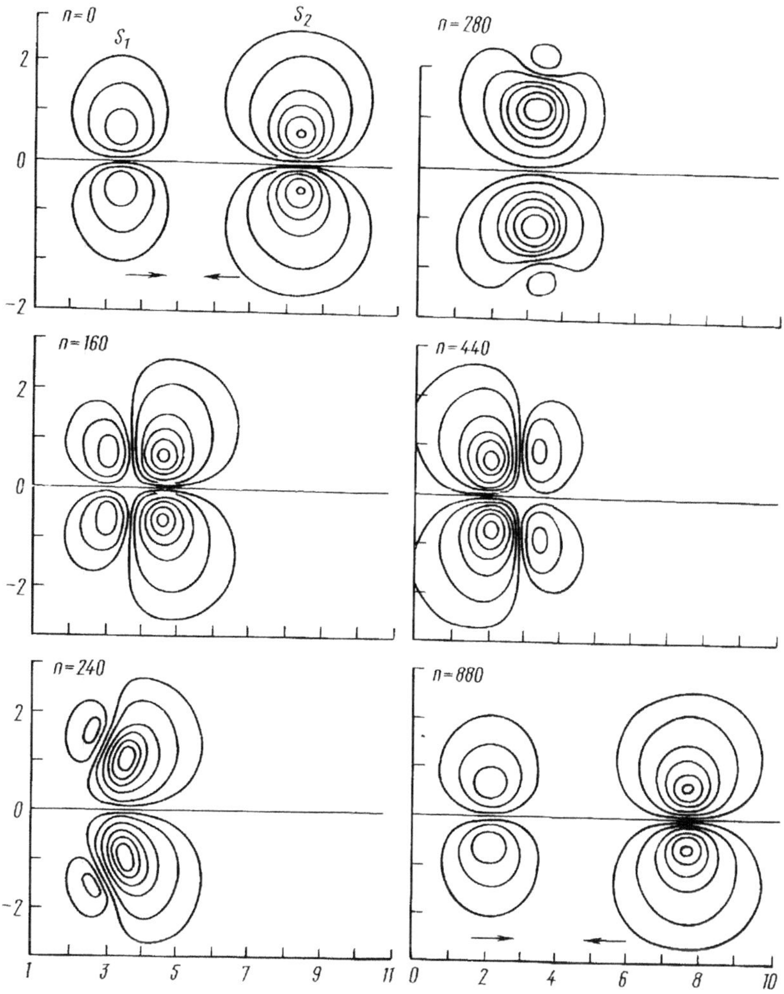

Fig. 2.4.5. Head-on collision of barotropic solitons according to the results of the numerical experiments of Larichev and Reznik (1982). The figure demonstrates streamlines; n is the number of time steps. For the sake of convenience, consideration was given to a problem with periodicity in x and y, the coordinate system moving with the velocity of the soliton S_1. After $n = 880$ steps the soliton S_2 travelled a distance approximately equal to the period with respect to x. The arrows indicate the velocities of the solitons in a fixed coordinate system. The unit line segments on the x and y axes correspond to the same distance

In the central collision of solitons propagating in the same direction (i.e. overtaking each other), the interaction pattern also depended on the parameters of the solitons. For some values of the parameters both solitons 'survived' (see Figure 2.4.6) and for other values only one of them survived. In some cases solitons 'merged' and a new soliton was formed.

Makino *et al.* (1981) also showed that if the antisymmetry axis of a soliton has an initial inclination to the x-axis, then the direction of propagation of the soliton

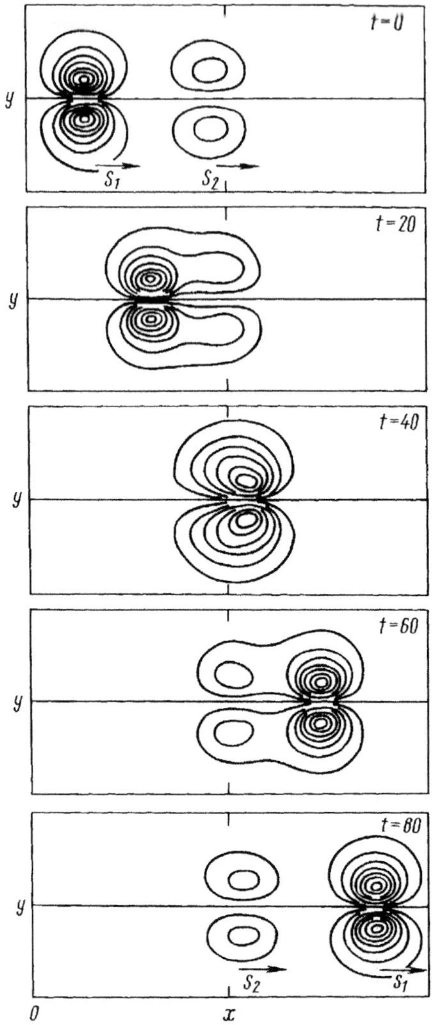

Fig. 2.4.6. Passage of a barotropic soliton S_1 through another barotropic soliton S_2 according to the results of the numerical experiments of Makino *et al.* (1981). The figure shows streamlines in a fixed coordinate system at different instants of conditional time t. The arrows indicate the velocities of the solitons.

oscillates about the x-axis and the intensities of the two eddies forming the soliton alternate.

Finally, to conclude this subsection we mention the Davis–Acrivos (1967) laboratory experiments on the excitation of internal-wave solitons in a fluid with density inhomogeneity. As was shown by Berestov and Monin (1980) (see also Berestov, 1981a), these experiments can be interpreted as modelling Rossby solitons in laboratory conditions.

4.4. Exact Solutions; a Two-Layer Ocean

We proceed from Equations (1.84), (1.85) without external forces ($\tau = 0$) ignoring the bottom friction ($A_z = 0$) and putting $b = 0$ (a constant depth). We seek the pressure disturbances p_1' and p_2' in the form of stationary waves

$$p_1' = f_0\varrho_0\psi_1(\xi, y); \quad p_2' = f_0\varrho_0\psi_2(\xi, y); \quad \xi = x - ct. \tag{4.32}$$

Substituting (4.32) into (1.84) and (1.85) we obtain

$$J\,[\psi_1 + cy,\ \Delta\psi_1 + \beta y + \Lambda_1\,(\psi_2 - \psi_1]) = 0; \tag{4.33}$$

$$J\,[\psi_2 + cy,\ \Delta\psi_2 + \beta y + \Lambda_2\,(\psi_1 - \psi_2]) = 0; \tag{4.34}$$

where Δ and J are given by formulas (4.20b).

We try to construct a class of exact solutions of Equations (4.33), (4.34) analogous to the class of solutions in the previous subsection. Let us now introduce the circle $r = a$ and consider the following systems of equations inside the circle ($r < a$) and outside it ($r > a$):

$$\Delta\psi_1 + \beta y + \Lambda_1\,(\psi_2 - \psi_1) = A_i\,(\psi_1 + cy); \tag{4.35a}$$

$$\Delta\psi_2 + \beta y + \Lambda_2\,(\psi_1 - \psi_2) = B_i\,(\psi_2 + cy); \tag{4.35b}$$

$$\Delta\psi_1 + \beta y + \Lambda_1\,(\psi_2 - \psi_1) = A_e\,(\psi_1 + cy); \tag{4.36a}$$

$$\Delta\psi_2 + \beta y + \Lambda_2\,(\psi_1 - \psi_2) = B_e\,(\psi_2 + cy), \tag{4.36b}$$

where A_i, B_i, A_e, and B_e are constants.

It can readily be verified that any solution ψ_1, ψ_2 of the system of equations (4.35) ((4.36)) satisfies the original equations (4.33) and (4.34) in the region $r < a$ ($r > a$).

We investigate Equations (4.36). We seek the solutions of these equations which are exponentially damped for large r (localized disturbances). The damping condition immediately implies that

$$A_e = B_e = \frac{\beta}{c}. \tag{4.37}$$

Let us introduce $\Psi_1 = \psi_1 - \psi_2$ and $\Psi_2 = \psi_1 + (\Lambda_1/(\Lambda_2)\,\psi_2$. Taking into account (4.37) we first subtract Equation (4.36b) from Equation (4.36a) and then add Equation (4.36b) multiplied by Λ_1/Λ_2 to Equation (4.36a). This yields

$$\Delta\Psi_1 - \left(\Lambda_1 + \Lambda_2 + \frac{\beta}{c}\right)\Psi_1 = 0; \tag{4.38a}$$

$$\Delta\Psi_2 - \frac{\beta}{c}\,\Psi_2 = 0. \tag{4.38b}$$

Thus, we have replaced the system of equations (4.36) by the two separate equations (4.38). Consider the solutions of Equations (4.38) having the form

$$\Psi_1(r, \theta) = \Phi_1(r) \sin \theta; \quad \Psi_2(r, \theta) = \Phi_2(r) \sin \theta. \tag{4.39}$$

Substituting (4.39) into (4.38) we find

$$\frac{d^2\Phi_1}{dr^2} + \frac{1}{r}\frac{d\Phi_1}{dr} - \left(\Lambda_1 + \Lambda_2 + \frac{\beta}{c} + \frac{1}{r^2}\right)\Phi_1 = 0; \tag{4.40a}$$

$$\frac{d^2\Phi_2}{dr^2} + \frac{1}{r}\frac{d\Phi_2}{dr} - \left(\frac{\beta}{c} + \frac{1}{r^2}\right)\Phi_2 = 0. \tag{4.40b}$$

Equations (4.40) present Bessel's equation. This is known to have an exponentially damped solution for large r on condition that

$$\Lambda_1 + \Lambda_2 + \frac{\beta}{c} > 0; \tag{4.41a}$$

$$\frac{\beta}{c} > 0. \tag{4.41b}$$

Taking into account the definitions of Ψ_1 and Ψ_2 and Equations (4.39) and (4.40) we obtain

$$\psi_1 = \left[D_0 K_1\left(r\sqrt{\frac{\beta}{c}}\right) + D_1 K_1 \times \times \left(r\sqrt{\Lambda_1 + \Lambda_2 + \frac{\beta}{c}}\right)\right] \sin \theta; \tag{4.42a}$$

$$\psi_2 = \left[D_0 K_1\left(r\sqrt{\frac{\beta}{c}}\right) - D_1 \frac{h_1}{h_2} K_1 \times \times \left(r\sqrt{\Lambda_1 + \Lambda_2 + \frac{\beta}{c}}\right)\right] \sin \theta, \tag{4.42b}$$

where h_1 and h_2 are the layer thicknesses in the model and D_0 and D_1 are arbitrary constants. For $D_1 = 0$, solution (4.42) will be called a barotropic soliton and for $D_0 = 0$ and $D_1 \neq 0$ it will be called a purely baroclinic soliton (irrespective of the character of the solution in the region $r < a$). In the general case when $D_0 \neq 0$ and $D_1 \neq 0$ the soliton under consideration is a linear combination of the barotropic and purely baroclinic solitons. If $D_0 \neq 0$, then, according to (4.41b), as in the previous subsection, we have $c > 0$ (we remind the reader that we use the rigid lid approximation, i.e. the first condition (1.36)). However, by virtue of (4.41a), for a purely baroclinic soliton ($D_0 = 0$) the velocity c must lie outside the range $(-\beta(\Lambda_1 + \Lambda_2)^{-1}, 0)$ of the possible phase velocities of the first baroclinic mode (see the dispersion relation (2.13)), i.e. a purely baroclinic soliton can propagate to the west as well.

Further, by analogy with the previous subsection, we must find the solution of

Equation (4.35) which is bounded for $r = 0$ and 'glue' it with solution (4.42) so that the functions ψ_1 and ψ_2 themselves, and their first and second derivatives with respect to r, are continuous at $r = a$. It is possible to perform this construction but the corresponding analysis is very cumbersome and we refer the reader to the paper by Flierl *et al.* (1980) where this solution was found. We only note that the construction of the complete solution shows that the velocity of a purely baroclinic soliton can be directed only to the west and it lies within a very narrow range. For the values $h_1/h_2 = 0.2$ and $(\Lambda_1 + \Lambda_2)^{-1/2} = 45$ km typical of the ocean we have

$$-4.98 \text{ cm/s} < c < -4.05 \text{ cm/s}; \quad 199 \text{ km} < a < 307 \text{ km}. \tag{4.43}$$

Thus, a purely baroclinic soliton can propagate to the west with velocities of the same order as the velocities of motion of synoptic eddies in the ocean. (For a barotropic ocean without a rigid lid we have $c < -g\beta H/f_0^2$; however, $g\beta H/f_0^2 = 2 \times 10^2$ m/s for $H = 5000$ m and $\varphi = 30°$ and, hence, these solitons can propagate to the west only with very large velocities.)

4.5. Exact Solutions; a Continuously Stratified Ocean

We consider the transfer equation (1.39a) for the quasigeostrophic potential vorticity and represent the pressure disturbance in the form

$$p' = f_0\varrho_0\psi(\xi, y, z), \quad \xi = x - ct. \tag{4.44}$$

Substituting (4.44) into the first two equations of (1.33) and Equation (1.39a) we obtain

$$J\left(\psi + cy, \quad \Delta\psi + \frac{\partial}{\partial z}\left(\frac{f_0^2}{N^2}\frac{\partial\psi}{\partial z}\right) + \beta y\right) = 0, \tag{4.45}$$

where Δ and J are given by formulas (4.20b).

First we present the solution obtained by Kizner (1984c). This solution is the sum of the Larichev–Reznik barotropic solution and an axisymmetric baroclinic mode. Consider the equations (cf. (4.21))

$$\Delta\psi + \frac{\partial}{\partial z}\left(\frac{f_0^2}{N^2}\frac{\partial\psi}{\partial z}\right) + \beta y = -k^2(\psi + cy) + CF(z) \quad \text{for } r < a, \tag{4.46a}$$

$$\Delta\psi + \frac{\partial}{\partial z}\left(\frac{f_0^2}{N^2}\frac{\partial\psi}{\partial z}\right) + \beta y = p^2(\psi + cy) \quad \text{for } r > a, \tag{4.46b}$$

where k, p, C, and a are arbitrary constants and $F(z)$ is an arbitrary function of z. It is clear that any function ψ satisfying Equation (4.46) also satisfies Equation (4.45) everywhere except, possibly, on the cylindrical surface $\sqrt{\xi^2 + y^2} = a$.

We introduce a cylindrical coordinate system r, θ, z and seek the solution ψ in the form

$$\psi = \psi_1(r, \theta) + \Phi(r)F(z), \tag{4.47}$$

where ψ_1 is determined by formulas (4.23) and (4.25), p is given by (4.22), and ka and $pa = q$ are related by (4.26); $\Phi(r)$ and $F(z)$ are the functions to be found.

Substituting (4.47) into (4.46) we obtain

$$\frac{d^2\Phi}{dr^2} + \frac{1}{r}\frac{d\Phi}{dr} + \left[\frac{1}{F(z)}\frac{d}{dz}\left(\frac{f_0^2}{N^2}\frac{dF}{dz}\right) + k^2\right]\Phi = C \quad \text{for } r < a, \tag{4.48a}$$

$$\frac{d^2\Phi}{dr^2} + \frac{1}{r}\frac{d\Phi}{dr} + \left[\frac{1}{F(z)}\frac{d}{dz}\left(\frac{f_0^2}{N^2}\frac{dF}{dz}\right) - p^2\right]\Phi = 0 \quad \text{for } r < a, \tag{4.48b}$$

Let $F(z)$ be an eigenfunction of the problem (2.7), (2.8) and let λ be the corresponding eigenvalue of this problem. For $\lambda = 0$ and $F = 1$ we obtain the solution mentioned earlier, which has a discontinuity of the vorticity at $r{=}a$. We show that for $\lambda > 0$ it is possible to construct a solution with continuous vorticity; assuming that $k^2 - \lambda > 0$ we find from (4.48) the expression

$$\Phi = \begin{cases} A_0 J_0(\sqrt{k^2 - \lambda}\, r) + C_0 & \text{for } r < a, \\ B_0 K_0(\sqrt{p^2 - \lambda}\, r) & \text{for } r < a, \end{cases} \tag{4.49}$$

where $C_0 = (k^2 - \lambda)^{-1}C$. Here A_0, B_0 and C_0 are constants; J_0 and K_0 are Bessel and Macdonald functions of the zeroth order, and hence Φ is damped for $r \to \infty$ faster than ψ_1 ($\Phi \sim r^{-1/2}\exp(-\sqrt{p^2 + \lambda}\, r)$).

The continuity conditions for Φ, $d\Phi/dr$, and $d^2\Phi/dr^2$ at $r = a$ make it possible to find two relations expressing B_0 and C_0 in terms of ka, q, $\lambda^{1/2}a$, and A_0 and the relationship between ka, $\lambda^{1/2}a$, and q:

$$-\frac{\sqrt{(ka)^2 - \lambda a^2}\, J_0(\sqrt{(ka)^2 - \lambda a^2})}{J_1(\sqrt{(ka)^2 - \lambda a^2})}$$

$$= \frac{\sqrt{q^2 + \lambda a^2}\, K_0(\sqrt{q^2 + \lambda a^2})}{K_1(\sqrt{q^2 + \lambda a^2})}. \tag{4.50}$$

We shall set parameter a and choose a definite baroclinic mode $F(z)$ and corresponding eigenvalue λ. Then relations (4.26) and (4.50) can be regarded as a system of equations for determining ka and q; this system was investigated in the paper by Kizner (1984c). After ka and q have been determined the function ψ_1 is found from (4.23) and (4.25) and the phase velocity c is calculated by (4.22); further, we find B_0 and C_0 and then determine the function Φ from (4.49). The parameter A_0 in this construction can be chosen arbitrarily. Finally, the desired solution having continuous pressure, velocity, and vorticity on the whole plane ξ, y is given by formula (4.47).

We confine ourselves to a single example. Let $N = 2 \times 10^{-3}\,\text{s}^{-1}$, $f_0 = 10^{-4}\,\text{s}^{-1}$, and $H = 5$ km. Consider the first baroclinic mode for which $\lambda_1^{1/2} = \pi f_0/HN$ and $F_1 = \cos(\pi z/H)$. Then for $\lambda^{1/2}a = 2.7$ ($a = 86$ km) the system of equations (4.26), (4.50) yields $ka = 4.13$ and $q = 2.38$, whence we find $p^{-1} = 36$ km and $c = 2.6$ cm/s ($\beta = 2 \times 10^{-8}\,\text{km}^{-1}\text{s}^{-1}$). This solution is presented in Figure 2.4.7.

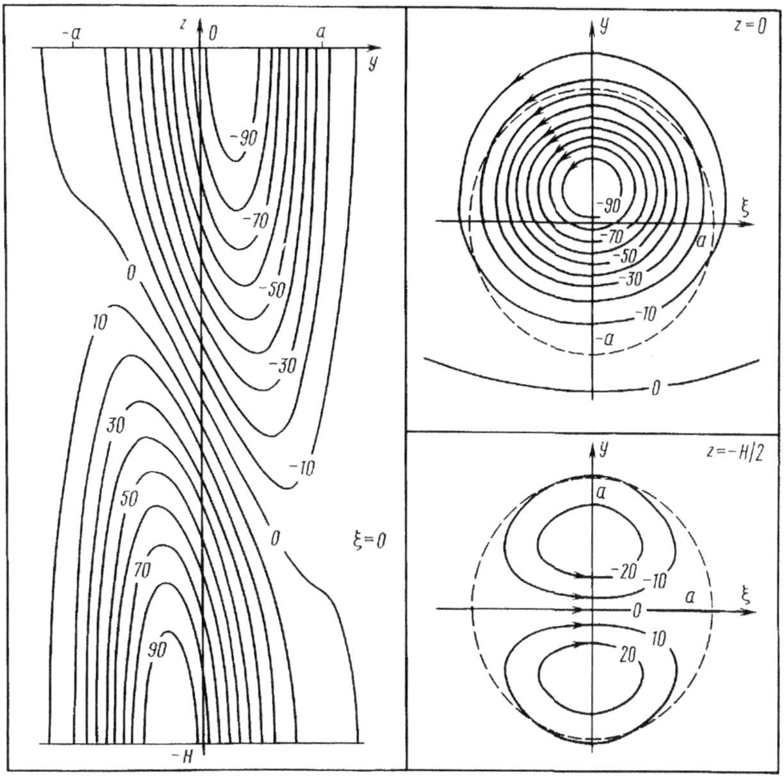

Fig. 2.4.7. Isolines of the stream function ψ (% of maximum value) in the vertical plane $\xi = 0$ and in the horizontal planes $z = 0$ and $z = -H/2$ for the Kizner soliton (4.47), (4.49). The constant $A_0 < 0$ is chosen so that max $\Phi = 3$ max ψ_1. The pattern of streamlines at $z = -H$ differs from that at $z = 0$ only in the direction of the arrows (after Kizner, 1984b).

Now we consider another soliton solution which was first constructed by Berestov (1979, 1981a, 1985). We assume the Brunt–Väisälä frequency to be constant and introduce the spherical coordinate system $\xi = r \sin\theta \cos\varphi$, $y = r \sin\theta \sin\varphi$, $z = (f_0/N) r \cos\theta$. We separate out the spatial region inside a sphere of radius a $(r < a)$ and the region outside the sphere $(r > a)$ and write down the following equations:

$$\Delta\psi + \frac{\partial}{\partial z}\left(\frac{f_0^2}{N^2}\frac{\partial\psi}{\partial z}\right) + \beta y = -k^2(\psi + cy), \quad r < a; \tag{4.51a}$$

$$\Delta\psi + \frac{\partial}{\partial z}\left(\frac{f_0^2}{N^2}\frac{\partial\psi}{\partial z}\right) + \beta y = p^2(\psi + cy), \quad r > a. \tag{4.51b}$$

Any solution of Equation (4.51a) ((4.51b)) satisfies Equation (4.45) in the region $r < a$ $(r > a)$. The condition of spatial localization of the disturbance implies

$$p^2 = \frac{\beta}{c}\,. \tag{4.52}$$

We seek ψ in the form $\psi = \Phi(r) \sin\theta \sin\varphi$; from (4.51) we easily find

$$\psi = \left[\frac{A}{\sqrt{r}} J_{3/2}(kr) - \frac{\beta + ck^2}{k^2} r \right] \sin\theta \sin\varphi, \quad r < a; \tag{4.53a}$$

$$\psi = \frac{B}{\sqrt{r}} K_{3/2}(pr) \sin\theta \sin\varphi, \quad r > a; \tag{4.53b}$$

where A and B are the constants to be determined.

We require that the function ψ and its derivatives $\partial\psi/\partial r$ and $\partial^2\psi/\partial r^2$ be continuous at $r = a$. It can readily be seen that these requirements are equivalent to the conditions

$$\psi + cy \,|_{r=a+0} = 0; \qquad \psi + cy \,|_{r=a-0} = 0;$$

$$\left.\frac{\partial\psi}{\partial r}\right|_{r=a+0} = \left.\frac{\partial\psi}{\partial r}\right|_{r=a-0}. \tag{4.54}$$

From the first two conditions (4.54) we obtain

$$A = \frac{\beta a\sqrt{a}}{k^2 J_{3/2}(ka)}, \qquad B = -\frac{\beta a\sqrt{a}}{p^2 K_{3/2}(pa)}, \tag{4.55}$$

and then the third condition (4.54) implies the following relation for the parameters k, p, and a:

$$\frac{J'_{3/2}(ka)}{J_{3/2}(ka)} = \frac{3}{2ka}\left(1 + \frac{k^2}{p^2}\right) - \frac{k}{p}\frac{K'_{3/2}(pa)}{K_{3/2}(pa)}.$$

Using the well-known relation $J'_{3/2}(z) = -J_{5/2}(z) + (3/2z)J_{3/2}(z)$ and the analogous relation for $K'_{3/2}(z)$, we bring this equation to a more compact form:

$$-\frac{1}{ka}\frac{J_{5/2}(ka)}{J_{3/2}(ka)} = \frac{1}{q}\frac{K_{5/2}(q)}{K_{3/2}(q)}, \quad q = a\sqrt{\frac{\beta}{c}}. \tag{4.56}$$

Since the Bessel functions involved in Equation (4.56) are simply expressed in terms of elementary functions, this equation can be rewritten in the form

$$q^3 + (1 + d)q^2 + 3dq + 3d = 0, \tag{4.57a}$$

where

$$d = -\frac{1}{(1 - ka \cot ka)^{-1} - 3(ka)^{-2}}. \tag{4.57b}$$

It can easily be shown that for the values of ka belonging to the interval $(\gamma_{3/2,\,1}, \gamma_{3/2})$ where $\gamma_{3/2,\,1}$ is the first zero of the function $z^{-3/2}J_{3/2}(z)$ for $z > 0$ and $\gamma_{3/2}$ is the minimum of this function we have $d \leqslant 0$ and, hence, Equation (4.57) has a single positive root q. On finding this root we determine all the parameters of the soliton. The corresponding dispersion curve is presented in Figure 2.4.8.

The function ψ and its first and second derivatives with respect to the spatial coordinates x, y, and z are continuous everywhere, but the third derivative $\partial^3\psi/\partial r^3$

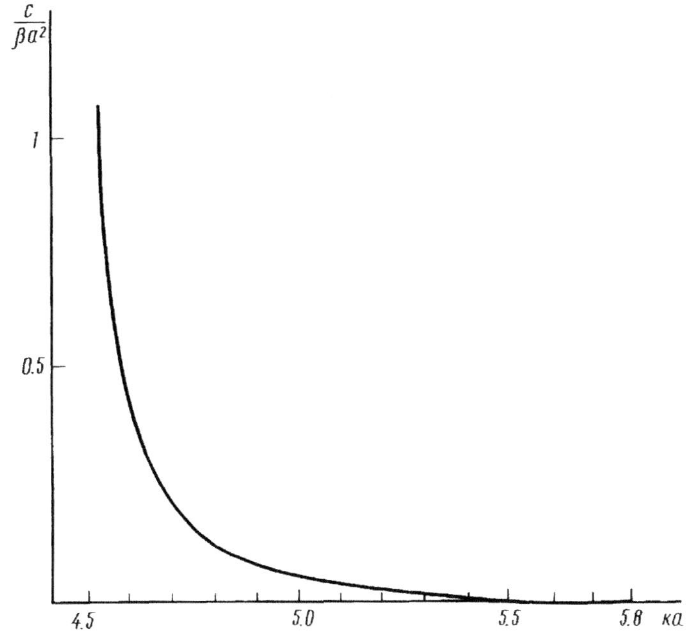

Fig. 2.4.8. Dispersion curve for the Berestov three-dimensional soliton (4.53), (4.55), (4.56).

suffers discontinuity at $r = a$. This means that the pressure, velocity, and vorticity are continuous everywhere but the vorticity gradient has a discontinuity at $r = a$. Since this discontinuity is finite, Equation (4.45) is satisfied throughout the whole space.

The solution we have constructed is represented in Figure 2.4.9. We see that, as in the barotropic case, the soliton consists of two eddies rotating in opposite directions. However, in contrast with the exact solutions considered earlier, which were localized in the plane ξ, y, the solution constructed here is localized in the space ξ, y, z; for large r we have $\psi \sim r^{-1} \exp(-\sqrt{\beta/c}\, r)$. By virtue of (4.52), we have $c > 0$, i.e. this soliton always propagates to the east.

According to Figure 2.4.8, the phase velocity c decreases with increasing dimensions of the soliton; for $ka = \gamma_{3/2}$ we have $c = 0$, i.e. there appears an immovable baroclinic soliton (Berestov, 1981a). It can easily be shown that, as in the barotropic case, the stronger the interaction of the two eddies forming the baroclinic soliton, the greater the velocity of its propagation.

Now we consider the Berestov (1985) 'three-layer' baroclinic soliton:

$$\psi = Ar^{-1/2}I_{3/2}(kr) \sin\theta \sin\varphi \quad \text{for } r < a, \tag{4.58a}$$

$$\omega = \left[Br^{-1/2}J_{3/2}(lr) - \frac{\beta + ck^2}{k^2} r \right] \sin\theta \sin\varphi \quad \text{for } a < r < b, \tag{4.58b}$$

$$\psi = Dr^{-1/2}K_{3/2}(pr) \sin\theta \sin\varphi \quad \text{for } r > b, \tag{4.58c}$$

where, in contrast to the 'three-layer' barotropic soliton (see (4.31)), different

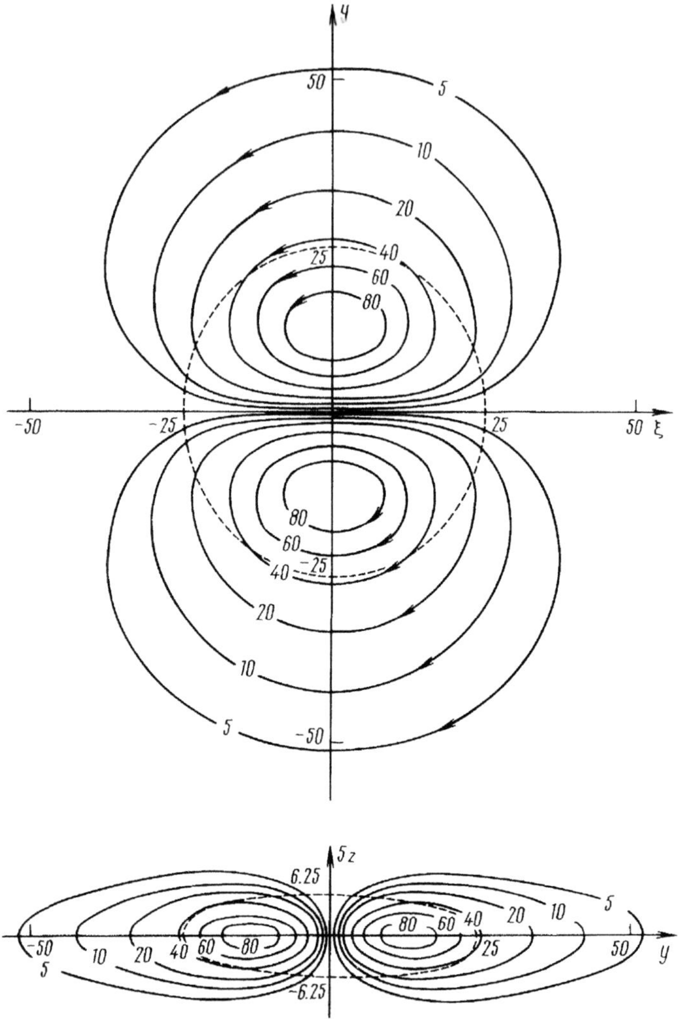

Fig. 2.4.9. Isolines of the stream function ψ (% of maximum value) in the planes ξ, y (above) and z, y (below) for the Berestov three-dimensional soliton (4.53), (4.55), (4.56). Here a = 25 km, $k = 1.8 \times 10^{-4}$ m^{-1}, p = 0.4×10^{-4} m^{-1}, c = 1.3 cm/s, max $\psi = 7 \times 10^2$ m^2/s, $N = 2 \times 10^{-3}$ s^{-1}, $f_0 = 10^{-4}$ s^{-1}; ξ, y, and z are measured in kilometers.

scales k^{-1}, l^{-1}, and p^{-1} are introduced in each of the regions $r < a$, $a < r < b$, and $r > b$. Here A, B, and D are the constants to be determined; the term with the coefficient E (analogous to the corresponding term in the barotropic solution (4.31b)) is not introduced here. The conditions of continuity of ψ, $\partial\psi/\partial r$, and $\partial^2\psi/\partial r^2$ at $r = a$ and $r = b$ yield six relations, three of which serve for determining A, B, and D from the parameters a, b, k, l, and p, and are not listed here. The other three relations yield a system of equations for determining la, lb, and pb for a given parameter ka; these equations have the form

$$-\frac{1}{la}\frac{J_{5/2}(la)}{J_{3/2}(la)} = -\frac{1}{ka}\frac{I_{5/2}(ka)}{I_{3/2}(ka)}, \tag{4.59}$$

$$-\frac{1}{lb}\frac{J_{5/2}(lb)}{J_{3/2}(lb)} = \frac{1}{pb}\frac{K_{5/2}(pb)}{K_{3/2}(pb)}, \tag{4.60}$$

$$\frac{(la)^{3/2}}{J_{3/2}(la)} = \frac{(lb)^{3/2}}{J_{3/2}(lb)}. \tag{4.61}$$

The system of equations (4.59)–(4.61) was analyzed in the paper by Berestov (1985). Figure 2.4.10 presents a dispersion curve for this soliton which was constructed in this paper for values of ka lying in the interval $(\gamma_{5/2,3}, \gamma_{5/2,5})$ where $\gamma_{5/2,n}$ is the nth zero of the function $J_{5/2}(z)$. When $b = a$, the 'three-layer' baroclinic soliton transforms into a 'two-layer' baroclinic soliton. For $qa = \gamma_{5/2,5}$ the phase velocity of the soliton vanishes and we obtain an immovable "three-layer' soliton for which Equation (4.58c) takes the form $\psi = 0$ for $r > b$. By analogy with solutions constructed earlier, the 'three-layer' baroclinic soliton can propagate only to the east.

We now present some estimates. In the coordinates ξ, y, z the sphere $r = a$ transforms into an oblate ellipsoid of revolution whose semiaxes and, consequently, typical horizontal and vertical scales L and L_v of the soliton are in the ratio $L/L_v \sim N/f_0 = 20$ ($N = 2 \times 10^{-3}$ s^{-1}; $f_0 = 10^{-4}$ s^{-1}). Now we place the center of the soliton at $z = -H/2$ and estimate the values of parameters of the soliton for which the vertical velocities at the sea surface ($z = 0$) and at the bottom ($z = -H$)

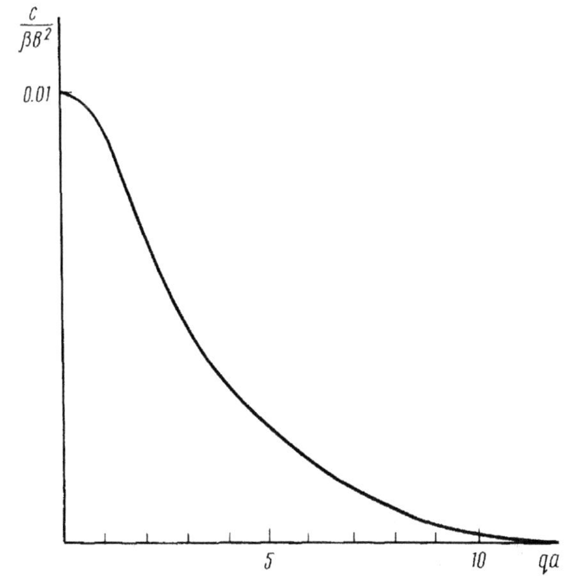

Fig. 2.4.10 Dispersion curve for the Berestov 'three-layer' baroclinic soliton (4.58)–(4.61).

are negligibly small. Taking into account the exponential character of the damping of the solution for large r we put $(N/f_0)p(H/2) = 2$ and $pL = 2$. For $H = 5$ km, the first relation yields $p = (1/25)\ \mathrm{km}^{-1}$ and the second relation gives $L = 50$ km. According to (4.52), the velocity of propagation of this soliton is equal to $c = \beta/p^2 = 1.3$ cm/s ($\beta = 2 \times 10^{-8}\mathrm{km}^{-1}\ \mathrm{s}^{-1}$). For the particle velocity in the soliton we have $U \sim c$ (since $Ki \sim \varepsilon_T$ and $T = L/c$; see (1.7)). Hence, a soliton with these values of parameters satisfies approximately the condition that vertical velocities at the sea surface and at the bottom are equal to zero, i.e. such a soliton can exist in a layer of a stratified fluid of finite thickness.

It is easy to show (Kizner, 1985) that the vertical velocity of solutions (4.53) and (4.58) vanishes in the plane $z = 0$, and hence the upper halves of these solutions can be placed near a plane ocean bottom and the lower halves can be placed near the sea surface. For these solutions the value of p can be decreased twice in comparison with the above estimates; accordingly, L_v and, consequently, L increase twice and c increases four times. In this case the estimates are closer to the estimates of the typical dimensions and velocity of propagation of synoptic eddies.

4.6. Large-Scale Solitons

Up to this point we have considered motions with a typical scale $L \sim L_R$. In conclusion, we mention soliton solutions whose scale substantially exceeds the Rossby deformation radius L_R. These solutions were first constructed by Mikhaĭlova and Shapiro (1980) for a two-layer fluid and by Petviashvili (1980) and Charney and Flierl (1981, §18.4.2) for a single-layer fluid. In the paper by Petviashvili (1980) it was suggested that such a soliton be regarded as a model of the Great Red Spot in the Jovian atmosphere. Later, in connection with the theory of the Great Red Spot, these results were developed by Nezlin (1981), Petviashvili and Yan'kov (1982), and Petviashvili (1983). For a discussion of these solutions see also the paper by Romanova and Tseitlin (1984). The indicated Rossby solitons were observed in laboratory experiments with a thin homogeneous layer of rotating fluid and were subjected to a thorough experimental study (see Antipov *et al.*, 1982, 1983; see also the bibliography in these papers and the paper by Antonova *et al.*, 1983). We note that to describe such Rossby solitons one should introduce some additional terms into the transfer equation for the quasigeostrophic potential vorticity which are important for motions with scales $L \gg L_R$; Flierl (1979) studied some solutions of a similar structure for the quasigeostrophic equations in the presence of a zonal current with a velocity shear.

CHAPTER 3

Theory of Ocean Eddies

1. BAROCLINIC INSTABILITY OF LARGE-SCALE CURRENTS

1.1. Energy Estimates

Observations show that isopycnic surfaces in large-scale currents in the ocean do not deviate strongly from horizontal planes (the average equilibrium positions). This indicates that, generally speaking, not the entire potential energy of the ocean but only the part called the *available potential energy* (a notion introduced by Lorenz, 1955) can be transformed into kinetic energy. Vulis and Monin (1975) used the existing data to estimate the average density of the available potential energy as 7×10^2 J/m^3 whereas the average potential energy density $(1/V)\int g\varrho z \, dV$ is estimated as 2×10^7 J/m^3.

We present a simple approximation formula for the available potential energy required for further estimates. Let a basin with vertical walls be filled with an incompressible stratified fluid and let $\partial\varrho/\partial z < 0$ everywhere (the z-axis is directed vertically upward). We introduce the heights $z = Z(x, y, \varrho, t)$ of isopycnic surfaces counted from a fixed level. Then, since the fluid particles do not intersect the isopycnic surfaces, the incompressibility condition implies

$$\frac{1}{S} \iint_S Z \, dx \, dy = \bar{Z} = \text{const}, \tag{1.1}$$

where S is the cross-section area of the basin at $z = \text{const}$.

The potential energy $\mathcal{E}$ of a fluid of volume V is given by the formula

$$\mathcal{E} = \int_V g\varrho z \, dV + \text{const}. \tag{1.2}$$

We transform formula (1.2) as follows:

$$\mathcal{E}(t) = \iiint g\varrho Z \frac{\partial Z}{\partial \varrho} \, dx \, dy \, d\varrho + \text{const}$$

$$= \iint \left[\frac{1}{2} g\varrho Z^2 \right]_{\varrho_b}^{\varrho_0} dx \, dy + \iint \left[\int_{\varrho_0}^{\varrho_b} \frac{1}{2} g Z^2 \, d\varrho \right] dx \, dy + \text{const}.$$

Let $Z = \bar{Z} + Z'$, where Z' is the displacement of an isopycnic surface relative to its equilibrium position and let ζ be the free surface level. Then

$$\mathcal{E}(t) = P(t) + \text{const}, \tag{1.3}$$

where

$$P = \frac{1}{2}\, g \int_{\varrho_0}^{\varrho_b} \left[\iint Z'^2 \, dx \, dy \right] d\varrho + \frac{1}{2} g\varrho_0 \iint \zeta^2 \, dx \, dy. \tag{1.4}$$

The quantity P(t) is called the *available potential energy*; $P(t) \geqslant 0$ and vanishes only in the undisturbed state. It is this part of the potential energy $\mathcal{E}(t)$ that can be transformed into kinetic energy.

Formula (1.4) holds in the case of an idealized problem; for a real ocean it will be regarded as an approximation formula and will be used for estimates. Consider fluid motion with a typical horizontal scale L. A typical scale for pressure variation $\mathcal{P}$ is found from the geostrophic relation $\mathcal{P} = \varrho_0 f_0 UL$, where U is the typical horizontal velocity. The quasistatic relation implies that the typical scale of density disturbances is $\varrho_* = \mathcal{P}/gH$; a typical scale ζ_0 for the elevation of a free surface is found from the boundary condition for pressure: $\zeta_0 = \mathcal{P}/g\varrho_0$ (cf. Section 1 of Chapter 2).

We estimate the displacement Z' of the isopycnic surface. We have $\varrho(z + Z') \simeq \varrho(z) + (d\varrho/dz)Z'$, whence

$$Z' \sim \frac{\varrho_*}{d\varrho/dz} \sim \frac{g\varrho_*}{\varrho_0 \bar{N}^2}.$$

Now, since $\varrho_b - \varrho_0 \sim (\varrho_0 \bar{N}^2/g)H$ and

$$\frac{g\varrho_0 \iint \zeta^2 \, dx \, dy}{g \int_{\varrho_0}^{\varrho_b} [\iint Z'^2 \, dx \, dy] \, d\varrho} \sim \frac{H\bar{N}^2}{g} \ll 1, \tag{1.5}$$

formula (1.4) implies an important estimate (see Gill *et al.*, 1974):

$$\frac{P}{K} \sim \left(\frac{L}{L_R} \right)^2, \tag{1.6}$$

where K is the resultant kinetic energy and $L_R = H\bar{N}/f_0$ is the Rossby deformation radius.

In the ocean there occurs a complicated interaction between large-scale currents and synoptic eddies. According to the estimate of Vulis and Monin (1975), the average density of the available potential energy is equal to 7×10^2 J/m^3 in the real situation. We mentally separate large-scale motions ($L \sim 1000$ km) and synoptic eddies ($L \sim L_R \sim 50$ km) from each other. Then, according to (1.6), we have $P \sim 400K$ for the large-scale motions and $P \sim K$ for the synoptic eddies. It was indicated in Chapter 1 that the average density of the kinetic energy of large-scale currents is equal to 1 J/m^3, whereas the average density P of the available potential energy of these motions must be of the order of 4×10^2 J/m^3, which is very close to the actual value 7×10^2 J/m^3 (taking into account the approximate character of the estimate). Further, the ratio of the average densities of kinetic energy of large-scale currents and synoptic eddies in the open ocean is equal to $1/20$ (see Wyrtki *et al*, 1976). Therefore, the average density of the available potential energy of synoptic eddies is equal to 20 J/m^3 and that of the resultant energy is equal to 40 J/m^3, i.e. amounts to 10% of 4×10^2 J/m^3.

Hence, we can assume that synoptic eddies in the open ocean obtain their energy from the available potential energy produced mainly by large-scale motions; since 10% of this energy suffices for the observed synoptic eddies, eddy formation does not result in the destruction of large-scale slopes of isopycnic surfaces. Indeed, hydrographic sections clearly show large-scale slopes of isopycnic surfaces with shorter-scale disturbances superimposed on them which seem to be caused by synoptic eddies (e.g. see Emery *et al.*, 1980).

It is useful to note that, according to (1.4), the available potential energy depends on the square of the *displacement* of isopycnic surfaces. At the same time, expressing the horizontal pressure gradient in terms of the slopes of isopycnic surfaces (by virtue of the quasistatic approximation), and using the geostrophic relations, we find that the kinetic energy depends on the square of the *slope* of isopycnic surfaces. Therefore, although the average slope Z'_e/L_e for synoptic eddies is greater than the average slope Z'_c/L_c for large-scale currents, the average displacement Z'_e is less than the average displacement Z'_c owing to the inequality $L_e < L_c$. It is this fact that accounts for the indicated distinctions between the available potential energies of large-scale and synoptic motions.

1.2. Statement of the Baroclinic Instability Problem

We shall show that large-scale currents can be unstable and can generate disturbances whose energy is primarily obtained from the stored available potential energy of these currents. It is this instability – usually called 'baroclinic instability' – that is likely to be one of the basic physical mechanisms causing the transformation of the available potential energy of the large-scale ocean circulation into the energy of synoptic eddies.

Let us consider the simplest zonal current in a channel $0 \leqslant y \leqslant \mathscr{L}_y$ of constant depth:

$$u = \bar{u}(y, z); \; v = 0; \; w = 0; \; p' = \bar{p}(y, z), \; \varrho' = \bar{\varrho}(y, z). \tag{1.7}$$

We can study its stability within the framework of the quasigeostrophic model (Equations (1.33)–(1.35) in Chapter 2). Substituting (1.7) into Equations (1.33)–(1.35) of Chapter 2, we find

$$f_0\bar{u} = -\frac{1}{\varrho_0}\frac{\partial\bar{p}}{\partial y}; \qquad \frac{\partial\bar{p}}{\partial z} = -g\bar{\varrho}. \tag{1.8}$$

We see that for *any* function $u(y, z)$ there are $\bar{p}$ and $\bar{\varrho}$ such that Equation (1.8) is satisfied. In other words, in a channel or in an ocean covering the entire Earth there can exist an arbitrary zonal current requiring no external forces for its maintenance.

Equations (1.8) imply

$$\frac{\partial \bar{u}}{\partial z} = \frac{g}{f_0 \varrho_0} \frac{\partial \bar{\varrho}}{\partial y} \,. \tag{1.9}$$

Let us now study the stability of solution (1.7) relative to disturbances with small amplitudes. Putting

$$u = \bar{u} + u_*; \ v = v_*; \ w = w_*; \ p' = \bar{p} + p_*; \ \varrho' = \bar{\varrho} + \varrho_* \tag{1.10}$$

and substituting (1.10) into Equations (1.33)–(1.35) of Chapter 2 we derive linearized equations for u_*, v_*, w_*, p_*, and ϱ_*:

$$u_* = -\frac{1}{f_0\varrho_0}\frac{\partial p_*}{\partial y}; \quad v_* = \frac{1}{f_0\varrho_0}\frac{\partial p_*}{\partial x}; \quad \varrho_* = -\frac{1}{g}\frac{\partial p_*}{\partial z}; \tag{1.11}$$

$$\left(\frac{\partial}{\partial t} + \bar{u}\frac{\partial}{\partial x}\right)\frac{1}{f_0\varrho_0}\Delta p_* + \left(\frac{1}{f_0\varrho_0}\frac{\partial^3 \bar{p}}{\partial y^3} + \beta\right) v_* - f_0\frac{\partial w_*}{\partial z} = 0; \tag{1.12}$$

$$\left(\frac{\partial}{\partial t} + \bar{u}\frac{\partial}{\partial x}\right)\varrho_* + v_*\frac{\partial \bar{\varrho}}{\partial y} - \frac{\varrho_0 N^2}{g} w_* = 0 \,. \tag{1.13}$$

Let us now consider the boundary conditions. Assuming the ocean depth to be constant, by virtue of the first condition (1.36) and conditions (1.37) and (1.38b) in Chapter 2, we find

$$w_* = 0 \quad \text{for } z = 0; \qquad w_* = 0 \quad \text{for } z = -H; \tag{1.14}$$

$$v_* = 0 \quad \text{for } y = 0; \qquad v_* = 0 \quad \text{for } y = \mathscr{L}_y; \tag{1.15}$$

$$\lim_{X\to\infty} \frac{1}{2X}\int_{-X}^{X} \frac{\partial u}{\partial t}\,dx = 0 \quad \text{for } y = 0 \text{ and } y = \mathscr{L}_y. \tag{1.16}$$

Moreover, we assume that

$$u_*, v_*, w_*, p_*, \text{ and } \varrho_* \text{ are periodic in } x \text{ with a period } \mathscr{L}_x. \tag{1.17}$$

When solving Cauchy's problem we must add initial conditions to the above boundary conditions.

Note that, as follows from (1.7) and (1.8), the choice of the channel width $\mathscr{L}_y$ does not relate to the large-scale current and is in fact quite arbitrary. The disturbance wavelengths along y are naturally determined by the choice of $\mathscr{L}_y$; varying $\mathscr{L}_y$ we can investigate disturbances with arbitrary wavelengths. The consideration of the problem for a channel is convenient because of the simplicity of the boundary conditions; the investigation of the problem for the whole plane is possible but, generally, requires setting additional radiation conditions (see Killworth, 1980). The scale $\mathscr{L}_x$ can also be chosen arbitrarily to investigate disturbances with an arbitrary wavelength along x.

We derive energy equations for the disturbances of a zonal current described by Equations (1.11)–(1.13) for the boundary conditions (1.14)–(1.17). Multiply Equation (3.12) by $(1/f_0)p_*$ and Equation (1.13) by $-(g^2/\varrho_0 N^2)\,\varrho_*$ and add together the results. Using the identity (1.58) in Chapter 2 and an analogous relation for $p_* \Delta\,(\partial p_*/\partial x)$ and also formulas (1.11) we obtain

$$\frac{\partial}{\partial t}\left(\varrho_0 \frac{u_*^2 + v_*^2}{2} + \frac{g^2}{\varrho_0 N^2} \frac{\varrho_*^2}{2}\right)$$

$$= \operatorname{div}\left[\frac{p_*}{\varrho_0 f_0^2} \nabla\left(\frac{\partial p_*}{\partial t}\right) + \frac{\bar{u} p_*}{\varrho_0 f_0^2} \nabla \times \left(\frac{\partial p_*}{\partial x}\right)\right] -$$

$$- \frac{\bar{u}}{2\varrho_0 f_0^2} \frac{\partial}{\partial x} (\nabla p_*)^2 + \frac{1}{2\varrho_0 f_0^2}\left(\frac{1}{f_0 \varrho_0} \frac{\partial^3 \bar{p}}{\partial y^3} + \beta\right) \times$$

$$\times \frac{\partial}{\partial x}(p_*^2) - \frac{g^2 \bar{u}}{2\varrho_0 N^2} \frac{\partial}{\partial x}(\varrho_*^2) - \frac{g^2}{\varrho_0 N^2} \varrho_* v_* \frac{\partial \bar{\varrho}}{\partial y} - \frac{1}{\varrho_0 f_0^2} \times$$

$$\times p_* \frac{\partial^2 p_*}{\partial x\, \partial y} \frac{\partial \bar{u}}{\partial y} - \frac{\partial}{\partial z}(p_* w_*). \tag{1.18}$$

Here and henceforth all the operators ∇, div, and Δ are understood to be plane operators with respect to x and y. We integrate (1.18) over the volume

$$V = \{ - X \leqslant x \leqslant X, 0 \leqslant y \leqslant \mathscr{L}_y, - H \leqslant z \leqslant 0 \}$$

and divide both sides of the resulting relation by $2X$. Taking, for instance, into account that, by virtue of condition (1.15) and the second of Equations (1.11), the function p_* is constant for $y = 0$ and $y = \mathscr{L}_y$ (see Figure 3.1.1), we write

$$\frac{1}{2X} \iint_S \operatorname{div}\left[p_* \nabla\left(\frac{\partial p_*}{\partial t}\right)\right] \mathrm{d}x\, \mathrm{d}y = \frac{1}{2X} \int_\Gamma p_* \frac{\partial^2 p_*}{\partial n\, \partial t}\, \mathrm{d}\Gamma$$

$$= \frac{1}{2X}\left\{ - p_*\big|_{y=0} \int_{\Gamma_1} \frac{\partial^2 p_*}{\partial y \partial t}\, \mathrm{d}x + \int_{\Gamma_2} p_* \frac{\partial^2 p_*}{\partial x\, \partial t}\, \mathrm{d}y \right.$$

$$\left. + p_*\big|_{y=\mathscr{L}_y} \int_{\Gamma_3} \frac{\partial^2 p_*}{\partial y\, \partial t}\, \mathrm{d}x - \int_{\Gamma_4} p_* \frac{\partial^2 p_*}{\partial x\, \partial t}\, \mathrm{d}y \right\},$$

where **n** is the outer normal.

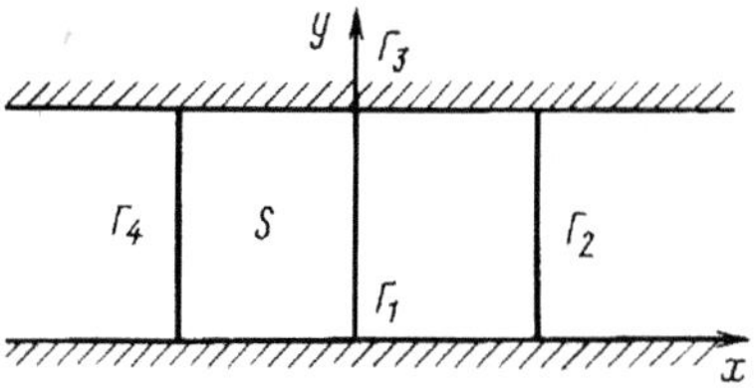

Fig. 3.1.1. To the derivation of Equation (1.19)

Making X tend to infinity and using the first formula (1.11), condition (1.16), and the boundedness of the function p_* for large $|x|$ (condition (1.17)) we find that in the limit the right-hand side of the above relation vanishes. Further,

$$\frac{1}{2X}\iint_S \frac{\partial \bar{u}}{y} p_* \frac{\partial^2 p_*}{\partial x\, \partial y}\, \mathrm{d}x\, \mathrm{d}y$$

$$= \frac{1}{2X}\int_0^{\mathcal{L}_y} \frac{\partial \bar{u}}{\partial y} p_* \frac{\partial p_*}{\partial y}\Bigg|_{x=-X}^{x=X} \mathrm{d}y - \frac{1}{2X}\iint_S \frac{\partial \bar{u}}{\partial y}\frac{\partial p_*}{\partial x}\frac{\partial p_*}{\partial y}\, \mathrm{d}x\, \mathrm{d}y.$$

It is obvious that for a function p_* bounded with respect to x, the first term on the right-hand side tends to zero for $X \to \infty$. The integrals of the other terms on the right-hand side of Equation (1.18) are also transformed in an analogous manner. Finally,

$$\frac{\partial}{\partial t}\iint \left(\varrho_0 \frac{\overline{u_*^2 + v_*^2}}{2} + \frac{g^2}{\varrho_0 N^2}\frac{\overline{\varrho_*^2}}{2} \right) \mathrm{d}y\, \mathrm{d}z$$

$$= \iint (-\varrho_0 \overline{u_* v_*}) \frac{\partial \bar{u}}{\partial y}\, \mathrm{d}y\, \mathrm{d}z + \iint \left(-\frac{g^2}{\varrho_0 N^2}\overline{v_* \varrho_*} \right) \frac{\partial \bar{\varrho}}{\partial y}\, \mathrm{d}y\, \mathrm{d}z, \tag{1.19}$$

where the overbar symbolizes the average value, which is defined for an arbitrary function a as

$$\bar{a} = \lim_{X \to \infty} \frac{1}{2X}\int_{-X}^{X} a\, \mathrm{d}x. \tag{1.20}$$

The integrals in (1.19) are taken over the cross-section area of the channel.

Equation (1.19) shows that in the case of an arbitrary zonal current there are two different energy sources for disturbances. The first source 'works' for $\partial \bar{u}/\partial y \neq 0$. Instability of this type is called *barotropic* since in this case the baroclinicity of the sea water is insignificant; the disturbances develop primarily owing to the kinetic energy of the large-scale current.

The other energy source 'works' for $\partial \bar{\varrho}/\partial y \neq 0$. This type of instability is called *baroclinic*. For this kind of instability the disturbances develop mainly as a result of the available potential energy of a large-scale motion.

We present the following estimate. Using formulas (1.9) and (1.11) and introducing the horizontal scales L and L_c of the eddies and the zonal current, respectively, we obtain

$$\varrho_0 u_* v_* \frac{\partial \bar{u}}{\partial y} \Big/ \frac{g^2}{\varrho_0 N^2} v_* \varrho_* \frac{\partial \bar{\varrho}}{\partial y} \sim \left(\frac{L_R}{L} \right)^2 \frac{L}{L_c}.$$

Assume that $L \sim L_R$ for the developing disturbances. Then if $L_R \ll L_c$, barotropic instability plays a secondary role in comparison with baroclinic instability, and vice versa for $L_R \gg L_c$. Usually $L_R \ll L_c$ for large-scale currents in the open ocean with the exception of narrow-jet currents. Therefore, globally, baroclinic instability processes most probably dominate over barotropic instability processes. This conclusion is also confirmed by the energy estimates presented in Section 1.

It is also easy to derive an equation for the kinetic energy of the disturbances. Multiplying (1.12) by $(1/f_0)p_*$ and repeating, essentially, the derivation of Equation (1.19) we find

$$\frac{\partial}{\partial t}\iint \varrho_0 \overline{\frac{u_*^2 + v_*^2}{2}}\, dy\, dz = \iint (-\varrho_0\overline{u_*v_*})\frac{\partial \bar{u}}{\partial y}\, dy\, dz + \iint (-g\overline{\varrho_* w_*})\, dy\, dz. \tag{1.21}$$

1.3. Physical Mechanism of Baroclinic Instability

Let us now study the physical mechanism of baroclinic instability. To this end assume that the zonal current velocity does not depend on y. Let the current be unstable and let the disturbances u_*, v_*, w_*, p_*, and ϱ_* grow with increasing time. This means that the right-hand sides of Equations (3.19) and (3.21) must be positive, i.e., on average, the quantities $\varrho_* v_* (\partial\bar{\varrho}/\partial y)$ and $\varrho_* w_*$ must be negative (the first terms on the right-hand sides of Equations (1.19) and (1.21) vanish for $\bar{u} = \bar{u}(z)$).

Density disturbances at a given point are caused by particle motion (it can be assumed that the density of a particle does not change during its motion). For definiteness, let $\partial\bar{\varrho}/\partial y > 0$. Consider Figure 3.1.2. If, on average, the product $\varrho_* w_*$ is negative, then a particle, when displaced upward ($w_* > 0$), is lighter than the surrounding medium ($\varrho_* < 0$), and when displaced downward ($w_* < 0$), is heavier than the surrounding medium ($\varrho_* > 0$). Further, on average, the product $\varrho_* v_*$ must also be negative, i.e. when a particle is displaced northward ($v_* > 0$), it is lighter than the surrounding medium ($\varrho_* < 0$), and when displaced southward ($v_* < 0$), is heavier than the surrounding medium ($\varrho_* > 0$). Hence, when the disturbances are increasing (the process of instability is developing), heavier particles must move southward and simultaneously sink, whereas lighter particles must move northward

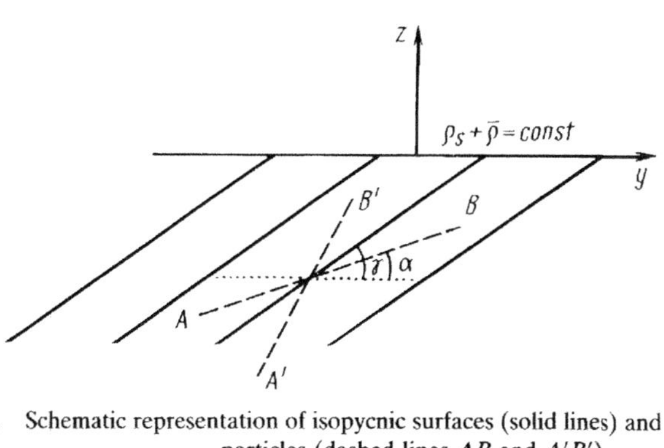

Fig. 3.1.2. Schematic representation of isopycnic surfaces (solid lines) and the possible trajectories of particles (dashed lines AB and $A'B'$).

and rise. Figure 3.1.2 clearly shows that these motions produce a tendency to decrease the stored available potential energy of the zonal current, and that the particles can move only within the limits of the angle γ, i.e.

$$0 < \tan \alpha < \tan \gamma. \tag{1.22}$$

Inequality (1.22) imposes a constraint on the disturbance scale L. Since $\tan\gamma = -(\partial\bar{\varrho}\,\partial y)/(\mathrm{d}\varrho_s/\mathrm{d}z)$, we conclude, by virtue of (1.9), that $\tan\gamma \sim f_0\bar{u}_0/(HN^2)$ where $\bar{u}_0$ is the typical value of the zonal current velocity. Further, for the disturbances we have $\tan\alpha \sim w_*/v_*$. The terms $\bar{u}(\partial/\partial x)(1/f_0\varrho_0)\ \Delta p_*$ and $f_0\,(\partial w_*/\partial z)$ in Equation (1.12) are of the same order and are significant. It follows that $w_*/v_* \sim (H/L)\bar{u}_0/(f_0 L)$, and inequality (1.22) takes the form

$$\left(\frac{L_R}{L}\right)^2 \lesssim 1. \tag{1.23}$$

Inequality (1.23) should be understood in the sense that the scale of the increasing disturbances of a baroclinically unstable current cannot be *much* smaller than the Rossby deformation radius.

1.4. Investigation of the Problem (1.11)–(1.17) by the Method of Normal Modes

Since the coefficients of Equations (1.11)–(1.13) do not depend on x and t, we consider solutions of the form

$$(u_*, v_*, w_*, p_*, \varrho_*) = \mathrm{Re}\,\{(u, v, w, \varrho_0 f_0 p, \varrho)\exp[ik_x(x - ct)]\}, \tag{1.24}$$

where the complex amplitudes u, v, w, p, and ϱ depend only on y and z, and k_x is a real parameter (a wavenumber). These solutions are called normal modes. Without loss of generality, we can assume that k_x is positive; the parameter c can be complex: $c = c_r + ic_i$ (c_r = Re c; c_i = Im c). The amplitude of a normal mode is proportional to exp $(k_x c_i t)$; if $c_i > 0$, the amplitude increases exponentially with time and solution (1.7) is unstable. If $c_i \leqslant 0$, the amplitude is exponentially damped as $t \to \infty$ or remains constant (a neutral oscillation); in this case solution (1.7) is stable with respect to disturbances of the form of (1.24).

The investigation of the stability of solution (1.7) by the method of normal modes consists in finding, for given k_x, *all* values of c for which problem (1.11)–(1.17) has nontrivial solutions of the form of (1.24). Then, if Im $c \leqslant 0$ for all c, the basic current (1.7) is stable with respect to disturbances with wavenumber k_x; however, if Im $c > 0$ for at least one value of c, the basic current (1.7) is unstable with respect to disturbances with wavenumber k_x. Carrying out these investigations for all the possible k_x, we obtain a solution of the problem of the stability of solution (1.7) with respect to arbitrary disturbances. Generally speaking, to investigate the stability completely it would be necessary to solve Cauchy's problem for Equations (1.11)–(1.13) with boundary conditions (1.14)–(1.17) and to consider the evolution of arbitrary initial disturbances with increasing time. When the normal modes form a complete system of functions, the two methods are equivalent, but the method of normal modes is simpler. As a rule, the analysis of the problem by the method of normal modes yields the solution to the problem of stability of a zonal current. However, there are some examples in which, prior to solving the stability problem, it is necessary to solve Cauchy's problem (e.g. see Dikiĭ, 1976, §§3 and 10).

Substituting (1.24) into Equations (1.11)–(1.13) and boundary conditions (1.14) and (1.15), and eliminating w, we obtain, by virtue of formula (1.9),

$$(\bar{u} - c)\left[\frac{\partial^2 p}{\partial y^2} + \frac{\partial}{\partial z}\left(\frac{f_0^2}{N^2}\frac{\partial p}{\partial z}\right) - k_x^2 p\right] + \frac{\partial Q}{\partial y} p = 0; \qquad (1.25)$$

$$(\bar{u} - c)\frac{\partial p}{\partial z} - \frac{\partial \bar{u}}{\partial z} p = 0 \quad \text{for } z = 0 \text{ and } z = -H; \qquad (1.26)$$

$$p = 0 \quad \text{for } y = 0 \text{ and } y = \mathscr{L}_y, \qquad (1.27)$$

where $\partial Q/\partial y$ is the meridional gradient of the quasigeostrophic potential vorticity of the basic current:

$$\frac{\partial Q}{\partial y} = \beta - \frac{\partial^2 \bar{u}}{\partial y^2} - \frac{\partial}{\partial z}\left(\frac{f_0^2}{N^2}\frac{\partial \bar{u}}{\partial z}\right). \qquad (1.28)$$

It is obvious that conditions (1.16) and (1.17) are satisfied identically for solutions of the form of (1.24).

If the basic zonal current is unstable, it is natural to seek k_{max} for which the increment of growth $k_x c_i$ of the disturbances is maximum. Within the framework of the problem under study we can assume that k_{max} determines the scale of developed disturbances under the assumption that it is the disturbances with the wavenumber k_{max} that develop faster than the other disturbances and receive the major part of the energy of the zonal current.

1.5. Necessary Conditions for Instability

Assume that the zonal current is unstable, i.e. there is a nontrivial solution of the problem (1.25)–(1.27) for which $c_i > 0$. Then $\bar{u} - c \neq 0$. We divide Equation (1.25) by $\bar{u} - c$, multiply the resulting equation by the complex conjugate function p^c, and integrate over the channel cross-section area. Taking into account boundary conditions (1.26) and (1.27), and separating the imaginary and real parts of the resulting expression, we find

$$\int_0^{\mathscr{L}_y}\left[\int_{-H}^{0}\frac{\partial Q}{\partial y}\frac{|p|^2}{|\bar{u} - c|^2}\mathrm{d}z\right]\mathrm{d}y - \int_0^{\mathscr{L}_y}\left[\frac{f_0^2}{N^2}\frac{\partial \bar{u}}{\partial z}\frac{|p|^2}{|\bar{u} - c|^2}\right]_{z=-H}\mathrm{d}y +$$

$$+ \int_0^{\mathscr{L}_y}\left[\frac{f_0^2}{N^2}\frac{\partial \bar{u}}{\partial z}\frac{|p|^2}{|\bar{u} - c|^2}\right]_{z=0}\mathrm{d}y = 0; \qquad (1.29)$$

$$\int_0^{\mathcal{L}_y}\left[\int_{-H}^{0}\bar{u}\frac{\partial Q}{\delta y}\frac{|p|^2}{|\bar{u}-c|^2}\,\mathrm{d}z\right]\mathrm{d}y-\int_0^{\mathcal{L}_y}\left[\frac{f_0^2}{N^2}\bar{u}\frac{\partial\bar{u}}{\partial z}\frac{|p|^2}{|u-c|^2}\right]_{z=-H}\mathrm{d}y\;+$$

$$+\int_0^{\mathcal{L}_y}\left[\frac{f_0^2}{N^2}\bar{u}\frac{\partial\bar{u}}{\partial z}\frac{|p|^2}{|\bar{u}-c|^2}\right]_{z=0}\mathrm{d}y=$$

$$=\int_0^{\mathcal{L}_y}\left\{\int_{-H}^{0}\left[\left|\frac{\partial p}{\partial y}\right|^2+\frac{f_0^2}{N^2}\left|\frac{\partial p}{\partial z}\right|^2+k_x^2\,|p|^2\right]\mathrm{d}z\right\}\mathrm{d}y. \qquad (1.30)$$

According to (1.29), at least one of the following three conditions holds:

(a) The meridional gradient of the quasigeostrophic potential vorticity $\partial Q/\partial y$ changes sign in the meridional cross-section $x = \text{const}$ $(0 \leqslant y \leqslant \mathcal{L}_y,\ -H \leqslant z \leqslant 0)$ of the channel. If this is not the case, then either

(b) there is a part of the ocean bottom such that the expression $(\partial\bar{u}/\partial z)_{-H}$, which is proportional to the slope of isopycnic surfaces at the bottom (see (1.9)), has the same sign as $\partial Q/\partial y$; or

(c) there is a part of the sea surface such that the expression $(\partial\bar{u}/\partial z)_0$, which is proportional to the slope of isopycnic surfaces at the sea surface (see (1.9)), and the derivative $\partial Q/\partial y$ have opposite signs.

Since the right-hand side of (1.30) is sure to be positive, at least one of the following three conditions must also hold:

(d) $\bar{u}(\partial Q/\partial y) > 0$ somewhere in the region $0 \leqslant y \leqslant \mathcal{L}_y$, $-H \leqslant z \leqslant 0$. If this is not the case then either

(e) there is a part of the bottom such that $[\bar{u}(\partial\bar{u}/\partial z)]_{-H} < 0$; or

(f) there is a part of the sea surface such that $[\bar{u}(\partial\bar{u}/\partial z)]_0 > 0$.

These are necessary conditions for instability. If none of the conditions of the group (a), (b), and (c) or of the group (d), (e), and (f) holds, the basic current is stable. Examples of application of the necessary conditions can be found in the paper by Gill, *et al.* (1974). We emphasize that, naturally, the fulfilment of at least one of the conditions of the group (a), (b), and (c) or of the group (d), (e), and (f) does not imply instability of the basic current; to prove instability it is necessary to investigate problem (1.25)–(1.27).

We now derive an important inequality (Pedlosky, 1979, §§5 and 7). Let $c_i > 0$; we perform the change of variables $p = (\bar{u} - c)^{1/2}\chi(y, z)$. Then the following problem is obtained for the function χ:

$$\frac{\partial}{\partial y}\left[(\bar{u}-c)\frac{\partial\chi}{\partial y}\right]+\frac{\partial}{\partial z}\left[\frac{f_0^2}{N^2}(\bar{u}-c)\frac{\partial\chi}{\partial z}\right]+\left\{-\frac{1}{4}(\bar{u}-c)^{-1}\times\right.$$

$$\times\left[\left(\frac{\partial\bar{u}}{\partial y}\right)^2+\frac{f_0^2}{N^2}\left(\frac{\partial\bar{u}}{\partial z}\right)^2\right]+\frac{1}{2}\frac{\partial^2\bar{u}}{\partial y^2}+\frac{\partial}{\partial z}\left(\frac{f_0^2}{N^2}\frac{\partial\bar{u}}{\partial z}\right)-$$

$$\left.-\,k^2(\bar{u}-c)+\frac{\partial Q}{\partial y}\right\}\chi=0, \qquad (1.31)$$

$$(\bar{u} - c)\frac{\partial \chi}{\partial z} - \frac{1}{2}\frac{\partial \bar{u}}{\partial z}\chi = 0 \text{ for } z = 0; -H, \tag{1.32}$$

$$\chi = 0 \text{ for } y = 0; \mathscr{L}_y . \tag{1.33}$$

Multiply (1.31) by the complex conjugate χ^c and integrate over the cross-section area of the channel. Using boundary conditions (1.32) and (1.33) and separating the imaginary part of the resulting expression we find

$$\iint \left[\left|\frac{\partial \chi}{\partial y}\right|^2 + \frac{f_0^2}{N^2}\left|\frac{\partial \chi}{\partial z}\right|^2 + k^2 |\chi|^2 \right] \mathrm{d}y\mathrm{d}z =$$

$$= \iint \left[\left(\frac{\partial \bar{u}}{\partial y}\right)^2 + \frac{f_0^2}{N^2}\left(\frac{\partial \bar{u}}{\partial z}\right)^2 \right] \frac{|\chi|^2}{4|\bar{u} - c|^2} \mathrm{d}y\mathrm{d}z.$$

Increasing the right-hand side of this expression by substituting c_i^2 for $|\bar{u} - c|^2$ and replacing the integrand in the square brackets by its maximum value, and dividing both sides of the resulting relation by $\iint |\chi|^2 \, dy \, dz$, we obtain

$$(2k_x c_i)^2 \leqslant \frac{k_x^2}{R_y^{-2} + R_z^{-2} + k_x^2} \left[\left(\frac{\partial \bar{u}}{\partial y}\right)^2 + \frac{f_0^2}{N^2}\left(\frac{\partial \bar{u}}{\partial z}\right)^2 \right]_{\max}, \tag{1.34}$$

where

$$R_y^{-2} = \frac{\iint |\partial \chi/\partial y|^2 \, dy \, dz}{\iint |\chi|^2 \, dy \, dz}; \quad R_z^{-2} = \frac{\iint (f_0^2/N^2) \, |\partial \chi/\partial z|^2 \, dy \, dz}{\iint |\chi|^2 \, dy \, dz}.$$

The scales R_y and R_z are obviously of the orders of the typical disturbance scale along y and the Rossby deformation radius, respectively.

Inequality (1.34) substantially clarifies the idea of barotropic ($\partial\bar{u}/\partial y \neq 0$) and baroclinic ($\partial\bar{u}/\partial z \neq 0$) sources of energy growth formulated on the basis of Equation (1.19). It is important to note that the increment $k_x c_i$ of growing disturbances is bounded above; given $\bar{u} = \bar{u}(y, z)$, the smaller the wavenumber k_x (or, in other words, the greater the disturbance wavelength), the lower the maximum possible increment of growth of these disturbances.

We consider the baroclinic instability process separately. By virtue of (1.23), the scale L of growing disturbances cannot be much smaller than the Rossby deformation radius L_R; at the same time, according to (1.34), the rate of growth of these disturbances seems to decrease as their length scale increases. Therefore, it is natural to expect that the scale of developed disturbances is of the order of the Rossby deformation radius L_R.

Thus we have drawn important qualitative conclusions for an arbitrary unstable zonal current. First, the baroclinic instability process obtains the required energy from the stored available potential energy of the basic current (which is the process we have sought) and, second, the scale of the generated disturbances is of the order of the Rossby deformation radius (which agrees with observations).

Before passing to a detailed analysis of problem (1.25)–(1.27) we note an interesting property of growing disturbances. Putting $p(y, z) = |p| \exp(i\theta)$ we write, according to (1.24), the expression

$$p_* = \varrho_0 f_0 |p| e^{k_x c_i t} \cos(k_x x - k_x c_r t + \theta). \tag{1.35}$$

Using formulas (1.11) we determine u_*, v_*, and ϱ_* and calculate the right-hand side of (1.19). After some simple transformations we obtain

$$\frac{k_x}{2}\left[\varrho_0 \iint |p|^2 \frac{\partial\theta}{\partial y}\frac{\partial\bar{u}}{\partial y}\,dy\,dz + gf_0 \iint |p|^2 \frac{\partial\theta}{\partial z}\frac{1}{N^2}\frac{\partial\bar{\varrho}}{\partial y}\,dy\,dz\right] e^{2k_x c_i t}.$$

It follows that the inflow of energy to disturbances is possible when the wave phase θ of the pressure disturbance p_* depends on y and z; in this case, barotropic instability 'works' for $\partial\theta/\partial y \neq 0$ and baroclinic instability 'works' for $\partial\theta/\partial z \neq 0$. For instance, let $\partial\theta/\partial y = 0$. Then, for disturbances growing with increasing time we have, on average,

$$\frac{\partial\theta}{\partial z}\frac{\partial\bar{\varrho}}{\partial y} > 0 \tag{1.36a}$$

or, taking into account (1.9),

$$\frac{\partial\theta}{\partial z}\frac{\partial\bar{u}}{\partial z} > 0. \tag{1.36b}$$

For decaying disturbances the sign of the inequalities changes to the opposite. This property makes it possible, in principle, to use the wave structure of an observed disturbance to find whether the disturbance grows or decays from the viewpoint of the development of the process of baroclinic instability (under the natural assumption that the basic current is zonal and the linear approximation applies to the description of real disturbances.). In Sections 1 and 2 of Chapter 5 it is shown that the eddies observed during the 'Polygon-70' and 'POLYMODE' expeditions have these vertical shifts of phase θ which indicate the important role played by the baroclinic instability of large-scale currents in the generation of eddies.

1.6. Two-Layer Model

In the foregoing sections the general properties of growing disturbances were elucidated but the conditions under which such disturbances actually appear were not investigated. To this end it is necessary to analyze the difficult problem (1.25)–(1.27). In this connection it is interesting to investigate the two-layer model which leads to much simpler equations (see Section 1 of Chapter 2). A detailed comparison of the solution for the two-layer model, with some known exact solutions for the continuously stratified model, shows that the former model provides a correct description of the most unstable modes (Gill *et al.*, 1974; Killworth, 1980; Pedlosky, 1979, §7.11).

Thus we consider Equations (1.84) and (1.85) of Chapter 2 for an ocean of constant depth ($b = 0$). Neglecting the effect of friction ($A_z = 0$) and external forces ($\tau = 0$) we study the stability of a current having velocities $\bar{u}_1(y)$ and $\bar{u}_2(y)$ in its upper and lower layers, respectively, and corresponding pressures $\bar{p}_1$ and $\bar{p}_2$. Putting $p_1' = \bar{p}_1 + p_1^*$ and $p_2' = \bar{p}_2 + p_2^*$ we obtain in the ordinary way equations for disturbances p_1^* and p_2^* with small amplitudes:

$$\left(\frac{\partial}{\partial t} + \bar{u}_1 \frac{\partial}{\partial x}\right)\left[\frac{1}{f_0\varrho_0}\Delta p_1^* + \frac{\Lambda_1}{f_0\varrho_0}(p_2^* - p_1^*)\right] + \\ + \frac{1}{f_0\varrho_0}\frac{\mathrm{d}Q_1}{\mathrm{d}y}\frac{\partial p_1^*}{\partial x} = 0; \tag{1.37a}$$

$$\left(\frac{\partial}{\partial t} + \bar{u}_2 \frac{\partial}{\partial x}\right)\left[\frac{1}{f_0\varrho_0}\Delta p_2^* + \frac{\Lambda_2}{f_0\varrho_0}(p_1^* - p_2^*)\right] + \\ + \frac{1}{f_0\varrho_0}\frac{\mathrm{d}Q_2}{\mathrm{d}y}\frac{\partial p_2^*}{\partial x} = 0, \tag{1.37b}$$

where Λ_1 and Λ_2 are given by formulas (1.86) in Chapter 2 and

$$\frac{\mathrm{d}Q_1}{\mathrm{d}y} = \beta - \frac{\mathrm{d}^2\bar{u}_1}{\mathrm{d}y^2} + \Lambda_1(\bar{u}_1 - \bar{u}_2); \tag{1.38a}$$

$$\frac{\mathrm{d}Q_2}{\mathrm{d}y} = \beta - \frac{\mathrm{d}^2\bar{u}_2}{\mathrm{d}y^2} + \Lambda_2(\bar{u}_2 - \bar{u}_1). \tag{1.38b}$$

The velocity disturbances (u_1^*, u_2^*) and (v_1^*, v_2^*) satisfy the geostrophic relations. The boundary conditions follow from the first condition (1.38b) and condition (1.59) of Chapter 2:

$$v_1^*, v_2^* = 0 \quad \text{for } y = 0 \text{ and } y = \mathscr{L}_y, \tag{1.39}$$

$$\lim_{X\to\infty} \frac{1}{2X}\int_{-X}^{X} \frac{\partial u_1^*}{\partial t}\,\mathrm{d}x = 0; \quad \lim_{X\to\infty} \frac{1}{2X}\int_{-X}^{X} \frac{\partial u_2^*}{\partial t}\,\mathrm{d}x = 0 \\ \text{for } y = 0; \mathscr{L}_y. \tag{1.40}$$

Moreover, the functions p_1^* and p_2^* are assumed to be periodic in x with a period $\mathscr{L}_x$ (see condition (1.17)).

We derive an equation for the energy of the disturbances. To this end we multiply Equations (1.37a, b) by $h_1(1/f_0)p_1^*$ and $h_2(1/f_0)p_2^*$, respectively, and add the results. After some simple transformations analogous to those in the derivation of (1.18) we obtain

$$\frac{\partial}{\partial t}\left[h_1\varrho_0 \frac{u_1^{*2} + v_1^{*2}}{2} + h_2\varrho_0 \frac{u_2^{*2} + v_2^{*2}}{2} + \frac{1}{g\delta\varrho}\frac{(p_2^* - p_1^*)^2}{2} \right]$$

$$= \operatorname{div}\left[h_1 \frac{p_1^*}{\varrho_0 f_0^2} \nabla\left(\frac{\partial p_1^*}{\partial t}\right) + h_2 \frac{p_2^*}{\varrho_0 f_0^2} \nabla\left(\frac{\partial p_2^*}{\partial t}\right) + h_1 \frac{\bar{u}_1 p_1^*}{\varrho_0 f_0^2} \nabla\left(\frac{\partial p_1^*}{\partial x}\right) + \right.$$

$$\left. + h_2 \frac{\bar{u}_2 p_2^*}{\varrho_0 f_0^2} \nabla\left(\frac{\partial p_2^*}{\partial x}\right)\right] - h_1 \frac{p_1^*}{\varrho_0 f_0^2}\frac{\partial^2 p_1^*}{\partial x \partial y}\frac{d\bar{u}_1}{dy} - h_2 \frac{p_2^*}{\varrho_0 f_0^2}\frac{\partial^2 p_2^*}{\partial x \partial y}\frac{d\bar{u}_2}{dy} -$$

$$- h_1 \frac{\bar{u}_1}{2\varrho_0 f_0^2}\frac{\partial}{\partial x}(\nabla p_1^*)^2 - h_2 \frac{\bar{u}_2}{2\varrho_0 f_0^2}\frac{\partial}{\partial x}(\nabla p_2^*)^2 + \frac{1}{g\delta\varrho}\left[(\bar{u}_1 - \bar{u}_2)p_1^* \frac{\partial p_2^*}{\partial x} + \right.$$

$$\left. + \bar{u}_2 \frac{\partial}{\partial x}(p_1^* p_2^*) - \bar{u}_1 \frac{\partial}{\partial x}\left(\frac{p_1^{*2}}{2}\right) - \bar{u}_2 \frac{\partial}{\partial x}\left(\frac{p_2^{*2}}{2}\right)\right] +$$

$$+ \frac{h_1}{\varrho_0 f_0^2}\frac{dQ_1}{dy}\frac{\partial}{\partial x}\left(\frac{p_1^{*2}}{2}\right) + \frac{h_2}{\varrho_0 f_0^2}\frac{dQ_2}{dy}\frac{\partial}{\partial x}\left(\frac{p_2^{*2}}{2}\right),$$

where $\delta\varrho = \varrho_2 - \varrho_1$. Operations analogous to those in the derivation of Equation (1.19) yield

$$\frac{\partial}{\partial t}\int_0^{\mathscr{L}_y}\left[h_1\varrho_0 \overline{\frac{u_1^{*2} + v_2^{*2}}{2}} + h_2\varrho_0 \overline{\frac{u_2^{*2} + v_2^{*2}}{2}} + \frac{1}{g\,\delta\varrho}\overline{\frac{(p_2^* - p_1^*)^2}{2}}\right] dy$$

$$= \int_0^{\mathscr{L}_y}\left[- h_1\varrho_0\overline{u_1^* v_1^*}\frac{d\bar{u}_1}{dy} - h_2\varrho_0\overline{u_2^* v_2^*}\frac{d\bar{u}_2}{dy} + \frac{1}{g\delta\varrho}\overline{p_1^*\frac{\partial p_2^*}{\partial x}}(\bar{u}_1 - \bar{u}_2)\right] dy, \tag{1.41}$$

where the averaging operator is determined by formula (1.20).

According to formula (1.42a) of Chapter 2, we have $p_2^* - p_1^* = g\ \delta\varrho\zeta_2^*$, where ζ_2^* is the interface displacement. We see that, as in the case of the problem with continuous stratification, the time variation of the total energy of the disturbances is caused by the action of two energy sources depending on the horizontal variations of velocity $\bar{u}_1$, $\bar{u}_2$ of the basic current (barotropic instability) and on the variations of this velocity along the vertical (baroclinic instability), respectively. The last term on the right-hand side of (1.41), corresponding to baroclinic instability, can be rewritten as

$$\frac{1}{g\delta\varrho}\overline{p_1^*\frac{\partial p_2^*}{\partial x}}(\bar{u}_1 - \bar{u}_2) = \frac{1}{g\delta\varrho}\overline{(p_1^* - p_2^*)\frac{\partial p_2^*}{\partial x}}(\bar{u}_1 - \bar{u}_2)$$

$$= -g\delta\varrho\overline{\zeta_2^* v_2^*}\frac{dZ_2}{dy},$$

where the displacement Z_2 of the interface in the basic current has been introduced and formula (1.42a) of Chapter 2 and the properties of the averaging operator (1.20) have been used.

Further, considering disturbances in the form of normal modes

$$(p_1^*, p_2^*) = \mathrm{Re}\{(p_1, p_2)\exp[ik_x(x - ct)]\}, \tag{1.42}$$

and repeating the derivation of Equation (1.25) and condition (1.27), we find

$$(\bar{u}_1 - c)\left[\frac{\mathrm{d}^2 p_1}{\mathrm{d}y^2} - k_x^2 p_1 - \Lambda_1(p_1 - p_2)\right] + \frac{\mathrm{d}Q_1}{\mathrm{d}y} p_1 = 0; \tag{1.43a}$$

$$(\bar{u}_2 - c)\left[\frac{\mathrm{d}^2 p_2}{\mathrm{d}y^2} - k_x^2 p_2 - \Lambda_2(p_2 - p_1)\right] + \frac{\mathrm{d}Q_2}{\mathrm{d}y} p_2 = 0; \tag{1.43b}$$

$$p_1 = p_2 = 0 \quad \text{for } y = 0 \text{ and } y = \mathcal{L}_y. \tag{1.44}$$

Now the stability problem for the zonal current $\bar{u}_1$, $\bar{u}_2$ can be stated as the problem of finding the eigenvalues c for which the system of ordinary differential equations (1.43) possesses a nontrivial solution (p_1, p_2) satisfying conditions (1.44). This problem is obviously much simpler than problem (1.25)–(1.27), which justifies the consideration of the two-layer model.

We see that the whole theory for a two-layer ocean is constructed by analogy with the case of a continuously stratified ocean. It is useful to note that Equations (1.37) for disturbances in the two-layer model can also be obtained from Equations (1.11)–(1.13) by introducing finite-difference approximations along the vertical of the type of (1.60a) in these equations. This fact reflects the general correspondence between the two-layer and two-level models (see Section 1 in Chapter 2).

Now we derive a necessary condition for instability.

Repeating the derivation of relation (1.29) we find

$$\int_0^{\mathcal{L}_y}\left[h_1 \frac{\mathrm{d}Q_1}{\mathrm{d}y}\frac{|p_1|^2}{|\bar{u}_1 - c|^2} + h_2 \frac{\mathrm{d}Q_2}{\mathrm{d}y}\frac{|p_2|^2}{|\bar{u}_2 - c|^2}\right]\mathrm{d}y = 0. \tag{1.45}$$

We see that if the basic current is unstable then: (1) at least one of the two functions $\mathrm{d}Q_1/\mathrm{d}y$ and $\mathrm{d}Q_2/\mathrm{d}y$ must change sign within the interval $(0, \mathcal{L}_y)$, or (2) if the functions $\mathrm{d}Q_1/\mathrm{d}y$ and $\mathrm{d}Q_2/\mathrm{d}y$ do not vanish anywhere, they must have opposite signs. The latter case is realized for a baroclinically unstable zonal current when $\bar{u}_1$, $\bar{u}_2$ and, consequently, $\mathrm{d}Q_1/\mathrm{d}y$ and $\mathrm{d}Q_2/\mathrm{d}y$ do not depend on y. Since $h_1 < h_2$ for the ocean, we have $\Lambda_1 > \Lambda_2$, and (1.38) and (1.45) immediately imply necessary conditions for instability:

$$U > \beta/\Lambda_2 \quad \text{if } U > 0; \tag{1.46a}$$

$$U < -\beta/\Lambda_1 \quad \text{if } U < 0, \tag{1.46b}$$

where

$$U = \bar{u}_1 - \bar{u}_2. \tag{1.47}$$

Later it will be shown that the instability conditions (1.46a, b) are not only necessary but also sufficient. We note that for $U < 0$ the minimum critical shear β/Λ_1 is less than the minimum critical shear β/Λ_2 for $U > 0$. Usually $|\bar{u}_1| > |\bar{u}_2|$, and therefore we can assume that in a zonal current of westward direction ($U < 0$) the instability develops more easily than in a zonal current of eastward direction ($U > 0$).

Now we consider in detail the baroclinic instability problem for a zonal current $\bar{u}_1$, $\bar{u}_2$ independent of y. In this case problem (1.43), (1.44) is considerably simplified since all the coefficients in Equations (1.43) are constant. We seek the solution of the system of equations (1.43) satisfying conditions (1.44) in the form

$$p_1 = A_1 \sin k_{jy}y;\ p_2 = A_2 \sin k_{jy}y;$$

$$k_{jy} = j\ \frac{\pi}{\mathscr{L}_y},\quad j = 1, 2, \ldots, \tag{1.48}$$

where A_1 and A_2 are constants.

The substitution of (1.48) into (1.43) yields an algebraic linear system of equations for A_1 and A_2:

$$[-(\bar{u}_1 - c)(k^2 + \Lambda_1) + \beta + \Lambda_1 (\bar{u}_1 - \bar{u}_2)] A_1 + \\ + \Lambda_1 (\bar{u}_1 - c) A_2 = 0; \tag{1.49a}$$

$$\Lambda_2 (\bar{u}_2 - c) A_1 + [-(\bar{u}_2 - c)(k^2 + \Lambda_2) + \beta + \\ + \Lambda_2 (\bar{u}_2 - \bar{u}_1)] A_2 = 0, \tag{1.49b}$$

where

$$k^2 = k_x^2 + k_{jy}^2. \tag{1.50}$$

The system (1.49) possesses a nontrivial solution only under the condition that its determinant should vanish. This condition implies a quadratic equation for c whose solutions are

$$c_{1,2} = \bar{u}_2 + \frac{Uk^2 (k^2 + 2\Lambda_2) - \beta (2k^2 + \Lambda_1 + \Lambda_2)}{2k^2 (k^2 + \Lambda_1 + \Lambda_2)} \pm \\ \pm \frac{[\beta^2 (\Lambda_1 + \Lambda_2)^2 + 2\beta U k^4 (\Lambda_1 - \Lambda_2) + k^4 U^2 (k^4 - 4\Lambda_1\Lambda_2)]^{1/2}}{2k^2 (k^2 + \Lambda_1 + \Lambda_2)}, \tag{1.51}$$

where U is determined according to (1.47).

We shall now show that the necessary stability conditions (1.46) are also sufficient conditions. To this end we investigate the radicand D in formula (1.51):

$$D = \beta^2 (\Lambda_1 + \Lambda_2)^2 + 2\beta U k^4 (\Lambda_1 - \Lambda_2) + k^4 U^2 (k^4 - 4\Lambda_1\Lambda_2).$$

We find the regions in the plane k^2, U where $D < 0$ (the instability regions). Putting $D = 0$ we obtain a quadratic equation for the critical value of the velocity shear $U_c(k^2)$ whose solution is written

$$U_c = \frac{\beta k^2 (\Lambda_1 - \Lambda_2) \mp 2\beta (\Lambda_1\Lambda_2)^{1/2} \sqrt{(\Lambda_1 + \Lambda_2)^2 - k^4}}{k^2(4\Lambda_1\Lambda_2 - k^4)}. \tag{1.52}$$

We see that the function $U_c(k^2)$ is real for $k^2 \leqslant \Lambda_1 + \Lambda_2$. Denote by $U_c^{(+)}$ the positive branch of $U_c(k^2)$ and by $U_c^{(-)}$ the negativè branch. These branches do not intersect the axis k^2; for $k^2 < 2\sqrt{\Lambda_1\Lambda_2}$ the two branches are single-valued functions and for $2\sqrt{\Lambda_1\Lambda_2} < k^2 < (\Lambda_1 + \Lambda_2)$ the lower branch is a two-valued function. It is clear that both branches have vertical asymptotes for $k^2 \to 0$ and $k^2 \to 2\sqrt{\Lambda_1\Lambda_2}$. Finally, differentiating the identity $D \equiv 0$ with respect to k^2 and putting $dU_c/dk^2 = 0$ we find that

$$\begin{aligned} \min U_c^{(+)} &= \beta/\Lambda_2 \quad \text{for } k^2 = [\Lambda_2(\Lambda_1 + \Lambda_2)]^{1/2}; \\ \max U_c^{(-)} &= -\beta/\Lambda_1 \quad \text{for } k^2 = [\Lambda_1(\Lambda_1 + \Lambda_2)]^{1/2}. \end{aligned} \tag{1.53}$$

Formulas (1.53) prove the sufficiency of conditions (1.46). The curves $U_c^{(+)}(k^2)$ and $U_c^{(-)}(k^2)$ are called curves of marginal stability; their behavior is demonstrated in Figure 3.1.3. We have $D = 0$ on the curves of marginal stability, and formula (1.51) yields a two-fold multiple root $c_1 = c_2$. In the regions where $D < 0$ the eigenvalue is equal to $c = c_r \pm ic_i$ ($c_i > 0$) and the zonal current $\bar{u}_1$, $\bar{u}_2$ is unstable. In the remaining part of the plane ($D > 0$) formula (1.51) yields two different real roots.

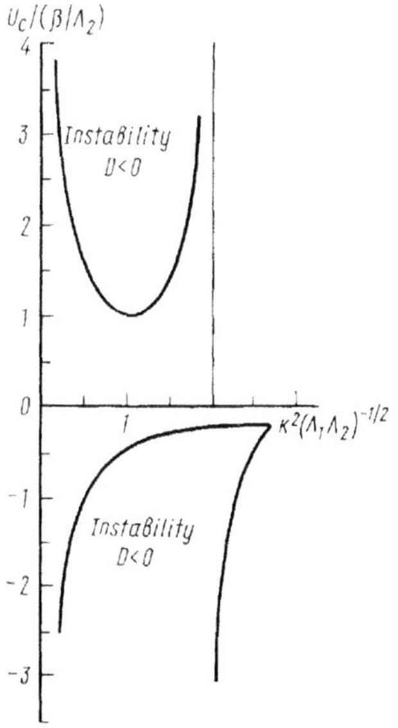

Fig. 3.1.3. Curves of marginal stability $U_c^{(+)}(k^2)$ and $U_c^{(-)}(k^2)$ for $h_1/h_2 = 0.2$ according to (1.52) (see Pedlosky, 1979, Fig. 7.11.3).

We remind the reader that, according to (1.86) and (2.21) in Chapter 2, $(\Lambda_1\Lambda_2)^{1/2}$ is of the order of L_R^{-2}. Indeed, $L_R^2(\Lambda_1\Lambda_2)^{1/2} = (h_1 + h_2)(h_1h_2)^{-1/2}$, and for $h_1 = 1000$ m and $h_2 = 4000$ m this product is equal to 2.5. Therefore, Figure 3.1.3 implies that the typical scales of increasing disturbances are of the order of the Rossby deformation radius; both very short-wave and very long-wave disturbances (in comparison with L_R) are damped with increasing time. It can easily be seen that the phase velocities of growing disturbances are of the same order as the phase velocities of Rossby waves. Hence, we have shown that within the framework of the two layer ocean model the baroclinic instability process generates disturbances with temporal and spatial scales typical of the synoptic eddies observed in the ocean. We note that this result has been established only for disturbances with small amplitudes; however we can expect that the temporal and spatial scales do not change strongly in the process of evolution of these disturbances and their transformation into disturbances with finite amplitudes.

Expressing A_2 in terms of A_1 with the aid of (1.49a) we find

$$A_2 = BA_1, \tag{1.54}$$

where

$$B = \frac{k^2 + \Lambda_1}{\Lambda_1} - \frac{\beta + \Lambda_1 U}{\Lambda_1} \frac{\bar{u}_1 - c_r}{|\bar{u}_1 - c|^2} - i\, c_i \frac{\beta + \Lambda_1 U}{\Lambda_1 |\bar{u}_1 - c|^2} . \tag{1.55}$$

According to (1.42) and (1.48), the pressure disturbances p_1^* and p_2^* can be represented in the form

$$\begin{aligned} p_1^* &= |A_1| \exp(k_x c_i t) \sin k_{jy} y \cos(k_x x - k_x c_r t + \theta_1), \\ p_1^* &= |A_1|\, |B| \exp(k_x c_i t) \sin k_{jy} y \cos(k_x x - k_x c_r t + \theta_1 + \vartheta), \end{aligned} \tag{1.56}$$

where $\theta_1 = \arg A_1$ and $\vartheta = \arg B$ and $|\theta_1| \leqslant \pi$, $|\vartheta| \leqslant \pi$. The expression (1.55) and conditions (1.53) imply that for growing disturbances ($c_i > 0$) we have

$$\begin{aligned} \vartheta &< 0 \text{ for } U > 0, \\ \vartheta &> 0 \text{ for } U < 0. \end{aligned} \tag{1.57}$$

For damped disturbances ($c_i < 0$) the signs of ϑ change to the opposite. It can readily be seen that the vertical phase variations of growing and decaying disturbances in a two-layer fluid have the same character as in the case of a continuously stratified fluid (see (1.36)).

Finally, we calculate the right-hand side of Equation (1.41) for the energy of the disturbances for the case under consideration. We have

$$\frac{1}{g\,\delta\varrho} \int_0^{\mathcal{L}_y} p_1^* \frac{\partial p_2^*}{\partial x} (\bar{u}_1 - \bar{u}_2)\, \mathrm{d}y = - \frac{k_x \mathcal{L}_y |B|\, |A_1|^2}{4g\,\delta\varrho} U \sin\vartheta\, \mathrm{e}^{2k_x c_i t} .$$

Since, according to (1.57), we have $U \sin \vartheta < 0$ in the instability region, as could be expected, the rate of growth of the disturbance energy is positive.

1.7. Stability of a Nonzonal Current

The above stability problem for a zonal current has a model character owing to the choice of the simplest current as a basic current. When studying baroclinic instability processes in a real ocean we should first decide how to construct the basic current whose stability we are going to study. This current must be determined by external forces (the wind stress and thermohaline effects), and (if the concept of the instability of the current is correct) it is its nonlinear interaction with growing disturbances that forms the currents observed in the ocean. However, generally speaking, it is not clear how the basic current can be reconstructed from the actually observed currents since the developed disturbances with finite amplitudes entirely change the character of the basic current. It seems that the basic current can be constructed theoretically by solving the corresponding mathematical problem; however, this problem is extremely difficult. Therefore, a detailed investigation of the stability of the simplest basic currents is an important stage in the development of the theory since it allows us to elucidate the general structure of growing disturbances and the character of those specific features of currents which cause their instability. (See also the discussion in the paper by Nikitin (1977) and the book by Pedlosky (1979, §7.1).)

Now we consider the stability of the simplest nonzonal geostrophic current (see Pedlosky, 1979, §7.13) within the framework of the two-layer model without friction and bottom relief (see (1.84) and (1.85) in Chapter 2). Assume that the curl $[1/(\varrho_0 h_1)]\mathrm{rot}_z \tau$ of the wind stress is constant; then the system of equations (1.84), (1.85) in Chapter 2 possesses the solution

$$\frac{1}{f_0 \varrho_0} p_1' = -\bar{u}_1 y + \bar{v}_1 x; \quad p_2' = 0; \quad \beta \bar{v}_1 = \frac{1}{\varrho_0 h_1} \mathrm{rot}_z \tau, \tag{1.58}$$

where $\bar{u}_1$, $\bar{v}_1$ are constants.

We emphasize once again that only a purely zonal current can exist without being maintained by external forces.

We consider the stability problem for solution (1.58) in the unbounded plane since, as we assume, typical disturbance scales are of the order of the Rossby deformation radius and, therefore, far from the coasts we can regard the ocean as being unbounded and the velocities of the basic current as being constant.

We seek the perturbed solution in the form

$$\frac{1}{f_0 \varrho_0} p_1' = -\bar{u}_1 y + \bar{v}_1 x + \mathrm{Re}\,\{A_1 \exp[i\,(k_x x + k_y y - \sigma t)]\}; \tag{1.59a}$$

$$\frac{1}{f_0 \varrho_0} p_2' = \mathrm{Re}\,\{A_2 \exp[i\,(k_x x + k_y y - \sigma t)]\}; \tag{1.59b}$$

where $\bar{u}_1$ and $\bar{v}_1$ are constant quantities; the wavenumbers k_x and k_y are real and the frequency σ can assume complex values. If $\mathrm{Im}\,\sigma > 0$, the disturbance amplitude grows; if $\mathrm{Im}\,\sigma < 0$, the amplitude is damped.

From the equations for disturbances we find

$$[-(U - c)\,(k^2 + \Lambda_1) + \tilde{\beta} + \Lambda_1 U]\,A_1 + \Lambda_1\,(U - c)\,A_2 = 0; \tag{1.60a}$$

$$-\Lambda_2 c A_1 + [c\,(k^2 + \Lambda_2) + \tilde{\beta} - \Lambda_2 U]\,A_2 = 0, \tag{1.60b}$$

where the notation

$$k^2 = k_x^2 + k_y^2; \quad \tilde{\beta} = \frac{k_x \beta}{k}; \quad U = \frac{\bar{u}_1 k_x + \bar{v}_1 k_y}{k}; \quad c = \frac{\sigma}{k} \tag{1.61}$$

has been introduced.

After the introduction of this notation Equations (1.60) completely coincide with Equations (1.49) if β, $\bar{u}_1$, and $\bar{u}_2$ in Equations (1.49) are replaced by $\tilde{\beta}$, U, and 0, respectively.

It is important that the orientation of the wave vector (k_x, k_y) can be chosen so that $k_x = 0$, and thus the stabilization effect caused by the latitudinal variation β of the Coriolis parameter f is eliminated. These disturbances increase with time for *any velocity shear* of the basic current (see (1.53)). Thus the nonzonal geostrophic current under study is *always unstable* and the general conclusions on the spatial and temporal disturbance scales remain valid in the case of a nonzonal current as well.

In conclusion we mention the review by Charney and Flierl (1981, §18.7.1) where some results relating to the investigation of instability of more complicated nonzonal basic currents are presented.

1.8. Some Concluding Remarks

The main conclusion that can be drawn from the results presented in the previous parts of this section is that the real oceanic currents produced by external forces are probably unstable for a rather broad range of conditions. The typical spatial scales of growing disturbances are of the same order as the corresponding scales for synoptic eddies actually observed in the ocean and, therefore, if we also take into account the energy estimates of Section 1.1, we see that baroclinic instability is probably one of the most important processes generating synoptic variability of the ocean. This is also confirmed by the results of processing the experimental data of the 'Polygon-70' and POLYMODE expeditions discussed in Sections 1 and 2 of Chapter 5. However, it should be noted that at present the role of the process of baroclinic instability in the dynamics of the real ocean has not been elucidated completely (owing, for instance, to the difficulty in determining the basic current; see Sections 4.4 and 7 of the present chapter).

Naturally, we have presented only the most general theoretical concepts and considered in detail only the simplest cases. In conclusion, we indicate some reviews containing extensive bibliography for the stability problem: the lectures by Charney (1973), the books by Pedlosky (1979) and Dikiĭ (1976), and the reviews by Kamenkovich and Reznik (1978, §6), Charney and Flierl (1981, §18.7.1), and Hart (1979).

2. GENERATION OF EDDIES BY BOTTOM RELIEF

In this section we discuss briefly some eddy generation mechanisms caused by the effect of a variable ocean depth.

Topographic Rossby waves or low-frequency waves in a rotating fluid with variable depth were discussed in Section 2 of Chapter 2. It is important to note that there exist topographic waves which can appear only in the presence of a bottom relief; when the disturbances of the mean ocean depth disappear, these waves are completely eliminated (e.g. this is the case for the zeroth topographic mode for $\gamma < 0$; see Section 2 in Chapter 2).

Eddies appearing in a flow over bottom irregularities. There is an interesting eddy generation mechanism that appears in deep ocean layers when a nonstationary current flows over an isolated seamount. This was studied by Huppert and Bryan (1976). Let a uniform current with velocity U_0 be generated far from a seamount. We shall consider the initial stage of motion. According to (1.39a) in Chapter 2, each fluid particle retains the initial value of the quasigeostrophic potential vorticity ω_*. Neglecting, for simplicity, the β-effect, assuming that N = const, and taking into account the quasistatics equation (the third of Equations (1.33) in Chapter 2), we write

$$\omega_* = \omega - \frac{gf_0}{\varrho_0 N^2} \frac{\partial \varrho'}{\partial z} , \qquad (2.1)$$

where ω is the relative vorticity ($\omega = [1/(f_0\varrho_0)]\,\Delta p'$), and the z-axis z is directed upward.

It is clear that for particles moving up the seamount the value of ω must decrease (owing to the upwelling of cold water and the increase of $-\partial\varrho'/\partial z$ produced by it); similarly, for particles moving down the seamount the value of ω must increase. As a result, an anticyclonic (cold) eddy must be formed above the seamount and a cyclonic (warm) eddy must be formed downstream behind the seamount. Naturally, the greater the number of particles flowing over the seamount (a part of the particles flows around the seamount), the more intensive these eddies are. Further, the anticyclonic eddy always remains 'linked' with the seamount; the cyclonic eddy can be kept by the anticyclonic eddy or can drift downstream depending on the velocity U_0 of the incident current and the intensities of the two eddies.

This qualitative consideration is confirmed by the results of numerical experiments demonstrating the possibility of the occurrence of both cases (Figure 3.2.1) and also by the approximate analytical theory developed by Huppert and Bryan (1976). The left-hand series of patterns in Figure 3.2.1 shows how a warm eddy gradually drifts by an incident flow, whereas a cold eddy remains almost fixed and its intensity remains almost unchanged. The right-hand series in Figure 3.2.1 shows a different situation – namely, the two eddies are closely linked and there is no tendency to separate the warm eddy from the cold one. We also note that the eddies in the right-hand series of patterns are less intensive than those in the left-hand series; accordingly, as is shown by calculations, the number of particles flowing over the seamount in the left-hand series considerably exceeds that in the right-hand series. A more detailed analysis shows that the distinction between the two cases of flow over a seamount depends not only on the velocity U_0 but also on the sea water stratification (N) and the height of the seamount (h_m) and is determined by the dimensionless ratio Nh_m/U_0 (Nh_m/U_0 = 4 for the left-hand series of patterns and Nh_m/U_0 = 20 for the right-hand series). The values typical of the real ocean are h_m = 500 m and $N = 10^{-3}\text{s}^{-1}$, and since the typical value of velocity is $U_0 < 5$ cm/s, the case demonstrated in the right-hand series of patterns is more characteristic of the real ocean. Probably this mechanism accounts for the energy-level increase near the bottom which was observed during the MODE expedition for a region with a rough bottom (in comparison with a region with a smooth bottom). In conclusion, we note that since the β-effect is not taken into account, this theory seems to apply only to a flow over small-scale irregularities of the bottom relief. For a thorough investigation of various models of formation of

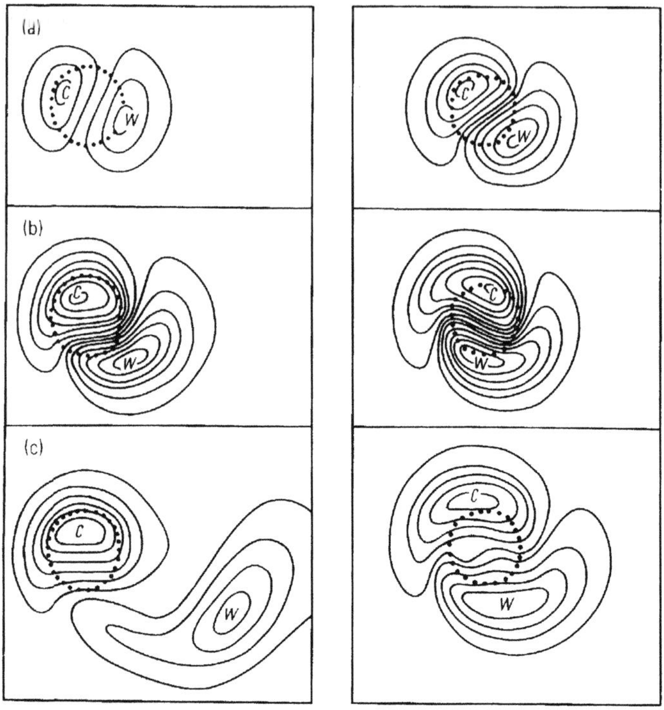

Fig. 3.2.1. Distribution of isopycnics at a depth of 3720 m after the beginning of the development of the current at time intervals (a) 4.6 days, (b) 9.3 days, and (c) 23.1 days (after Huppert and Bryan, 1976, Figs 5 and 8). Left-hand patterns: $U_0 = 5.1$ cm/s, $h_m = 200$ m, and $CI = 16 \times 10^{-6}\ \varrho_0$; right-hand patterns: $U_0 = 1$ cm/s, $h_m = 200$ m, and $CI = 4 \times 10^{-6}\ \varrho_0$. C: the cold anomaly; W: the warm anomaly; CI: the contour intervals at which the ϱ isolines are drawn. The dots show a circle with a radius of 20 km. The total ocean depth is 4 km; the seamount elevation is $h = h_m \exp[-(x^2 + y^2)/L^2]$; $L = 20$ km; $f = 10^{-4}\text{s}^{-1}$; $N = 10^{-3}\ \text{s}^{-1}$.

topographic eddies see the paper by Zyryanov (1981), the monograph by Kozlov (1983), and the bibliography presented in these works.

When a current flows over bottom irregularities, there appear wave disturbances downstream behind the obstacle which were thoroughly investigated by meteorologists. These waves can also be important for the ocean (e.g. see Charney and Flierl, 1981, §18.5.2).

Effect of bottom relief on the stability of currents. In Section 1 we investigated thoroughly the stability problem for a zonal current in an ocean of constant depth. If we assume that there is a bottom depth disturbance $b = b(y)$, then the basic current does not change and the bottom relief effect leads only to a change of the coefficients in the problem for disturbances. For example, in the simplest case of the barotropic problem it is in fact the parameter β that changes (see Equation (1.39b) in Chapter 2), which, as we know, can substantially affect the stability conditions for the basic current (see conditions (1.46) and Pedlosky, 1979, Chapter 7).

Let us now indicate another mechanism of the effect of bottom relief on the stability of the basic current (Charney and Flierl, 1981, §18.7.3). For simplicity, consider the barotropic problem, and let $b = b_0 \sin(kx)$. We introduce the stream function $\psi = [1/(f_0\varrho_0)]p'$ and study the stability of the following basic current:

$$\bar{\psi} = -\bar{u}y + A \sin(kx), \quad \bar{u} = \text{const}. \tag{2.2}$$

Substituting (2.2) into Equation (1.39b) of Chapter 2 we find

$$A = \frac{f_0}{H} \frac{b_0}{k^2 - k_u^2}, \quad k_u^2 = \frac{\beta}{\bar{u}}.$$

Denoting the disturbance stream function by $\psi'(x, y, t)$ we derive in a usual way the equation describing the disturbance evolution:

$$\frac{\partial}{\partial t}\Delta\psi' + J\left(-\frac{\beta}{k_u^2}y + A \sin(kx), \Delta\psi' + k_u^2\psi'\right) = 0, \tag{2.3}$$

where J is the Jacobian with respect to the variables x and y.

Equation (2.3) also describes the evolution of disturbances in the problem of stability of a barotropic Rossby wave (see Gill, 1974 and Lorenz, 1972). The authors of these works studied the triad interaction of disturbance waves leading to the instability of the basic current and the mechanism of shear instability of the basic current; some important instability mechanisms were also studied by Charney and Flierl (1981, §18.7.3). Here we only indicate a specific feature of the effect of the bottom relief on the stability of the basic current. If $b_0 = 0$ (a constant depth), then the basic current $\bar{\psi} = -\bar{u}y$ ($\bar{u}$ = const) is stable for any $\bar{u}$; the wave disturbances turn out to be neutral. When a bottom relief is introduced in the problem ($b_0 \neq 0$), the basic current itself changes and turns out to be stable for $\bar{u} < 0$ (a westward current) and unstable for some $\bar{u} > 0$ (an eastward current). We remind the reader that the analysis of the two-layer problem for an ocean of constant depth implies that in westward zonal currents the instability develops more easily than in eastward currents (see Section 1.6).

3. GENERATION OF EDDIES BY DIRECT FORCING BY THE ATMOSPHERE

3.1. Statement of the Model

We divide the whole ocean into three zones: Ekman boundary layers at the surface of the ocean and at its bottom where the vertical turbulent friction is significant, and the interior region where the friction is small. Accordingly, the functions u, v, w, p', ϱ' are represented in the form

$$u = u_T(x, y, \zeta_T, t) + u_g(x, y, z, t) + u_B(x, y, \zeta_B, t) \tag{3.1}$$

(expressions for the other variables are written in a similar way). Here u_T and u_B are boundary-layer correcting functions at the sea surface and at the bottom; $\zeta_T = -z/D_T$ and $\zeta_B = (z + H)/D_B$ are the 'stretched' boundary-layer variables; D_T and D_B are the thicknesses of the Ekman boundary layers at the surface and the bottom.

We consider quasigeostrophic motions for which the parameters Ki, ε_l, and ε_L are small (conditions (1.7) in Chapter 2). Ignoring the effect of horizontal turbulent friction, we derive equations for the Ekman surface layer:

$$fD_T v_T = \frac{1}{\varrho_0}\frac{\partial R_{xz}}{\partial \zeta_T}; \tag{3.2a}$$

$$-fD_T u_T = \frac{1}{\varrho_0}\frac{\partial R_{yz}}{\partial \zeta_T}; \tag{3.2b}$$

$$-D_T\left(\frac{\partial u_T}{\partial x} + \frac{\partial v_T}{\partial y}\right) + \frac{\partial w_T}{\partial \zeta_T} = 0, \tag{3.2c}$$

with boundary conditions

$$R_{xz} = \tau_x; \quad R_{yz} = \tau_y \quad \text{for} \quad \zeta_T = 0; \tag{3.2d}$$

$$w_T + w_g = \frac{\partial \zeta}{\partial t}; \tag{3.2e}$$

$$p'_g = g\varrho_0\zeta + p'_a \quad \text{for} \quad \zeta_T = 0; \tag{3.2f}$$

$$u_T, v_T, w_T, R_{xz}, R_{yz} \to 0 \quad \text{for} \quad \zeta_T \to \infty, \tag{3.2g}$$

where R_{xz} and R_{yz} are the components of the Reynolds stress tensor, τ_x and τ_y the components of the wind stress, ζ is the sea surface level, and p'_a is the variable part of the atmospheric pressure.

These relations are readily derived by the ordinary methods of boundary-layer theory (e.g. see Kamenkovich, 1977, Ch. 5, §5.3 and Ch. 7, §7.1).

Integrating (3.2c) with respect to ζ_T from 0 to ∞ we express $w_T|_{\zeta_T=0}$ in terms of the integrals $\int_0^\infty u_T\, d\zeta_T$ and $\int_0^\infty v_T\, d\zeta_T$ which, in turn, are found by integrating (3.2a, b) with respect to ζ_T from 0 to ∞, taking into account the boundary conditions (3.2d) and (3.2g). Finally,

$$w_T|_{\zeta_T=0} = -\frac{1}{f_0\varrho_0}\,\mathrm{rot}_z\boldsymbol{\tau}, \tag{3.3}$$

where, in accordance with the quasigeostrophic approximation, the variable Coriolis parameter f is replaced by its average value f_0 at the given latitude.

Equations (3.3) and (3.2e, f) imply the boundary conditions for geostrophic motions at $z = 0$:

$$w_g - \frac{1}{g\varrho_0}\frac{\partial p'_g}{\partial t} = \frac{1}{f_0\varrho_0}\,\mathrm{rot}_z\boldsymbol{\tau} - \frac{1}{g\varrho_0}\frac{\partial p'_a}{\partial t}. \tag{3.4}$$

To derive the boundary conditions at $z = -H$ we consider the Ekman bottom layer (also see Chapter 2, Section 1.5). By virtue of the nonslip conditions at the bottom, $U_B \sim U_g$, where U_B and U_g are typical horizontal velocities in the Ekman bottom layer and in the geostrophic region. The continuity equation implies that the scale of the vertical velocity in the Ekman bottom layer is $W_B \sim (D_B/L)\,U_g$, where L is the horizontal scale of motion. By virtue of formula (1.12a) in Chapter 2, for the scale of the geostrophic vertical velocity we have $W_B/W_g \sim D_B H^{-1} \varepsilon_T^{-1}(L/L_R)^2$. Confining ourselves to motions for which the Ekman number D_B/H is small relative to the parameter $\varepsilon_T(L/L_R)^2$ (in other words, neglecting bottom friction), we find the boundary condition at the bottom:

$$w_g = 0 \quad \text{for} \quad z = -H. \tag{3.5}$$

Neglecting the nonlinear effects and introducing $\psi = [1/(f_0\varrho_0)]p'_g$, according to formulas (1.39a) and (1.35) and the third of Equations (1.33) in Chapter 2, we obtain

$$\frac{\partial}{\partial t}\left(\Delta\psi + \frac{\partial}{\partial z}\frac{f_0^2}{N^2}\frac{\partial\psi}{\partial z}\right) + \beta\frac{\partial\psi}{\partial x} = 0; \tag{3.6}$$

$$w_g = -\frac{f_0}{N^2}\frac{\partial^2\psi}{\partial z\,\partial t}. \tag{3.7}$$

Taking (3.7) into consideration, we reduce the boundary conditions (3.4) and (3.5) to the form

$$\frac{1}{N^2}\frac{\partial^2\psi}{\partial z\,\partial t} + \frac{1}{g}\frac{\partial\psi}{\partial t} = \frac{1}{gf_0\varrho_0}\frac{\partial p'_a}{\partial t} - \frac{1}{\varrho_0 f_0^2}\operatorname{rot}_z\boldsymbol{\tau}$$

$$= \frac{Q}{g} \quad \text{for } z = 0; \tag{3.8a}$$

$$\frac{\partial^2\psi}{\partial z\,\partial t} = 0 \quad \text{for} \quad z = -H. \tag{3.8b}$$

Problem (3.6)–(3.8) describes in linear approximation the generation of quasi-geostrophic motions under the action of wind stress and the time-varying atmospheric pressure; the statement of the problem does not take into account the buoyancy flux at the sea surface (formula (2.2) in Chapter 1) and the variation of the sea surface level owing to moisture exchange with the atmosphere (for this approximation see below).

It is important to note that exactly the same problem for the interior region arises when a more realistic model is used which includes, in addition to the Ekman surface layer, the upper mixed layer and the intermediate layer (Müller, 1982).

We seek the solution of problem (3.6)–(3.8) for the zero initial conditions

$$\psi = 0 \quad \text{for} \quad t = 0, \tag{3.9}$$

in the form

$$\psi = \psi_0 + \int_0^t Q\,\mathrm{d}t; \tag{3.10a}$$

$$\psi_0 = \sum_{n=0}^{\infty}\int \tilde{\psi}_n^0(\mathbf{k}, t)\,\Phi_n(z)\exp[i(k_x x + k_y y)]\,\mathrm{d}\mathbf{k}, \tag{3.10b}$$

where $\Phi_n(z)$ are eigenfunctions of the following Sturm–Liouville problem (cf problem (2.7), (2.8) in Chapter 2):

$$\frac{\partial}{\partial z}\left(\frac{f_0^2}{N^2}\frac{\partial \Phi}{\partial z}\right) + \lambda\Phi = 0; \tag{3.11a}$$

$$\frac{1}{N^2}\frac{\partial \Phi}{\partial z} + \frac{1}{g}\Phi = 0 \text{ for } z = 0;$$

$$\frac{\partial \Phi}{\partial z} = 0 \text{ for } z = -H. \tag{3.11b}$$

The system of eigenfunctions Φ_n ($n = 0, 1, 2, \ldots$) is complete and orthonormal, i.e.

$$\int_{-H}^{0} \Phi_m \Phi_n \, dz = \delta_{mn}, \tag{3.12}$$

where δ_{mn} is the Kronecker delta.

By virtue of conditions (3.11b), an expansion of the form of (3.10) satisfies the boundary conditions (3.8).

Substituting expansion (3.10) into Equation (3.6) we obtain an equation for $\tilde{\psi}_n^0$; coming back to the variables $\tilde{\psi}_n = \tilde{\psi}_n^0 + \int_0^t \tilde{Q}_n \, dt$, where $\tilde{\psi}_n(\mathbf{k}, t)$ and $\tilde{Q}_n(\mathbf{k}, t)$ are the amplitudes in the expansions of the form of (3.10b) of the functions ψ and Q, we write

$$\frac{\partial \tilde{\psi}_n}{\partial t} + i\sigma_n \tilde{\psi}_n = \frac{1}{1 + k^2\lambda_n^{-1}}\tilde{Q}_n = \frac{s_n}{1 + k^2\lambda_n^{-1}}\hat{Q}, \tag{3.13}$$

where k is the modulus of the wave vector $\mathbf{k}$, $k^2 = k_x^2 + k_y^2$; σ_n is the Rossby wave frequency corresponding to the nth mode (see formula (2.9) in Chapter 2); and $\hat{Q}(\mathbf{k}, t)$ is the Fourier transform of the function Q with respect to the coordinates x and y. Since Q does not depend on z, we have $Q_n = s_n \hat{Q}$ and

$$s_n = \int_{-H}^{0} \Phi_n \, dz = \frac{f_0^2}{g\lambda_n}\Phi_n(0). \tag{3.14}$$

Equation (3.13) was derived by Frankignoul and Müller (1979a) in a somewhat different manner. Setting the spectral characteristics of wind stress and atmospheric pressure we can use (3.13) to obtain information about the spectrum of the ocean response ψ, the energy distribution over the vertical modes, etc.

3.2. Characterization of Atmospheric Forcing Fields

We consider the following three kinds of force in the ocean: wind stress; time- and space-varying atmospheric pressure; and buoyancy flux across the ocean–atmosphere boundary owing to evaporation and precipitation processes, absorption and emission of various forms of radiation, and also owing to the variation of the sea surface level produced by the moisture exchange with the atmosphere. Frankignoul and Müller (1979b) showed that the combined effect of all these factors leads to a system of equations of the form of (3.6)–(3.8) in which, however, a modified function Q consisting of three terms is used:

$$Q = Q_\tau + Q_p + Q_M$$

These terms are due to the wind stress, the variation of the atmospheric pressure, and the buoyancy flux, respectively.

Observation data were used to construct model spectra $\tilde{Q}_\tau(\mathbf{k}, \sigma)$, $\tilde{Q}_p(\mathbf{k}, \sigma)$, and $\tilde{Q}_M(\mathbf{k}, \sigma)$ for Q_τ, Q_p, and Q_M analogous to the spectrum (3.20) presented below, and the ratios

$$R_{\tau p}(\mathbf{k}, \sigma) = \frac{\tilde{Q}_p(\mathbf{k}, \sigma)}{\tilde{Q}_\tau(\mathbf{k}, \sigma)}; \quad R_{\tau M}(\mathbf{k}, \sigma) = \frac{\tilde{Q}_M(\mathbf{k}, \sigma)}{\tilde{Q}_\tau(\mathbf{k}, \sigma)}.$$

were considered.

The investigation showed that in the range of the space–time scales of synoptic eddies $R_{\tau p}(k, \sigma) \lesssim R_{\tau M}(k, \sigma) \sim 10^{-3}$, i.e. the effect of atmospheric pressure and buoyancy flux can be neglected. The atmospheric pressure proves significant for substantially greater frequencies and wavelengths, whereas the buoyancy flux is significant for substantially smaller spatial scales. These conclusions are confirmed by the calculations performed by Magaard (1977), who estimates separately the contribution of each of the three factors, and also by the results obtained by Müller (1982), who used a more detailed description of the dynamics of the upper ocean. Thus, in what follows we approximately put

$$Q = -\frac{g}{\varrho_0 f_0^2} \operatorname{rot}_z \boldsymbol{\tau}. \tag{3.15}$$

The determination of wind stress $\boldsymbol{\tau}$ (and, more particularly, of $\operatorname{rot}_z \boldsymbol{\tau}$) is a very complicated problem, particularly for the region above the ocean where the measurements are scarce and, basically, irregular both in space and time. Nevertheless, in recent years a number of works generalizing measurement data obtained mainly over the North Atlantic and the North Pacific have appeared. They give an idea of the space–time characteristics of wind velocity and stress. A detailed review of the existing results can be found in the papers by Frankignoul and Müller (1979a), Willebrand (1978), and Müller (1982).

In middle latitudes the dominant periods and wavelengths in the spectra of atmospheric factors are of the order of several days and from 3000 to 6000 km, respectively. They primarily relate to cyclones and anticyclones moving eastward. However, these spectra are rather broad and possess a substantial energy at lower frequencies and greater wavenumbers, and we describe below some important properties of the spectra for these scales.

At periods greater than 10–20 days the frequency spectra of atmospheric characteristics for middle and high latitudes are approximately 'white' (see Figures 3.3.1 and 3.3.2; a simple physical explanation of this fact was given by Frankignoul and Müller, 1979a). The wavenumber–frequency spectrum $F(\mathbf{k}, \sigma)$ for this frequency range can be approximately written in the form

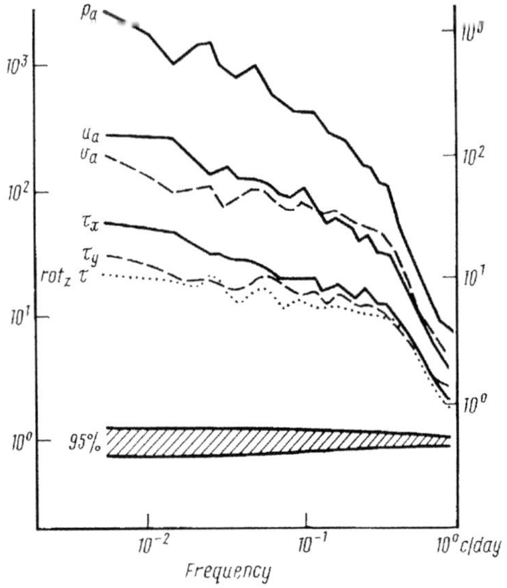

Fig. 3.3.1. Longitude-averaged spectra of atmospheric characteristics over the North Pacific at 43°N (after Willebrand, 1978). p_a, atmospheric pressure; u_a and v_a, zonal and meridional wind-velocity components at the sea surface; τ_x and τ_y wind-stress components. The scales on the left: $(m/s)^2/(c/day)$ and $(10^{-7}\ N/m^3)^2/(c/day)$; the scales on the right: $(mbar)^2/(c/day)$ and $(10^{-1}\ N/m^{-2})/(c/day)$.

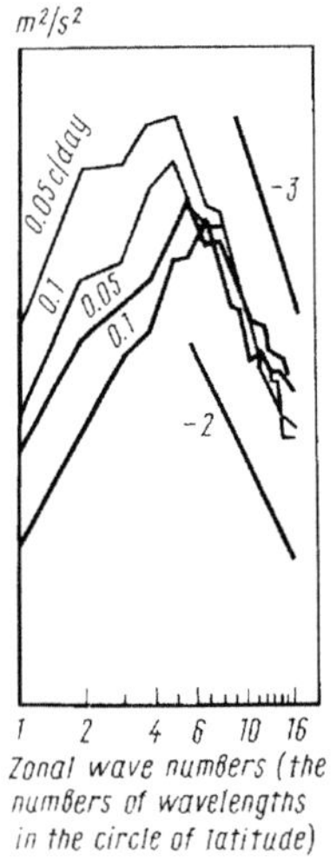

Fig. 3.3.2. Zonal wavenumber spectra of the meridional geostrophic wind for low frequencies at 30°N (thick lines) and 50°N (thin lines) at the 1000 mbar level (after Frankignoul and Müller, 1979a). The line segments indicate spectral slopes.

$$F(\mathbf{k}, \sigma) \approx F(0)\, S(\mathbf{k}), \tag{3.16}$$

where $F(0) = \int F(\mathbf{k}, 0)\, d\mathbf{k}$ is the white noise level; $S(\mathbf{k})$ is the normalized wavenumber spectrum; $\int S(\mathbf{k})\, d\mathbf{k} = 1$.

Since $F(\mathbf{k}, \sigma) = F(-\mathbf{k}, -\sigma)$, relation (3.16) implies that atmospheric spectra are approximately symmetric in the wavenumber space for sufficiently small frequencies (see also Willebrand, 1978). The presence of symmetry in the spectra is very important since it indicates the existence of atmospheric disturbances moving westward which can produce resonant excitation of Rossby waves in the ocean.

Frankignoul and Müller (1979a, Table 2) suggested the following estimate of the white noise level of wind stress based on a comparison of extensive experimental data:

$$F_\tau(0) = 10^4 \mathrm{N}^2(\mathrm{m}^4\ \mathrm{Hz}). \tag{3.17}$$

There exists almost no information about the wavenumber spectrum of wind stress. However, we can assume (see Frankignoul and Müller, 1979a, §6) that this spectrum is similar to the wind velocity spectrum, which has been studied much more thoroughly. In middle latitudes the latter spectrum has a clearly marked maximum at a wavelength $\lambda \sim 5000$ km and decreases somewhat steeper than k^{-2} at the level of 1000 mbar. Typical wind-velocity wavenumber spectra for low frequencies are shown in Figure 3.3.2. It is also important that the wind spectrum can be regarded as being isotropic for $\lambda < 3000$ km.

Thus we assume that the random atmospheric forcing function Q is statistically stationary and homogeneous. Further, we assume that the wind field is isotropic in the range of scales under study, i.e. the spectrum is given by

$$F_{\tau_\alpha \tau_\beta} = F_\tau(0)\, S_\tau(k) \left(\delta_{\alpha\beta} - \frac{k_\alpha k_\beta}{k^2} \right) \tag{3.18a}$$

and

$$S_\tau(k) = \frac{1}{2\pi k} \times \begin{cases} \frac{3}{4} k_b^{-3} k^2 & \text{for} \quad 0 \leqslant k \leqslant k_b\,, \\ \frac{3}{4} k_b k^{-2} & \text{for} \quad k_b \leqslant k \leqslant k_c\,, \\ 0 & \text{for} \quad k > k_c\,, \end{cases} \tag{3.18b}$$

where $\alpha = 1, 2$; $\beta = 1, 2$; $\tau_1 \equiv \tau_x$ and $\tau_2 \equiv \tau_y$; and $\delta_{\alpha\beta}$ is the Kronecker delta.

Formulas (3.18a, b) are in good agreement with the behavior of a real atmospheric spectrum for middle latitudes in its short-wave part. For $k < k_b$ the assumption of isotropy is inadequate. However, here it is not very significant because this range of scales mainly relates to atmospheric disturbances moving eastward and producing no resonant response.

It can easily be shown that, by virtue of the isotropy, the spectrum of $\mathrm{rot}_z \tau$ is related to the spectrum of τ in the following way:

$$F_{\mathrm{rot}}(\mathbf{k}, \sigma) = F_\tau(0)\, k^2 S_\tau(\mathbf{k}). \tag{3.19}$$

Finally, the spectrum of the atmospheric forcing function Q has the form

$$\langle Q(\mathbf{k}, \sigma) Q(\mathbf{k}', \sigma') \rangle = F(\mathbf{k}, \sigma)\delta(\mathbf{k} - \mathbf{k}')\delta(\sigma - \sigma'), \tag{3.20a}$$

$$F(\mathbf{k}, \sigma) = \frac{g^2}{\varrho_0^2 f_0^4}\, k^2 S_\tau(\mathbf{k})\, F_\tau(0)\,, \tag{3.20b}$$

where $\langle \cdot \cdot \cdot \rangle$ denotes averaging over the ensemble and $F_\tau(0)$ is given by (3.17).

Hence, in the range of wavenumbers $k > k_0$ in which we are interested, the spectrum $F(\mathbf{k}, \sigma)$ of the atmospheric forcing function is approximately 'white' in both k and σ.

3.3. The Ocean Response

We solve Equation (3.13) with respect to the normal amplitudes $\tilde{\psi}_n$:

$$\tilde{\psi}_n = \frac{s_n}{1 + k^2\lambda_n^{-1}} \int \hat{Q}(\mathbf{k}, \sigma)\, \Delta\,(\sigma - \sigma_n) \exp\,(-i\sigma_n t)\, \mathrm{d}\sigma. \tag{3.21}$$

Here $\hat{Q}(\mathbf{k}, \sigma)$ is the total Fourier transform of the function Q; $\Delta(\sigma) = [1 - \exp\,(-i\sigma t)]/(i\sigma)$.

Formula (3.21) makes it possible to find spectra of the normal amplitudes for $t \to \infty$:

$$\langle \tilde{\psi}_n\,(\mathbf{k}, t)\, \tilde{\psi}_n^*\,(\mathbf{k}', t)\rangle = 2\pi t\, \frac{s_n^2}{(1 + k^2\lambda_n^{-1})^2}\, F(\mathbf{k}, \sigma_n)\, \delta\,(\mathbf{k} - \mathbf{k}'). \tag{3.22}$$

Here the asterisk (*) designates the complex conjugate, and the relation $\lim_{t\to\infty} \Delta(\sigma)\, \Delta^*(\sigma) = 2\pi t\delta(\sigma)$ has been used.

Equation (3.6) and boundary conditions (3.8) readily provide an expression for the total energy of a water column with unit cross-section area:

$$\bar{E}\,(t) = \frac{1}{2}\,\varrho_0 \int_{-H}^{0} \left[(\nabla\psi)^2 + \frac{f_0^2}{N^2}\left(\frac{\partial\psi}{\partial z}\right)^2 + \frac{f_0^2}{gH}\psi^2\,\big|_{z=0} \right] \mathrm{d}z. \tag{3.23a}$$

The first term in (3.23a) is equal to the kinetic energy, and the sum of the second and third terms is equal to the available potential energy of quasigeostrophic motions. Substituting the expansion of ψ of the form of (3.10b) into (3.23a), using (3.11a, b) and (3.12), and averaging $\bar{E}$ over the ensemble we find, after some simple transformations, the expression

$$\langle \bar{E}\,(t)\rangle = \frac{1}{2}\,\varrho_0 \int \mathrm{d}\mathbf{k} \int_{-H}^{0} \mathrm{d}z \sum_{n=0}^{\infty} \int \langle \tilde{\psi}_n\,(\mathbf{k}, t)\, \tilde{\psi}_n^*\,(\mathbf{k}', t)\rangle \times$$
$$\times\, (\mathbf{k} \cdot \mathbf{k}' + \lambda_n)\, \Phi_n^2\,(z)\, \mathrm{d}\mathbf{k}' = \int \mathrm{d}\mathbf{k} \int_{-H}^{0} \bar{E}\,(\mathbf{k}, z, t)\, \mathrm{d}z, \tag{3.23b}$$

$\bar{E}(\mathbf{k}, z, t)$ is the spectral density of the total energy:

$$\bar{E} = \sum_{n=0}^{\infty} \bar{E}^n\,(\mathbf{k}, z, t);$$
$$\bar{E}^n = \frac{1}{2}\,\varrho_0 \Phi_n^2\,(z) \int \langle \tilde{\psi}_n\,(\mathbf{k}, t)\, \tilde{\psi}_n^*\,(\mathbf{k}', t)\rangle\, (\mathbf{k} \cdot \mathbf{k}' + \lambda_n)\, \mathrm{d}\mathbf{k}', \tag{3.24}$$

where $\tilde{E}^n$ is the contribution of the nth mode to the energy spectrum.

Using (3.22) we find that, for $t \to \infty$,

$$\frac{\partial \tilde{E}}{\partial t} = \sum_{n=0}^{\infty} \frac{\partial \tilde{E}^n}{\partial t} = \pi \varrho_0 \sum_{n=0}^{\infty} \frac{s_n^2 \Phi_n^2 (z)}{\lambda_n^{-1} (1 + k^2 \lambda_n^{-1})} F(\mathbf{k}, \sigma_n). \tag{3.25}$$

Here $\partial \tilde{E}/\partial t$ is the energy flux to the wavenumber $\mathbf{k}$ at depth z produced by the atmospheric forcing function Q. It is seen from (3.25) that for sufficiently long times the major contribution to the energy flux comes from resonant harmonics in the spectrum of the source. Further, the resonant excitation results in a linear increase of the energy spectrum in time. This is due to the stochastic character of the external action[1]; in the case of a deterministic external force with a continuous spectrum, the ocean response remains finite for $t \to \infty$ (see Kamenkovich and Reznik, 1978, §3).

The energy flux to the spectral density $\partial \tilde{E}^n/\partial t$ of the nth mode varies with depth proportionally to $\Phi_n^2 (z)$, i.e. it is determined by the form of the $N(z)$ profile. The plots of the functions $\Phi_n^2 (z)$ for various n in the MODE region are presented in Figure 2.2.1. The figure shows that for the first baroclinic mode the energy of the response is mainly concentrated in the upper 1000 m layer; as the index of the mode increases, the response energy becomes more and still more 'diffused' over the depth.

The energy flux integrated over the depth, taking into account (3.12), is less sensitive to the details of the $N(z)$ profile (and therefore it is more suitable for estimation):

$$\frac{\partial E}{\partial t} = \int_{-H}^{0} \frac{\partial \tilde{E}}{\partial t} \mathrm{d}z = \sum_{n=0}^{\infty} \frac{\partial E^n}{\partial t} = \pi \varrho_0 \sum_{n=0}^{\infty} \frac{s_n^2 F (\mathbf{k}, \sigma_n)}{\lambda_n^{-1} (1 + k^2 \lambda_n^{-1})} . \tag{3.26}$$

Analogous formulas for the spectral density of the kinetic energy can be derived in a completely similar way:

$$\frac{\partial K}{\partial t} = \sum_{n=0}^{\infty} \frac{\partial K^n}{\partial t} = \pi \varrho_0 \sum_{n=0}^{\infty} \frac{s_n^2 k^2 F (\mathbf{k}, \sigma_n)}{(1 + k^2 \lambda_n^{-1})^2} . \tag{3.27}$$

Stratification exerts an effect on the energy fluxes (3.26) and (3.27) by virtue of the parameters $\Phi_n(0)$ and λ_n. Frankignoul and Müller (1979a, Table 1) compared the values of these parameters for the cases of a constant N and real stratification in the MODE region. The constant Brunt–Väisälä frequency N can be chosen so that the values of $\lambda_n^{-1/2}$ differ very slightly (the maximum difference is 20% for $n > 1$). The effect of replacing $N(z)$ by a constant N on $\Phi_n(0)$ is stronger; namely, for $n > 1$ the value of $\Phi_n(0)$ in the case of $N = N(z)$ is approximately twice as great as that in the case of a constant N.

It is natural to split the total energy fluxes (3.26) and (3.27) into barotropic

[1] Similar effects are described in the Phillips theory of generation of wind waves by random fluctuations of atmospheric pressure.

$$\frac{\partial E^{\mathrm{bt}}}{\partial t} = \frac{\partial E^{0}}{\partial t}; \qquad \frac{\partial K^{\mathrm{bt}}}{\partial t} = \frac{\partial K^{0}}{\partial t} \tag{3.28a}$$

and baroclinic

$$\frac{\partial E^{\mathrm{bc}}}{\partial t} = \sum_{n=1}^{\infty} \frac{\partial E^{n}}{\partial t}; \qquad \frac{\partial K^{\mathrm{bc}}}{\partial t} = \sum_{n=1}^{\infty} \frac{\partial K^{n}}{\partial t} \tag{3.28b}$$

parts. Since the quantities $\lambda_0^{-1/2} \simeq \sqrt{gH}/f_0$ and $\Phi_0(0) \simeq 1/H^{-1/2}$ do not significantly depend on N, stratification exerts no effect on the barotropic response. As to the magnitude of the baroclinic response, its dependence on stratification is stronger; for instance, replacing the real frequency $N(z)$ in the MODE region by $N =$ const results in a four-fold underestimation of the baroclinic energy flux.

All the numerical estimates presented below are obtained for the constant frequency $N = N_0 = 2 \times 10^{-3}\mathrm{s}^{-1}$, $H = 5$ km, and $f_0 = 7 \times 10^{-5}\mathrm{s}^{-1}$. It can easily be verified that for $N =$ const the following relations hold:

$$s_0 = \frac{f_0^2}{g\lambda_0 \sqrt{H}}; \quad s_n = \frac{\sqrt{2}\, f_0^2}{g\lambda_n \sqrt{H}}, \quad n \geq 1; \tag{3.29a}$$

$$\lambda_0^{-1/2} = \frac{\sqrt{gH}}{f_0}; \quad \lambda_n^{-1/2} = \frac{HN_0}{\pi n f_0}, \quad n \geq 1. \tag{3.29b}$$

In the frequency range under consideration, $F(\mathbf{k}, \sigma) \simeq F(\mathbf{k}, 0)$; taking into account relations (3.29a, b) it is possible to calculate the sums of the infinite series in (3.26) and (3.27). This results in

$$\frac{\partial E^{bt}}{\partial t} = \frac{\pi \varrho_0 f_0^4 F(\mathbf{k}, 0)}{g^2 H (k^2 + \lambda_0)} = T^{\mathrm{bt}}_{\mathrm{tot}} F(\mathbf{k}, 0); \tag{3.30a}$$

$$\frac{\partial K^{\mathrm{bt}}}{\partial t} = \frac{\pi \varrho_0 f_0^4 k^2 F(\mathbf{k}, 0)}{g^2 H (k^2 + \lambda_0)^2} = T^{\mathrm{bt}}_{\mathrm{kin}} F(\mathbf{k}, 0); \tag{3.30b}$$

$$\frac{\partial E^{\mathrm{bc}}}{\partial t} = \frac{\pi \varrho_0 f_0^4}{g^2 H k^2} (kL_{\mathrm{R}} \coth kL_{\mathrm{R}} - 1) F(\mathbf{k}, 0) = T^{\mathrm{bc}}_{\mathrm{tot}} F(\mathbf{k}, 0); \tag{3.30c}$$

$$\frac{\partial K^{\mathrm{bc}}}{\partial t} = \frac{\pi \varrho_0 f_0^4 L_{\mathrm{R}}^2}{2g^2 H} \left(\coth^2 kL_{\mathrm{R}} + \frac{1}{kL_{\mathrm{R}}} \coth kL_{\mathrm{R}} - 1 - \frac{2}{(kL_{\mathrm{R}})^2} \right) F(\mathbf{k}, 0)$$
$$= T^{bc}_{\mathrm{kin}} F(\mathbf{k}, 0), \tag{3.30d}$$

where $L_{\mathrm{R}} = HN_0/f_0$.

Thus the energy fluxes from the atmosphere to the ocean relate linearly to the spectrum of the atmospheric forcing function $F(\mathbf{k}, 0)$, the proportionality coefficients being equal to the corresponding transfer functions $T^{\mathrm{bt}}(k)$ and $T^{\mathrm{bc}}(k)$. The plots of the transfer functions are shown in Figure 3.3.3. Of course, main specific properties of the behavior of these functions relate to the substantial difference between the values of λ_0 and λ_1; for instance, for the values of H and f_0 indicated above we have $\lambda_0^{-1/2} \simeq 3200$ km and $\lambda_1^{-1/2} \simeq 45$ km. The two barotropic transfer functions $T_{\text{tot}}^{\text{bt}}$ and $T_{\text{kin}}^{\text{bt}}$ are primarily 'concentrated' in the vicinity of $k = \lambda_0^{1/2}$, rapidly decrease as k recedes from this value, and are damped like k^{-2} for $k \to \infty$. The baroclinic functions $T_{\text{tot}}^{\text{bc}}$ and $T_{\text{kin}}^{\text{bc}}$ behave much more 'smoothly' and are damped like k^{-1} as $k \to \infty$. This damping characteristic is due to the fact that in the calculation of $T_{\text{tot}}^{\text{bc}}$ and $T_{\text{kin}}^{\text{bc}}$ all the vertical modes with $n \geqslant 1$ are taken into account. The total transfer functions $T_{\text{tot}} = T_{\text{tot}}^{\text{bt}} + T_{\text{tot}}^{\text{bc}}$ and $T_{\text{kin}} = T_{\text{kin}}^{\text{bt}} + T_{\text{kin}}^{\text{bc}}$ behave approximately like barotropic transfer functions; however, they are dampled like $\mathbf{k}^{-1}$ for $\mathbf{k} \to \infty$ (cf. Kamenkovich and Reznik, 1978, §3).

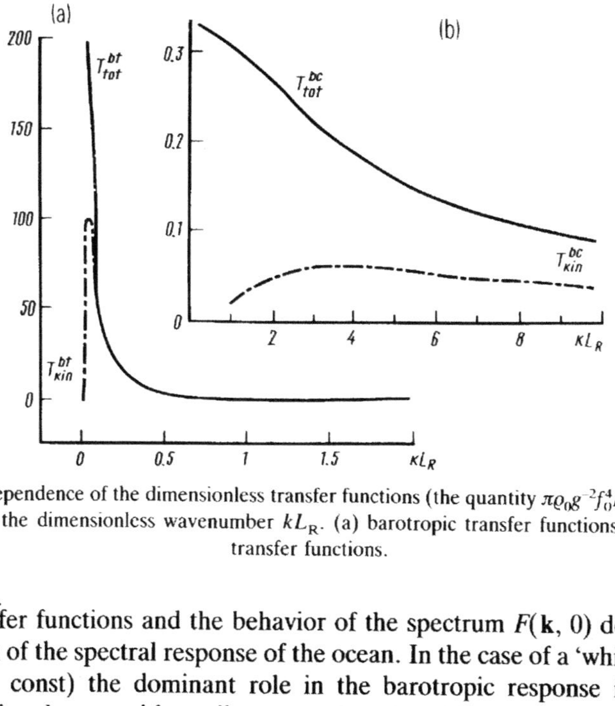

Fig. 3.3.3. Dependence of the dimensionless transfer functions (the quantity $\pi\varrho_0 g^{-2} f_0^4 H^{-1} L_R^2$ is chosen as a scale) on the dimensionless wavenumber kL_R. (a) barotropic transfer functions; (b) baroclinic transfer functions.

The transfer functions and the behavior of the spectrum $F(\mathbf{k}, 0)$ determine the composition of the spectral response of the ocean. In the case of a 'white' spectrum ($F(\mathbf{k}, 0)$ = const) the dominant role in the barotropic response is played by long-wave disturbances with small wavenumbers $k < \lambda_1^{1/2}$, whereas in the baroclinic response (particularly in the kinetic energy spectrum) the major contribution comes from disturbances with shorter wavelengths of the order of $\lambda_1^{-1/2}$. If $F(\mathbf{k}, 0)$ is damped for large k, this exerts no substantial effect on the dominant scale of the barotropic response since, for large k, the barotropic transfer functions are very small in comparison with their values in the neighborhood of $k = 0$. The dominant scale of the baroclinic response has a greater change (it is shifted towards longer wavelengths in comparison with the case $F(\mathbf{k}, 0)$ = const) since the baroclinic transfer functions are damped much more smoothly than barotropic ones. These qualitative conclusions are confirmed by the calculations presented below.

Formulas (3.30) make it possible to derive integral estimates for various energy fluxes. (Such estimates are least sensitive to the choice of model; see also Frankignoul and Müller, 1979a.) We are interested in the range of wavenumbers from $k_m = 2\pi/4000\ \text{km}^{-1}$ to $k_m = 2\pi/50\ \text{km}^{-1}$. First we consider the barotropic response. Taking into account relations (3.17), (3.18), and (3.20) we find from (3.30a, b) the integral fluxes of the total and kinetic energies[2]:

$$\frac{\partial \bar{E}^{\mathrm{bt}}}{\partial t} = \int_{\pi/2}^{3\pi/2} \mathrm{d}\theta \int_{k_m}^{k_M} \frac{\partial E^{\mathrm{bt}}}{\partial t} k\,\mathrm{d}k = 2.90 \times 10^{-4}\ \mathrm{W/m^2}; \tag{3.31a}$$

$$\frac{\partial \bar{K}^{\mathrm{bt}}}{\partial t} = \int_{\pi/2}^{3\pi/2} \mathrm{d}\theta \int_{k_m}^{k_M} \frac{\partial K^{\mathrm{bt}}}{\partial t} k\,\mathrm{d}k = 2.85 \times 10^{-4}\ \mathrm{W/m^2}. \tag{3.31b}$$

Hence, the major part of the total barotropic energy flux ($\approx$ 99%) is expended on an increase of kinetic energy. The energy is mainly transferred to long waves; calculations show that 50% of the energy flux is concentrated in wavelengths > 2000 km and 75% in wavelengths > 1000 km. It is also important to note that integral fluxes are not very sensitive to the form of the spectrum $F(\mathbf{k}, 0)$; for instance, when the law k^{-2} is replaced by k^{-3} for $k > k_b$, the variation in the fluxes is approximately 35%.

Expressions for the baroclinic fluxes are obtained in a similar manner:

$$\frac{\partial \bar{E}^{\mathrm{bc}}}{\partial t} = \int_{\pi/2}^{3\pi/2} \mathrm{d}\theta \int_{k_m}^{k_M} \frac{\partial E^{\mathrm{bc}}}{\partial t} k\,\mathrm{d}k = 1.5 \times 10^{-4}\ \mathrm{W/m^2}; \tag{3.32a}$$

$$\frac{\partial \bar{K}^{\mathrm{bc}}}{\partial t} = \int_{\pi/2}^{3\pi/2} \mathrm{d}\theta \int_{k_m}^{k_M} \frac{\partial K^{\mathrm{bc}}}{\partial t} k\,\mathrm{d}k = 0.5 \times 10^{-4}\ \mathrm{W/m^2}. \tag{3.32b}$$

The estimates (3.32a, b) imply that the major part of the energy flux to baroclinic modes (about 70%) is expended on increasing the available potential energy and only 30% is expended on increasing the kinetic energy. The integral baroclinic response turns out to be sensitive to the choice of model; for instance, the use of the law k^{-3} instead of k^{-2} in expression (3.18) ($S_\tau(\mathbf{k})$) for $k > k_b$ decreases the integral response approximately eight times. This replacement also changes the scale of the baroclinic response; namely, for the law k^{-2} the energy is mainly transferred to wavenumbers greater than or of the order of $\lambda_1^{1/2}$, whereas for the law k^{-3} it is transferred to intermediate wavenumbers ($k_m \leqslant k < \lambda_1^{1/2}$).

3.4. Comparison with Observation Data

Let eddies be generated in a certain ocean region primarily by atmospheric forcing. Then the wavenumber spectrum of the response must increase in the vicinity of small wavenumbers (see, above, consideration of the behavior of transfer functions). Unfortunately, there are no reliable data on the space spectrum of the

[2] Here, and in formulas (3.32a, b), integration is performed taking into consideration the dispersion relation for Rossby waves, and only positive frequencies are considered.

depth-averaged energy of synoptic eddies and therefore at present a comparison of theoretical and experimental spectra is impossible.

A crude estimate of the efficiency of atmospheric forcing can be obtained by comparing the observed eddy energy with theoretical influxes (3.31a, b) and (3.32a, b). (For more detail, see Frankignoul and Müller, 1979a.) It can be assumed that the energy at a depth of 4000 m approximately coincides with the barotropic eddy energy. Then, according to W. J. Schmitz Jr's data (Frankignoul and Müller, 1979a), the barotropic eddy energy varies from a value less than 5×10^2 J/m^2 near 28°N, 55–65°W (the smallest value ever registered in the North Atlantic) to 5×10^3 J/m^2 in the region 28°N, 70°W and further to the north nearer the Gulf Stream. Dividing these values by the corresponding energy influxes we obtain a typical time τ_0 characterizing the efficiency of atmospheric forcing (during time τ_0 the energy of synoptic eddies increases from 0 to the observed value). The smallest value is $\tau_0 \approx 19$ days; since the friction affects long waves comparatively weakly, atmospheric forcing can play a dominant role in the generation of barotropic eddies. In the MODE region and nearer the Gulf Stream τ_0 increases ten times and more; here the role of atmospheric forcing is substantially less significant. The values of τ_0 obtained for baroclinic eddies are much greater. For instance, according to Schmitz' estimate, the kinetic energy of baroclinic disturbances varies from 2.5×10^3 J/m^2 in the vicinity of 28°N, 55°W to 10^4 J/m^2 and more in the MODE region and nearer the Gulf Stream. If these values are divided by the energy flux (3.32b), the smallest value of τ_0 is 600 days at the point 28°N, 55°W, and in the MODE region $\tau_0 = 2400$ days. Thus, the atmospheric action is more effective in the generation of barotropic than baroclinic eddies.

It is also advisable to compare the energy fluxes to synoptic eddies produced by the instability of large-scale circulation with those caused by atmospheric forcing. Gill *et al.* (1974) estimated the flux produced by instability as 10^{-3} W/m^2, which agrees in order of magnitude with (but is slightly smaller than) the values obtained in a number of numerical experiments (Frankignoul and Müller, 1979a, Table 3). The total energy flux produced by atmospheric forcing, by virtue of (3.31a) and (3.32a), is 0.45×10^{-3} W/m^2, i.e. the fluxes under consideration are comparable in magnitude. However, it should be noted that, according to the results of the numerical modelling of synoptic eddies (see Section 4 of the present chapter), the main eddy-formation zones resulting from instability are strongly localized in the vicinity of jet and counter currents. The magnitude of the energy flux to synoptic eddies in these zones substantially exceeds the above estimates.

Thus the direct generation of oceanic eddies by atmospheric forcing seems to play a significant role far away from intensive currents in regions with low eddy activity. Moreover, resonant atmospheric forcing is mainly responsible for barotropic eddies.

The central part of the Pacific is a region of low eddy activity lying far away from intensive currents and topographic irregularities. The pycnocline is very shallow here and the Rossby deformation radius is small ($\lambda_1^{-1/2} \simeq 10$ km). Accordingly, the resonant excitation of baroclinic modes can occur only at periods of the order of several years and more; at shorter periods the baroclinic response of the ocean is nonresonant.

Using Equation (3.13) we find that in the nonresonant case

$$\psi_n(\mathbf{k}, \sigma) = \frac{i}{\sigma - \sigma_n} \frac{s_n}{1 + k^2\lambda_n^{-1}} Q(\mathbf{k}, \sigma), \tag{3.33}$$

where $\psi_n(\mathbf{k}, \sigma)$ and $Q(\mathbf{k}, \sigma)$ are Fourier transforms with respect to time of the functions $\tilde{\psi}_n(\mathbf{k}, t)$ and $\hat{Q}(\mathbf{k}, t)$.

Expression (3.33) implies that the amplitude spectra of the normal modes are

$$\langle \psi_n(\mathbf{k}, \sigma)\, \psi_m^*(\mathbf{k}', \sigma') \rangle = \frac{1}{(\sigma - \sigma_n(\mathbf{k}))\,(\sigma - \sigma_m(\mathbf{k}))} \times$$

$$\times \frac{s_n s_m}{(1 + k^2\lambda_n^{-1})\,(1 + k^2\lambda_m^{-1})} F(\mathbf{k}, \sigma)\, \delta(\mathbf{k} - \mathbf{k}')\, \delta(\sigma - \sigma'). \tag{3.34}$$

According to (3.34), the spectra of the characteristics of synoptic eddies (with the exception of the vertical velocity spectrum) behave like σ^{-2} for $\sigma \gg \sigma_{m,\,n}$. This behavior is in qualitative agreement with the behavior of the frequency spectra of vertical displacements constructed by R. L. Bernstein and W. B. White for the center of the North Pacific (Frankignoul and Müller, 1979a; see also Figure 3.3.4). There also exists a correlation between the oscillations of the energy level of the frequency spectra and the seasonal oscillations of the intensity of the wind field above the North Pacific (see also Müller, 1982). Magaard (1977) showed that wind stress over the central parts of the Pacific can generate long and low-frequency Rossby waves (with wavelengths from 1200 to 1700 km and periods from 1 to 2 yr); they were observed by Emery and Magaard (1976) in this region. Hence, direct atmospheric forcing can be one of the basic mechanisms for generating synoptic disturbances in the central part of the Pacific.

The above estimates relate to middle latitudes (approximately from 15 to 35°). For higher latitudes the decay of the spectrum for large k becomes stronger and the white noise level $F_\tau(0)$ increases approximately three times. Accordingly, the estimates of the energy fluxes also change; namely, the flux to the barotropic mode increases twice and the flux to the baroclinic mode decreases by more than an order of magnitude. Such estimates were not constructed for the tropical zone of the ocean because there are no detailed data on the spectrum of $\mathrm{rot}_z\tau$. It can be assumed (Fedorov *et al.*, 1978) that in this region an important role is played by tropical typhoons and hurricanes which are rather intensive cyclonic formations with a typical scale close to the Rossby deformation radius for the ocean.

If the atmospheric action on the ocean plays a significant role in a certain region, then a correlation between atmospheric and oceanic fields must be observed there. One can assume that the (one-point) correlation is maximum for nonresonant excitation since the Rossby waves forming the resonant response tend to leave the generation region (Willebrand *et al.*, 1980). The existence of this correlation (mainly at periods $\lesssim$ 10 days) was demonstrated in a number of works (e.g. see Ivanov and Novitskiĭ, 1975; Frankignoul and Müller, 1979a, §8; Willebrand *et al.*, 1980).

The role of such important factors as the existence of boundaries, the bottom relief, and nonlinearity was investigated numerically by Willebrand *et al.* (1980)

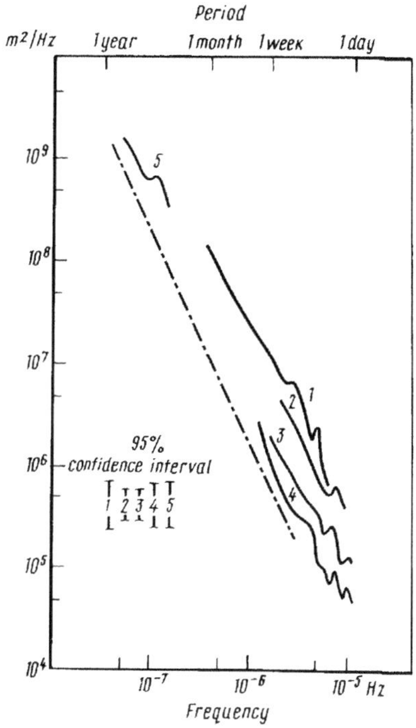

Fig. 3.3.4. Spectra of vertical displacements obtained from measurements at the center of the North Pacific in a region between 30 and 43°N, 165 and 140°W at depths of (1–4) 150 m and (5) 200 m (after Frankignoul and Müller, 1979a). Spectra 1, 2, and 4 were measured in winter, spectrum 3 was measured in summer, and spectrum 5 was obtained on the basis of measurements over a period of 9 years. The dotted line demonstrates the spectrum predicted by theory.

who demonstrated the strong influence of the bottom topography (see also Müller, 1982). The reaction of a bounded ocean to a deterministic atmospheric forcing was also studied by Ivanov and Novitskiĭ (1975 – a rectangular ocean) and Resin, (1980 – a zonal channel).

3.5. The Inclusion of Dissipation

The inclusion of turbulent friction and buoyancy diffusion in the consideration of the geostrophic region makes it possible, in principle, to calculate not only the energy influx rate from the atmosphere to the ocean but the energies of the generated barotropic and baroclinic modes themselves, the relationships between these energies, etc. Here the major difficulty is that the mechanisms of turbulent mixing have scarcely been studied and, in particular, the coefficients of turbulent friction and diffusion are almost unknown. Nevertheless, some specific qualitative features of the generation of synoptic eddies relating to the inclusion of dissipation are of interest, and we shall describe them briefly. (For a more detailed discussion see Reznik, 1983, and Müller, 1982.)

If, for simplicity, only horizontal turbulent exchange is taken into account in the geostrophic region, and if vertical exchange is neglected, the resulting vorticity equation is written as

$$\frac{\partial}{\partial t}\left(\Delta\psi + \frac{\partial}{\partial z}\frac{f_0^2}{N^2}\frac{\partial\psi}{\partial z}\right) + \beta\frac{\partial\psi}{\partial x} -$$

$$- \Delta\left(A_H\Delta\psi + A'_H\frac{\partial}{\partial z}\frac{f_0^2}{N^2}\frac{\partial\psi}{\partial z}\right) = 0, \qquad (3.35a)$$

where A_H and A'_H are the coefficients of horizontal turbulent friction and diffusion, respectively (see Reznik, 1983). The boundary conditions (3.8a, b) take the following form:

$$\frac{1}{N^2}\frac{\partial^2\psi}{\partial z\,\partial t} - \frac{A'_H}{N^2}\frac{\partial\Delta\psi}{\partial z} = \frac{Q}{g} \quad \text{for } z = 0, \qquad (3.35b)$$

$$\frac{\partial^2\psi}{\partial z\,\partial t} - A'_H\frac{\partial\Delta\psi}{\partial z} = 0 \quad \text{for } z = -H. \qquad (3.35c)$$

Here, for the sake of simplicity, the free surface is replaced by a rigid lid.

The solution of problem (3.35) is sought in the form

$$\psi = \sum_{n=0}^{n=\infty}\Phi_n(z)\int\tilde{\psi}_n(\mathbf{k},\sigma)\exp\left[i(k_x\mathbf{x} + k_y y - \sigma t\right]\,d\mathbf{k}\,d\sigma. \qquad (3.36)$$

After some simple transformations, we find

$$\tilde{\psi}_n = \frac{i f_0^2\tilde{Q}(\mathbf{k},\sigma)\,\Phi_n(0)}{g(\mathbf{k}^2 + \lambda_n)(\sigma - \sigma_n + iR_n)}, \qquad (3.37a)$$

where

$$R_n = \frac{k^2(\lambda_n A'_{\mathrm{H}} + k^2A_H)}{k^2 + \lambda_n}, \qquad (3.37b)$$

and $\tilde{Q}(\mathbf{k},\sigma)$ is the Fourier transform of the function $Q(x, y, t)$. Substituting the representation (3.36) for ψ into (3.23a), and averaging over the ensemble, we obtain expressions for the kinetic ($\bar{K}$) and potential ($\bar{P}$) energies:

$$\bar{K} = \sum_{n=0}^{\infty}\bar{K}^n, \qquad \bar{P} = \sum_{n=0}^{\infty}\bar{P}^n, \qquad (3.38a)$$

$$\bar{K}^n = \tfrac{1}{2}\varrho_0\int k^2F_n(\mathbf{k},\sigma)\,d\mathbf{k}\,d\sigma, \qquad \bar{P}^n = \tfrac{1}{2}\varrho_0\int\lambda_n F_n(\mathbf{k},\sigma)\,d\mathbf{k}\,d\sigma, \quad (3.38b)$$

$$F_n(\mathbf{k},\sigma) = \frac{f_0^4\,\Phi_n^2(0)}{g^2(k^2 + \lambda_n)[(\sigma - \sigma_n)^2 + R_n^2]}\,F_n(\mathbf{k},\sigma). \qquad (3.38c)$$

Accordingly, the spectral densities of the kinetic, potential, and total energies of the nth mode are equal to

$$\hat{K}^n = \tfrac{1}{2}\varrho_0 k^2F_n(\mathbf{k},\sigma), \quad \hat{P}^n = \tfrac{1}{2}\varrho_0\lambda_0 F_n(\mathbf{k},\sigma), \hat{E}^n = \hat{K}^n + \hat{P}^n. \qquad (3.39)$$

Now we find the distribution over the wavenumbers of the energy contained in the frequency range $0 \leqslant |\sigma| \leqslant \sigma_{\max}$. The spectral densities (3.39) integrated over this range are given, to within small terms, by the formulas (Reznik, 1983)

$$E^n(\mathbf{k}) = \frac{\varrho_0 \pi f_0^4 \Phi_n^2(0) F_0}{2k^2 g^2 (A'_H \lambda_n + A_H k^2)}, \tag{3.40a}$$

$$K^n(k) = \frac{k^2}{k^2 + \lambda_n} E^n, \tag{3.40b}$$

$$P^n = \frac{\lambda_n}{k^2 + \lambda_n} E^n, \tag{3.40c}$$

where it is taken into account that $F(\mathbf{k}, \sigma) = F_0 = \text{const}$ in the range of spacetime scales under study. Since $\lambda_n \sim n^2$ and

$$\Phi_n^2(0) \simeq \frac{2\, e^{-f_0^2/N^2|_{z=0}}}{\int_{-H}^{0} e^{-f_0^2/N^2}\, dz}, \quad n \gg 1, \tag{3.41}$$

for large n (this can be shown using the WKB method), the energy of the nth mode decreases proportionately to n^{-2} as the index n increases. For $N = \text{const}$ the spectral densities of the barotropic ($n = 0$) and baroclinic ($n \geqslant 1$) modes are written particularly simply:

$$E^{\text{bt}} = E^{0} = \frac{\varrho_0 \pi f_0^4 F_0}{2A_H g^2 k^4 H}, \tag{3.42a}$$

$$E^n = \frac{\varrho_0 \pi f_0^4 F_0}{A_H k^2 g^2 [(A'_H/A_H) \lambda_n + k^2] H}, \quad n \geqslant 1. \tag{3.42b}$$

The total baroclinic energy is equal to

$$E^{\text{bc}} = \sum_{n=1}^{\infty} E^n = E^{\text{bt}} (q \coth q - 1), \quad q = \sqrt{\frac{A_H}{A'_H}} \frac{N_0 H}{f_0} k. \tag{3.43}$$

Since for synoptic eddies in the ocean $(N_0 H/f_0)k \sim 1$, the relationship between the baroclinic and barotropic energies depends predominantly on the ratio A_H/A'_H. If the bouyancy diffusion is more intensive than the momentum exchange, i.e. $A_H/A'_H \ll 1$, then we also have $q \ll 1$. In this case, $E_{\text{tot}}^{\text{bc}} \simeq (1/3)\, q^2 E_{\text{tot}}^{\text{bt}}$, i.e. the baroclinic energy amounts to only a small fraction of the barotropic energy. As A_H/A'_H increases, the baroclinic energy increases, and for $q \sim 1$ it is of the same order as the barotropic energy. Finally, for $A'_H \to 0$ (i.e. $q \to \infty$) the baroclinic energy tends to ∞. This follows from formula (3.42) according to which, for $A'_H = 0$ (in the absence of buoyancy diffusion), the quantities E^n do not depend on n for $n \geqslant 1$, i.e. the total energy is distributed homogeneously over all baroclinic modes. In this case the total energy of each baroclinic mode is twice as great as the total energy of the barotropic mode. It readily follows from (3.40a) and (3.41) that for $N \neq$ const the resultant baroclinic energy is also equal to ∞ for $A'_H = 0$; in this case the homogeneous distribution of energy over the modes holds (approximately) only for higher modes when $\Phi_n^2(0)$ depends weakly on the index n.

One may think that this result contradicts the calculations in Section 3, according to which, in the absence of friction, the rate of energy influx to the barotropic mode $\partial E^{bt}/\partial t$ is at least twice as great as the resultant rate of influx to baroclinic modes $\Sigma_{n=1}^{\infty} \partial E^n/\partial t$, and therefore it seems that, irrespective of the magnitude of A'_H, when the friction is taken into consideration, the barotropic energy transferred to the ocean must exceed the baroclinic energy. The role of buoyancy diffusion can easily be elucidated if the solution $\bar{\psi}$ of the homogeneous system corresponding to (3.35a–c) is represented in the form

$$\bar{\psi} = \sum_{n=0}^{\infty} \Phi_n(z) \int \bar{\psi}_n(\mathbf{k}, t)\, e^{i\mathbf{kr}}\, d\mathbf{k}.$$

For $\bar{\psi}^n$ we obtain the equation

$$\frac{\partial \bar{\psi}^n}{\partial t} + (i\sigma_n + R_n)\, \bar{\psi}^n = 0,$$

which shows that the coefficient of friction for the nth mode is equal to $R_n = k^2(\lambda_n A'_H + k^2 A_H)/(k^2 + \lambda_n)$. For $A'_H = 0$ the coefficient R_n decreases proportionately to n^{-2} as n increases and thus, as the index n of the mode increases, the effect of friction on the mode decreases. It follows that, although the energy influx to the baroclinic mode with index n is proportional to n^{-2} (see(3.26)), the energy accumulated in the mode during a sufficiently long time is of the same order of magnitude as the barotropic energy. However, if $A'_H \neq 0$, then $R_n \simeq k^2 A'_H$ for large n, i.e. the effect of friction on the mode does not depend on its index but decreases together with A'_H. Accordingly, the baroclinic energy of the forced motions decreases when A'_H increases, and vice versa.

It is also possible to give a somewhat different interpretation of the above result. Assume that the buoyancy diffusion is negligibly small and consider the action of a low-frequency wind component (a harmonic with $\sigma \to 0$) on the ocean. According to the boundary condition (3.35b), in this case very large values of the density $\varrho \sim \partial\psi/\partial z \sim \sigma^{-1}$ are generated in the vicinity of the upper boundary. If we assume that the vorticity equation (3.35a) holds for this region, it can easily be shown that the thickness h of the region is proportional to $\sqrt{\sigma}$. Then the available potential energy accumulated here (proportional to $\psi_z^2 h$) is of the order of $\sigma^{-3/2}$, whence it follows that the total available potential energy of the response is infinite (since the spectrum of the external force does not tend to zero for $\sigma \to 0$; see above). When buoyancy diffusion is taken into consideration the situation changes because, by virtue of (3.35b), the density ϱ tends to a finite limit for $\sigma \to 0$; accordingly, the available potential energy is also finite. It also follows from (3.35b) that the limiting value of ϱ increases as the diffusion coefficient A'_H decreases, which leads to an increase of stored baroclinic energy.

Hence the energy stored by the ocean depends on the relationship between the intensities of the turbulent momentum exchange and the turbulent buoyancy diffusion, and it increases when the buoyancy exchange decreases. It is natural to assume that this conclusion does not depend on the specific parametrization of the turbulent exchange.

In our consideration we have intentionally simplified the problem by assuming that the motion in the geostrophic region always satisfies the vorticity equation (3.35a) for all values of the frequency σ of the harmonic external force. However, it can easily be seen that in the absence of turbulent buoyancy diffusion, and for very small frequencies σ, the thickness of the region of sharp density gradients which is proportional to $\sqrt{\sigma}$ (see above) is smaller than or of the order of the thickness D_T of the Ekman layer. For these small frequencies there is a change in both the structure of the Ekman layer and the dynamics of the geostrophic region where the vertical exchange becomes significant, and the vorticity equation (3.35a) no longer applies (see Reznik, 1983). Thus the suggested theory holds only for $\sigma/f \gg D_T/H$, and it can be shown that the baroclinic response energy remains finite even in the absence of the buoyancy diffusion. However, in this case the baroclinic energy is many times as great as the barotropic energy so that, qualitatively, all the above conclusions remain valid.

Thus, to form an ultimate judgement upon the role of turbulent mixing in the generation of synoptic eddies by atmospheric forcing, it is first necessary to have an indication of the relative intensity of the processes of turbulent friction and turbulent buoyancy diffusion at synoptic scales.

4. EDDY-RESOLVING NUMERICAL MODELS

In Sections 1 to 3 we studied some mechanisms of generating synoptic eddies in the ocean within the framework of simplified models (the barotropic and the baroclinic instability of large-scale currents, the effect of irregularities of the bottom relief, and direct atmospheric forcing). What are the relative contributions of these mechanisms? What is the character of the interaction of mature eddies with a large-scale current? To gain a fundamental understanding of the part played by synoptic eddies in the ocean, both theoretical and experimental studies of these aspects are necessary. At present the available observation data are obviously insufficient even for an approximate investigation of these problems and their analytical study is extremely difficult. In this connection, numerical models of the motion of ocean waters have acquired importance since they provide a means of studying the dynamics of synoptic eddies and their interaction with large-scale currents (eddy-resolving models). To present the general idea of this modelling, and to discuss some methods of interpreting the calculation results, we begin with a detailed consideration of Holland's (1978) quasigeostrophic model.

4.1. Holland's Model

We consider a two-level model of quasigeostrophic motions (Chapter 2, Sections 1.2 and 1.3) for an ocean of constant depth ($b = 0$). We introduce a vertical velocity $w_{3/2}$ at the interface between the layers as an unknown function. Taking into account (1.64), (1.67), and (1.86) in Chapter 2 and introducing the terms F_1 and F_2 specified by the horizontal turbulent friction in the upper and lower layers into the transfer equation for vorticity, we write (1.84), (1.85), and (1.64) from Chapter 2 in the form

$$\frac{\partial}{\partial t} \Delta\psi_1 + J(\psi_1, \Delta\psi_1 + \beta y) + \frac{f_0}{h_1} w_{3/2} = \frac{1}{\varrho_0 h_1} \operatorname{rot}_z \tau + F_1 ; \tag{4.1}$$

$$\frac{\partial}{\partial t} \Delta\psi_2 + J(\psi_2, \Delta\psi_2 + \beta y) - \frac{f_0}{h_2} w_{3/2} = F_2 - \varepsilon \Delta\psi_2; \tag{4.2}$$

$$\frac{\partial}{\partial t}(\psi_2 - \psi_1) + J(\psi_1, \psi_2 - \psi_1) = \frac{g'}{f_0} w_{3/2}, \tag{4.3}$$

where $\psi_1 = [1/(f_0\varrho_0)]p_1'$ and $\psi_2 = [1/(f_0\varrho_0)]p_2'$ are the stream functions for the upper and lower layers having thicknesses h_1 and h_2; $-\varepsilon\Delta\psi_2$ is the bottom friction; $\varepsilon = (f_0/h_2)[A_z/(2f_0)]^{1/2}$; τ is a given wind stress; and J denotes the Jacobian with respect to x and y.

Equations (4.1)–(4.3) are solved in the region $0 \leqslant x \leqslant L_x, 0 \leqslant y \leqslant L_y$ under the impermeability condition at the boundary of the region:

$$\psi_1 |_\Gamma = \psi_{1b}(t), \qquad \psi_2 |_\Gamma = \psi_{2b}(t), \tag{4.4}$$

the condition

$$\int_0^{\mathscr{L}_x} \int_0^y w_{3/2} \, dx \, dy = 0, \tag{4.5}$$

implied by (1.58a) in Chapter 2 for $k = 2$, taking into account the formulas $H_2 = h_2 + \zeta_2$ and $\zeta_2 = (f_0/g')(\psi_2 - \psi_1)$, Equation (4.3), and conditions (4.4) (condition (4.5) is also directly implied by condition (1.31) in Chapter 2 since $w_{5/2} = 0$), and also some other additional boundary conditions determined by the structure of the expressions for F_1 and F_2.

In Holland's numerical experiments the following two versions of setting F_1 and F_2 were used:

$$(F_1, F_2) = A_H^{(2)} \Delta\Delta(\psi_1, \psi_2); \tag{4.6a}$$

$$(F_1, F_2) = -A_H^{(4)} \Delta\Delta\Delta(\psi_1, \psi_2). \tag{4.6b}$$

The version (4.6a) follows from the usual formula describing the horizontal turbulent friction with the aid of a constant exchange coefficient $A_H^{(2)}$ when the friction force is proportional to the Laplacian of the velocity components. The version (4.6b) requires a special explanation.

The matter is that, so far, dissipative processes in the ocean have been very poorly studied. However, the form in which they are taken into consideration in numerical models affect greatly the calculation results. We shall proceed from the following physical scheme which is in agreement with the basic concepts of quasigeostrophic turbulence (see Section 5). We assume that, generally, the vertical turbulent friction is negligibly small in the ocean and that it plays a significant role only in the Ekman bottom and surface boundary layers. The surface boundary layer provides the transfer of momentum from the wind to the ocean and makes almost no contribution to the dissipation; almost all the dissipation produced by the vertical turbulence occurs in the bottom boundary layer (Chapter 2, Section 1.5.). The horizontal turbulent friction substantially depends on the scales of the processes under study. For scales of the order of L_R the horizontal turbulence is entirely due to the interaction between synoptic eddies and large-scale currents. This is the process we are going to study in detail by means of a numerical model and, therefore, no parametrization of the horizontal turbulence for scales of the order of and greater than L_R should be introduced into the model. However, the horizontal turbulent friction cannot simply be excluded from consideration as it makes an important contribution to motions of smaller scales (of the order of the grid size) and provides the dissipation of enstrophy for these scales. The dissipative terms F_1 and F_2 are introduced into the model to parametrize this effect. In this approach, the energy dissipation must be mainly specified by the bottom friction if the model is to attain a quasistationary regime.

The realization of this physical scheme by means of a numerical model seems highly expedient. In this connection, expression (4.6b) has certain advantages over (4.6a). Namely choosing the coefficient $A_H^{(4)}$ so that the two expressions yield the same damping decrement for a wave with wavelength equal to twice the grid size we find that expression (4.6b) yields a substantially smaller damping decrement for longer waves. However, expression (4.6b) requires the introduction of some additional boundary conditions of a purely mathematical character which are not quite clear from the physical viewpoint. For example, to describe the friction with the aid of expression (4.6a), it is usually required that, in addition to conditions (4.4), the slip conditions $\Delta\psi_1 = 0$ and $\Delta\psi_2 = 0$ be fulfilled on the boundary Γ; to describe the friction with the aid of expression (4.6b) it is additionally required that $\Delta\Delta\psi_1 = 0$ and $\Delta\Delta\psi_2 = 0$ on Γ (see also the discussion on the parametrization of dissipative processes in the paper by Harrison, 1980).

We note that from the physical viewpoint the correct condition on a rigid boundary is the nonslip condition (the tangential velocity component must vanish) and not the slip condition. The nonslip condition is usually replaced by the slip condition proceeding from the assumption that this replacement excludes from consideration only thin, near-coast, viscous boundary layers (which are difficult to resolve from the viewpoint of finite-difference approximations) and does not affect strongly the general circulation pattern in a closed basin. This assumption seems to disagree with reality; for instance, it was shown that the replacement of a nonslip condition by a slip condition in the investigation of the barotropic problem substantially affects the general circulation pattern (see the discussion of this problem in the papers by Holland, 1977, and Kamenkovich *et al.*, 1985a).

We now pass to the consideration of energy transformations in the model. Let us introduce $\psi_1^* = \psi_1 - \psi_{1b}$ and $\psi_2^* = \psi_2 - \psi_{2b}$ and multiply (4.1) by ψ_1^*, (4.2) by ψ_2^*, and (4.3) by $\psi_2 - \psi_1$. After some simple transformations, taking into account (4.5), we obtain (cf. Section 1.3 in Chapter 2)

$$\frac{\partial K_1}{\partial t} = (K_1, \tau) + (K_1, P) + (K_1, K_2) + (K_1, F_1); \tag{4.7}$$

$$\frac{\partial K_2}{\partial t} = (K_2, P) - (K_1, K_2) + (K_2, F_2) + (K_2, B); \tag{4.8}$$

$$\frac{\partial P}{\partial t} = -(K_1, P) - (K_2, P). \tag{4.9}$$

The quantities K_1, K_2 and P in relations (4.7)–(4.9) are the kinetic energy of the upper layer, the kinetic energy of the lower layer, and the available potential energy per unit area of the basin:

$$K_1 = \frac{1}{2S} \varrho_0 h_1 \int_S\int (\nabla\psi_1)^2 \,dx\,dy; \quad K_2 = \frac{1}{2S} \varrho_0 h_2 \int_S\int (\nabla\psi_2)^2 \,dx\,dy;$$

$$P = \frac{\varrho_0 f_0^2}{2g'S} \int_S\int (\psi_2 - \psi_1)^2 \,dx\,dy = \frac{g'\varrho_0}{2S} \int_S\int \zeta_2^2 \,dx\,dy, \tag{4.10}$$

where $S = L_x L_y$. The energy fluxes (per unit time) on the right-hand sides of (4.7)–(4.9) are written as

$$(K_1, \tau) = -\frac{1}{S}\int_S\int \psi_1^* \operatorname{rot}_z \tau \,dx\,dy;$$

$$(K_1, K_2) = \frac{f_0\varrho_0}{S}\int_S\int w_{3/2}\psi_{3/2}^* \,dx\,dy;$$

$$(K_1, P) = \frac{f_0\varrho_0}{hS}\int_S\int h_1 w_{3/2} (\psi_1^* - \psi_2^*) \,dx\,dy;$$

$$(K_2, P) = \frac{f_0\varrho_0}{hS}\int_S\int h_2 w_{3/2} (\psi_1^* - \psi_2^*) \,dx\,dy;$$

$$(K_1, F_1) = -\frac{\varrho_0}{S}\int_S\int h_1\psi_1^* F_1 \,dx\,dy;$$

$$(K_2, F_2) = -\frac{\varrho_0}{S}\int_S\int h_2\psi_2^* F_2 \,dx\,dy;$$

$$(K_2, B) = \frac{\varepsilon\varrho_0}{S}\int_S\int h_2\psi_2^* \,\Delta\psi_2 \,dx\,dy; \tag{4.11}$$

where

$$\psi_{3/2}^* = h^{-1}(h_2\psi_1^* + h_1\psi_2^*), \quad h = h_1 + h_2.$$

The quantity (K_1, τ) describes the energy flux due to wind stress; (K_1, P) and (K_2, P) are the energy fluxes produced by the work of buoyancy forces in the upper and lower layers, respectively; (K_1, K_2) is the energy flux due to the work of forces of pressure at the interface between the layers resulting in the transfer of kinetic energy from the upper to the lower layer; and (K_1, F_1), (K_2, F_2), and (K_2, B) are the energy sinks caused by horizontal turbulent friction in the upper and lower layers and the bottom friction, respectively.

Usually Cauchy's problem is solved for Equations (4.1)–(4.3) with the indicated boundary conditions, the external force (wind stress) being regarded as time-invariable. Here we do not discuss the corresponding numerical schemes; we only indicate that these schemes are based on finite-difference approximations of second-order accuracy and also make some remarks on the choice of the horizontal grid size. It is usually assumed (e.g. see Kreiss and Oliger, 1973) that when finite-difference schemes of second-order accuracy are applied, it is necessary to have approximately 20 grid points per typical wavelength to describe correctly the process under study. On average, the wavelength of synoptic motions is equal to $2\pi \times 50$ km; therefore this grid size must be approximately equal to 16 km. The grid size in the paper by Holland (1978) was equal to 20 km, which practically satisfies the above condition. Of course, it is clear that such a small grid size imposes some requirements on the computer since dimensions of the basin are equal to 1000 or even to 2000 km (for a discussion of the specific properties of the numerical model, see also Haidvogel, 1979).

The results of the numerical experiments allow us to assert that the problem under consideration possesses a stationary solution which proves to be stable, e.g. for sufficiently large values of the coefficients of horizontal friction (in these cases there appeared no eddies; see Holland and Lin, 1975). As the coefficients of friction are decreased, the indicated solution of the problem seems to become unstable; at a certain stage of the evolution of currents there appear eddies with parameters typical of synoptic eddies in the ocean. So far it has not been possible to investigate the stability of such a stationary solution (see the discussion at the beginning of Section 1.7). For an attempt to carry out an approximate analysis based on the replacement of some parts of the time-averaged calculated quasi-stationary current by a model stationary zonal current, see Haidvogel and Holland (1978). We note that the analysis of the $K_1(t)$, $K_2(t)$, and $P(t)$ plots at the stage of the development of the current makes it possible to draw some conclusions on the character of this instability (see, below, the analysis of Figure 3.4.1).

In those cases when the evolution of currents is accompanied by the appearance of eddies, the system eventually attains a quasistationary regime in which the energies $K_1(t)$, $K_2(t)$, and $P(t)$ usually oscillate relative to some average values. Therefore, it is natural to define the large-scale current as the time average over a time interval T and to regard the deviations from the large-scale current thus defined as synoptic eddies. This approach allows us to study the interaction between the synoptic eddies and the large-scale current using the model by calculating, for instance, various energy transformations (see the energy diagrams in Figures 3.4.2, 3.4.4., 3.4.6, 3.4.8, and 3.4.14). We present the following formulas for calculating the kinetic and available potential energies of the large-scale current ($\bar{K}_1$, $\bar{K}_2$, and $\bar{P}$) and the time-averaged energies of the eddy current (K_1', K_2', and P') per unit area of the basin:

$$\bar{K}_1 = \frac{1}{2S} \varrho_0 h_1 \iint_S (\nabla\overline{\psi_1})^2 \, dx \, dy; \quad K_1' = \frac{1}{2S} \varrho_0 h_1 \iint_S \overline{(\nabla\psi_1')^2} \, dx \, dy;$$

$$\bar{K}_2 = \frac{1}{2S} \varrho_0 h_2 \iint_S (\nabla\overline{\psi_2})^2 \, dx \, dy; \quad K_2' = \frac{1}{2S} \varrho_0 h_2 \iint_S \overline{(\nabla\psi_2')^2} \, dx \, dy;$$

$$\bar{P} = \frac{f_0^2}{2g'S} \iint_S (\overline{\psi_2} - \overline{\psi_1})^2 \, dx \, dy; \quad P' = \frac{f_0^2}{2g'S} \iint_S \overline{(\psi_2' - \psi_1')^2} \, dx \, dy,$$

($S = L_x L_y$), where, for instance, (4.12)

$$\overline{\psi_1} = \frac{1}{T} \int_{t_0}^{t_0+T} \psi_1 \, dt \tag{4.13}$$

(the other averaged quantities are expressed analogously); t_0 is the time the system attains the quasistationary regime.

Based on definitions (4.11) and (4.12) and Equations (4.1)–(4.3) and (4.7)–(4.9) we can readily derive expressions for various energy transitions between, for instance, $\bar{K}_1$ and $\bar{K}_1'$, etc. The corresponding formulas are presented in the paper by Holland (1978); these fluxes are indicated by arrows in the figures (see below, Figure 3.4.2, etc.).

We now briefly consider the results of Holland's (1978) three experiments; the values $\tau_x = -\tau_0 \cos(\pi y/L)$, $\tau_y = 0$, $L = 1000$ km, $\beta = 2 \times 10^{-11} \text{m}^{-1}\text{s}^{-1}$, $g' = 0.02$ m/s^2, $h_1 = 1$ km, and $h_2 = 4$ km are used in all the experiments. There is one large-scale gyre for $L_y = L$ and two large-scale gyres for $L_y = 2L$ (Table 3.4.1).

Experiment A. Figure 3.4.1 shows that, first, for $t < 500$ days, accumulation of the available potential energy P takes place and then, for $500 < t < 800$ days, a sharp decrease occurs. The analysis of the fields ψ_1 and ψ_2 shows that, for approximately $t > 500$ days, eddies begin to form mainly in the part of the gyre

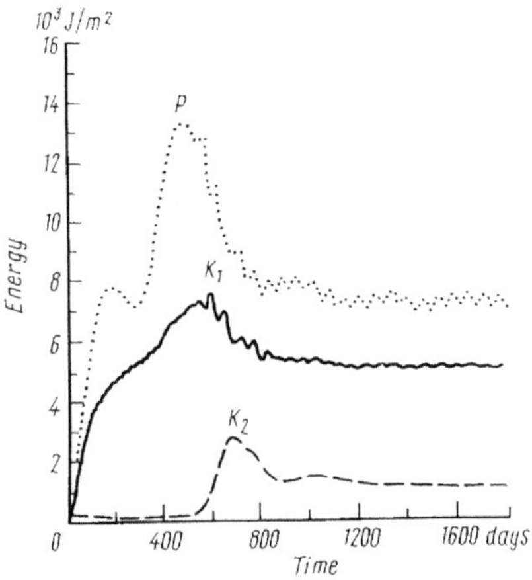

Fig. 3.4.1. Kinetic energies K_1 and K_2 of the upper and lower layers and the available potential energy p per unit area (formulas (4.10)) as functions of t for experiment A (after Holland, 1978, Fig. 3).

TABLE 3.4.1

Experiment	Number of gyres	$\tau_0 \times 10^{-4}$, m^2/s^2	f_0, s^{-1}	$A_H^{(2)}$, m^2/s	$A_H^{(4)}$, m^4/s	ε, s^{-1}	L_y	L_x	L_R, km; see (2.21) in Chapter 2
A	One	1	8.3×10^{-5}	330	—	—	L	L	120
B	Two	1	9.3×10^{-5}	100	—	—	$2L$	L	108
C	Two	1	9.3×10^{-5}	—	8×10^9	1×10^{-7}	$2L$	L	108

where the current flows westward. The fact that almost simultaneously the available potential energy sharply decreases and the amplitude of the eddy increases indicates that the eddies are probably generated as a result of baroclinic instability in the basic current (more precisely, the baroclinic character of the basic current plays a substantial part in the evolution of the disturbances; see Section 1).

Approximately at $t = 1500$ days the system attains a quasistationary regime; Figure 3.4.2 presents the energy diagram for this regime. We note that usually the averaging period T in (4.12) is chosen so that all the quantities $\bar{K}_1$, $\bar{K}_2$, $\bar{K}'_2$, $\bar{P}$, and P' do not vary with time; then the sum of all the fluxes must be equal to zero for each of the squares, which holds to a sufficient degree of accuracy. It is seen from Figure 3.4.2 that 78% of the energy imparted to the system by the wind dissipates owing to horizontal turbulent friction, and only 22% takes part in the eddy generation process. Shown in Figure 3.4.3 are the time-averaged fields $\bar{\psi}_1$ and $\bar{\psi}_2$ ($T = 300$ days during the quasistationary regime) and examples of instantaneous fields ψ_1 and ψ_2. The fields $\bar{\psi}_1$ and $\bar{\psi}_2$ clearly demonstrate boundary currents near the western and northern walls.

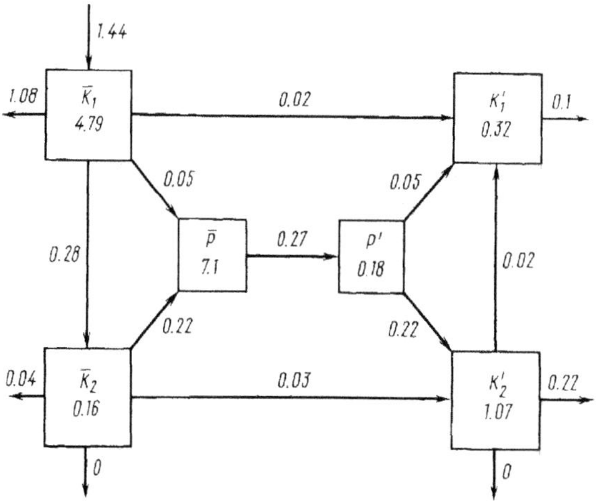

Fig. 3.4.2. Energy diagram for the quasistationary regime in experiment A (after Holland, 1978, Fig. 4). Indicated in the squares are the energies (see (4.12)) in 10^3 J/m^2. The arrows (and their figures) joining the squares indicate the directions and magnitudes of the energy fluxes in 10^{-3} J/m^2·s. The figures at arrows not joining the squares indicate dissipative fluxes owing to horizontal turbulent friction (horizontal arrows) and bottom friction (vertical arrows). The vertical arrow directed towards $\bar{K}_1$ indicates the energy flux imparted to the ocean by wind stress.

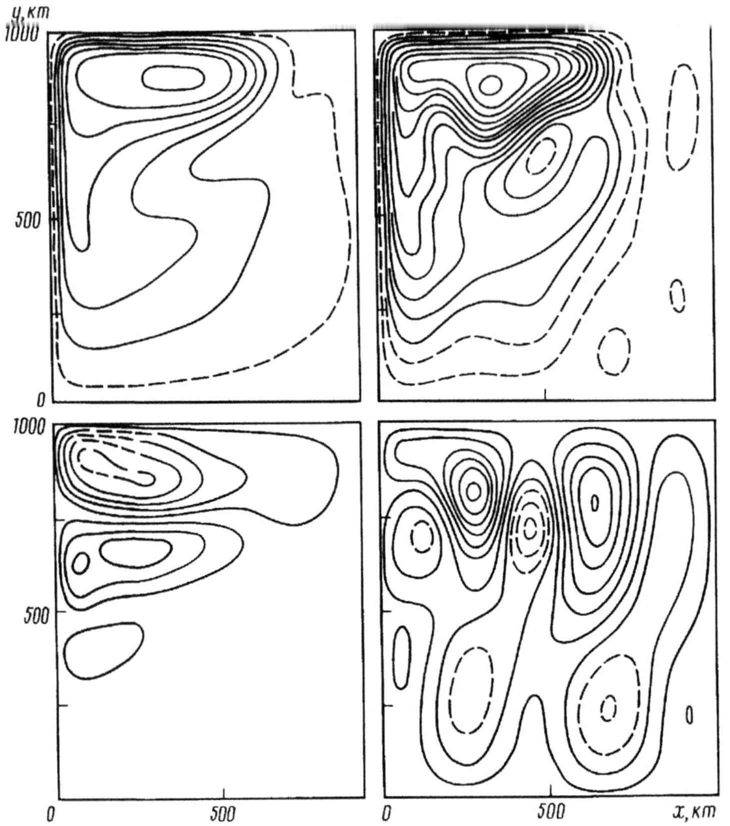

Fig. 3.4.3. Stream functions ψ_1 (above) and ψ_2 (below) for experiment *A* (after Holland, 1978, Fig. 5). On the left: the average fields $\bar{\psi}_1$ ($CI = 4000$ m^2/s) and $\bar{\psi}_2$ ($CI = 500$ m^2/s). On the right: an example of instantaneous fields $\bar{\psi}_1$ ($CI = 2000$ m^2/s) and ψ_2 ($CI = 1000$ m^2/s). *CI* are the contour intervals at which the isolines are drawn.

Experiment B. The size of the basin along *y* is increased twice in comparison with experiment A, and this results in the formation of two large-scale gyres and an intensive narrow jet between them (an analog of the continuation of the Gulf Stream). This jet seems to be barotropically unstable since the formation of eddies takes place primarily in the region of the jet. Figure 3.4.4 demonstrates the energy diagram for the quasistationary regime (the averaging period is $T = 900$ days). Note that now 71% of the energy imparted to the system by the wind takes part in the generation of eddies and 29% dissipates owing to horizontal turbulent friction. Moreover, the energy values themselves, which are indicated in the squares, have increased substantially. All this is undoubtedly related to the decrease (in comparison with experiment A) of the horizontal turbulent exchange coefficient. Figure 3.4.5 presents the time-averaged fields $\bar{\psi}_1$ and $\bar{\psi}_2$ and examples of instantaneous fields ψ_1 and ψ_2. In the fields $\bar{\psi}_1$ and $\bar{\psi}_2$ we clearly see boundary currents at the western wall and a narrow meandering eastern jet separating the two large-scale gyres.

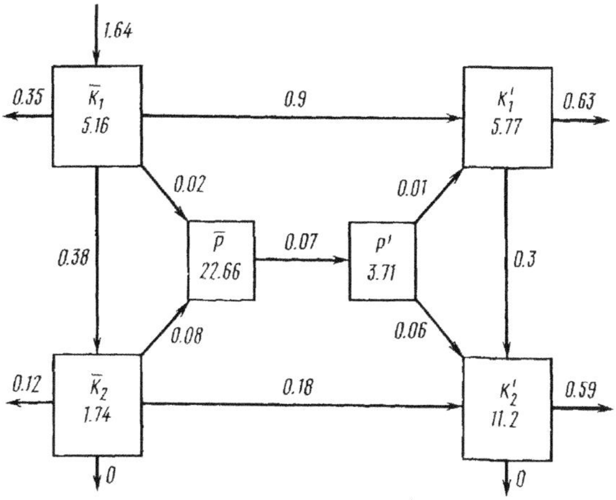

Fig. 3.4.4. Energy diagram for the quasistationary regime in experiment *B* (after Holland, 1978, Fig. 12). The notation is the same as in Figure 3.4.2.

Experiment C. The difference between experiment C and experiment B is that, in the former, horizontal turbulent friction is described by expression (4.6b) and the model takes bottom friction into account. Qualitatively, the fields ψ_1 and ψ_2 look like those in experiment *B*; the major role in the generation of eddies again seems to be played by the barotropic instability of the eastern jet. However, in this case a more realistic energy diagram (Figure 3.4.6) is obtained. We see that 86% of the energy imparted to the system by the wind takes part in the eddy generation process, 10% dissipates as the result of bottom friction, and only 4% dissipates by the action of horizontal turbulent friction. In this experiment the above-mentioned physical concept of the role of dissipative processes in the ocean is realized most completely.

A thorough analysis of experiments A, B, and C (Haidvogel and Holland, 1978) shows that the mature eddies move to the west and have the following periods, diameters (half-wavelengths), and phase velocities. Experiment A–60 days, 200 km, and 9 cm/s; experiment B–43–50 days, 250–350 km (the upper layer), 250–500 km (the lower layer), and 15–25 cm/s; experiment C–50 days, 220 km (the upper layer), 310 km (the lower layer), and 18 cm/s.

We note that in the above experiments the motion in the lower layer in the quasistationary regime ($\bar{K}_2 \neq 0$) is entirely due to the effect of eddies. Indeed, assume that there are no eddies and that the system has attained a stationary state. Then, according to (4.3) and (4.4),

$$\frac{g'}{f_0} w_{3/2} = J(\psi_1, \psi_2); \quad \iint_S w_{3/2} (\psi_1^* - \psi_2^*), \, dx \, dy = 0$$

and, by virtue of the corresponding formulas (4.11), we have $(K_1, K_2) = 0$ and $(K_2, P) = 0$ for the energy fluxes. In other words, the energy sources in Equations (4.8) disappear and only the sinks remain; these sinks are due to dissipation, which means that there must be $\bar{\psi}_2 = 0$ in the stationary regime and, hence, $\bar{w}_{3/2} = 0$. At present it is difficult to decide the extent to which this result has a general

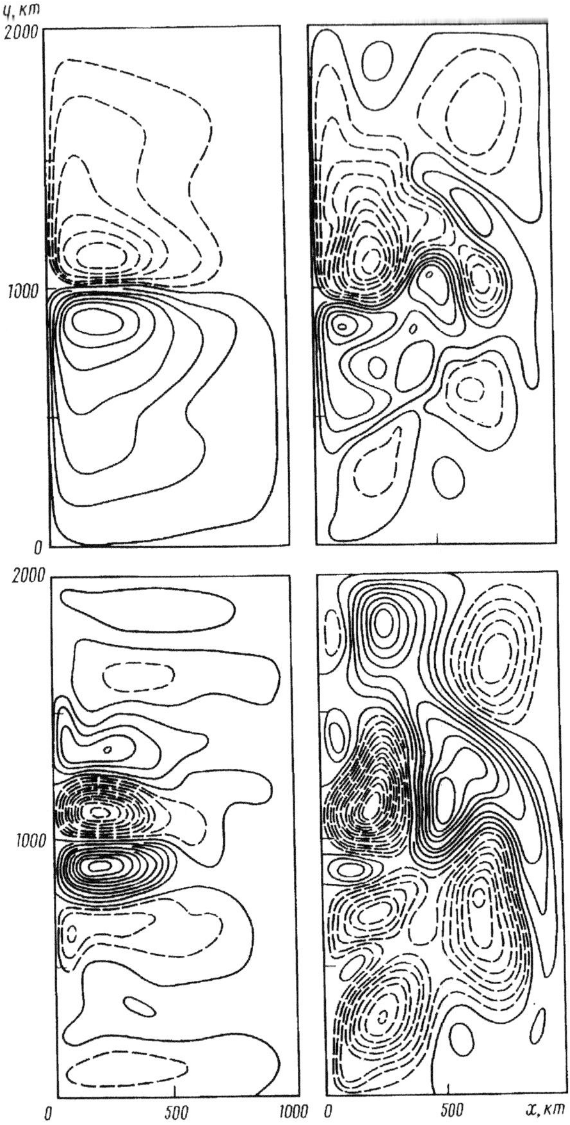

Fig. 3.4.5. Stream functions ψ_1 (above) and ψ_2 (below) for the experiment *B* (after Holland, 1978, Figs 13 and 15). On the left: the average fields $\bar{\psi}_1$ ($CI = 5000\ m^2/s$) and $\bar{\psi}_2$ ($CI = 1000\ m^2/s$). On the right: examples of instantaneous fields ψ_1 ($CI = 5000\ m^2/s$) and ψ_2 ($CI = 2000\ m^2/s$). *CI* are the contour intervals at which the isolines are drawn.

character and does not depend on the properties of the two-level quasigeostrophic model (for a discussion see Harrison, 1979). For a further investigation of the problem of the generation of mean currents due to the effect of synoptic eddies, see Holland and Rhines (1980).

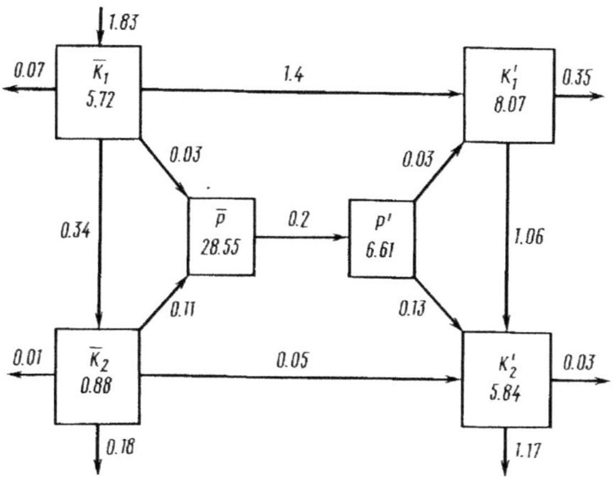

Figure 3.4.6. Energy diagram for the quasistationary regime in experiment C (after Holland, 1978, Fig. 22). The notation is the same as in Figure 3.4.2.

4.2. Seidov's Model

By now several eddy-resolving models have been constructed on the basis of both the quasigeostrophic equations and the primitive equations of the ocean hydrothermodynamics, taking into account the effect of mass flux at the sea surface.[3] A common feature of all these models is the detailed analysis of the quasistationary regime by calculating various time-averaged characteristics (first of all the characteristics of energy transformation). Here we shall not compile the results of such experiments and shall not analyze them comparatively (e.g. see Harrison, 1979). We discuss only some of the results of numerical experiments on the generation of eddies in a continuously stratified fluid carried out on the basis of the primitive equations of motion along the horizontal, the quasistatics equation, the continuity equation, and the transfer equation for density.

We begin with Seidov's eddy-resolving model (1978, 1980, 1981, and 1985) where the wind was chosen so that the time-averaged circulation had two gyres separated by an intensive eastern jet. The model also involved an external heat influx proportional to the difference between the sea surface temperature and a given atmospheric temperature. We consider the experiment where the horizontal turbulent friction was described by expression (4.6a) with $A_L = 700$ m^2/s; the coefficient of horizontal turbulent temperature conductivity was also 700 m^2/s. Accordingly, the coefficients of vertical turbulent exchange of momentum and heat were 10^{-3} and 2.5×10^{-4} m^2/s. The depth of the basin was 2 km; its horizontal dimensions were 780 × 1170 km; the grid size was 65 km; the number of grid points was 13 × 13 × 5. The eddies obtained in the model had a diameter (half-wavelength) of 300 km and periods of 60–100 days.

[3] We note that, as was shown by special comparison experiments, the quasigeostrophic model reproduces quite satisfactorily the basic dynamical features of the models with the primitive equations (Sempter and Holland, 1978). For a further discussion, see Harrison (1979).

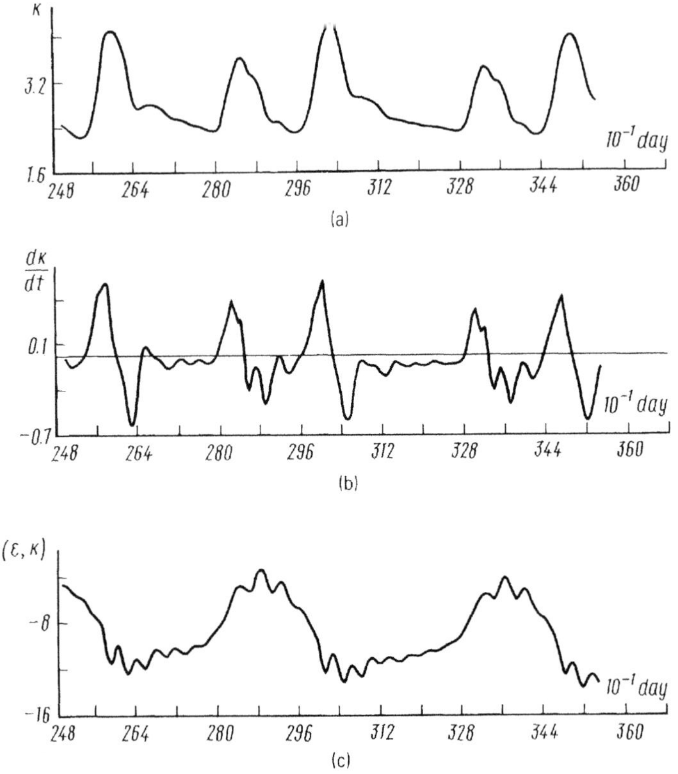

Fig. 3.4.7. Time variation of the volume-averaged (a) density K of the kinetic energy (J/m^3), (b) the time rate $\mathrm{d}K/\mathrm{d}t$ (in 10^{-6} J/m^3·s), and (c) the density of the work of the buoyancy forces per unit time ($\mathcal{E}$, K) (in 10^{-6} J/m^3·s) for the quasistationary regime of currents (after Seidov, 1985).

Figure 3.4.7 demonstrates the time variation of some energy characteristics of the quasistationary regime. (It can be seen from Figure 3.4.7(b) that, on average, $\mathrm{d}K/\mathrm{d}t = 0$, i.e. there is no trend to the variation of K.) We note that the 'peaks' on the K plot are narrower than the 'troughs'; the same is also characteristic of the ($\mathcal{E}$, K) plot. Generally this fact is in agreement with the POLYMODE observations. The energy diagrams for the experiment under consideration are presented in Figure 3.4.8. The first of these diagrams (Figure 3.4.8(a)) shows that the kinetic energy of the barotropic component of motion (depth-averaged motion) grows at the expense of the kinetic energy of the baroclinic component (a tendency towards the barotropization of currents; see Section 5). We give particular attention to the negative viscosity effect discovered by D. G. Seidov (Figure 3.4.8(b)): the resultant energy flux ($\bar{K}$, K') is directed from the eddy motion to the mean motion. The distribution of this flux in the plane is demonstrated in Figure 3.4.9(a). Figure 3.4.9(b) also shows the distribution of the kinetic energy of eddies in the plane. We see that the eddy activity is concentrated in the region of the eastern jet, to the south of it, in the neighborhood of the western recirculation. This distribution of kinetic energy is characteristic of many experiments with various eddy-resolving models. To give a more detailed indication of the negative viscosity effect we

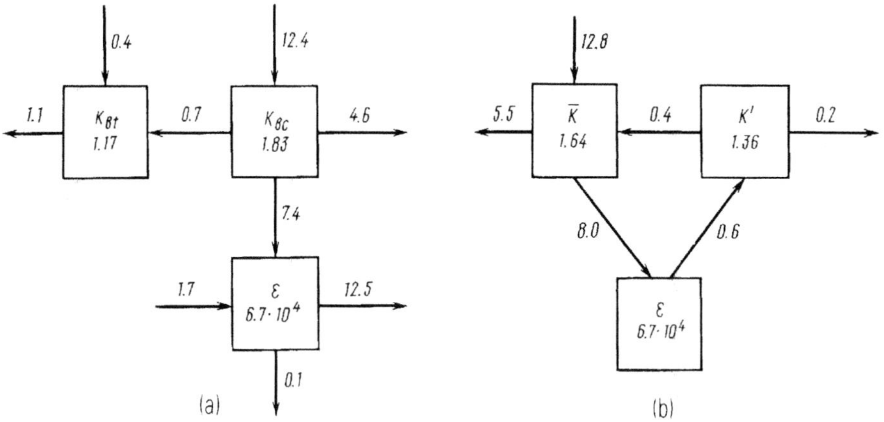

Fig. 3.4.8. Energy diagram for the quasistationary regime (after Seidov, 1985); (a) the characteristics of the energy balance averaged over a period of 960 days; (b) the interaction between the mean and eddy motions. K is the average density of kinetic energy in J/m^3; $\mathcal{E}$ is the average density of potential energy in J/m^3. The energy fluxes at the arrows are given in 10^{-6} J/m^3·s. K_{bt} is the kinetic energy of the barotropic component; K_{bc} is the kinetic energy of the baroclinic component; $\bar{K}$ is the kinetic energy of the mean motion (averaged over a period of 960 days); K' is the kinetic energy of eddies averaged over a period of 960 days. The horizontal arrows and the downward arrow at $\mathcal{E}$ indicate the variation of the potential energy due to heat fluxes.

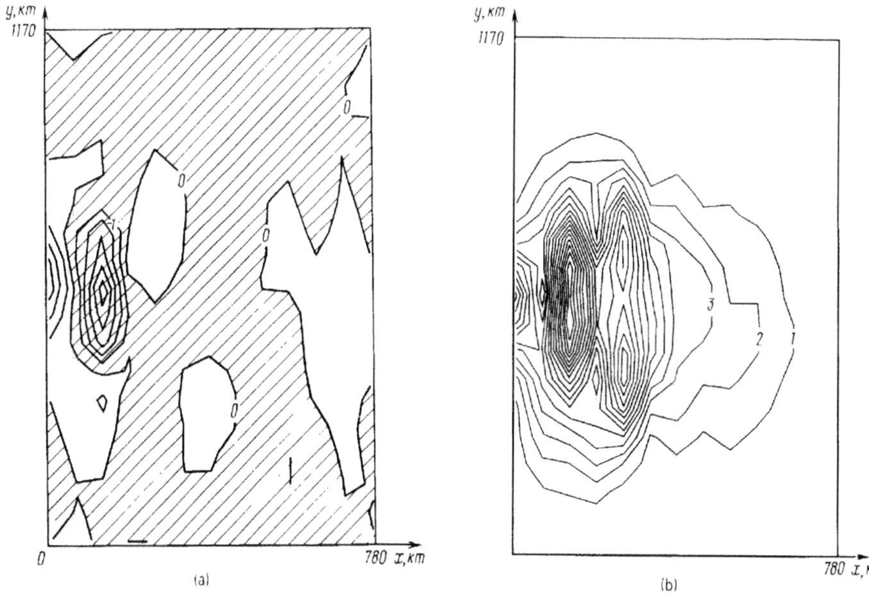

Fig. 3.4.9. Distribution in the horizontal plane of (a) the depth-averaged energy flux from the kinetic energy of the mean motion to the kinetic energy of eddies (in 2×10^{-6} J/m^3·s) and (b) the depth-averaged density of the kinetic energy of eddies (in J/m^3) for the quasistationary regime (after Seidov, 1981). The regions of negative values in Figure 3.4.9(a) are shaded.

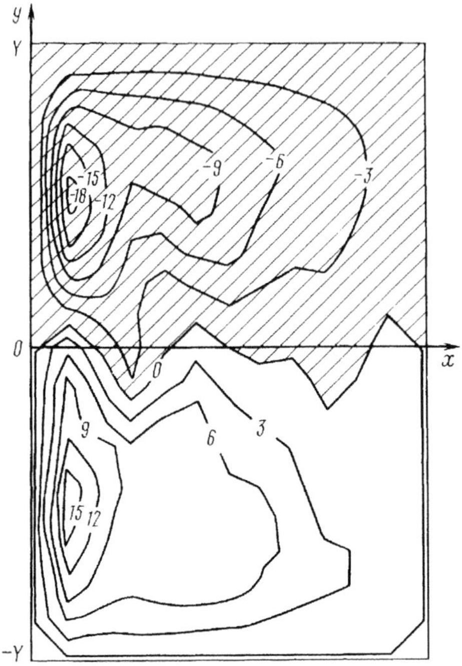

Fig. 3.4.10. Isolines of the total flow (transport) stream function averaged over a period of 960 days (after Monin and Seidov, 1982). The indicated values are given in 10^6 m^3/s. The region of the cyclonic circulation is shaded. $Y = 585$ km.

present Figure 3.4.10 where the isolines of the total flow (transport) stream function[4] are shown, and Figure 3.4.11 demonstrating meridional profiles of the energy flux from the mean motion to the eddy motion. It follows from Figure 3.4.11 that, at a distance of 130 km from the western coast, the negative viscosity manifests itself on both sides of the jet axis at $y = 0$ whereas at a distance of 260 km from the western coast the negative viscosity manifest itself only south of the jet axis and does not manifest itself north of the jet. The model considerations of Monin and Seidov (1982) show that this distribution is characteristic of a free jet and must lead to its narrowing and to the strengthening of its intensity (an analogous mechanism is also responsible for the formation of atmospheric jet currents).

It is also advisable to point out another experiment with Seidov's model where the coefficients of the horizontal and vertical exchange of momentum and heat remained as before and the basin depth was not changed either, but some other horizontal dimensions of the basin were taken (960 × 1440 km) and the horizontal resolution was finer (the number of grid points was 25 × 37 × 5), which corresponded to a horizontal grid size of 40 km and made it possible to study the eddy motion dynamics in more detail. In this case the characteristics of the eddies and the energy distributions did not change substantially in comparison with the

[4] We remind the reader that the total flow vector $\mathbf{S} = (S_x, S_y)$ is equal to the integral over the ocean depth of the horizontal velocity vector: $(S_x, S_y) = \int_{H}^{0} (u, v)\,dz$.

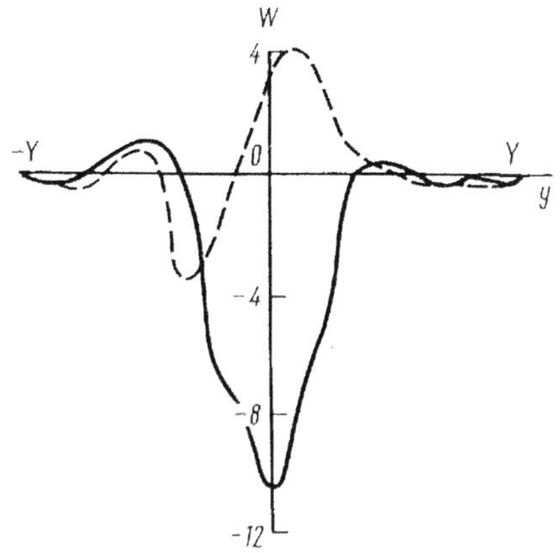

Fig. 3.4.11. Meridional profiles of the depth-averaged density of work of the Reynolds stresses per unit time W (the energy flux from the kinetic energy of the mean motion to the kinetic energy of eddies) at different distances from the western coast (after Monin and Seidov, 1982). The solid line: 130 km from the coast. The dashed line: 260 km from the coast. The values of W indicated on the axis of ordinates are given in 10^{-6} J/m^3·s. $Y = 585$ km.

above-mentioned experiment. We present the consecutive patterns of isolines (taken at 6-day intervals) of the transport stream function for a quasistationary regime. These patterns clearly demonstrate the character of the variability of the currents (Figure 3.4.12).

4.3. Neelov–Chalikov Model

Now we describe experiments with the Neelov–Chalikov model based on the primitive equations (Neelov and Chalikov, 1981; Neelov, 1982) which were performed proceeding from another standpoint. First, the authors of the model considered a basin open in the zonal direction by setting periodicity conditions at the x-boundaries of the basin. There appears no western boundary current in this model and it seems that the analysis of the model can be regarded as an investigation of the dynamics of synoptic eddies in the open ocean. Second, there was no wind at the sea surface and the role of external forcing exciting the current was played by a given latitude-dependent mass flux at the sea surface. In this setting of the problem the authors of the model wanted to investigate in a pure form the mutual transformations of the potential and kinetic energies. Formula (4.6a) was taken to describe the horizontal turbulent exchange; the values of the coefficients of friction and diffusion were 20 m^2/s and 200 m^2/s, respectively. The vertical friction was assumed to be proportional to the square of the velocity difference between the layers; the bottom friction was described by the Ekman formulas. A finite-difference scheme was constructed with the aid of the box-method, and had first-order accuracy with respect to time and second-order accuracy with respect to the spatial coordinates.

We describe the experiment with a three-level version of the model where the impermeability and slip conditions were set on the northern and southern bound-

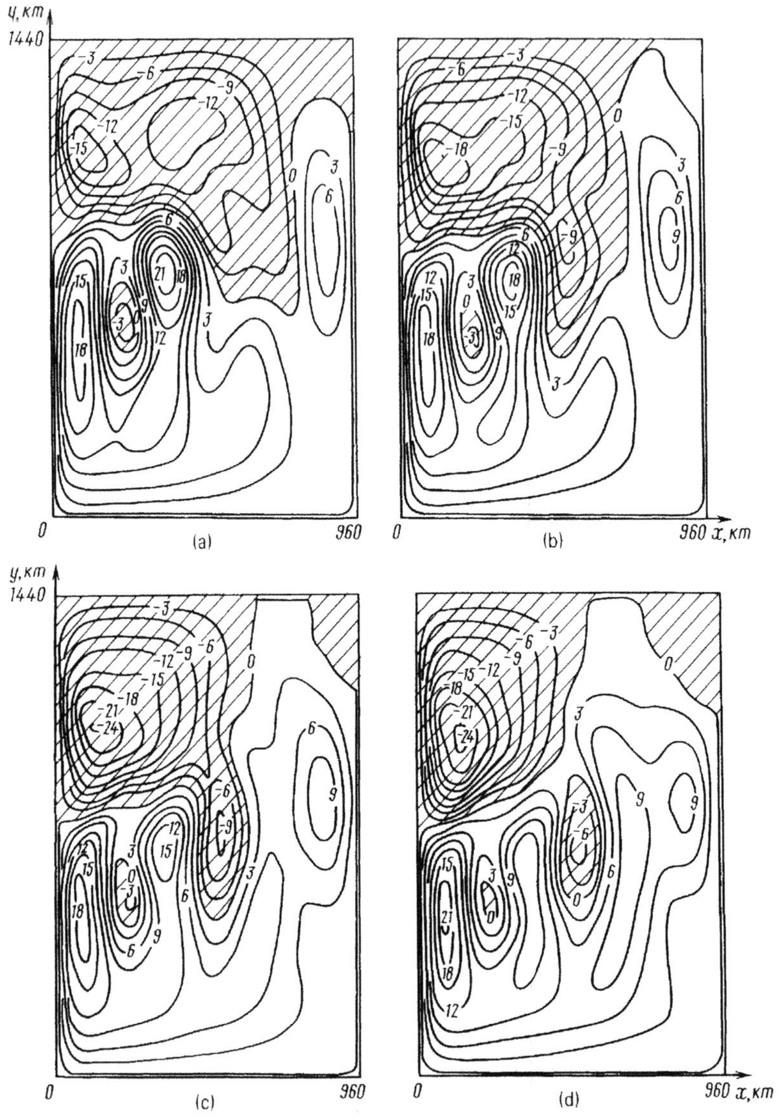

Fig. 3.4.12. Isolines of the total flow (transport) stream function for the quasistationary regime at different times: (a) 2233 days after the beginning of motion; (b) 6 days later; (c) another 6 days later; (d) another 6 days later (after Monin and Seidov, 1983). The values at the isolines are given in 10^6 m^3/s. The regions of the cyclonic circulation are shaded.

aries of the basin. The horizontal size of the region was 400 km; the layer thicknesses (from the sea surface to the bottom) were 200 m, 1000 m, and 2400 m. The equations of the model were integrated with a time step of 10 min and with a step of 20 km along the horizontal coordinates.

The results of the numerical experiment yielded a circulation characterized by a meandering jet at the center of the region and eddies, located in the meanders, with time-varying dimensions from 150 to 300 km. The orbital velocities of the eddies in the upper layer were about 30 cm/s. The analogous velocities in the lower layer were about 5 cm/s. The eddies propagated westward with velocities from 5 to 12 cm/s.

Figure 3.4.13 demonstrates the time variation of the basic energy fluxes for the case of the zonal averaging. It is seen that during almost the entire period of integration the kinetic energy of the nonzonal (eddy) motion increased at the expense of the potential energy of the zonal current, i.e. the process of baroclinic instability was observed (curve 1). It is interesting that, as a rule, the major part of the energy imparted to the eddies was not dissipated but was tranformed into kinetic energy of the mean current (curve 2). In other words, the work of the Reynolds stresses was negative, which indicated a manifestation of the negative viscosity effect. During the periods of 220–270 and 1020–1050 days the work of the Reynolds stresses became positive, which was accompanied by a rapid increase in the intensity of the eddies. It can be seen from the figure that the increase of kinetic energy in the mean current owing to the variation of potential energy in the current is relatively small (curve 3).

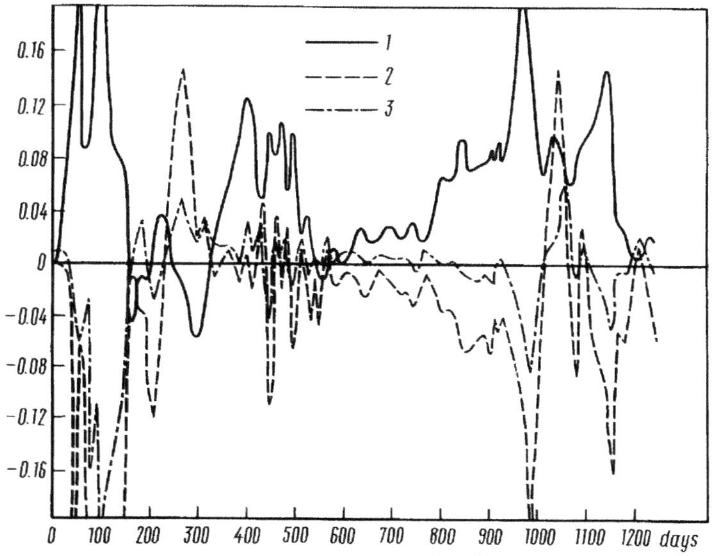

Fig. 3.4.13. Time variation of the basic energy fluxes in 10^{-6} J/m^3·s (after Neelov and Chalikov, 1981). (1) the rate of increase of the kinetic energy of the nonzonal motion at the expense of the potential energy of the zonal motion; (2) the energy flux from the kinetic energy of the zonal motion to the kinetic energy of the nonzonal motion owing to the work of Reynolds stresses; (3) the rate of increase of the kinetic energy of the zonal motion at the expense of the potential energy of the zonal motion.

Figure 3.4.14 presents an energy diagram calculated for the quasistationary period of the evolution of currents. We indicate some of the most typical specific features of the diagram. First, we observe the transformation of the potential energy of the mean current into the kinetic energy of eddies in the middle and upper layers, which can be interpreted as a manisfestation of the process of

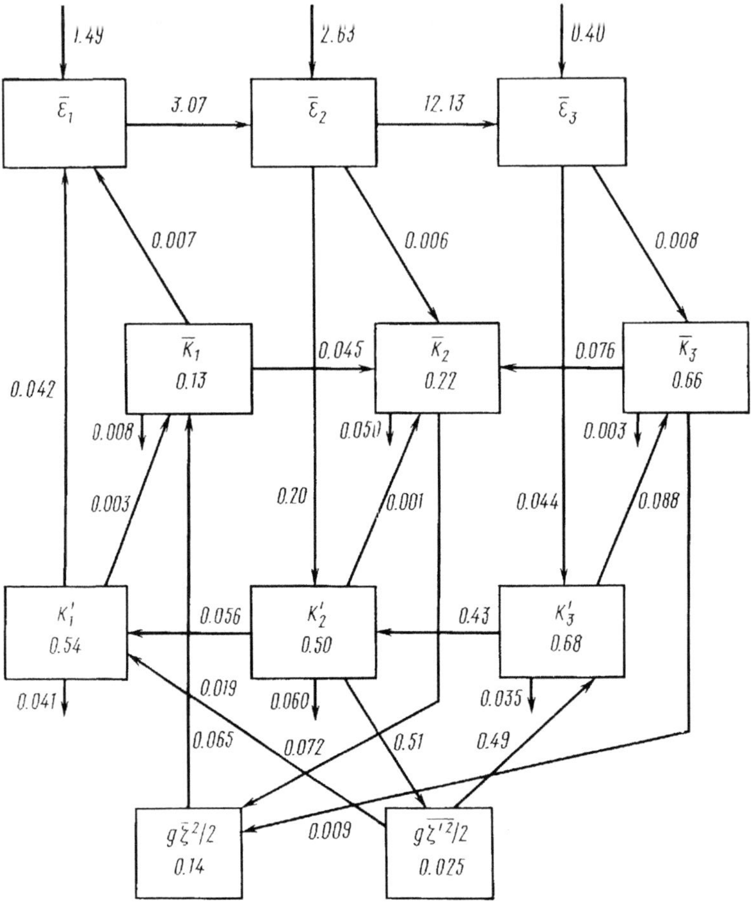

Fig. 3.4.14. Energy diagram for the quasistationary regime (according to the results of Neelov, 1982). Indicated in the squares are the average energy densities of the layers in 10^3 J/m^2. The magnitudes of the corresponding average energy fluxes per unit time are given in 10^{-3} J/m^2·s. The layers are numbered from the bottom to the sea surface. The bar denotes time averaging over a period of 795 days (from day 444 to day 1239); the prime indicates the eddy component. The values in the lower squares express that part of potential energy that is related to displacement ζ of the free surface.

baroclinic instability. Second, in all the layers there is transformation of the kinetic energy of the eddies into the kinetic energy of the mean current (the coefficient of negative viscosity for the upper layer is estimated as 600 m^2/s). Third, we also note that the potential energy associated with the displacement of the free surface plays an important role in the general energy exchange.

4.4. Some Concluding Remarks

We note that the density of kinetic energy in the mean current calculated in the eddy-resolving models is close to that estimated for the real ocean (see Section 1.1); however, the densities of the available potential energy $\bar{P}$ and the eddy kinetic energy K' are underestimated (in comparison with the estimates for the real ocean; see Section 1.1).[5] The underestimation of $\bar{P}$ is probably due to the relatively small scales of the resulting mean currents (owing to the fact that the dimensions of the basin are smaller than those of the real ocean; see the estimate (1.6)) and the underestimation of K' is likely to relate to the smoothing effect of the expressions for the horizontal turbulent friction which are used in the models. According to Harrison's (1980) review of American papers devoted to eddy-resolving models, the ratio $K'/\bar{K}$ varies from 0.3 to 6.3, and $K'/\bar{K} < 1$ in all the cases when the parametrization (4.6a) is used (with the exception of experiment B above, where $K'/\bar{K} = 2.5$; however, this experiment was performed for $A_L = 100$ m^2/s) and $K'/\bar{K} > 2$ when the parametrization (4.6b) is applied.

Finally, we make one more remark about the usually accepted interpretation of the transformation of the kinetic energy of the mean motion into the kinetic energy of eddies and of the potential energy of the mean motion into the kinetic energy of eddies as a manifestation of the respective processes of barotropic and baroclinic instability of the time-averaged current. The matter is that, generally, the time-averaged solution does not satisfy the *original equations of the model* and therefore there is hardly any sense in speaking about its instability. It is probably better to speak about the barotropic ($\bar{K} \rightleftarrows K'$) and baroclinic ($\bar{P} \rightleftarrows K'$) interaction between the synoptic eddies and the time-averaged current; these interactions are based on different physical mechanisms requiring further studies.

Thus we have described a number of numerical experiments with different eddy-resolving models. Our aim has been to state the general principles of this modelling, to discuss the methods used to interpret the calculation results, and to emphasize the most important of the resulting physical conclusions (the negative viscosity effect, etc.). This trend of investigations is undoubtedly very promising; however, in our opinion the results that have been published by now do not allow us to give definite answers to the questions posed at the beginning of this section.

In conclusion, we mention eddy-resolving models for analyzing the dynamics of eddies in an open bounded region (i.e. a region with boundaries through which the fluid can flow freely). These models are of interest from the viewpoint of studying the possibility of local oceanographic forecasts (e.g. see Grachev *et al.*, 1984, and Kamenkovich *et al.*, 1985b; see also the bibliography in these papers).

5. STATISTICAL DYNAMICS OF OCEAN EDDIES

The eddy-resolving models considered in the foregoing section (particularly the nongeostrophic models) make it possible, in principle, to describe all individual synoptic eddies (although this objective has not yet been accomplished). However, this description is inevitably very cumbersome. For many cases where it is not the influence of an individual eddy but the effect of a large collection of eddies that is important (e.g. when calculating the total horizontal transfer of heat) such a

[5] We note that in the construction of the energy diagrams in Figures 3.4.8 and 3.4.14 the available potential energy was not separated out from the total amount of stored potential energy.

description is not needed and only some statistical characteristics of the collection are required. Therefore, it is necessary to develop statistical dynamics of synoptic eddies regarded as a special kind of ocean macroturbulence (see the reviews by Mirabel' and Monin, 1979, 1980).

5.1. Quasi-Two-Dimensional Character of Ocean Macroturbulence

We note that large-scale ocean turbulence (with horizontal scales exceeding the ocean thickness) is quasi-two-dimensional and not two-dimensional because (1) the hydrodynamic fields describing it vary substantially with height, (2) in its dynamics a significant role is played by vertical motions interrelating horizontal motion fields at different heights, and (3) for adiabatic processes the vorticity is not conserved in fluid particles (and, by contrast, stretching of the vortex lines takes place). Nevertheless, it is natural to expect that some properties characteristic of two-dimensional turbulence manifest themselves in large-scale ocean turbulence. In this case the role of the 'two-dimensional' vorticity is played by the so-called *potential vorticity* (see Ertel, 1942; Obukhov, 1962; Monin, 1969) which, under the assumption that the large-scale three-dimensional motions in the ocean are nondivergent, has the form $I = \omega_a \nabla\eta$, where η is the entropy, $\omega_a = \text{rot}\,\mathbf{v} + 2\omega_e$ is the vorticity of the absolute velocity of motion, and ω_e is the angular velocity vector of the Earth's rotation (for the quantity η in the ocean, the so-called pseudoentropy should be taken, i.e. the entropy which is isopycnically and isobarically reduced to a constant salinity; see Monin, 1973). Adiabatic processes are characterized by the equations

$$\frac{d_h\eta}{dt} + w\frac{\partial\eta}{\partial z} = 0; \qquad \frac{d_h I}{dt} + w\frac{\partial I}{\partial z} = 0, \tag{5.1}$$

where d_h/dt is the symbol of the individual derivative relative to the horizontal motion; z is the height and w is the vertical velocity.

Eliminating w from (5.1) it is possible to derive, for large-scale processes, a certain conservation law for horizontal motions which is completely analogous to the law of conservation of vorticity in two-dimensional hydrodynamics. Taking into consideration that, in the case of large-scale motions, the field $\eta(\mathbf{x})$ is quasihorizontal and, hence, the vector $\nabla\eta$ is approximately vertical and that the part $\eta' = \eta - \eta_0$ of the field η depending on horizontal coordinates and time is small in comparison with the stationary value η_0, and assuming that the large-scale processes are quasihydrostatic, we can derive the conservation law

$$\frac{d_h}{dt}\left(\ln\omega_{az} + \frac{\partial}{\partial z}\frac{1}{\varrho_0 N^2}\frac{\partial\eta'}{\partial z}\right) = 0\,. \tag{5.2}$$

Moreover, if we take into account that the large-scale processes are quasigeostrophic, i.e. $|\text{rot}_z\,\mathbf{v}| \ll |2\omega_{ez}| = f$, $u = -\partial\psi/\partial y$, $v = \partial\psi/\partial x$, and $p' \approx \varrho_0 f\psi$, then Equation (5.2) can approximately be written as

$$\frac{d_h\omega_*}{dt} = \frac{\partial\mathscr{L}\psi}{\partial t} + \frac{\partial(\psi, \mathscr{L}\psi)}{\partial(x, y)} + \beta\frac{\partial\psi}{\partial x} = F; \qquad \omega_* = \mathscr{L}\psi + f;$$

$$\mathscr{L} = \Delta + \frac{\partial}{\partial z}\frac{f^2}{N^2}\frac{\partial}{\partial z}, \tag{5.3}$$

where F describes the possible sources and sinks of the quantity ω_* (cf. formula (1.39a)). This quantity can also be called the potential vorticity, and then $\Omega_* = \frac{1}{2}\omega_*^2$ is the *potential enstrophy*. Thus there appears profound analogy with two-dimensional turbulence (see Gavrilin *et al.*, 1972; Charney, 1971).

5.2. Spectrum of Two-Dimensional Turbulence

Two-dimensional turbulence possesses some specific properties determining its distinctions from three-dimensional turbulence. The most important of these distinctions is that in two-dimensional flows of incompressible nonviscous fluid not only the kinetic energy but also the vorticity ω in fluid particles is conserved (and hence its square is also conserved; half the square of the vorticity $\omega^2/2$ is called the *enstrophy*). In a viscous fluid the total kinetic energy $E = \frac{1}{2}\overline{|\mathbf{v}|^2}$ and the total enstrophy $\Omega = \frac{1}{2}\overline{\omega^2}$ degenerate according to the laws

$$\frac{\partial E}{\partial t} = -2\nu\Omega = -\varepsilon; \quad \frac{\partial \Omega}{\partial t} = -\nu\overline{|\nabla\omega|^2} = -\varepsilon_\omega. \tag{5.4}$$

It follows that for any bounded initial values of $\overline{|\mathbf{v}|^2}$ and $\overline{\omega^2}$ the value of $\overline{\omega^2}$ can only decrease, and for $\nu \to 0$ we have $\varepsilon \to 0$; hence, for large Reynolds numbers the kinetic energy is approximately constant and a cascade transfer of a significant amount of energy from small to large wavenumbers is impossible. In this case, according to Batchelor (1969), the energy spectrum in the energy-containing subrange of scales must have the form

$$E(k, t) = E^{3/2} t\, \mathcal{F}(E^{1/2} kt), \tag{5.5}$$

by virtue of which the energy and the enstrophy degeneration obeys the laws

$$E = E_\infty - A\nu t^{-1}; \quad \Omega = \frac{1}{2}At^{-2} \quad \text{and} \quad \varepsilon_\omega = At^{-3}, \tag{5.6}$$

where $A = \int_0^\infty x^2\mathcal{F}(x)\, dx$ is a numerical constant.

At the same time, a cascade enstrophy transfer from small to large wavenumbers and the corresponding nonzero limit of the quantity ε_ω for $\nu \to 0$ are not excluded.

Therefore, generally, in the inertial subrange of wavenumbers k the energy spectral density $E(k)$ is determined by the parameters ε and ε_ω. Here the constant length scale $L_\omega = (\varepsilon/\varepsilon_\omega)^{1/2}$ appears, and dimensional considerations show that the energy spectral density must have the form

$$E(k) = C_1(kL_\omega)\, \varepsilon^{2/3} k^{-5/3}, \tag{5.7}$$

where $C_1(kL_\omega)$ is a function of the dimensionless wavenumber kL_ω.

If we assume that at one of the ends of the inertial subrange of the spectrum only the parameter ε is significant, while at the other end only ε_ω is significant, then at the former end $C_1(kL_\omega) \sim$ const and the energy spectrum satisfies the ordinary five-thirds law, and at the latter end $C_1(kL_\omega) \sim (kL_\omega)^{-4/3}$ and the energy spectrum obeys the minus-three law

$$E(k) = C_\omega \varepsilon_\omega^{2/3} k^{-3}, \tag{5.8}$$

where C_ω is a numerical constant. (We note that in the range of applicability of this law the turbulent diffusion coefficient $\mathcal{K}(l)$ must satisfy not Richardson's four-thirds law, $\mathcal{K} \sim l^{4/3}$, but the anomalous 'square law'

$$\mathcal{K}(l) = C_{\mathcal{K}}\, \varepsilon_\omega^{1/3} l^2, \tag{5.9}$$

for which the ordinary parabolic diffusion equation does not provide any scale length.)

Finally, in the inertial-convective subrange of the spectrum the spectral density $E_T(k)$ of the passive scalar field is determined by the parameters ε, ε_ω, and ε_T, and dimensional considerations show that it must have the form

$$E_T(k) = B_1(kL_\omega)\, \varepsilon_T \varepsilon^{-1/3} k^{-5/3}, \tag{5.10}$$

where $B_1(kL_\omega)$ is a function of kL_ω.

At the end of the inertial-convective subrange, where the parameter ε_ω is insignificant, we have $B_1(kL_\omega) \sim$ const, and for $E_T(k)$ the ordinary five-thirds law is obtained. At the end where the parameter ε is insignificant, $B_1(kL_\omega) \sim (kL_\omega)^{2/3}$, and $E_T(k)$ satisfies the minus-one law

$$E_T(k) = B_\omega \varepsilon_T \varepsilon_\omega^{-1/3} k^{-1}, \tag{5.11}$$

where B_ω is a numerical constant (Gavrilin *et al.*, 1972; Saunders, 1972; Kraichnan, 1974).

The minus-three law was first discovered by Ogura (1952) from data on spectra of large-scale meteorological fields (see also Ogura, 1958, 1962). The explanation of this law as a consequence of enstrophy spectral transfer was first given by Kraichnan (1967) (see also Kraichnan, 1971, 1975) and then also by Batchelor (1969), who presented this explanation together with the results of preliminary numerical experiments on the evolution of two-dimensional turbulence which were performed by his post-graduate student R. W. Bray in 1961–1964 and defended as a thesis at Cambridge in 1966.

We note that the enstrophy spectral transfer turns out to be nonlocal, namely the major contribution to the rate of enstrophy transfer through a point k on the wavenumber axis is made by spectral components with wavenumbers $p \ll k \sim q$ and $q \ll k \sim p$. Therefore, the spectrum (5.8) should be corrected by introducing a logarithmic factor. (In the case of the minus-three law the total enstrophy $\int_{k_1}^{k_{\max}} k^2 E(k)\, dk$ is logarithmically divergent for both $k_1 \to 0$ and $k_{\max} \to \infty$.) Kraichnan (1971) suggested that the logarithmic correction should be introduced using the following argument. The energy spectral flux $\Pi(k)$ can be represented in its order of magnitude as a product of the total energy $kE(k)$ of motion with wavenumbers of the order of k by the root-mean-square velocity gradient in large-scale motions (with wavenumbers less than k) determined by the formula

$$\omega(k) \sim \left[\int_0^k k_1'^2\, E(k_1')\, \mathrm{d}k_1' \right]^{1/2}. \tag{5.12}$$

Then the condition $\Pi(k)$ = const can only be guaranteed by the five-thirds law, and in this case the major contribution to the integral in (5.12) is made by wavenumbers $k_1' \sim k$, which demonstrates the local character of the spectral transfer of energy. Similarly, the enstrophy spectral flux can be represented as $Q(k) \sim k^3E(k)\omega(k)$. In the case of the minus-three law we would obtain $\omega(k) \sim [\ln(k/k_1)]^{1/2}$, and $Q(k)$ would increase with k. To guarantee the condition $Q(k)$ = const the minus-three law must be replaced by

$$E(k) = C_\omega \varepsilon_\omega^{2/3} k^{-3} \left(\ln \frac{k}{k_1} \right)^{-1/3}. \tag{5.13}$$

Here the major contribution to the integral in (5.12) for $k \gg k_1$ is made by wavenumbers k' belonging to the subrange $k_1 \ll k' \leqslant k$, and hence the process of spectral transfer of enstrophy turns out to be substantially less localized than the process of spectral transfer of energy. Using (5.13) and the so-called almost-Markovian Galilean-invariant model of turbulence, Kraichnan derived for C_ω the estimate[6] $C_\omega = 1.65$ (whereas the analogous calculation of the coefficient C_1 in the five-thirds law gave the estimates $C_1 = 1.40$ and $C_1 = 6.69$ in the three- and two-dimensional cases, respectively). Kraichnan (1975) also put forward the idea that in the subrange of enstrophy spectral transfer there appears intermittency of spatial derivatives of enstrophy, which increases with k, but not of the enstrophy itself since it is conserved in each fluid particle provided that the viscosity effect is negligibly small; therefore, this intermittency does not change the form of the spectrum (5.13). By contrast, in the subrange of energy spectral transfer there can appear intermittency of nonnegative hydrodynamic characteristics which increases with decreasing k. This intermittency leads to the replacement of the five-thirds law by a power law with a smaller exponent and to a tendency of large-scale eddies of the same sign to form groups in the plane.

5.3. Spectral Transfer of Energy and Enstrophy

Kinetic energy spectra of low-frequency oscillations of current velocity (see the typical example in Figure 1.1.3) show that in the subrange of several-day periods separating the synoptic and mesoscale subranges there is a deep minimum of the energy spectral density. It must substantially hamper the possibility of continuous cascade transfer of energy across the spectrum from the synoptic subregion to the mesoscale subregion. On the contrary, as we are going to show, in the inertial subrange of the spectrum of two-dimensional turbulence, energy is transferred in the opposite direction, i.e. from small to large scales. This probably accounts for the existence of the indicated spectral minimum.

In three-dimensional turbulence to the absolute statistical equilibrium at zero viscosity there corresponds a uniform energy distribution over the degrees of freedom (over the wavenumbers k), i.e. a constant three-dimensional spectrum $F(k)$ = const and an energy spectral density $E(k) = 4\pi k^2 F(k) \sim k^2$, so that for the spectrum $E(k) \sim k^{-5/3}$ of the inertial subrange large wavenumbers are much lower

[6] Taking into consideration the correction suggested by Leith and Kraichnan (1971, footnote on p. 1054)

in their energy levels than small wavenumbers. It is natural to expect that interactions between the wavenumbers must tend to restore the absolute statistical equilibrium, i.e. to produce a spectral flux of energy from small to large wavenumbers ($\varepsilon > 0$).

In two-dimensional isotropic turbulence the quantity $\int_0^\infty (\alpha + \beta k^2)F(k)\, dk$ (where α and β are constants) is conserved for zero viscosity, and to the absolute statistical equilibrium or the uniform energy distribution over the degrees of freedom there corresponds a two-dimensional spectrum $F(k) \sim 1/(\alpha + \beta k^2)$ and an enstrophy spectral density $k^2E(k) = \pi k^3 F(k) \sim k^3/(\alpha + \beta k^2)$. Hence, for the spectrum $k^2E(k) \sim k^{-1}$ large wavenumbers are much lower in their enstrophy levels than the absolute equilibrium with small wavenumbers. Interactions between the wavenumbers must tend to restore the absolute equilibrium, i.e. to produce a spectral flux of enstrophy from small to large wavenumbers ($\varepsilon_\omega > 0$).

The same can also be established on the basis of considerations relating to the conservation of vorticity or potential vorticity ω_*. Indeed, in this case the fluid flows along the contours $\omega_* = \text{const}$. The deformation of any of these contours or surfaces produced by turbulence, when the material volumes are preserved, leads to formation of small-scale details in the field ω_* and hence to an increase of local vorticity gradients $|\nabla \omega_*|$; hence, the deformation generates the transfer of ω_* to the small-scale subrange.

Below we shall see that the spectral flux of energy must be of opposite sign, i.e. $\varepsilon < 0$. The possibility of this anomalous direction of energy spectral transfer was indicated by Onsager (1949) and Fjortoft (1953). This effect, called 'negative viscosity' (see the book by Starr, 1968), is characteristic of atmospheric and oceanic macroturbulence (see the above-mentioned book by Starr, 1968, and the book by Monin *et al.*, 1977). The phenomenon of negative viscosity was observed in laboratory experiments on convection carried out by Starr and Long (1953) and was found by Ward (1966) when processing statistical data on the solar photosphere.

Thus if, in the neighborhood of a wavenumber k_1, there is energy influx from outside with rate ε and enstrophy influx with rate ε_ω (and if k_1 lies in the corresponding inertial subrange), then we should expect that the major part of the incoming enstrophy is transferred towards large wavenumbers according to the minus-three law and dissipates under the viscosity effect for $k \gg k_1$, while the major part of the incoming energy is transferred towards small wavenumbers according to the five-thirds law and attains wavenumbers $k \ll k_1$ in time $t \sim (\varepsilon k^2)^{-1/3}$ (when the volume occupied by the fluid is bounded and the minimum wavenumber corresponding to it is $k_0 \ll k_1$, the energy accumulates in the turbulence components with wavenumber k_0).

We introduce the average wavenumbers of two-dimensional turbulence $k_1(t)$ and $k_2(t)$ for the energy and enstrophy spectra, respectively, by putting

$$k_1(t) = E^{-1} \int_0^\infty kE(k,t)\, dk; \quad k_2(t) = \Omega^{-1} \int_0^\infty k^3 E(k,t)\, dk. \tag{5.14}$$

Representing the quantities $kE(k)$ and $k^2E(k)$ as the products $[kE^{1/2}(k)] \times [E^{1/2}(k)]$ and $[k^{1/2}E^{1/2}(k)]\ [k^{3/2}E^{1/2}(k)]$ and applying the Cauchy–Bunyakovskiĭ–

Schwarz inequality to the integrals $\int_0^\infty kE(k)\,dk$ and $\int_0^\infty k^2E(k)\,dk$, we can prove the inequalities

$$k_1(t)\,k_2(t) \geqslant k_*^2 = \Omega/E; \quad k_1(t) \leqslant k_*; k_2(t) \geqslant k_*. \tag{5.15}$$

(see Novikov, 1978; Chefranov, 1979), which show that an enstrophy transfer towards large k (an increase of $k_2(t)$) must be accompanied by an energy transfer towards small k (a decrease of $k_1(t)$) and that, in the cascade processes, energy cannot accumulate in the small-scale subrange while enstrophy cannot accumulate in the large-scale subrange.

An alternative way of determining the direction of spectral transfer of energy and enstrophy in two-dimensional turbulence is to consider a subrange of the spectrum where similarity hypotheses are applicable. To this end we proceed, according to Kraichnan (1967), from the spectral equation for the energy of two-dimensional isotropic turbulence:

$$\left(\frac{\partial}{\partial t} + 2\nu k^2\right) E(k) = \frac{1}{2}\int_0^\infty\int_0^\infty T(k, p, q)\,dp, dq. \tag{5.16}$$

Here the function $T(k, p, q)$ is expressed in terms of the Fourier transform of the third-order two-point correlation function of the velocity field and possesses the following properties: it is nonzero only when k, p, and q form the sides of a triangle, is symmetric relative to the second and third arguments, and satisfies the identities

$$T(k, p, q) + T(p, q, k) + T(q, k, p) = 0;$$
$$k^2\,T(k, p, q) + p^2\,T(p, q, k) + q^2\,T(q, k, p) = 0. \tag{5.17}$$

These identities express the laws of conservation of energy and enstrophy for the interaction of any three wave components of turbulence with wave vectors **k**, **p**, and **q**. Consider the subrange of the spectrum where the spectral functions $E(k)$ and $T(k, p, q)$ possess the following similarity properties:

$$\frac{E(ak)}{E(k)} = a^{-n}; \quad \frac{T(ak, ap, aq)}{T(k, p, q)} = a^{-(3n+1)/2} \tag{5.18}$$

for an arbitrary a. The exponent chosen for the second of conditions (5.18) is the same as that for the function $[E(k)]^{3/2}k^{-1/2}$ whose dimension coincides with the dimension of T. Using the enumerated properties of the function $T(k, p, q)$ we derive the following expressions for the rates of transfer $\Pi(k)$ and $Q(k)$ of kinetic energy and enstrophy from wavenumbers less than k to those exceeding k:

$$\Pi(k) = \frac{1}{2}\int_k^\infty dk' \int_0^k\int_0^k T(k', p, q)\,dp\,dq -$$

$$-\frac{1}{2}\int_0^k dk' \int_k^\infty\int_k^\infty T(k', p, q)\,dp\,dq \tag{5.19}$$

$$= k^{(5-3n)/2}\int_0^1 dv \int_1^\infty T(1, v, w)\,W_1(v, w, n)\,dw;$$

$$W_1(v, w, n) = -(w^2 - v^2)^{-1}\left[(1 - v^2)\int_1^{\omega} u^{(3n-7)/2}\,du - \right.$$

$$\left. - (w^2 - 1)\int_v^1 u^{(3n-7)/2}\,du\right];$$

$$Q(k) = \frac{1}{2}\int_k^{\infty} k'^2\,dk' \int_0^k\int_0^k T(k', p, q)\,dp\,dq -$$

$$-\frac{1}{2}\int_0^k k'^2\,dk' \int_k^{\infty}\int_k^{\infty} T(k', p, q)\,dp\,dq \qquad (5.20)$$

$$= k^{(9-3n)/2}\int_0^1 dv \int_1^{\infty} T(1, v, w)\,W_2(v, w, n)\,dw;$$

$$W_2(v, w, n) = -(w^2 - v^2)^{-1}\left[(1 - v^2)\,w^2\int_1^{w} u^{(3n-11)/2}\,du - \right.$$

$$\left. - (w^2 - 1)v^2\int_v^1 u^{(3n-11)/2}\,du\right].$$

It follows that in the subrange of energy spectral transfer where the five-thirds law applies, i.e. $n = \frac{5}{3}$, the quantity $\Pi(k)$ does not depend on k (it is equal to ε) and $W_2(v, w, \frac{5}{3})$ is equal to zero so that $Q(k) = 0$. By contrast, in the subrange of enstrophy spectral transfer where the minus-three law applies, i.e. $n = 3$, the quantity $Q(k)$ does not depend on k (it is equal to ε_ω) and $W_1(v, w, 3)$ is equal to zero, and hence $\Pi(k) = 0$. Moreover, we can prove that $W_1(v, w, \frac{5}{3}) > 0$ and $W_2(v, w, 3) < 0$, and thus the sign of ε in the subrange of energy spectral transfer and the sign of ε_ω in the subrange of enstrophy spectral transfer are opposite.

The above predictions of the relationships between the processes of spectral transfer of energy and enstrophy were confirmed by the calculations of the functions (5.19) and (5.20) performed by Kraichnan (1967) for the quasi-Markovian model of two-dimensional turbulence. These calculations result in

$$T(k, p, q) = \pi k^2\{[b(k, p, q) + b(k, q, p)]\,\Theta(k, p, q)\times$$
$$\times\, F(p)\,F(q) + b(k, p, q)\,\Theta(p, q, k)\,F(k)\times$$
$$\times\, F(q) - b(k, q, p)\,\Theta(q, k, p)\,F(k)\,F(p)\}, \qquad (5.21)$$

Here $F(k) = E(k)/(\pi k)$; $b(k, p, q) = 2k^{-4}(k^2 - q^2)(p^2 - q^2)(1 - \alpha^2)^{1/2}$, where α is the cosine of the angle opposite to the side k in a triangle with sides k, p, and q; finally, $\Theta(k, p, q)$ is the relaxation time, i.e. the integral with respect to time of Green's function of the operator $\partial/\partial t + \nu(k^2 + p^2 + q^2) + \eta(k, p, q)$ where η is the relaxation coefficient of friction which must be determined additionally. In this case the energy equation (5.16) becomes analogous to Boltzmann's kinetic equation. Introducing energy injection at wavenumber k_1 in this equation, Pouquet *et al.* (1975) found its stationary solution of the form of (5.13). For the constant C_ω in this solution the value $C_\omega \approx 1.74$ was obtained.

Leith (1969) carried out a simplified calculation of the functions $\Pi(k)$ and $Q(k)$ in the so-called diffusion approximation, which gave

$$\Pi(k) = -bk^{-1}\frac{\partial}{\partial k}k^{9/2}E^{3/2}(k);\ Q(k) = -bk^3\frac{\partial}{\partial k}k^{5/2}E^{3/2}(k), \qquad (5.22)$$

where b is a positive constant. Hence, in the case of the five-thirds law we have $\Pi = \text{const} < 0$ and $Q = 0$, and in the case of the minus-three law, $Q = \text{const} > 0$ and $\Pi = 0$.

5.4. Numerical Experiments on Two-Dimensional Turbulence

Until recently it has been impossible to calculate an individual realization of a turbulent flow (on a sufficiently fine spatial grid or with a sufficiently large number of spectral components or 'fluid particles') and then determine statistical characteristics of the turbulence, say, by means of time averaging. At the present time the appearance of high-speed computers and the development of highly accurate methods of numerical integration of equations of hydromechanics open up possibilities for such calculations which have already begun to be realized, naturally beginning with the two-dimensional case which is less cumbersome.

Even the first preliminary numerical experiments on two-dimensional turbulence performed by Bray (see Batchelor, 1969) revealed manifestations of the minus-three law (5.8). More definite results were obtained by Lilly (1969) who integrated numerically the two-dimensional vorticity equation

$$\frac{\partial}{\partial t}\Delta\psi + \frac{\partial(\psi, \Delta\psi)}{\partial(x, y)} = \nu\Delta\Delta\psi + F. \qquad (5.23)$$

The value F_n of the exciting function F at the nth time step was set as the solution of the first-order finite-difference equation of a Markovian process $F_{n+1} = R_n F_n + (1 - R_n^2)^{1/2}\hat{F}_{n+1}$, where $R_n = (1 - \Delta t/2\tau)/(1 + \Delta t/2\tau)$ and $\hat{F}_{n+1}$ is determined by a random sample from the Gauss distribution of the amplitudes of all the Fourier components of the field F with wave vectors $\mathbf{k} = (k_x, k_y)$ on the sides of a square of dimension $2k_1$ (statistically stationary excitation at wavenumber k_1). In the main experiment, $k_1 = 2\pi$, $\nu = 2.5 \times 10^{-4}$, $\tau = \frac{1}{2}$, and $\Delta x = \frac{1}{8}$ in dimensionless form, and a 64×64 spatial grid of points was taken. The spectrum $E(k)$ turned out to satisfy the five-thirds law for $k < k_1$ with a constant $C_1 \approx 6.2$ after 2200 time steps and the minus-three law with a constant $C_\omega \approx 4$ for $k > k_1$ within the limits from one to three octaves. The estimate $C_\omega \approx 4$ differs from the above-mentioned estimate of Kraichnan. However, Lilly (1971, 1972) obtained the estimate $C_\omega \approx 2$ (in experiments on the degeneration of two-dimensional turbulence where it was assumed that $F = 0$ after 50 time steps), which is close to the theoretical estimate. These works also confirmed the laws of degeneration of turbulence (5.6) indicated by Batchelor. The minus-three law was also confirmed in the numerical experiments whose results were published in the papers by Gavrilin *et al.* (1972), Gavrilin and Mirabel' (1972), and Mirabel' (1974). The last of these papers presents the results of experiments on the degeneration of two-dimensional turbulence (with an initial field ψ in the form of a Gauss random field with a spectrum $(k_0^2/\pi)(k^2 + k_0^2)^{-2}$ which gave the estimate $C_\omega \approx 1.54$) and experiments on the evolution of a two-dimensional field of a passive scalar ϑ' described by the equation

$$\frac{\partial\vartheta'}{\partial t} + \frac{\partial(\psi, \vartheta')}{\partial(x, y)} = \chi\Delta\vartheta' + \frac{\partial(\bar{\vartheta}, \psi)}{\partial(x, y)}. \tag{5.24}$$

At the first 500 time steps a constant gradient of the averaged concentration $\bar{\vartheta}$ was set after which the gradient was equated to zero; the initial field $\vartheta'(x, y, 0)$ was equal to zero; after 1500–2000 time steps a spectrum of the form of (5.11) with a coefficient $B_\omega \approx 1.56$ was obtained for the field $\vartheta'(x, y, t)$.

However, all these facts confirming the laws of spectral transfer of enstrophy were called into question in the paper by Herring *et al.* (1974), where the most detailed experiments on the degeneration of two-dimensional turbulence were carried out. The initial field ψ was taken in the form of a Gauss periodic random field in a $2\pi \times 2\pi$ square with a spectrum $v_0^2(k/k_0)\, e^{-k/k_0}$; the viscosity coefficient was equal to 5×10^{-3}, 2.5×10^{-3}, and 1×10^{-3}; the spatial grid contained 64×64 or 128×128 points (the Reynolds numbers $Re = E\eta^{-1/3}\nu^{-1}$ varied within the limits $50 \leqslant Re \leqslant 100$); the vorticity equation (5.2) was integrated by both the Arakawa second-order finite-difference method and the Orszag method (1971; also see the papers by Fox and Orszag, 1973a, b). This calculation was an extension of the works on mathematical modelling of three-dimensional turbulence of Orszag and Patterson (1972), and Herring *et al.* (1973) to the two-dimensional case. The main result of the calculation is that the correct modelling of the inertial range of the spectrum of two-dimensional turbulence, i.e. the resolution of the scales making notable contributions to the dissipation rate of the enstrophy ε_ω for $\nu = 1 \times 10^{-3}$, requires a spatial grid of 512×512 points. From this viewpoint all the above-mentioned numerical experiments lacked the necessary resolution (in the Lilly experiments the Reynolds number Re was equal to 315, 411, and 537). On the other hand, the calculation showed that the large-scale components of the turbulence with $k \leqslant 10$ are almost independent of Re in the range $Re = 150$–1100 and therefore they can be satisfactorily modelled for small Re as well.

We also mention the numerical experiments on the appearance of absolute statistical equilibrium with positive or negative temperatures performed in a number of works. A typical work of this kind is the paper of Seyler *et al.* (1975), devoted to numerical integration of a truncated two-dimensional spectral equation for vorticity in an inviscid fluid with smooth initial data. In particular, we note that a typical configuration in the states with negative temperatures ($\alpha < 0$) is the one with pair of large eddies of opposite sign.

We note that earlier Manabe *et al.* (1970) revealed the minus-three law in the range of zonal wavenumbers $k = (2\pi a \cos\theta)/l = 8$–$20$ (where a is the Earth's radius, θ is the complement of the latitude, and l is the wavelength) in the kinetic energy spectra of hydrodynamical fields obtained by numerical integration of the equations of the physico-mathematical model of the general circulation of the

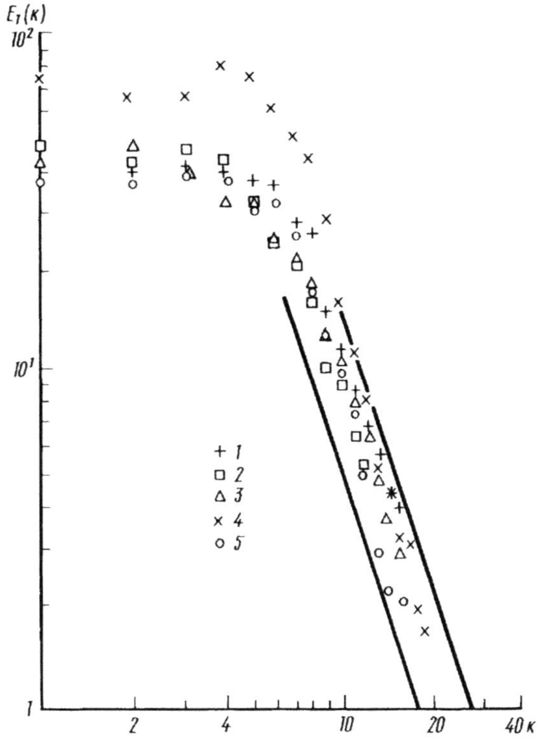

Fig. 3.5.1. One-dimensional wind velocity spectra E_1 (k) (in 0.31 m^2/s^2) according to empirical data of different authors (see Leith, 1971); k is the spherical wavenumber (the number of wavelengths in the circle of latitude). (1) the data of Saltzman and Fleisher (1962); (2) the data of Horn and Bryson (1963); (3) the data of Kao and Wendell (1970); (4) the data of Julian *et al.* (1970); (5) the data of Wiin-Nielsen (1967).

atmosphere. The minus-three law in the range of scales $l \sim$ 1200–4000 km is also found in the empirical spectra of large-scale meteorological fields whose summary can, for instance, be found in the above-mentioned paper by Gavrilin *et al.* (1972). As an example, Figure 3.5.1 presents one-dimensional kinetic energy spectra of large-scale meteorological fields according to various data, demonstrating visually the existence of subranges with the minus-three law in these spectra.

5.5. Geostrophic Eddies and Rossby Waves

Let us return to Equation (5.3) describing ocean macroturbulence. Its major distinction from the two-dimensional vorticity equation (5.23), which was subjected to a special study by Rhines (1975; see also Rhines, 1973, 1977), is the presence of the term $\beta(\partial\psi/\partial x)$, owing to which the solutions of the equation include not only eddies carrying the water involved in them but also Rossby waves travelling over the water which are not contained among the solutions of the two-dimensional equation (5.23). As to the replacement of the operator Δ by the three-dimensional operator $\mathscr{L}$, which leads to the appearance of baroclinic modes in addition to barotropic ones, this is not so important since the basic specific properties of Equation (5.3) which we are going to discuss here are also retained in the barotropic case, i.e. for $\mathscr{L} = \Delta$. We begin our consideration with this case.

If the initial field ψ for (5.3) is taken in the form of close-packed eddies with a narrow-band spectrum having a peak in the vicinity of a wavenumber k_0, then for the measure of relative contribution of turbulence and Rossby waves we can take the ratio of the nonlinear term on the left-hand side of (5.3) to the term $\beta(\partial\psi/\partial x)$. This ratio is equal to $\delta = U/c_\varphi = 2k_0^2U/\beta$ where U is the root-mean-square velocity of motion ($U^2/2 = E$) and $c_\varphi = \beta/(2k_0^2)$ is the phase velocity of Rossby waves with wavenumber k_0. For $\delta > 1$, eddies dominate in the dynamics of motion, and for $\delta < 1$, Rossby waves dominate. The wavenumber corresponding to the boundary $\delta = 1$ is equal to

$$k_\beta = \left(\frac{\beta}{2U}\right)^{1/2}. \tag{5.25}$$

For typical conditions in the ocean at latitude 30° for $U = 5$ cm/s we obtain $k_\beta^{-1} = 70$ km, to which eddy diameters about 220 km correspond. This is in good agreement with the observation data on synoptic eddies in the open ocean.

If the typical fluid velocity $\hat{U}(k)$ for a wavenumber k is introduced with the aid of the relation $\hat{U}^2(k)/2 = kE(k)$, then to the boundary between turbulence and waves there corresponds the relation $\hat{U}(k) = c_\varphi(k)$ or $[2kE(k)]^{1/2} = \beta/(2k^2)$, whence the boundary spectrum is obtained:

$$E_\beta(k) = c_1\beta^2k^{-5}, \tag{5.26}$$

where $c_1 = 1/8$.

To the condition $E(k) > E_\beta(k)$ there corresponds turbulence and to the condition $E(k) < E_\beta(k)$ there correspond waves. The boundary spectrum is very steep since the Rossby wave frequencies $\sigma = -\beta k_x k^{-2}$ rapidly increase together with the wavelengths $2\pi/k$. (The possibility of 'saturation' of the spectrum up to the value (5.26) for $k \ll k_1$ under continuous energy influx in the neighborhood of the wavenumber k_1 is ambiguous because the 'saturated' spectrum, besides its dependence on β and k, can also depend on the position of the spectral peak $\hat{k}$; the velocity shear which acts on eddies with scales of the order of k^{-1} is primarily determined by eddies with scales of the order of $\hat{k}^{-1}$.)

5.6. Growth of Eddies

Let $k_0 \gg k_\beta$ so that the initial field ψ is formed by turbulence. Naturally, the initial narrow spectral peak will broaden, i.e. $(\partial/\partial t)\int(k - k_1)^2E(k)\,dk > 0$, where k_1 is the energy-averaged wavenumber that was used previously (see (5.14)). Since the total enstrophy cannot increase, i.e. $(\partial/\partial t)\int k^2E(k)\,dk \leqslant 0$, and the total energy does not change, we obtain $\partial k_1/\partial t < 0$. More precisely, the employment of Batchelor's self-similar spectrum (5.5) in the energy-containing subrange of scales yields

$$\frac{\partial}{\partial t}\frac{1}{k_1} = TU, \tag{5.27}$$

where $T = \int \mathcal{F}(\xi)\, d\xi [\int \xi \mathcal{F}(\xi)\, d\xi]^{-1}$ is a numerical constant.

Thus the dominant wavenumber k_1 of eddies and their typical frequency $\sigma_1 \sim Uk_1$ decrease, and in time of the order of $(TUk_\beta)^{-1} \sim [T(\beta U)^{1/2}]^{-1}$, which is independent of k_0, the wavenumber k_1 attains the value k_β for which the eddies are transformed into Rossby waves. k_1 then continues to decrease owing to the weak interactions between Rossby waves, but this decrease is much slower since the wave interactions require both superposition of triples of waves in space and resonance between their wavenumbers and frequencies. The resultant effect of these interactions on the evolution of the spectrum $F(\mathbf{k}, \tau)$ of the two-dimensional wave field is described by the kinetic equation (see Section 3 of Chapter 2 and also Longuet-Higgins and Gill, 1967; Kenyon, 1964):

$$\begin{aligned}\frac{\partial F}{\partial \tau} &= 8\pi \iint (D_{\mathbf{k}_1\mathbf{k}_2} F_1 F_2 + D_{\mathbf{k}\mathbf{k}_1} FF_1 + D_{\mathbf{k}\mathbf{k}_2} FF_2) \times \\ &\times \left[\frac{1}{(k^2+1)(k_1^2+1)(k_2^2+1)}\right] \times \\ &\times \delta(\mathbf{k} + \mathbf{k}_1 + \mathbf{k}_2)\, \delta(\sigma + \sigma_1 + \sigma_2)\, D_{\mathbf{k}_1\mathbf{k}_2}\, d\mathbf{k}_1\, d\mathbf{k}_2,\end{aligned} \tag{5.28}$$

where dimensionless quantities are used (such that the Rossby deformation radius is taken as unit measure), $\tau = \delta^2 t$ is slow time (δ is the small parameter in the nonlinear terms of the vorticity equation), $\sigma(k) = -k_x(k^2+1)^{-1}$ is the Rossby wave frequency, and

$$D_{\mathbf{k}_1\mathbf{k}_2} = \tfrac{1}{2}(k_{1x}k_{2y} - k_{1y}k_{2x})(k_2^2 - k_1^2).$$

The frequency decrease resulting from these interactions is due to the fact that the wave with the greatest frequency in a resonant triad is unstable relative to the energy increase in the other two waves (Hasselmann, 1967; Rhines, 1975).

The numerical experiments on integrating the barotropic equation for the potential vorticity (i.e. Equation (5.3) for $\mathcal{L} = \Delta$), carried out by Rhines (1975) using a method similar to that of Herring *et al.* (1974), confirmed the above predictions and gave the estimate $T \approx 3 \times 10^{-2}$ for T in the turbulent regime. The values of T rapidly decreased with increasing $\delta^{-1} = \beta/(2k_1^2U)$ and became five times as small as that estimate (i.e. $T \approx 6 \times 10^{-3}$) when the wave regime appeared (i.e. for $\delta = 1$). We note that these results are in good agreement with the theoretical estimates obtained for the effect of resonant interaction by Kenyon (1967) ($T \sim 4 \times 10^{-3}$), Lorenz (1972), Longuet-Higgins and Gill (1967), and Gill (1974) ($T \sim (3\text{–}6) \times 10^{-3}$). Further, the typical width $\langle k^2\rangle \langle k\rangle^{-2} - 1$ of the developed spectrum and its form in the numerical experiments performed by Rhines were 0.34 and $k^{-4.3}$ without inclusion of the β-effect and 0.1 and $k^{-5.4}$ with inclusion of the β-effect. We also mention the numerical experiments of Holloway and Hendershott (1977), whose results are in good agreement with the theoretical calculations of these authors, carried out with the aid of the 'Markovian quasinormal' closure of equations of two-dimensional turbulence with the β-effect. (In these experiments a spectrum with a slope between k^{-3} and k^{-4} was obtained for $k_\beta = 3$–5 in the subrange $k = 10$–60.)

5.7. Anisotropization

As a consequence of the dispersion relation $\sigma = -\beta k_x k^{-2}$ for Rossby waves, their appearance results in tendency towards westward motion of the inhomogeneities of the field ψ, and their frequency decrease leads to anisotropization of these inhomogeneities, mainly to the increase of their meridional wavenumbers k_y. This leads to a tendency of eddies to stretch along circles of latitude and 'organize' zonal currents (and, in the limit, stationary zonal currents in narrow latitudinal strips with westward or eastward direction changing from strip to strip; for some reasons this limit is not attained in the ocean). This tendency was observed in the trajectories of labelled fluid particles during the MODE experiment (Rossby *et al.*, 1975; Freeland *et al.*, 1975) and also in the numerical experiments (Williams, 1978, 1979) on computer modelling of some turbulent regimes observed in the Jovian atmosphere which can provide an explanation of its strips (zonal jets with average velocities of the order of 50 m/s and alternating vortex strips mainly of oval form whose dimensions increase towards the equator and are of the order of $L_\beta = \pi k_\beta^{-1} \sim (1$–$2) \times 10^4$ km).

It is the tendency of wave interactions to preferably transfer the energy – not to other wavelengths but to other directions in space – that leads to a sharp decrease of T in (5.27) in the transition from turbulence to waves. As a result of this sharp decrease of T for $k_1 \sim k_\beta$, eddies do not grow up to the global dimensions, which could have happened if the radiation of Rossby waves had not appeared and the turbulence had remained unaffected. The decrease of T owing to the β-effect, i.e. the deceleration of the cascade transfer of energy towards large scales, results in the deceleration of the cascade transfer of enstrophy towards small scales. In the presence of dissipation of enstrophy the spectrum must lose its small-scale structure and become narrower and therefore steeper.

The influence of the β-effect on small scales is not confined to decelerating the rate of spectral transfer of enstrophy. This effect substantially influences the scales $k^{-1} > k_\beta^{-1}$ and also serves as a source of anisotropization for $k > k_\beta$. It is interesting that the meridional anisotropy for small k excites the zonal anisotropy for large k (Holloway and Hendershott, 1977). In other words, the ordinary isotropization process of small-scale components in the case of two-dimensional turbulence, in contrast to three-dimensional turbulence, has a nonlocal character (Herring, 1975), which leads to the effect of the large-scale anisotropy on the small scale anisotropy.

5.8. Barotropization

We come back to the baroclinic equation for the potential vorticity (5.3). Here there appears a distinction between the behavior of waves with wavelengths less than the Rossby deformation radius $L_R = NH/f$ for which $\mathscr{L} \approx \Delta$ (i.e. the vertical interaction between different fluid layers plays a minor role and these layers undergo evolution approximately independently of one another) and the behavior of waves with wavelengths $L > L_R$ for which different fluid layers strongly interact along the vertical (i.e. behave as a single layer) so that for these waves the fluid turns out to be effectively barotropic. Hence, if $k_R = 2\pi/L_R$, then, as the average wavenumber $k_1(t)$ decreases and attains the value k_R, there must appear a tendency towards the 'barotropization' of eddies, i.e. towards a decrease of the distinctions between the motions at different levels. The numerical experiments with the two-layer model performed by Rhines (1977) confirmed these predictions and showed that in the absence of disturbing factors (which will be discussed below), and for not too loose-packed eddies, the barotropization process is total and very fast.

The interaction between barotropic and baroclinic components can be studied in the simplest way by considering the example of the two-layer model (and with neglect of the β-effect which does not influence the barotropization directly). Without loss of generality, we assume that the layer thicknesses are equal. We introduce the half-sum ψ and the half-difference φ of the stream functions at middle levels in the upper and lower layers. Then (5.3) implies the following equations for the Fourier components $\psi_{\mathbf{k}}$ and $\varphi_{\mathbf{k}}$ of ψ and φ:

$$\frac{d\psi_{-\mathbf{k}}}{dt} = \sum_{\mathbf{k}+\mathbf{p}+\mathbf{q}} a_{\mathbf{kpq}}(\psi_{\mathbf{p}}\psi_{\mathbf{q}} + \varphi_{\mathbf{p}}\varphi_{\mathbf{q}}), \tag{5.29}$$

$$\frac{d\varphi_{-\mathbf{k}}}{dt} = \sum_{\mathbf{k}+\mathbf{p}+\mathbf{q}} \left[b_{\mathbf{kpq}}(\psi_{\mathbf{p}}\varphi_{\mathbf{q}} + \psi_{\mathbf{q}}\varphi_{\mathbf{p}}) + c_{\mathbf{kpq}}(\psi_{\mathbf{p}}\varphi_{\mathbf{q}} - \psi_{\mathbf{q}}\varphi_{\mathbf{p}}) \right], \tag{5.30}$$

where the interaction coefficients are calculated by the formulas

$$a_{\mathbf{kpq}} = \frac{1}{2}(\mathbf{p} \times \mathbf{q})_z(q^2 - p^2)\, k^{-2},$$

$$b_{\mathbf{kpq}} = \frac{1}{2}(\mathbf{p} \times \mathbf{q})_z(q^2 - p^2)(k^2 + k_R^2)^{-1}, \tag{5.31}$$

$$c_{\mathbf{kpq}} = \frac{1}{2}(\mathbf{p} \times \mathbf{q})_z k_R^2(k^2 - k_R^2)^{-1}.$$

Multiplying the relations (5.29) and (5.30) by $\psi_{\mathbf{k}}$ and $\varphi_{\mathbf{k}}$ respectively, adding them to their complex conjugates, and performing the operation of averaging (denoted by $\langle \cdots \rangle$) we obtain nonclosed equations for the spectral moments $\langle |\psi_{\mathbf{k}}|^2 \rangle$ and $\langle |\varphi_{\mathbf{k}}|^2 \rangle$:

$$\frac{\partial}{\partial t}\langle|\psi_{\mathbf{k}}|^2\rangle = \sum_{\Delta} a_{\mathbf{kpq}}\, 2\,\mathrm{Re}(\langle\psi_{\mathbf{k}}\,\psi_{\mathbf{p}}\,\psi_{\mathbf{q}}\rangle + \langle\psi_{\mathbf{k}}\,\varphi_{\mathbf{p}}\,\varphi_{\mathbf{q}}\rangle),$$

$$\frac{\partial}{\partial t}\langle|\varphi_{\mathbf{k}}|^2\rangle = \sum_{\Delta}\Big[b_{\mathbf{kpq}}\, 2\,\mathrm{Re}(\langle\psi_{\mathbf{p}}\,\varphi_{\mathbf{q}}\,\varphi_{\mathbf{k}}\rangle + \langle\psi_{\mathbf{q}}\,\varphi_{\mathbf{p}}\,\varphi_{\mathbf{k}}\rangle) +$$

$$+ c_{\mathbf{kpq}}\; 2\;\mathrm{Re}(\langle\psi_{\mathbf{p}}\,\varphi_{\mathbf{q}}\,\varphi_{\mathbf{k}}\rangle - \langle\psi_{\mathbf{q}}\,\varphi_{\mathbf{p}}\,\varphi_{\mathbf{k}}\rangle)\Big]. \tag{5.32}$$

Here the summation extends only to those triads of wave vectors **k**, **p**, and **q** which form triangles. The laws of conservation of the total energy and the potential enstrophy in the spectral form are obtained from (5.32):

$$\frac{\mathrm{d}}{\mathrm{d}t}\sum_{\mathbf{k}}(U_{\mathbf{k}} + T_{\mathbf{k}}) = 0; \tag{5.33}$$

$$\frac{\mathrm{d}}{\mathrm{d}t}\sum_{\mathbf{k}}[k^2U_{\mathbf{k}} + (k^2 + k_{\mathrm{R}}^2)\,T_{\mathbf{k}}] = 0, \tag{5.34}$$

where $U_{\mathbf{k}} = k^2\langle|\psi_{\mathbf{k}}|^2\rangle$ is the energy of the barotropic component with wave vector **k**; $T_{\mathbf{k}} = (k^2 + k_{\mathrm{R}}^2)\langle|\varphi_{\mathbf{k}}|^2\rangle$ is the energy of the baroclinic component with wave vector **k**.

The conditions (5.33) and (5.34) are fulfilled when, for each of the interacting triads satisfying the homogeneity condition $\mathbf{k} + \mathbf{p} + \mathbf{q} = 0$, the rates of the spectral transfer of energy and enstrophy satisfy the conditions

$$\frac{\mathrm{d}}{\mathrm{d}t}(E_{\mathbf{k}} + E_{\mathbf{p}} + E_{\mathbf{q}}) = 0; \tag{5.35}$$

$$\frac{\mathrm{d}}{\mathrm{d}t}(\Omega_{\mathbf{k}} + \Omega_{\mathbf{p}} + \Omega_{\mathbf{q}}) = 0, \tag{5.36}$$

where $E_{\mathbf{k}} = U_{\mathbf{k}} + T_{\mathbf{k}}$; $\Omega_{\mathbf{k}} = k^2U_{\mathbf{k}} + (k^2 + k_{\mathrm{R}}^2)T_{\mathbf{k}}$.

Now we analyze the possible processes of energy transfer between elementary interaction triads. First we note that the structure of the terms involved in the right-hand sides of Equations (5.32) admits only of interactions between three barotropic components and between one barotropic and two baroclinic components. (For the sake of brevity, we shall call them interactions of types I and II, respectively, and denote them by superscripts 1 and 2 in parentheses). For these two types of interaction the conservation laws (5.35) and (5.36) are rewritten thus (Salmon, 1978):

$$\frac{\mathrm{d}}{\mathrm{d}t}[U_{\mathbf{k}}^{(1)} + U_{\mathbf{p}}^{(1)} + U_{\mathbf{q}}^{(1)}] = 0; \tag{5.37a}$$

$$\frac{\mathrm{d}}{\mathrm{d}t}[U_{\mathbf{k}}^{(2)} + T_{\mathbf{p}}^{(2)} + T_{\mathbf{q}}^{(2)}] = 0; \tag{5.37b}$$

$$\frac{\mathrm{d}}{\mathrm{d}t}[k^2U_{\mathbf{k}}^{(1)} + p^2U_{\mathbf{p}}^{(1)} + q^2U_{\mathbf{q}}^{(1)}] = 0; \tag{5.38a}$$

$$\frac{d}{dt}[k^2 U_{\mathbf{k}}^{(2)} + (p^2 + k_R^2)\, T_{\mathbf{p}}^{(2)} + (q^2 + k_R^2)\, T_{\mathbf{q}}^{(2)}] = 0. \tag{5.38b}$$

The processes of energy and enstrophy transfer between barotropic modes satisfying the conservation laws (5.37a) and (5.38a) are completely analogous to transfer processes in two-dimensional turbulence. Here the cascades between local modes with $k \approx p$ and nonlocal modes with $k \ll p \sim q$ (forming an isosceles triangle) are forbidden and those between modes with $p \ll k \sim q$ and $q \ll k \sim p$ are allowed. The energy in an elementary triad can propagate from a mode with a medium wavenumber to the two extreme modes, and ultimately the transfer of energy acquires a statistically irreversible tendency towards low wavenumbers. For interactions of type II (with the effect of stretching of the vortex lines) there appears a distinction between the behavior of turbulent components with wavenumbers $k \gg k_R$ and $k \ll k_R$. In the former case the interaction between the fluid layers plays an insignificant role; these layers undergo evolution approximately independently of each other (more precisely, the motions in the layers are uncorrelated), and the energy flux is directed predominantly towards large scales. To facilitate the analysis of the more complicated case $k \ll k_R$ we make use of the following relations:

$$\frac{\dot{U}_{\mathbf{k}}}{\dot{T}_{\mathbf{p}}} = \frac{q^2 - p^2}{k^2 - q^2 - k_R^2}; \quad \frac{\dot{T}_{\mathbf{q}}}{\dot{T}_{\mathbf{p}}} = \frac{p^2 - k^2 + k_R^2}{k^2 - q^2 - k_R^2};$$

$$\frac{\dot{U}_{\mathbf{k}}}{\dot{T}_{\mathbf{q}}} = \frac{q^2 - p^2}{k_R^2 - k^2 + p^2}. \tag{5.39}$$

Relations (5.39) are implied by (5.37b) and (5.38b) and describe the relative contributions of the triad interactions to energy transfer processes (here $\dot{U} \equiv dU/dt$, etc.). If $p \sim q$ or $p \ll q$, then (5.39) shows that the dominant role is played by the processes of energy exchange between the two baroclinic modes with a tendency to transfer energy towards large wavenumbers. To estimate the resultant contribution of the interactions to the energy transfer from or towards a wavenumber p satisfying the inequality $p < q < k = k_R(1 - \delta)$, where $\delta \ll 1$, we reduce the last relation in (5.39) to the form

$$\gamma \equiv \frac{\dot{U}_{\mathbf{k}}}{\dot{T}_{\mathbf{q}}} = \frac{1 + 2\alpha \cos\theta}{\alpha^2 + 2\delta} + O(\delta^2), \tag{5.40}$$

where $\alpha = k/p$ and θ is the angle between the wave vectors $\mathbf{k}$ and $\mathbf{p}$.

According to Merilees and Warn (1975), the desired quantity in the α, θ plane (where $0 \leqslant \alpha \leqslant 1$; $\arccos(-\alpha/2) < \theta < \pi/2 - \delta$) is proportional to the ratio s_1/s of the areas described by the end of the vector $\mathbf{k}$ (for a fixed $\mathbf{p}$) in the case $\gamma > 1$ and in all the possible cases $\gamma \geqslant 0$, respectively. A simple calculation shows that about 63% of interactions in (5.32) lead to energy exchange with the barotropic mode, and, as a result, there appears a tendency towards the 'barotropization' of the current, i.e. to a decrease of the distinction between the motions at different levels. This phenomenon can also be illustrated by the form of the spectra of equilibrium ensembles of geostrophic turbulence constructed by Salmon *et al.* (1976) which are

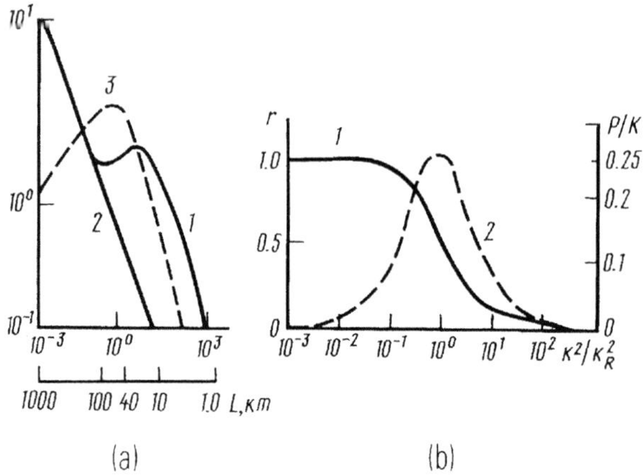

Fig. 3.5.2 (a) Kinetic energy spectra (per unit depth) of motions (1) in the upper layer and (2) in the lower layer, and (3) the spectrum of the available potential energy in the case $h_1/h_2 = 1/7$ (after Salmon *et al.*, 1976). The scales along the axis of ordinates are given in arbitrary units; the Rossby deformation radius is equal to 35 km. (b) (1) The correlation coefficient r for values of the stream function in the upper and lower layers and (2) the ratio of the available potential energy P to the total kinetic energy K.

shown in Figure 3.5.2. The available potential energy level sharply decreases in the subrange of wavenumbers $k < k_R$, where there occurs superposition of spectral curves corresponding to each of the levels, and this subrange becomes a region where the current tends towards a barotropic state.

Finally, the last possible case of interaction of type II for which the triples of wavenumbers satisfy the condition $k \approx q \gg p \to 0$ and the initial energy is entirely concentrated in the mode with wave vector $\mathbf{p}$ relates to the appearance of baroclinic instability specifying processes of energy transfer from the basic current to the baroclinic modes with $k \sim k_R$.

For the sake of visuality, it is advisable to sum up the cases considered above using the schematic diagram in Figure 3.5.3. The plots in Figure 3.5.3 represent the state of the fluid at different times, and the arrows between them indicate the

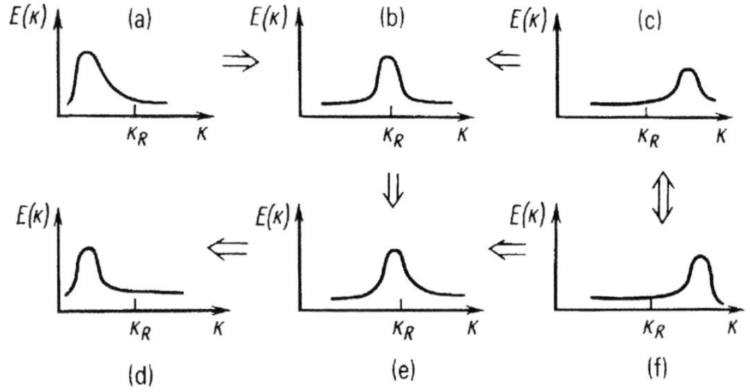

Fig. 3.5.3. Schematic diagram of states of the two-layer quasigeostrophic model according to the results of the numerical experiments performed by Rhines (see Salmon, 1977).

transitions from one state to another. We shall describe the possible transitions demonstrated in Figure 3.5.3. Let the energy of baroclinic currents be concentrated in large scales $L \gg k_R^{-1}$ (state A). Owing to baroclinic instability the transition to state B is possible; further, the rapid 'barotropization' performs the transition to state E, and, finally, the gradual increase of the scales of energy-containing eddies (as in the case of two-dimensional turbulence) reduces the system to state D. If the energy in the two-layer model is introduced in small scales $L \ll k_R^{-1}$, then in some time the motions in the layers become uncorrelated and reverse cascades appear in each of the layers, i.e. the transitions C $\rightarrow$ B and F $\rightarrow$ E are possible.

5.9. Statistical Regimes of Geostrophic Turbulence in the Ocean

We shall now enumerate some tendencies opposite to barotropization effects, eddy growth, westward motion, and anisotropization (stretching along the west-to-east direction). The first is the excitation of small eddies by both external forces and internal processes of baroclinic instability of big eddies. The numerical experiments performed by Rhines showed that very long baroclinic Rossby waves ($L \gg k_R^{-1}$) are unstable (the motions in the meridional direction are particularly unstable) and rapidly disintegrate into eddies with dimensions of the order of L_R which then undergo evolution, as was described above. However, if the baroclinic zonal currents develop sufficiently fast, the vertical velocity shear ΔU retains stability (under the condition $\beta L_R^2/\Delta U > 1$) so that complete barotropization does not necessarily occur. Second, owing to the integral condition (Rhines, 1975).

$$\frac{\partial}{\partial t}\int_{\Sigma}\int |\Delta\psi|^2\, dx\, dy = -\frac{\beta}{2}\oint_{\Gamma} |v^2|^2 \sin\theta\, dl, \tag{5.41}$$

where Σ is the area and Γ is the coastal line of the ocean (θ is the angle between Γ and the x-axis), the reflection of Rossby waves from western coasts of the ocean is accompanied by transformation of large-scale motions into small-scale motions, and the enstrophy increases (by contrast, at eastern coasts it decreases). Further, since small-scale eddies move more slowly, both the energy and the enstrophy accumulate at western coasts and dissipate at eastern coasts. Third, the flow over irregularities on the ocean bottom results in a scattering of Rossby waves and the generation of enstrophy (in the form of small-scale topographic eddies distributed inhomogeneously in space) including the 'baroclinization' of motion (see Rhines and Bretherton, 1973; Rhines, 1977). The typical bottom relief spectrum is rather planar, namely $E_h(k) \sim k^{-3/2}$ or $E_h(k) \sim k^{-2}$ (the latter is the 'white noise' for bottom slope spectra). This spectrum has a tendency to smooth the kinetic energy spectrum of currents. When the bottom irregularities are of a larger scale than the initial eddies, they strengthen the β-effect, namely the f/h isolines play the role of circles of latitude, the wavenumber $k_h = [h|\nabla_h(f/h)|/(2U)]^{1/2}$ substitutes for k_β, and the anisotropization manifests itself in a tendency to flow along the f/h isolines. Finally, the interaction between excitation, dissipation, and the above-mentioned opposite tendencies must produce still more complicated statistical regimes of synoptic eddies in different regions of the world ocean.

CHAPTER 4

Eddies of Western Boundary Currents

1. GULF STREAM EDDIES

Gulf Stream eddies are types of synoptic oceanic eddies that were the first to be studied experimentally (Iselin and Fuglister, 1948; Fuglister and Worthington, 1951). The Gulf Stream is a very strong jet current carrying its waters to the north, north-east, and east from the Florida Straits to the southern extremity of the Newfoundland Grand Banks (Figures 4.1.1 and 4.1.2). On passing the Cape Hatteras traverse (35°N), the Gulf Stream goes away from the North American continental slope and flows into the open ocean, retaining its jet character. With a width of 70–90 km, the Gulf Stream penetrates almost down to the ocean bottom (Fuglister, 1963; Schmitz *et al.*, 1970). The current velocity in the surface layer attains very large values, up to 3.5 m/s. At depths of the order of 1000–1500 m it decreases to 10–20 cm/s. The Gulf Stream transport in the layer above 4000 m in the region 60–70°W is of the order of 100–120 × 10^6 m^3/s (Baranov, 1971).

The Gulf Stream is a geostrophic current. This property, and the fact that the current velocity rapidly decreases with increasing depth, specify the frontal character of the Gulf Stream which separates the cold (and somewhat less saline) slope water in the west and north from the warmer (and more saline) Sargasso Sea water in the east and south (Figure 4.1.2). For instance, the temperature drop across the Gulf Stream at a depth of 300 m is around 9°, namely it varies from 8–9 °C on the left of the current to 17–18 °C on the right of the current. A particularly sharp transverse temperature gradient which, in some cases, attains 0.2–0.5 °C/km (Stommel, 1965) is observed on the left periphery of the current (the so-called 'cold wall' of the Gulf Stream). In the surface ocean layer about 200 m thick, the above-mentioned temperature distribution is accompanied by the effect of heat advection by the current, owing to which the temperature at the Gulf Stream axis is by two or three degrees higher than in the Sargasso Sea waters adjoining it on the right; e.g. this effect can be seen in the right upper part of Figure 4.1.4.

A great number of observations show (e.g. Hansen, 1970) that, along its whole extent, the Gulf Stream undergoes wave-shaped transverse oscillations characterized by a wavelength of 300–400 km and a phase velocity of 6–10 cm/s directed to the east. On passing the Cape Hatteras traverse, these waves become unstable, their amplitude tending to increase in the course of the eastward motion, and are transformed into the so-called 'meanders' whose dimension in the direction from south to north can attain 600 km and even more (Baranov, 1971). For instance,

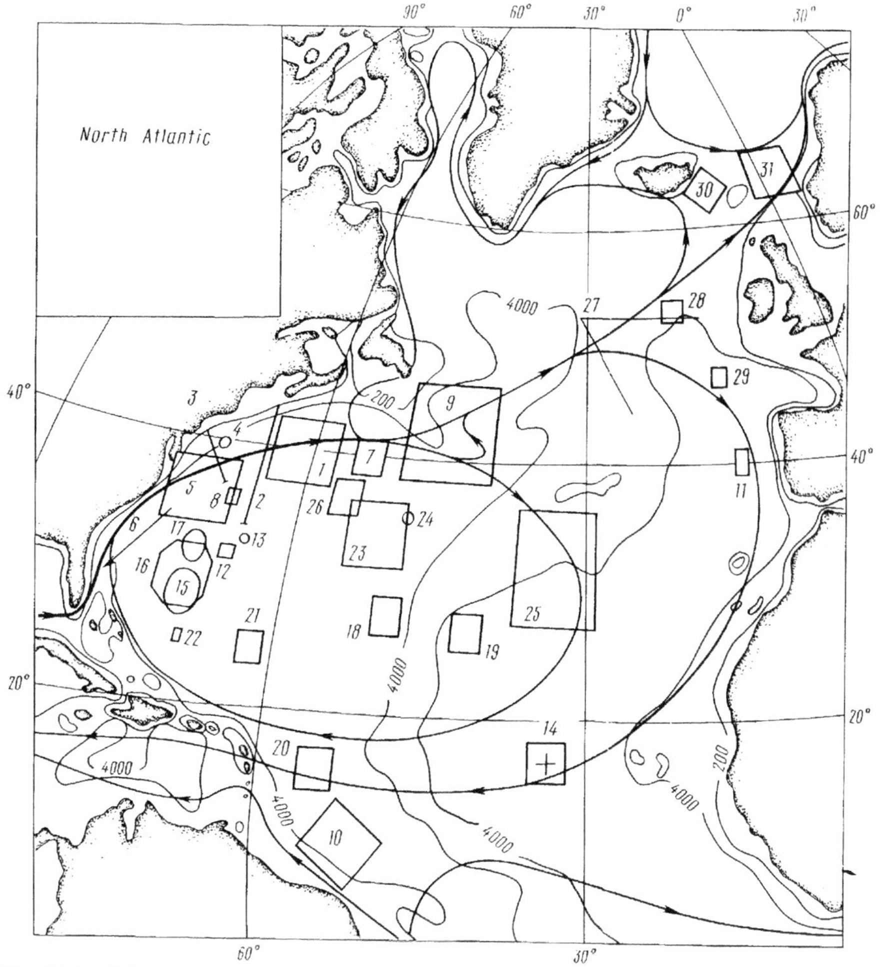

Fig. 4.1.1. Scheme of large-scale circulation of the North Atlantic upper layer waters (of thickness of the order of 1 km). The symbols and numbers in the figure indicate the places of expedition observations whose results are discussed in Chapters 4 and 5.

well-developed meanders can be observed in Figure 4.1.3 constructed from the synchronous survey data obtained by several Soviet research vessels in 1963. Further, in some cases the separation of meanders from the main current jet occurs, and thus their transformation into cold cyclonic (warm anticyclonic) eddies located south (north) of the main jet takes place. The process of formation of a meander and its transformation into an eddy penetrates into the whole current depth, and the entire cycle from the appearance of a wave-shaped bending of the jet to the formation of an eddy lasts about two or three months.

The classical example of Gulf Stream meander separation accompanied by formation of a cold cyclonic eddy, which still remains the best in the oceanographic literature, is demonstrated in Figure 4.1.4 (symbol 1 in Figure 4.1.1). In this case the surface current velocity in the cyclonic meander just before the meander

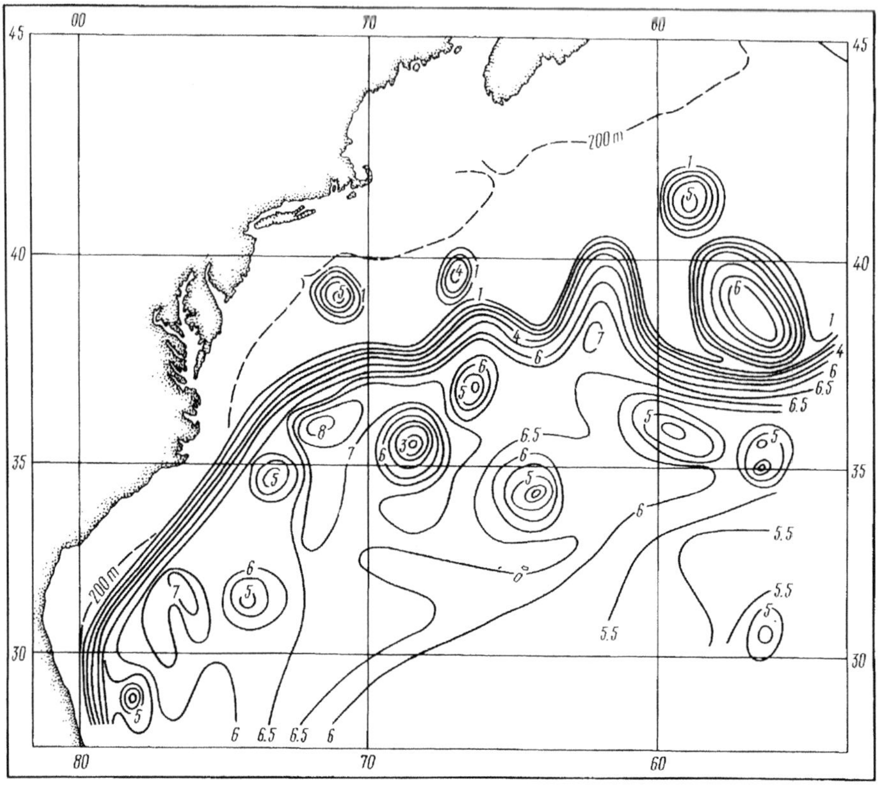

Fig. 4.1.2. Topography (in hundreds of meters) of the 15 °C isothermal surface according to the data of XBT, CTD, and routine hydrographical observations performed during the period March 15 to July 9, 1975 (after Richardson *et al.*, 1978). The elongated band-shaped bundle of isotherms corresponds to the Gulf Stream. South (north) of the Gulf Stream cold cyclonic (warm anticyclonic) eddies are seen.

separation attains 3 m/s. Owing to the presence of closed rings of very warm Gulf Stream surface waters in upper parts of young Gulf Stream eddies, these eddies are sometimes called 'rings'.

Cyclonic Gulf Stream eddies are extremely concentrated formations. The 'diameter' of a young Gulf Stream cyclone – defined as the total transverse dimension of the cold anomaly region in the main thermocline (i.e. at depths of 500–1200 m in the northern part of the Sargasso Sea; see Figure 4.1.5) – is 200–300 km (Parker, 1971; Fuglister 1972, Titov *et al.*, 1979; Richardson, 1980 etc.). The Rossby scale $L_R = (\Delta\varrho\varrho_0^{-1} gh)^{1/2} f^{-1}$, where $\Delta\varrho$ is the density drop across the main pycnocline, h is the depth of the lower boundary of the main pycnocline, f is the Coriolis parameter, g is the acceleration of gravity, and ϱ_0 is the average density,[1] in the oceanic region under consideration is equal to 40 km. Putting $\mathscr{L} = 2\pi L_R$, where $\mathscr{L}$ is the disturbance wavelength, and identifying the quantity $\mathscr{L}$ with the total transverse dimension of a Gulf Stream cyclone we obtain good agreement between

[1] Throughout Chapters 4,5, and 6, the Rossby scale means the quantity L_R defined in this way.

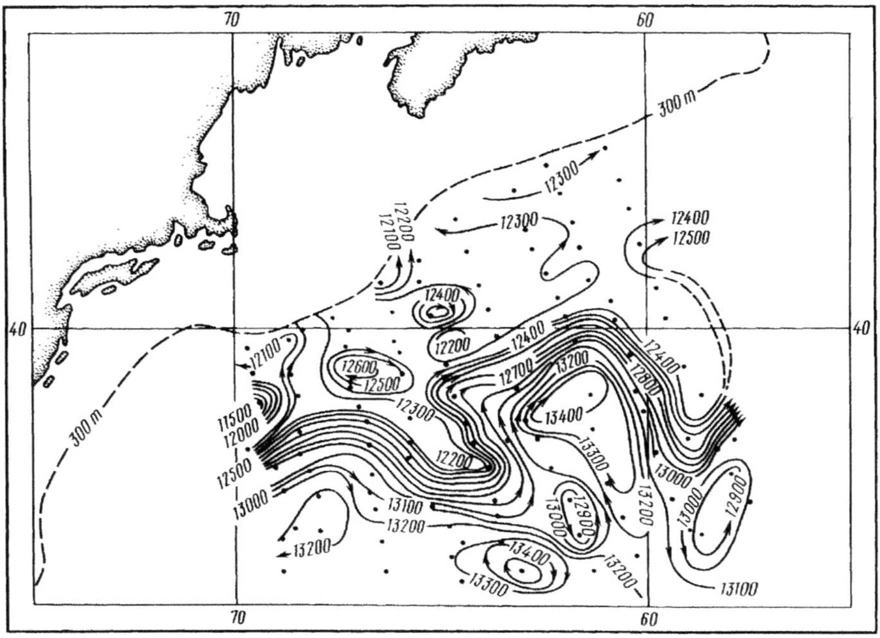

Fig. 4.1.3. Dynamic topography of the 100 dbar isobaric surface relative to the 4000 dbar reference surface in the Gulf Stream region according to the survey data of March 21–April 3, 1963 (after Baranov, 1971). The points show the positions of hydrographical stations. The current velocity in the region of the main Gulf Stream jet shown in the figure varies from 100 to 250 cm/s.

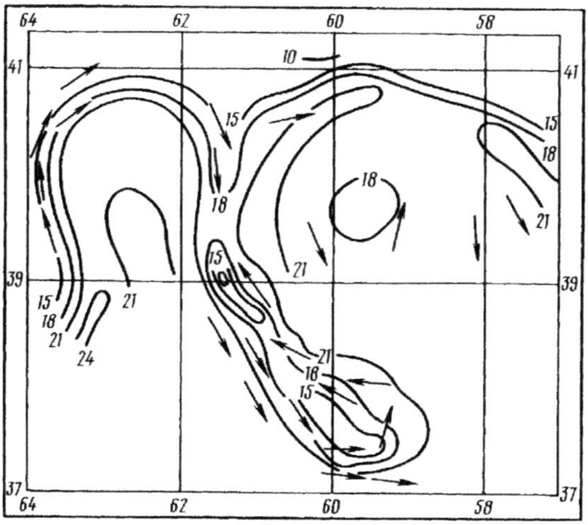

Fig. 4.1.4. Separation of a Gulf Stream meander and formation of a cold cyclonic eddy (after Fuglister and Worthington, 1951). The curves demonstrate the average temperature distribution (°C) over an upper ocean layer of thickness of 200 m in January 17, 1950. The arrows indicate the directions of surface currents. (See symbol 1 in Figure 4.1.1.)

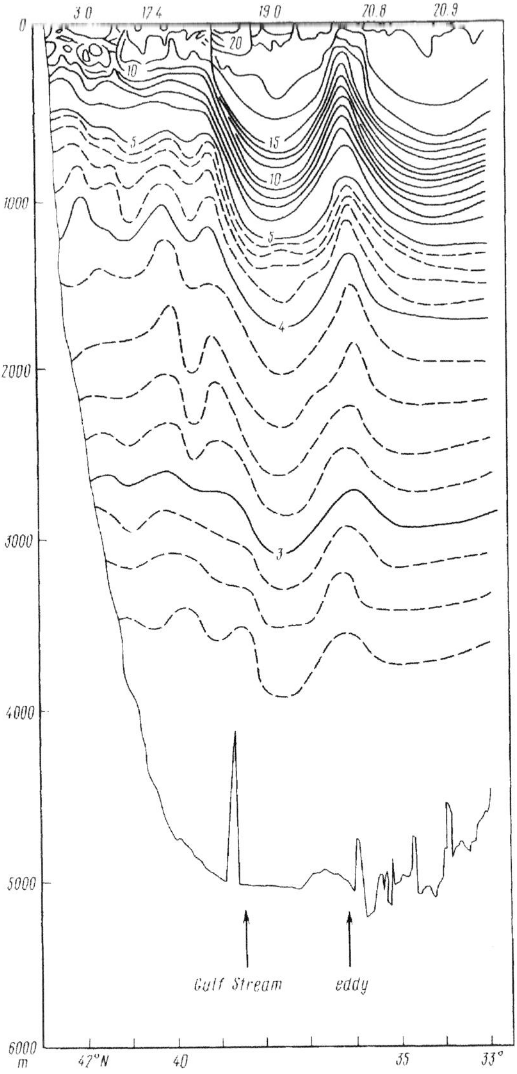

Fig. 4.1.5. Temperature distribution (°C) in the section along 64°30′W in April 1960 (after Fuglister, 1963). The section goes across the Gulf Stream and a cyclonic eddy generated by it. (See symbol 2 in Figure 4.1.1.)

experimental data and theory, suggesting that synoptic oceanic eddies are generated owing to baroclinic instability of large-scale currents (Chapter 3). The horizontal temperature difference between the center of a young Gulf Stream cyclone and the surrounding Sargasso Sea water in the main thermocline can attain 10–12 °C, which corresponds to the isothermal surface height difference between the eddy and the surrounding water in the temperature range from 6 to 17 °C attaining a very large value of 600–700 m (Parker, 1971; Fuglister, 1972; Richardson *et al.*, 1979). As is clearly seen from Figure 4.1.5 (the symbol 2 in Figure 4.1.1), young Gulf Stream cyclones, like the current generating them, penetrate the maximum ocean depths. The rotational water motion velocity in the upper part of a Gulf Stream cyclone can attain 3 m/s. According to the geostrophic relation, the velocity decreases with increasing depth and is of the order of 10 cm/s at depths of 1000–2000 m. Some interesting results were obtained in current measurements using autonomous buoy stations at the end of 1976 in a Gulf Stream cyclone in the region 35°N, 70°W. They showed that the southward current in the upper ocean in the western part of the eddy was replaced at a depth of about 800 m by a northward counter current whose velocity attained 40 cm/s(!) at 3000 m (Titov *et al.*, 1979).

Gulf Stream cyclones, which are extremely concentrated formations, store large amounts of excessive potential and kinetic energies. Henceforward, by the synoptic potential energy anomaly (PEA) of a vertical water column with unit cross-section area located between the reference level $z = 0$ and the sea surface $z = D$ we shall mean the quantity (measured in erg/cm^2)

$$E_p = g \int_0^D z\,(\varrho - \bar{\varrho})\,\mathrm{d}z, \qquad (1.1)$$

where $\varrho\ (x, y, z, t)$ is the synoptic value of the density and $\bar{\varrho}(z)$ is its average distribution in the Sargasso Sea. Table 4.1.1 presents the amount of PEA (the integral of (1.1) over the eddy area), the kinetic energy (KE), and some other parameters of several Gulf Stream cyclones. Eddies 3 and 8 were young in the observation period, the age of eddies 2 and 4 exceeded one year, and the remainder had intermediate ages. The eddy diameter in Table 4.1.1 is defined as the average transverse dimension of the region contoured by the 15 °C isotherm at a depth of 500 m, which is less than the total transverse dimension of the negative temperature anomaly region. The large diameter of eddy number 6 is due to its generation in the eastern part of the Gulf Stream (see below). Taking into account that the undisturbed 15 °C isotherm depth in the northern part of the Sargasso Sea is approximately equal to 600 m, we can estimate the degree of elevation of isothermal surfaces in the various eddies from the data in column (4) of Table 4.1.1. The numbers in the parentheses in column (5) are the reference surface depths relative to which the current transports in the eddy field were determined. The energy values in columns (6) and (7) were also calculated relative to the same reference surfaces $z = 0$ (see formula (1.1)). It should be noted that the PEA density per 1 cm^3 in Gulf Stream cyclone fields is primarily concentrated in the main thermocline and is maximum at depths of 500–800 m (Barrett, 1971; Cheney and Richardson, 1976; Hagan *et al.*, 1978). By contrast, the KE density is maximum in the subsurface layer, rapidly decreases with increasing depth, and attains small values at the lower boundary of the main thermocline.

The formation process of warm Gulf Stream anticyclones described, for instance, by Saunders (1971) is in many aspects analogous to the cyclone formation process. In the initial period after the formation, the advective effect is very clearly marked in the upper layer of Gulf Stream anticyclones. It manifests itself in the drawing of the cold slope and shelf waters north of the Gulf Stream in the anticyclonic motion (Saunders, 1971). The typical total transverse dimension of young Gulf Stream anticyclones is 150–200 km (Saunders, 1971; Fornshell and Criess, 1979), which is

TABLE 4.1.1

Characteristics of several Gulf Stream cyclones (after Richardson *et al.*, 1979)

(1) Observation period	(2) Coordinates		(3) Diameter, km (15 °C, at depth of 500m)	(4) Minimum depth (m) of the 15 °C isotherm	(5) Transport, 10^6 m^3/s	(6) Potential energy anomaly, 10^{15} J	(7) Kinetic energy, 10^{15}
	Latitude, °N	Longitude, °W					
May 1960	36	65	150	175	45 (2000 m)	120	
October 1963	28	74	90	300	25 (2000 m)	25	
June 1967	36	66	115	0	40 (3000 m)	55	
April 1970	31	70	65	355	15 (1000 m)	15	
October 1971	32	73	160	190	60 (3500 m)	93	6
March 1975	34	63	260	215	45 (2000 m)		
June 1975	36	68	120	150	73 (3000 m)		
December 1975	36	58	140	120	70 (3000 m)		
December 1975	35	65	160	170	46 (3000 m)		
December 1976	36	66	165	275	61 (4700 m)	134	3

less than that of Gulf Stream cyclones. This can probably be accounted for by the smaller value of Rossby scale for the slope water region. The temperature difference between the center of a young anticyclone and the surrounding slope waters can attain 9–10 °C at depths of 200–500 m, and the corresponding isothermal surface depth difference in the main thermocline is 400–500 m (Figure 4.1.6; symbol 3 in Figure 4.1.1). The rotational water motion velocity in the upper parts of young Gulf Stream anticyclones can exceed 1 m/s (Fornshell and Criess, 1979). The velocity decreases comparatively quickly with increasing depth so that, as a rule, the effective vertical extent of Gulf Stream anticyclones across the ocean depth is less than that of cyclones. The young Gulf Stream anticyclone described by Fornshell and Criess (1979) possessed amounts of PEA and KE in the layer above 3000 m equal to 9×10^{15} J and 0.6×10^{15} J, respectively.

The initial period of life of both Gulf Stream cyclones and anticyclones is characterized by a temperature contrast with the surrounding waters not only in the main thermocline but also in the upper layer, including the sea surface. The contrast can attain several degrees. That is why young Gulf Stream cyclones and, particularly, anticyclones are so clearly marked in satellite infrared pictures (Legeckis, 1978). However, in the course of time the surface temperature signal is weakened owing to the direct atmospheric effect and completely disappears in several months (and sometimes even in several weeks), whereas the temperature contrast in the main thermocline can still remain very strong. For instance, a situation of this kind is seen in Figure 4.1.5. Figure 4.1.7 was mainly constructed from the data of satellite temperature surveys of the sea surface and also from

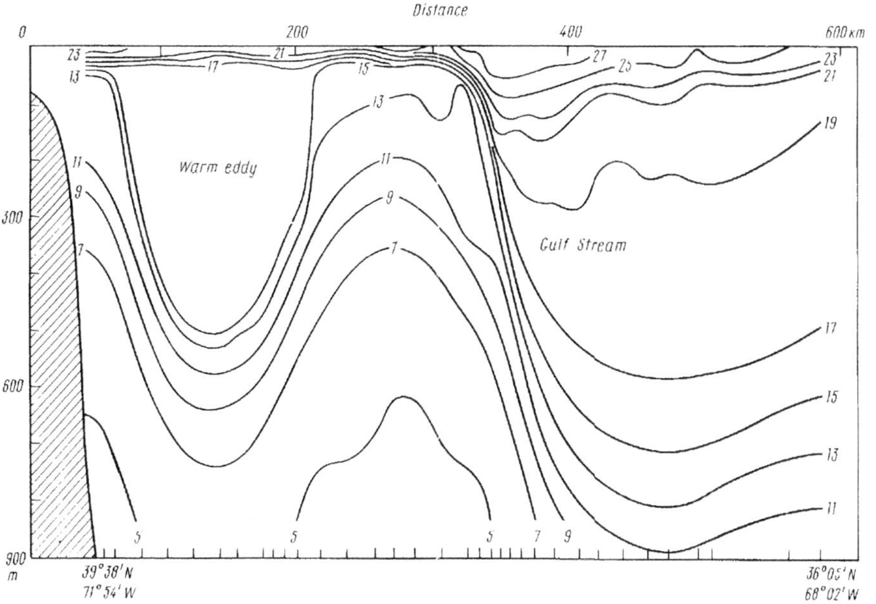

Fig. 4.1.6. Temperature distribution (°C) in the section across the Gulf Stream and an anticyclonic eddy generated by it; June 29, 1975 (after Csanady, 1979). The marks indicate the places of thermal soundings of the ocean. (See symbol 3 in Figure 4.1.1.)

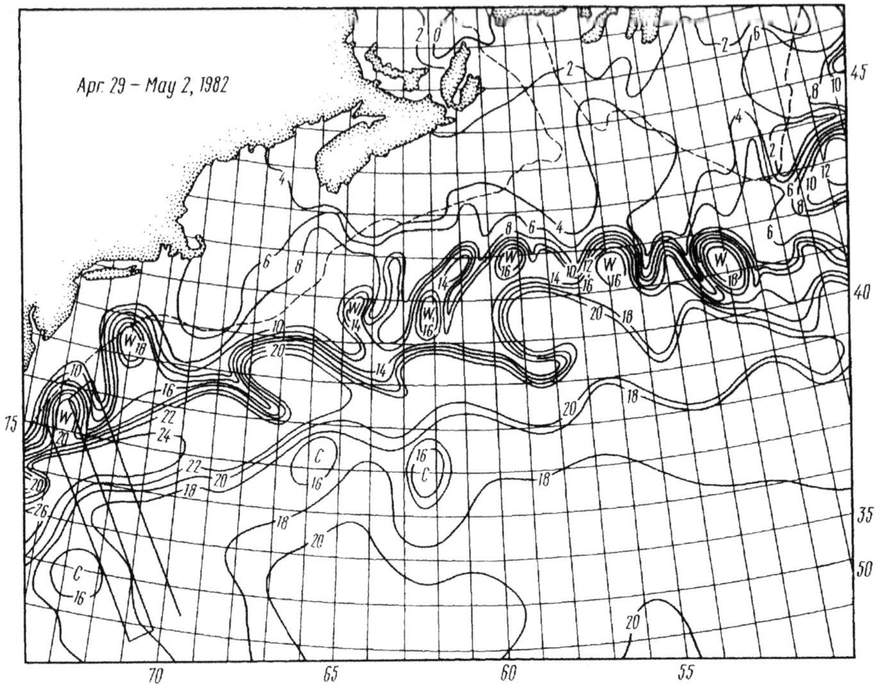

Fig. 4.1.7. Temperature distribution (°C) at the sea surface in the Gulf Stream region in April 29–May 2, 1982 (presented by V. A. Burkov). W and C are the centers of warm and cold regions with closed isotherms. The dashed line is the 1000 m isobath. Shown in the western part of the chart are the positions of the three hydrographic sections made by the research vessel *Vityaz'* in May 1982.

limited ship measurement data. To correctly interpret the figure we should take into account that at the sea surface the Gulf Stream core coincides with the axis of the warm water tongue (the 26–20 °C isotherms in the figure) while the main frontal zone, represented in the case under consideration by two clearly marked thermal fronts, is located north of the current. Figure 4.1.7 demonstrates three cold cyclonic eddies immediately south of the Gulf Stream and a series of warm anticyclonic centers north of the main current jet which were probably at different stages of their life cycles.

According to Fuglister's (1972) estimate, along the whole extent of the Gulf Stream (from Cape Hatteras to the Newfoundland Grand Banks), five to eight cold cyclonic eddies are formed every year. As to the warm Gulf Stream anticyclones, their annual number (judging, for instance, from Figure 4.1.7) seems to be much greater. However, owing to the smaller horizontal dimensions and vertical thicknesses of Gulf Stream anticyclones in comparison with the cyclones, and also, probably, because of their more frequent reverse absorption by the Gulf Stream, the total annual amounts of warm and cold waters transported by the eddies northward and southward across the Gulf Stream turn out to be approximately equal.

After separating from the current generating them, Gulf Stream cyclones begin to move to the west and south-west (i.e. approximately parallel to the Gulf Stream but in the opposite direction) with an average velocity of 4–5 cm/s (Lai and

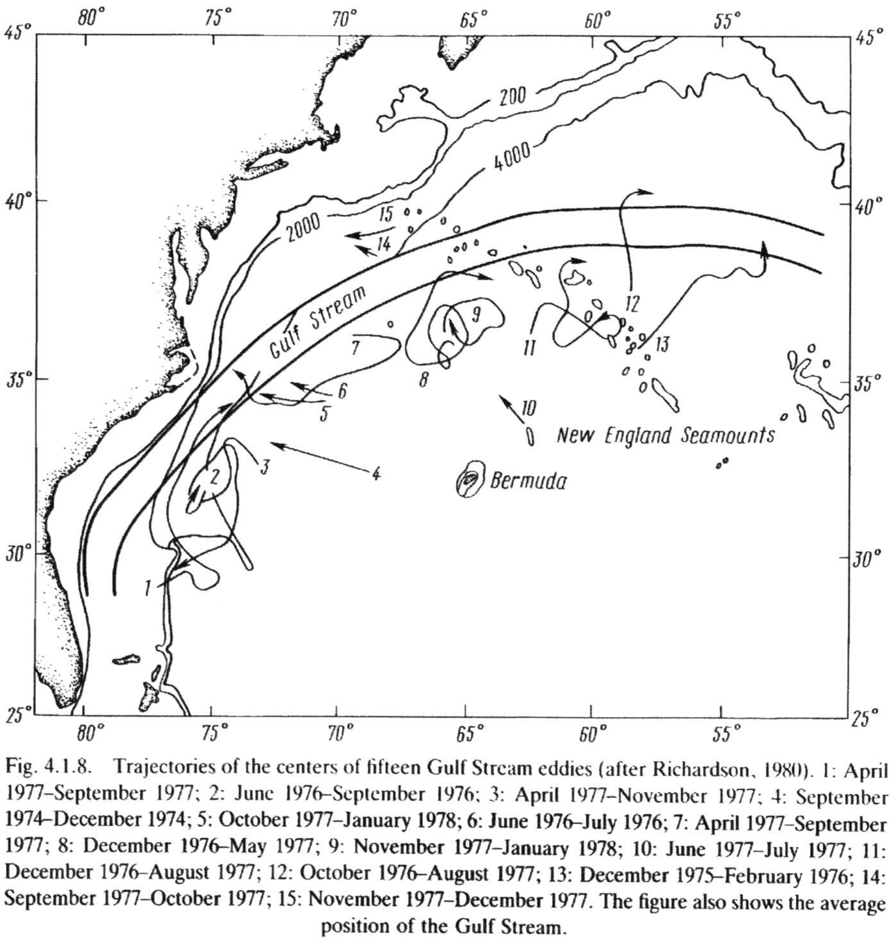

Fig. 4.1.8. Trajectories of the centers of fifteen Gulf Stream eddies (after Richardson, 1980). 1: April 1977–September 1977; 2: June 1976–September 1976; 3: April 1977–November 1977; 4: September 1974–December 1974; 5: October 1977–January 1978; 6: June 1976–July 1976; 7: April 1977–September 1977; 8: December 1976–May 1977; 9: November 1977–January 1978; 10: June 1977–July 1977; 11: December 1976–August 1977; 12: October 1976–August 1977; 13: December 1975–February 1976; 14: September 1977–October 1977; 15: November 1977–December 1977. The figure also shows the average position of the Gulf Stream.

Richardson, 1977; Richardson, 1980). The latest data (Richardson, 1980) show that the majority of the cyclones do not travel far to the south of the Gulf Stream but, on drifting several hundred kilometers westward, they again flow into the Gulf Stream (Figure 4.1.8). This fact is very important for a correct estimate of the integral heat exchange across the Gulf Stream relating to the generation of its eddies. The eddy center trajectories in Figure 4.1.8 were mainly constructed from the data of surface current measurements inside the eddies using free-drifting satellite-tracked buoys. The eastern parts of trajectories 2 and 3, and the southern parts of trajectories 7, 8, and 9, corresponded to those phases of motion when the eddies had no direct contact with the Gulf Stream. By contrast, the western and northern parts of these trajectories, which revealed the northward and eastward eddy motion, related to those life periods of the eddies when they again came into contact with the Gulf Stream, and the process of their trapping by the current began. Trajectories 4 and 10 show that some individual cyclones can nevertheless pass great distances away from the Gulf Stream and drift in the Sargasso Sea for a long time.

Gulf Stream anticyclones, on separating from the main current, also move to the west and south-west parallel to the Gulf Stream with a typical velocity of 4–7 cm/s (Lai and Richardson, 1977). The life of all of them, almost without exception, ends in reverse merging with the Gulf Stream mainly in the region west of 68°W. The spectrum analysis of the results of two-year current measurements performed at four levels in a layer from 100 to 2000 m at the point 39°12′N, 70°03′W lying in the continental slope region north of the Gulf Stream (symbol 4 in Figure 4.1.1) showed very clearly marked predominance of oscillations with a period around 30 days (Thompson, 1971). It can be assumed (Thompson, 1971; Rhines, 1971) that these oscillations were produced by the joint effect of the generation of quasibarotropic Rossby waves by the meandering Gulf Stream and the passage of Gulf Stream anticyclones through the above-mentioned point.

The chart in Figure 4.1.9 sums up the results of all observations of Gulf Stream

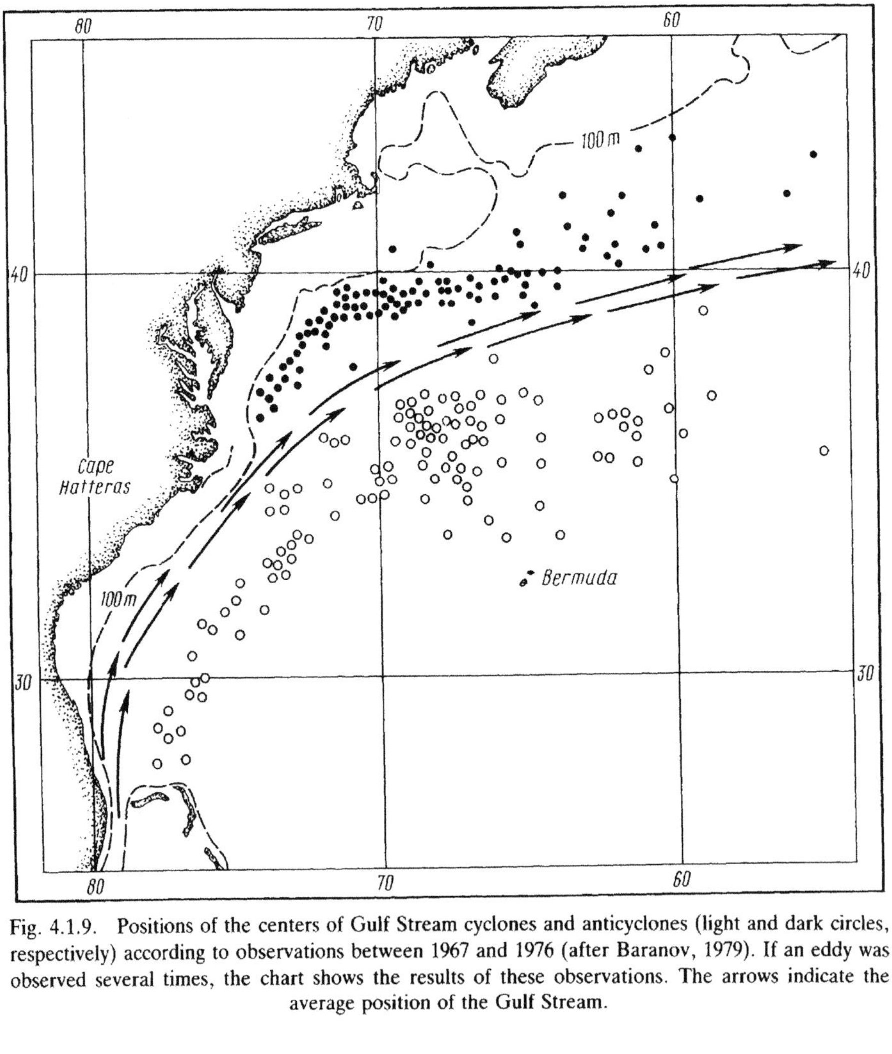

Fig. 4.1.9. Positions of the centers of Gulf Stream cyclones and anticyclones (light and dark circles, respectively) according to observations between 1967 and 1976 (after Baranov, 1979). If an eddy was observed several times, the chart shows the results of these observations. The arrows indicate the average position of the Gulf Stream.

eddies (which were mainly performed using XBT and CTD) during 1967–1976. Although the chart does not distinguish between single registrations of individual eddies and series of several registrations of the same eddies, on the whole it provides valuable information about Gulf Stream cyclone and anticyclone propagation areas.

Direct long-term current measurements (e.g. Schmitz, 1977), together with geostrophic current calculations from the large-scale sea water density distribution known from observations (Stommel *et al.*, 1978) and deep-ocean circulation schemes constructed from distributions of various oceanographic characteristics (Worthington, 1976), make us assume that the mean large-scale currents in the whole ocean thickness in both Gulf Stream cyclone and anticyclone propagation areas (Figure 1.4.9) are directed to the west and south-west, i.e. approximately opposite the Gulf Stream. The depth-averaged velocities of these currents turn out to be close to the above-mentioned average westward drift velocities of Gulf Stream eddies. Some data of direct current measurements performed during surveys of Gulf Stream eddies also agree with the above result. For example, the average direction and velocity of motion of the Gulf Stream anticyclone observed in October 1969–January 1970 in the region 69–72°W (Saunders, 1971) were close to the corresponding parameters of the average velocity vector obtained by averaging (during a two-month period of observation of the eddy) the current measurement data at a depth of 2000 m at the point 39°12′N, 70°03′W (see symbol 4 in Figure 4.1.1) located in the region of motion of this anticyclone. The average direction of motion of a Gulf Stream cyclone in November 1974–March 1975 in the region 33–34°N, 71–74° was almost identical with the direction of motion of a SOFAR float which was drifting in the same period at a depth of 1100 m in the region immediately south of that eddy (Cheney *et al.*, 1976). As to the velocity of the eddy and the velocity of the float, the former was one and a half times as great as the latter (3 cm/s for the eddy and 2 cm/s for the float). The following result obtained in April 1976 in the region 37°N, 66°W (Richardson *et al.*, 1979) is still more remarkable: the average velocity vector over the whole ocean thickness (6 cm/s in the direction to north-north-east) obtained by measuring twice the vertical current distribution by a velocity profilometer at the very center of an intensive cyclonic eddy, turned out to be close to the eddy drift velocity vector (4 cm/s to north-east) in the same period.

The above-mentioned facts allow us to draw a natural conclusion about a substantial contribution to translational motion of Gulf Stream eddies made by large-scale currents. At the same time, hydrodynamical models of isolated oceanic eddies (Warren, 1967; Korotaev, 1980) suggest that, owing to the effect of the latitudinal variation of the Coriolis parameter, Gulf Stream eddies must move with a western velocity component of several centimeters per second *relative to the surrounding water*. It seems that the actual translational motion of Gulf Stream eddies results from both the large-scale current effect and the dynamics of the eddies themselves. Another cardinal question relating to the above fact is whether the eddy motion is predominantly of advective or wave character. The most evident indication of the advective character of the motion of Gulf Stream eddies in their upper parts is the above-mentioned fact that young eddies retain a temperature

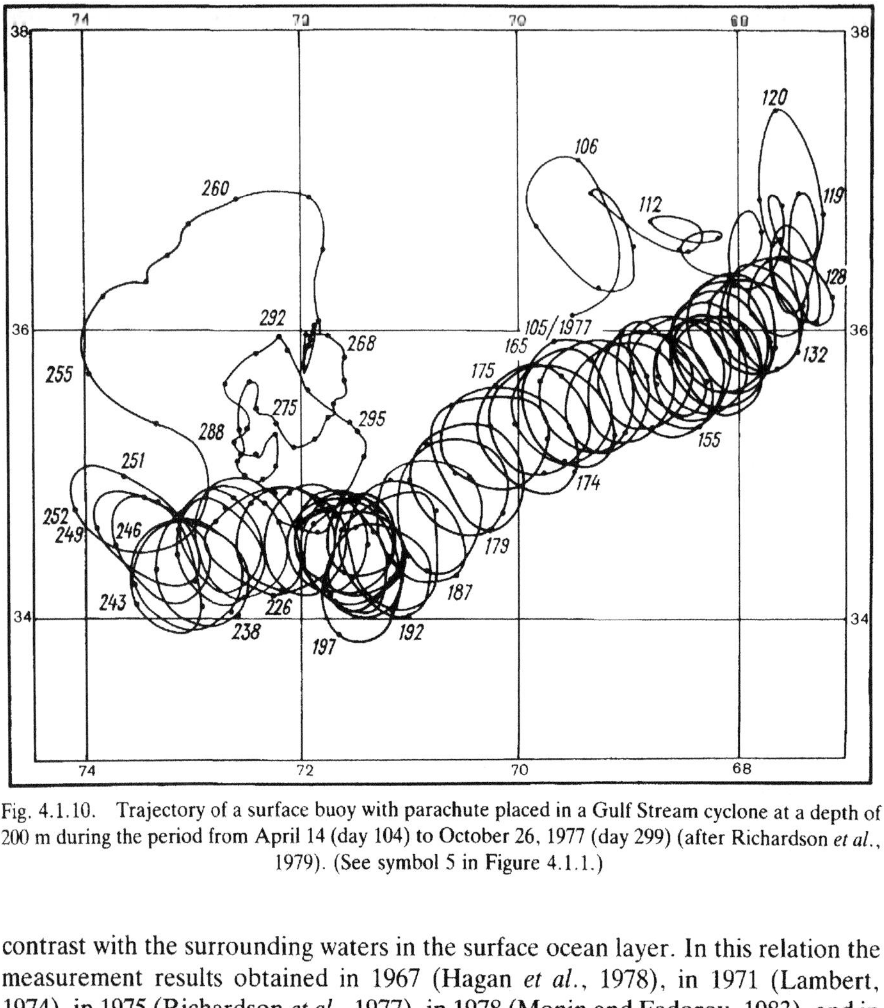

Fig. 4.1.10. Trajectory of a surface buoy with parachute placed in a Gulf Stream cyclone at a depth of 200 m during the period from April 14 (day 104) to October 26, 1977 (day 299) (after Richardson *et al.*, 1979). (See symbol 5 in Figure 4.1.1.)

contrast with the surrounding waters in the surface ocean layer. In this relation the measurement results obtained in 1967 (Hagan *et al.*, 1978), in 1971 (Lambert, 1974), in 1975 (Richardson *et al.*, 1977), in 1978 (Monin and Fedorov, 1983), and in 1979 (Richardson *et al.*, 1979) are not less convincing. These were temperature, salinity, and dissolved oxygen concentration measurements in Gulf Stream cyclones which were separated from the Gulf Stream by distances of 300–500 km at the time of measurement. They all showed that, to a depth of 500–600 m, the central parts of these eddies consisted of 'slope water' whose source was located north of the Gulf Stream. This clearly indicates the advective character of eddy motion, at least to depths around 600 m.

Current measurements in eddies by free-drifting floats are particularly important for elucidating the physical nature of Gulf Stream eddy motion. The most vivid example of such a measurement in the upper ocean is demonstrated in Figure 4.1.10 (symbol 5 in Figure 4.1.1). The position of the buoy was tracked by a satellite from the radio signal emitted from the surface float of the buoy. The whole period of observation, the results of which are shown in Figure 4.1.10, lasted 195 days of which 131 days corresponded to circular rotation of the buoy in the eddy field. The diameter of the rotation circle was 80 km and the rotation velocity varied from 150 cm/s at the beginning of May to 100 cm/s at the beginning of September 1977, which indicated decay of the eddy. In the period under consideration the eddy moved to the south-west and west with an average velocity of 5.6 cm/s. However, the drift velocity was extremely nonuniform; it varied from 12 cm/s at the beginning of July (days 175–185; Figure 4.1.10) to almost zero in the second half of July (days 190–200). The initial part of the buoy trajectory in Figure 4.1.10 corresponds to the period of formation of the eddy from a Gulf Stream meander and the final part corresponds to the reverse trapping of the eddy by the Gulf Stream. A great number of current measurements in Gulf Stream eddies were carried out by free-drifting floats during the period 1975–1978 (Richardson *et al.*, 1977; Richardson, 1980). They all gave results which, in principle, agreed with those described above; this confirmed the advective character of motion of eddies in their upper parts.

The results of current measurements in a Gulf Stream cyclone performed in October–December 1974 in the region 32–34°N, 68–74°W *at depths of* 750–1080 m (Cheney *et al.*, 1976) provided results which substantially differ from the data described above. These measurements were performed using acoustic SOFAR floats whose positions were registered by coastal tracking stations. The above-mentioned cyclone drifted westward during the observation period with an average velocity of 6 cm/s. One of the three floats launched in the center of the cyclone at the beginning of October 1974 immediately left the limits of the eddy while the other two did the same after one or two rotations; the velocities of the floats were about 7–11 cm/s. The measurement results showed that at depths of 700–1100 m the water 'flowed through the eddy' and hence the westward motion of the eddy at these depths was not of advective but of wave character. Here it should be noted that this eddy was comparatively weak; namely, the difference between the 15 °C isotherm depths across the eddy was about 200 m, which probably indicated that the age of the eddy was about one year.

On the whole, the observation data demonstrate a high degree of stability of Gulf Stream eddies. For instance, the observations of two Gulf Stream cyclones in 1965–1966 (Fuglister, 1972) and in 1967 (Parker, 1971) were in very good agreement and gave the value of 0.5–0.6 m/day for the average subsidence rate of the main thermocline at the center of a decaying eddy, which corresponds to several years of life of an eddy not coming into secondary contact with the Gulf Stream or other eddies. The stability of Gulf Stream cyclones seems to relate to their thermohaline and kinematic structure. For instance, the zonal CTD section across the central part of a Gulf Stream cyclone in the region 34°30′N, 72°30′W in August 1977 showed that in the upper layer of thickness about 700 m the central eddy core with radius around 40 km rotated almost as a solid, which naturally did not contribute to the destruction of the eddy owing to lateral friction. The distribution of the potential vorticity $P = (\partial\varrho/\partial z)\ [(v/r) + (\partial v/\partial r) + f]$ (where r is the distance from the eddy center, v is the water rotation velocity, and f is the Coriolis parameter) in this section, which is shown in Figure 4.1.11, was still more peculiar. Within the framework of the geostrophic approximation P is a Lagrangian invariant (see Chapter 2). Figure 4.1.11 shows that this property hampers the isopyc-

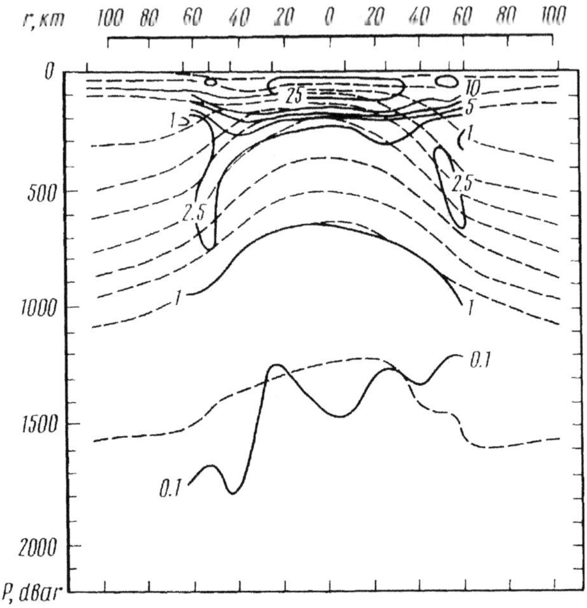

Fig. 4.1.11. Distribution of potential vorticity (in 10^7 kg/m^4·s; solid lines) and potential density (dashed lines) in the radial zonal section across a Gulf Stream cyclone in August 1977 in the region 34°30′N, 72°30′W (after Olson, 1980). The marks at the top of the figure show the places of CTD soundings of the ocean.

nic mixing of eddy water with the surrounding water in the main pycnocline. On the other hand, in the layers above and below the main pycnocline the P and water-density isolines are parallel, and the water exchange between the eddy and the surrounding ocean must take place in these very layers, which must lead to damping of the eddy. In application to the uppermost of these layers this conclusion is, for instance, confirmed by the multiple temperature, salinity, and dissolved oxygen concentration measurements performed in a Gulf Stream cyclone in 1971–1972 in the region 29–33°N, 73–78°W (Lambert, 1974). The observation revealed the eddy's complex structure in its upper part (above 500 m): layers of the water formed on the continental slope north of the Gulf Stream alternated with layers of the water characteristic of the Sargasso Sea. Typical thicknesses of the layers were 25–30 m, which suggests that the process of mixing eddy waters with surrounding sea waters, which is responsible for eddy decay, is of a layer-wise character. The corresponding effective turbulent mass flux coefficient was 10^3 cm^2/s. The Gulf Stream cyclone survey data (symbol 8 in Figure 4.1.1) obtained in September 1978 (Figure 4.1.12) are also in agreement with the above results. The patches of higher and lower temperature and salinity values in the section across the upper part of the eddy demonstrated the mutual isopycnic intrusion of the slope water forming the central part of the eddy and the surrounding Sargasso Sea water. Attention should also be paid to the sharp salinity front which surrounded the eddy core in the 100–500 m layer. Figure 4.1.13, constructed from the same data as Figure 4.1.12, demonstrates the isolines of the isopycnic temperature

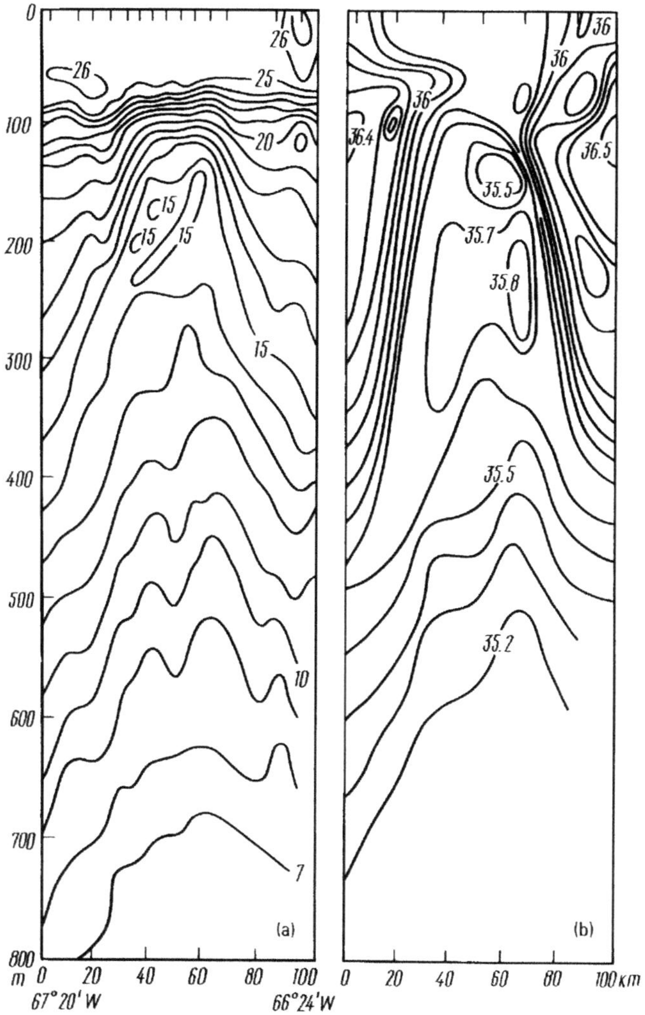

Fig. 4.1.12. Distributions of (a) temperature (°C) and (b) salinity (‰) in the section across a Gulf Stream cyclone along 35°N in the region 67°W in September 1978 (after Fedorov *et al.*, 1979). The marks at the top of the figure show the places of temperature and salinity measurements. (See symbol 8 in Figure 4.1.1.)

anomaly, i.e. the difference between the temperature observed at a given point of the section and the temperature of the surrounding Sargasso Sea waters corresponding to the same water-density value. These isolines contour perfectly well the core of the cyclonic eddy formed by the slope water whose T, S characteristics sharply differ from those of the Sargasso Sea water.

The longest observation cycle of one and the same Gulf Stream cyclone was performed during the period from March 1971 to April 1972 (Cheney and Richardson, 1976). In this period, when five CTD and XBT surveys were carried out, the eddy travelled a path of 750 km in the western part of the Sargasso Sea from 33°30′N, 73°W to 28°30′N, 78°20′W (symbol 6 in Figure 4.1.1) moving to the

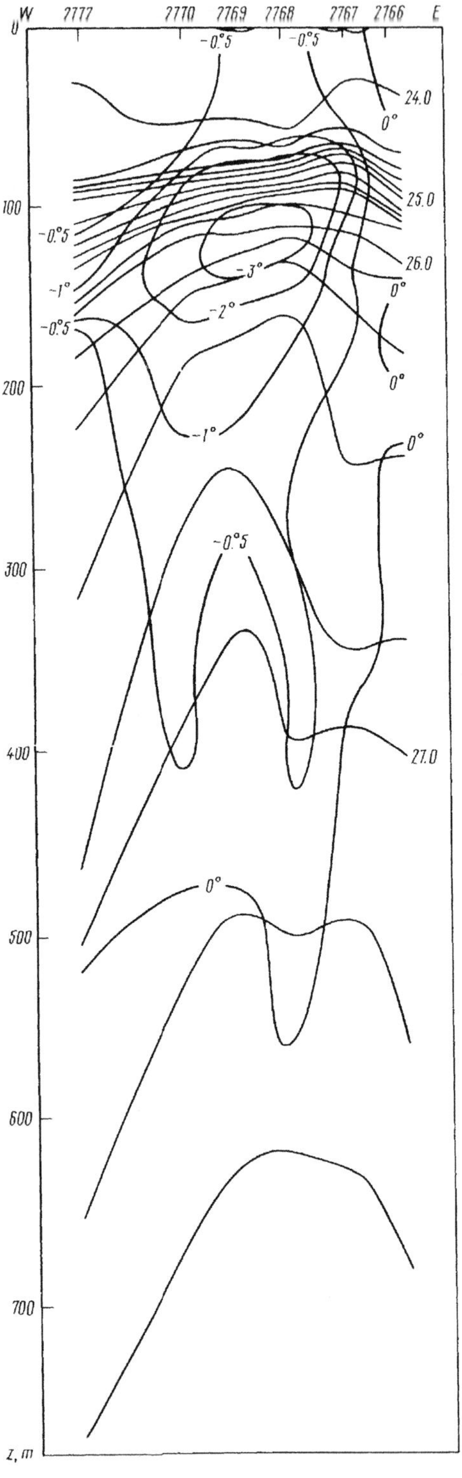

Fig. 4.1.13. Water density (σ_t; thin lines) and isopycnic temperature anomaly (°C; thick lines) distributions in the same section across the Gulf Stream cyclone to which Figure 4.1.12 corresponds (after Monin and Fedorov, 1983). The marks and figures at the top indicate the locations and numbers of hydrographic stations. (See symbol 8 in Figure 4.1.1.)

south-west with an average velocity about 2 cm/s. The data in Table 4.1.2 demonstrate the eddy evolution along this path. The eddy had a slightly elongated shape, and by the eddy 'radius' in Table 4.1.2 is meant the half-sum of the semi-major and semi-minor axes of the 'ellipse' contoured by the 15 °C isotherm at a depth of 500 m. The undisturbed 15 °C isotherm depth in this ocean region is 550–600 m. The 'potential energy anomaly' (PEA) in Table 4.1.2 is understood as a quantity calculated by formula (1.1) for values corresponding to the reference level $z = 0$ (these values are compiled in column (6) of the table) and then integrated over the eddy area. As in all Gulf Stream cyclones, the amount of PEA in the eddy under consideration was concentrated in the main thermocline with a density maximum at a depth about 700 m (Figure 4.1.14). The observations showed that at the time of the third survey (Table 4.1.2) the eddy came close to the Blake Plateau, after which its damping sharply increased (Figure 4.1.14). Namely, before that time the average rates of decrease of the amounts of PEA and KE in the eddy in the layer above 1000 m were approximately 12×10^{13} and 0.6×10^{13} J/day, and after the eddy came into contact with the ocean bottom these rates were 31×10^{13} and 3.3×10^{13} J/day, respectively. Taking into account the initial value of the PEA of the eddy in the layer above 1000 m (73×10^{15} J) and the rate of its decrease in the open ocean (12×10^{13} J/day), and assuming that the age of the eddy at the time of the

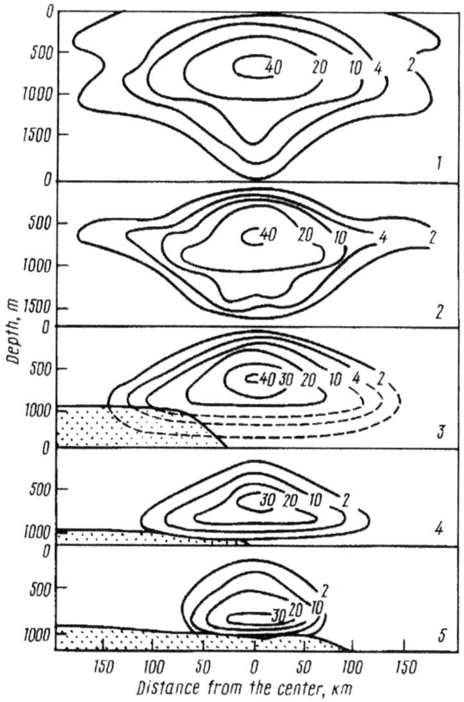

Fig. 4.1.14. Azimuth-averaged potential energy anomaly distributions $g\varrho_0^{1}z\ (\varrho - \bar{\varrho})$ (see formula (1.1)) in 10^3 erg/g in the radial section of a Gulf Stream cyclone in (1) May and (2) October 1971, and in (3) January, (4) February, and (5) April 1972 in the region 28-33°N, 73-78°W (see Table 4.1.2) (after Cheney and Richardson, 1976). (See symbol 6 in Figure 4.1.1.)

TABLE 4.1.2

Characteristics of a Gulf Stream cyclone in May 1971–April 1972 (after Cheney and Richardson, 1976)

(1) Survey No.	(2) Observation time	(3) Coordinates of the eddy center		(4) Radius, km	(5) Minimum depth (m) of the 15 °C isotherm	(6) Reference level, m	(7) Potential energy anomaly, 10^{15} J	(8) Transport, 10^6 m^3/s	(9) Kinetic energy, 10^{15} J
		Latitude, N	Longitude, W						
1	May 9–11, 1971	32°50′	73°10′	79	190	1000	73	36	2.4
						3500	93	60	5.6
2	October 21–24, 1971	31°00′	75°00′	71	150	1000	46	32	2.2
						1500	52	40	3.5
3	January 8, 1972	29°50′	76°40′	70	190	1000	42	32	1.9
4	February 29, 1972	29°20′	77°40′	55	300	1000	22	23	1.1
5	April 8–9, 1972	28°30′	78°20′	55	305	1000	14	19	0.9

beginning of the observation was about half a year (Cheney and Richardson, 1976), we obtain an estimate of 2.5 yr for the lifetime of a free-drifting eddy in the Sargasso Sea. However, this estimate has a conditional character since, as indicated above, the majority of the Gulf Stream eddies come into secondary contact with the Gulf Stream and are absorbed by it (Richardson, 1980). The maximum surface current velocity during the period of observation of this eddy changed very slightly and was approximately 1 m/s.

In two cases, one of which relates to December 1974 (Cheney *et al.*, 1976) and the other to December 1976 (Richardson *et al.*, 1979), XBT surveys and satellite observations made it possible to register the disintegration of a Gulf Stream cyclone into two cyclonic eddies of different power. In the first case the bifurcation process lasted about 10 days, and in the second case it was not possible to estimate the time. In 1974 the total transverse dimensions of the daughter eddies were around 200 and 120 km and the 15 °C isotherm anomaly depths were 250 and 100 m, respectively; in 1976 the corresponding parameters had values of 220 and 160 km and 350 and 200 m. The 1976 observations showed that after disintegration the daughter eddies went in different directions.

The damping process of a Gulf Stream anticyclone was tracked by performing two CTD–XBT surveys of an anticyclonic eddy. The first was carried out at the end of September 1977 when the eddy center was in the region 40°40′N, 65°W, and the other was performed at the beginning of December 1977 when the center was at 39°40′N, 68°W (Fornshell and Criess, 1979). The most remarkable change in the eddy structure during the two months was a sharp decrease of the total depth of the temperature disturbance from 3000 m in the first observation period to 1600 m in the second period. During the same two months the eddy diameter and maximum current velocity in the surface layer decreased from 190 to 150 km and from 103 to 73 cm/s, and the total amounts of PEA and KE in the layer above 1600 m decreased from 7×10^{15} to 3×10^{15} J and from 6×10^{14} to 2×10^{14} J, respectively. As in the above-mentioned case of a Gulf Stream cyclone, the rapid damping of the eddy and the decrease of its thickness probably related to the fact that the eddy came into contact with the continental slope. Moreover, in the period between the surveys the eddy travelled over the New England Seamount chain whose tops are at depths of 1160–2000 m below the sea surface. Slower damping of a Gulf Stream anticyclone was described by Saunders (1971); the amount of PEA of the eddy in a layer above 1000 m which was equal to 10^{16} J in the first observation period decreased by one third over two months.

To estimate correctly integral meridional energy fluxes of different kinds relating to the phenomenon of formation and motion of Gulf Stream eddies, attention should be primarily paid to investigating the process of secondary interaction between these eddies and the current generating them. Many of the details of this process have been elucidated quite recently (Richardson *et al.*, 1979; Richardson, 1980). As was indicated above, after a several hundred kilometer drift to the west and south-west, all the Gulf Stream anticyclones (without exception) – and the vast majority of its cyclones – again come into contact with the Gulf Stream. The immediate consequence of a contact of a Gulf Stream cyclone with the Gulf Stream is usually that the eddy gains an additional amount of the warm and saline Gulf

Stream waters predominantly in the upper ocean (to a depth of 400 m) and that these waters are drawn in the cyclonic circulation around the eddy (Vukovich and Crissman, 1978). This results in an increase in the amounts of APE and KE of the eddy. Moreover, as a rule, after an eddy comes into contact (and, usually, partly merges) with the Gulf Stream, the westward motion of the eddy is replaced by an eastward motion because it is drawn into the Gulf Stream. The course of the process then develops in accordance with one of the three 'scenarios' described below (Richardson, 1980).

When a Gulf Stream eddy is sufficiently large and intensive, as it was, for instance, in the situation represented in Figure 4.1.15 (symbol 7 in Figure 4.1.1), it can completely merge with the Gulf Stream after a repeated approach and become a meander. This process can be thought of as a reversed motion picture of the formation of an eddy owing to meander separation from the Gulf Stream. In Figure 4.1.15 it is seen how a surface float, which was in the field of a cyclonic eddy during the process of its merging with the Gulf Stream, made two additional rotations around the eddy center and then, after the merging process ended, was carried away by the Gulf Stream in the north-easterly direction. Another process of reverse transformation of a Gulf Stream cyclone into a meander was registered in September 1977 in the region 74°W (Richardson, 1980).

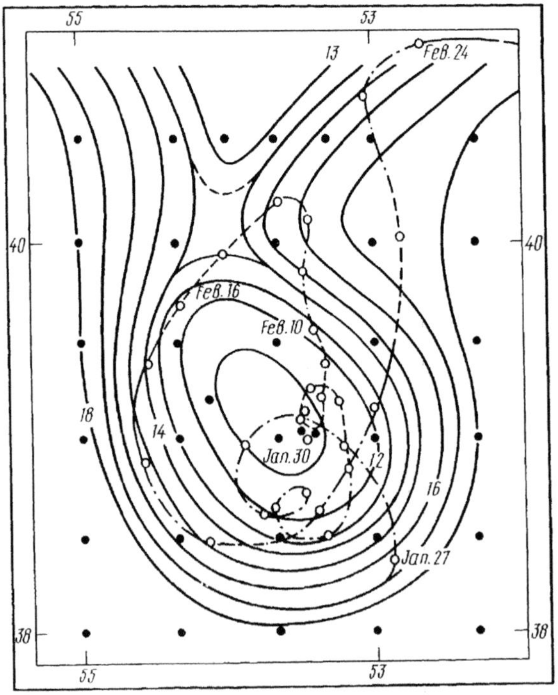

Fig. 4.1.15. Temperature distribution (°C) at a depth of 300 m in the region of merging of a Gulf Stream cyclone with the Gulf Stream in February 16, 1976 and the trajectory of a surface satellite-tracked float in the period from January 27 to February 24, 1976 (after Richardson *et al.*, 1977). The points show the places of AXBT temperature soundings. (See symbol 7 in Figure 4.1.1.)

One more possible result of a secondary contact of a Gulf Stream eddy with the current generating it can be a short-time but intensive exchange of water, momentum, energy, etc., between the eddy and the Gulf Stream, after which the eddy again leaves the Gulf Stream, this time in a transformed state. Such a case was registered at the end of February to the beginning of March 1976 in the region 66°W (Richardson *et al.*, 1979). After contact with the Gulf Stream, which lasted about a week, a sharp concentration of a clearly marked Gulf Stream cyclone took place, owing to which its diameter decreased from 220 to 70 km. The difference between the isotherm depths in the main thermocline remained the same (about 350 m) while in the upper ocean it even increased owing to the drawing of the Gulf Stream water in the cyclonic circulation on the periphery of the eddy. As a result of these changes, the total amount of PEA in the eddy notably decreased while the amount of KE somewhat increased. On changing its shape the eddy left the Gulf Stream to the south, made a clockwise circulation, and again approached the Gulf Stream at a distance of 250 km to the west. Here the Gulf Stream finally trapped the eddy, and, since in the case under consideration the eddy diameter at the time of trapping turned out to be even smaller than the total Gulf Stream width (70 and 100 km, respectively), the character of the trapping differed from that in the above-mentioned cases of the eddies in February 1976 (Figure 4.1.15) and September 1977. That is, the eddy was not transformed into a meander but was trapped as a whole by the Gulf Stream jet and then transported by the jet to the east until it dissipated in the surrounding flow. This is the third of the possible versions of secondary interaction of Gulf Stream eddies with the current generating them. The satellite temperature surveys of the sea surface, the current measurements by surface buoys, and the XBT measurements performed in February–May 1977 in the region 66–72°W registered one more process of partial merging of a Gulf Stream cyclone with the Gulf Stream, secondary departure of the transformed eddy from the Gulf Stream, and, finally, its complete merging with the current after their repeated approach (Richardson, 1980). However, in contrast to the above-mentioned case of February–April 1976, the eddy retained its large dimension and substantial potential energy after the first contact with the Gulf Stream, owing to which the course of its final merging with the Gulf Stream was like the one shown in Figure 4.1.15.

Current measurements by free-drifting surface floats and satellite temperature surveys of the sea surface (Richardson, 1981) show that the phenomenon of formation of a very large cyclonic meander in the region where the Gulf Stream intersects the New England Seamount chain (62–66°W; Figure 4.1.8) and transformation of this meander, which was elongated in the south-east direction (i.e. along the seamount chain) into an extensive cyclonic eddy (a process which is represented in Figure 4.1.4), has a quasipermanent character. After the passage of the region of the New England Seamount chain, the current width increases approximately two-fold (Baranov, 1971; Richardson, 1981) and the velocity reduces accordingly. Further, when the current approaches the southern extremity of the Newfoundland Grand Banks (51–52°W; Figure 4.1.1), its bifurcation occurs (Mann, 1967; Clarke *et al.*, 1980; Mountain and Shuhy, 1980), and, as a result, its northern branch, which is mainly concentrated in the upper ocean, goes north-east, intersects the

Newfoundland Ridge, and, finally, generates the North Atlantic Current. The other deeper and, as a rule, more powerful branch flows to the south-east along the south-west periphery of the Newfoundland Ridge and bifurcates (Figure 4.1.1) near the end of the ridge (43–45°W). One of the new branches goes around the ridge on its north-east side and then flows into the North Atlantic Current. The other branch, which has mainly been studied quite recently (Kirwan *et al.*, 1976; Richardson, 1981), retains its jet character, continues to go south-east, and intersects the axis of the North Atlantic Ridge (Figure 4.1.1) approximately in the region 35°N, 35°W. Observations show that the whole Atlantic region, from the New England Seamount chain to the North Atlantic Ridge described here, is an area of intensive eddy formation which seems to be substantially affected by the above-mentioned elements of the ocean bottom relief. However, owing to the slightly diffuse character of the jets of the current generating the eddies, Gulf Stream cyclones formed in the region east of 60°W are somewhat different from those generated west of 60°W. That is, as a rule, they have larger dimensions but their isothermal surface height difference is less than in the western cyclones. For instance, the cyclone discovered in March 1975 in the region 34°N, 63°W, which was probably formed in the eastern part of the Gulf Stream, had a total transverse dimension of 360 km and was characterized by a temperature drop at a depth of 600 m between the eddy center and the surrounding water of 7 °C (McCartney *et al.*, 1978).

The intensive character of eddy-formation process in the region of the Gulf Stream bifurcation near the southern extremity of the Newfoundland Grand Banks was confirmed in June 1976 by the results of a detailed CTD survey of this region (Figure 4.1.16; symbol 9 in Figure 4.1.1). It simultaneously revealed three cyclonic eddies which were probably produced in succession by cyclonic meanders of the beginning of the North Atlantic Current. The fact that the centerline of the eddies is stretched along the axis of the Newfoundland Ridge (Figure 4.1.16) indicated the effect of the ridge on the eddy-formation process. The isothermal surface height difference in the main thermocline in the field of the most developed cyclone with center at 39°30′N, 45°W (Figure 4.1.16) attained 600 m.

The chart in Figure 4.1.2 cannot be even conditionally called synoptic since it is constructed from the data of observations which lasted about four months. Nevertheless, it seems to provide certain information about the possible pattern of simultaneous distribution of Gulf Stream cyclones and anticyclones in the region west of 55°W. Figure 4.1.2 demonstrates three anticyclones, one anticyclone-meander of vast dimensions, and nine cyclones in the southern part of the region. As was noted by the authors themselves (Richardson *et al.*, 1978), one or two cyclones might have been omitted. All the anticyclones in Figure 4.1.2 are rather intensive, which can be understood if we take into account that the typical time from the formation of an anticyclone to its reverse absorption by the Gulf Stream seems to be not more than several months (e.g. see Lai and Richardson, 1977). Only one of the cyclones in Figure 4.1.2 was actually strong, namely the one with center at 35°N, 68°30′W. However, it should be taken into consideration that for four of the nine cyclones we cannot be sure that the sections passed through the eddy centers, and therefore their intensities can be underestimated. Attention

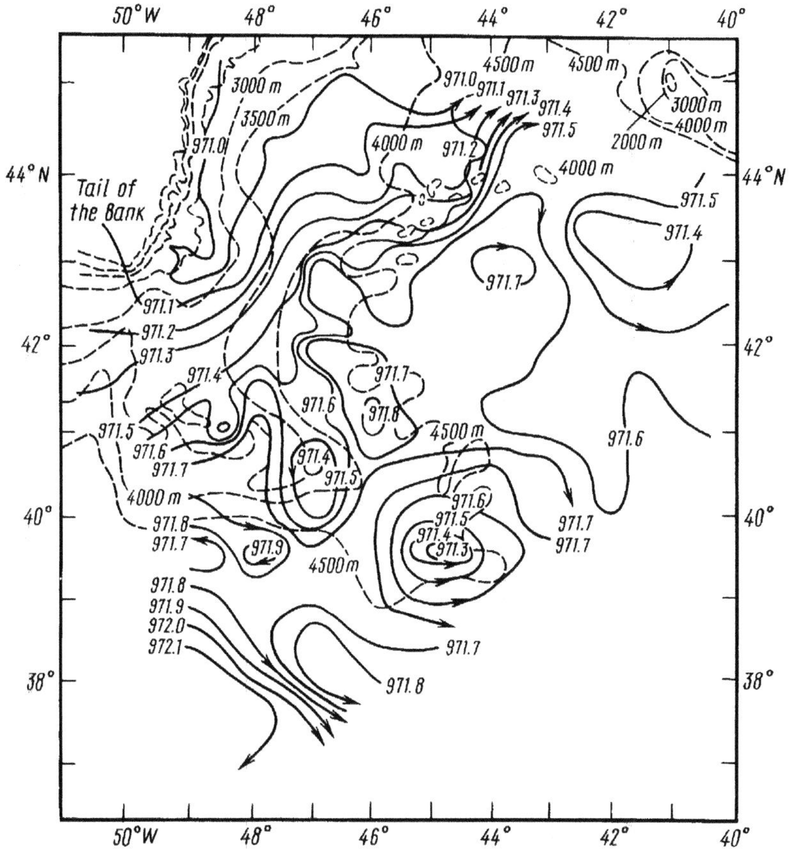

Fig. 4.1.16. Dynamic topography (in dynamic meters) of the ocean surface relative to the 1000 dbar reference surface according to the data of CTD surveys of the region south-east of the Newfoundland Grand Banks in June 1976 (after Mountain and Shuhy, 1980). The dashed lines show the isobaths; the depth is measured in meters. (See symbol 9 in Figure 4.1.1.)

should be paid to the cyclone at 30°30′N, 56°30′W which went 800 km off the Gulf Stream at the time of observation. The cyclone west of Florida was apparently at the initial stage of its reverse absorption by the Gulf Stream. Lai and Richardson (1977) mentioned eleven Gulf Stream cyclones which were simultaneously observed in the Sargasso Sea west of 55°W in November 1971.

Besides the big eddies described above, which are formed as a result of their separation from Gulf Stream meanders, surface cyclonic eddies of smaller dimensions which seem to be of frictional origin are also formed on the left (western and northern) side of the Gulf Stream. Such eddies are observed particularly frequently near the Florida coasts in the region from 28 to 31°N (Vukovich *et al.*, 1979) where they have an elliptic shape with a major axis ranging from 80 to 200 km which is stretched along the current. They are characterized by a clearly marked heat advection effect in their upper parts, penetrate the ocean depth at least to 100 m, and are transported northward by the Gulf Stream with an average velocity of about 30 km/day. The results of the joint satellite and ship observations of the upper ocean temperature performed in March 1978 in the Gulf Stream region at 72–74°N (Sklyarov and Fedorov, 1980) turned out to be very remarkable. The observations showed the formation of a narrow and long surface cold slope-water filament in the northern part of the Gulf Stream which was stretched eastward along the current and then twisted to form a cold cyclonic eddy with a diameter about 50 km, a depth about 500 m, and a temperature drop about 2 °C between the water that formed the eddy core and the Gulf Stream water which surrounded the eddy. The eddy was carried away by the Gulf Stream to the east–north-east with a velocity of the order of 40 km/day.

Surface current measurements by satellite-tracked floats (Richardson, 1981) and some other measurements also indicate the formation of small cyclonic and anti-cyclonic eddies in the Gulf Stream near seamounts, e.g. over some individual tops of the New England Seamount chain or of the Corner Rise Seamounts (36°N, 52°W). These eddies have transverse dimensions comparable with the seamount scale (10–30 km). They can be located both over the top itself and on its 'lee' side, are characterized by an increase in the water rotation velocity with depth down to the ocean bottom, and seem to be the so-called 'Taylor columns', i.e. rotating water cylinders appearing when a uniform current flows over an individual sea floor rise.

Although this section is devoted to the Gulf Stream, it is natural to mention anticyclonic eddies which are periodically formed in the eastern part of the Gulf of Mexico owing to the separation of anticyclonic meanders of the Yucatan Current which is a source of the Gulf Stream (Figure 4.1.17). For instance, during the period from October 1966 to September 1967 successive hydrographical surveys of the gulf registered three such separations (Elliott, 1982). The eddy radius varied from 100 to 200 km and the orbital velocity of water motion in one of these eddies attained 150 cm/s. After separation these eddies drift westward with an average velocity of 2–2.5 km/day until they have reached, albeit in a weakened form, the western coast of the gulf where they thus form a northward boundary current with periodically varying power (Nowlin *et al.*, 1982).

2. EDDIES OF THE KUROSHIO SYSTEM

Eddies of the Kuroshio–Oyashio frontal region (Figures 4.2.1 and 4.2.2) began to be studied particularly intensively by Japanese oceanographers in the mid-1950s; the papers by Masuzawa (1955), Ichiye (1956), and Sugiura (1957) can be regarded as classical works in this field. The results of these and some other investigations whose review can be found in the work by Kawai (1972) provided, on the whole, a correct indication of the structure of the frontal zone and the character and intensity of the eddy formation process in that zone. In the 1960s and 1970s Soviet oceanographers carried out an extensive series of studies in the Kuroshio–Oyashio frontal region which was primarily devoted to investigating the seasonal and year-to-year variability of the position and intensity of the main fronts and branches of a large-scale oceanic circulation (Bulgakov *et al.*, 1972; Muromtsev, 1972; Pavlychev, 1975; Pokudov and Vel'yaots, 1976; Pokudov *et al.*, 1978; Peregudin, 1982). In recent years special studies on the structure and dynamics of the

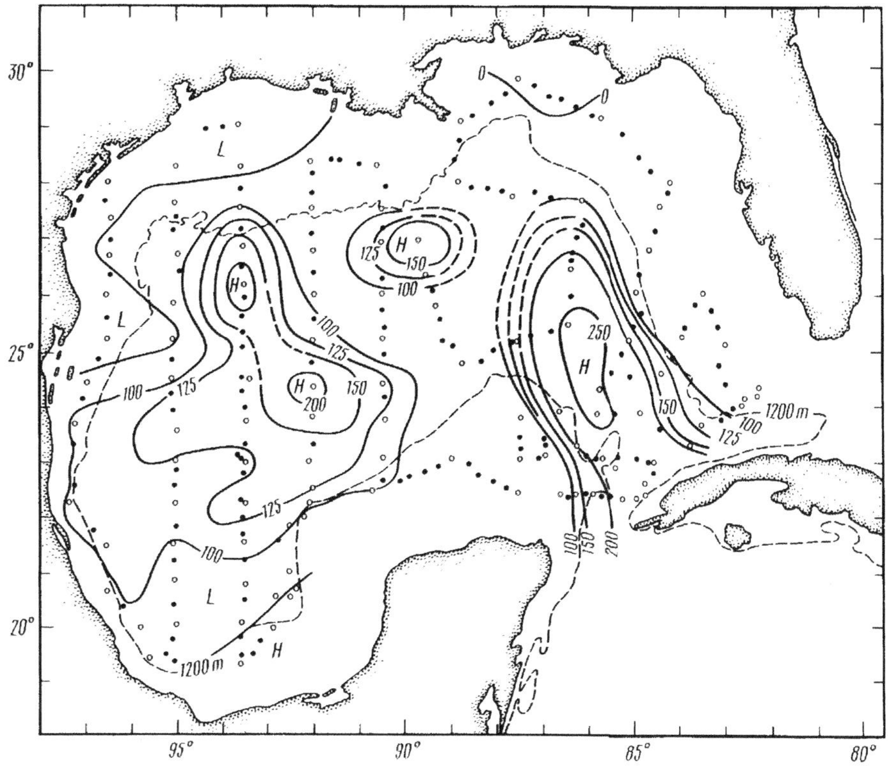

Fig. 4.1.17. The 20 °C isotherm depth (in meters) according to the data of the combined hydrographic and XBT survey of the Gulf of Mexico (the light and dark circles, respectively) in February 21–March 31, 1967 (after Elliott, 1982). H and L show the centers of high and low pressure.

Kuroshio and Oyashio eddies have been intensified (Kitano, 1975; Byshev *et al.*, 1976; Bernstein and White, 1977, 1981, 1982; Cheney, 1977, 1978b; Wilson and Dugan, 1978; Pokudov and Vel'yaots, 1978, 1979; Burkov and Pavlova, 1979; Bulatov, 1980; Nishiyama *et al.*, 1980; Kawai, 1980; Pokudov and Tunegolovets, 1983).

The basic specific structural features of the Kuroshio–Oyashio frontal region are clearly seen in Figure 4.2.2. Owing to the short-time character of the aircraft survey (13 days), the charts in the figure can be regarded as 'shots' of a sharply nonstationary synoptic situation. The geostrophic character of synoptic-scale currents allows us to identify the Kuroshio and Oyashio fronts in Figure 4.2.2 with the main jets of these currents; the current is directed so that the warm water remains on its right. Naturally, the same law also applies to the Kuroshio and Oyashio eddies which are cut-off meanders of the currents generating them. Thus the region south of the Kuroshio is an area of propagation of cold Kuroshio cyclones and the region north of the Oyashio is an area of warm Oyashio anticyclones[2]; and the region between

[2] In this case the terms 'warm' and 'cold' indicate only the temperature contrast between an eddy and the surrounding water. In reality, owing to the origin of the 'cold' Kuroshio cyclones and the 'warm' Oyashio anticyclones, the temperature of the former is higher than that of the latter (Figure 4.2.2).

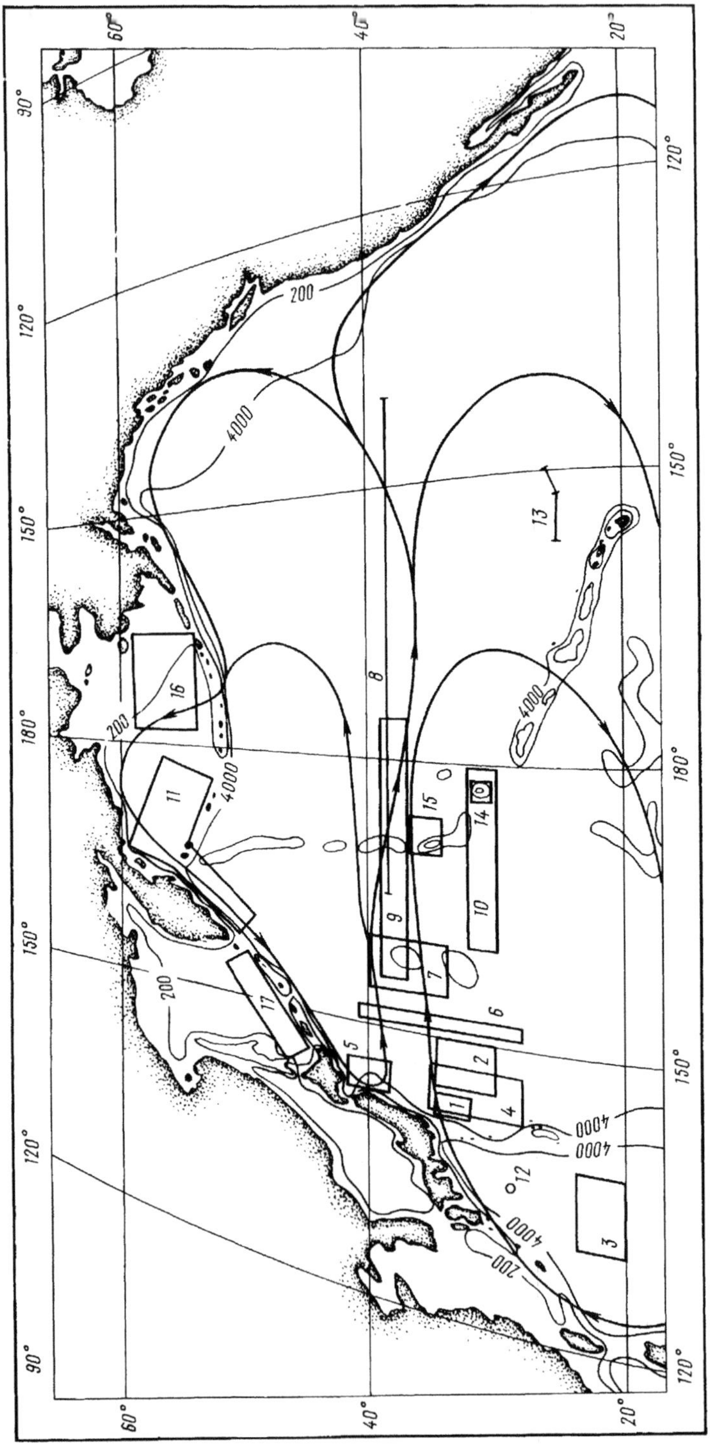

Fig. 4.2.1. Scheme of the large-scale circulation of waters of the upper (about 1 km thick) layer of the North Pacific. The symbols and numbers indicate the areas of expedition experiments whose results are discussed in Chapters 4 and 5.

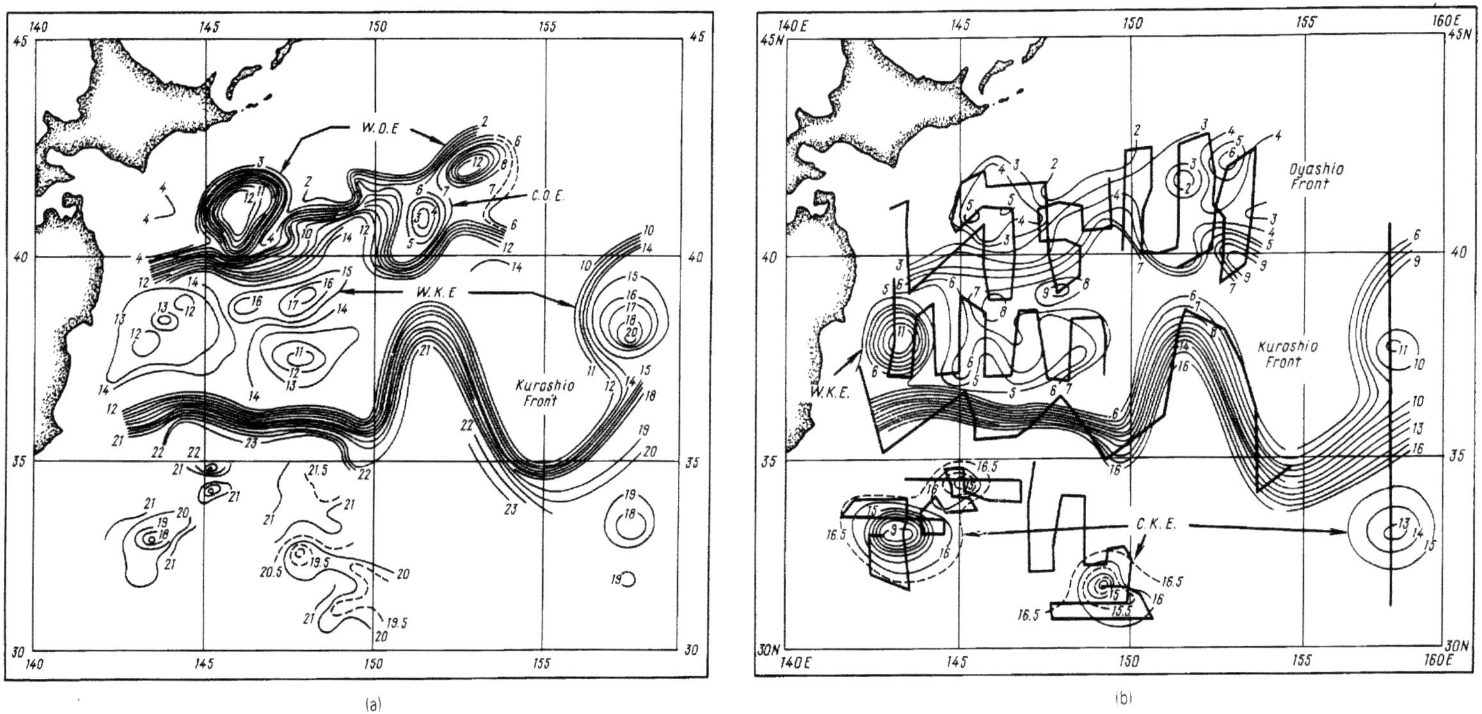

Fig. 4.2.2. Temperature distribution (°C) at depths (a) 100 m and (b) 300 m in the Kuroshio–Oyashio frontal region on October 9–22, 1976 (after Cheney, 1977). C.K.E, W.K.E, C.O.E., and W.O.E denote the cold and warm Kuroshio eddies and the cold and warm Oyashio eddies. The broken lines in (b) show the aircraft tracks from which the XBT soundings of the ocean were carried out; the distance between the soundings along the tracks was around 60 km.

the main jets of these currents contains warm Kuroshio anticyclones and cold Oyashio cyclones (Figure 4.2.2). The current velocity at the sea surface in the field of the main jet of the Kuroshio and the young eddies of this current can attain 2 m/s and more, and in the field of the Oyashio and its eddies it attains 1 m/s (e.g. see Kawai, 1972). This velocity decreases with increasing depth and, as can easily be seen from a comparison of Figures 4.2.2(a) and 4.2.2(b), for the Oyashio and its eddies this decrease is substantially sharper than for the main jet of the Kuroshio and its eddies. The velocity in the field of the Kuroshio main jet at a depth of 1000–1500 m is 10–15 cm/s.

The eddies shown in Figures 4.2.2(a) and 4.2.2(b) were of different power and age. The Oyashio anticyclone with center at 41°N, 146°E seemed to be the youngest formation. Among all the Kuroshio cyclones and anticyclones in Figures 4.2.2(a) and 4.2.2(b) the two extreme western eddies were the strongest. The Kuroshio anticyclone with center at 38°N, 143°E was characterized by an isothermal surface height difference between its center and periphery in the main thermocline layer attaining 320 m, and also by the presence in its central part of an isothermal water layer (with temperature from 11 to 12 °C) between depths of 100 and 400 m. In this relation the structure of this anticyclone was analogous to that of Gulf Stream anticyclones (Figure 4.1.6). It is remarkable that the Oyashio anticyclone with center at 41°N, 146°E did not contain such an isothermal layer. The isothermal surface height difference in the main thermocline (temperature from 16 to 5 °C; and water layer outside an eddy from 300 to 800 m) in the field of the Kuroshio cyclone with center at 33°N, 143°E also attained 300 m. Assuming that the 15 °C isothermal surface in a newly formed Kuroshio cyclone is elevated in its central part up to a depth of 75 m and that, by analogy with Gulf Stream cyclones (Parker, 1971; Cheney and Richardson, 1976) the subsidence rate of this surface is 0.5 m/day, Cheney (1977) obtained estimates of 4, 12, and 9 months for the age of the Kuroshio cyclones in Figure 4.2.2 occupying positions at 143, 149, and 158°E, respectively.

On the whole, Kuroshio cyclones have been studied less thoroughly than Gulf Stream cyclones. However, as early as 1939 the results of simultaneous operation of 30(!) research ships allowed Japanese oceanographers to obtain the first synoptic chart of the distribution of Kuroshio cyclones in the region bounded by the Kuroshio, 22°N, 126°E, and 162°E (Figure 4.2.3). As is seen from Figure 4.2.3, altogether 13 cyclonic eddies with diameters from 100 to approximately 230 km were registered and, as could be expected, on average, bigger (and hence, probably, younger) eddies occupied positions closer to the Kuroshio. Attention should be paid to the chain of five eddies, of gradually decreasing dimensions, which stretch from 33°N, 155°E to the passage in the Izu–Ogasawara Ridge at 23°N and probably indicate one of the predominant routes of Kuroshio cyclones in the ocean.

Masuzawa (1957) described the results of the surveys of a cold Kuroshio cyclone in February and May 1956 in the region 34–36°N, 142–148°E. The eddy had an elliptic shape with a major axis of 300 km oriented along the parallel and a minor axis about 160 km. According to the current measurements by a geoelectromagnetic kinetograph, the velocity of the surface current in the eddy field attained 2 m/s. During a three-month period the eddy travelled 170 km to the west-north-west,

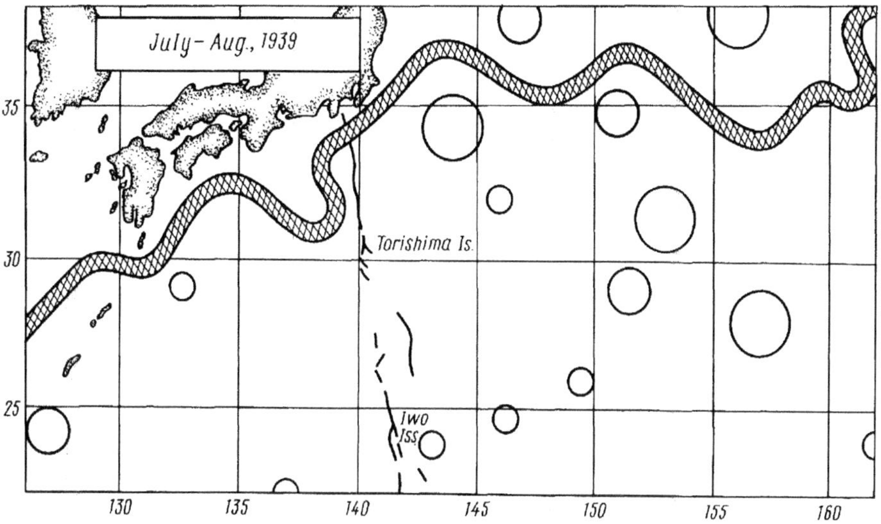

Fig. 4.2.3. Distribution and dimensions (the area contoured by the 16 °C isotherm at a depth of 200 m) of Kuroshio cyclones in the region south of the Kuroshio in July–August 1939 (after Kawai, 1980). The shaded strip shows the position of the Kuroshio at the indicated time. Shown in the region 139–143°E are the positions of the parts of the Izu–Ogasawara Ridge with depths less than 1500 m.

which corresponded to an average velocity of motion of 2 cm/s; during the same time the temperature difference between the eddy center and the surrounding waters at a depth of 100 m decreased from 8 to 7 °C.

A series of observations, lasting almost two months, of a cold Kuroshio cyclone with the aid of XBT and CTD soundings and current measurements by a satellite-tracked drifting buoy was carried out in October–November 1976 in the region 32–35°N, 142–144°E (Cheney, 1978b; symbol 1 in Figure 4.2.1). The eddy was of a circular shape with a diameter around 250 km and a radial temperature drop of 9 °C at a depth of 400 m. A maximum surface current velocity of 80 cm/s was attained at a distance of 60 km from the eddy center. The maximum current velocity at a depth of 1200 m was equal to 10 cm/s. The total kinetic energy and total available potential energy (see Section 1) of the eddy counted from a depth of 3000 m were estimated as 3.3×10^{15} and 1.3×10^{17} J, respectively. During the observation period the eddy first moved to the north-west with an average velocity of 5 cm/s and then moved to the north (towards the Kuroshio) with a velocity of 8 cm/s. The life of the eddy ended when it merged with the Kuroshio. A still longer series of observations of a Kuroshio cyclone was performed in the period from January to October 1972 (Pokudov and Vel'yaots, 1978). During this time the eddy located immediately south of the Kuroshio, and not contacting it, moved 1000 km to the west from 155 to 143°E with an average velocity about 4 km/day. In 1973–1974 the formation of eddies from the cyclonic Kuroshio meander at 147–148°E was registered at least twice (Pokudov and Vel'yaots, 1978) (symbol 2 in Figure 4.2.1). However, in contrast with the above-mentioned case, 2–3 months after their formation the eddies again merged with the current that had generated them.

The enormous quasistationary Kuroshio meander south of Japan at 135 139°E (Figure 4.2.4) has been known since the mid-1930s (Shoji, 1972). Multiyear observations (Shoji, 1972; Pokudov and Vel'yaots, 1979) show that this meander undergoes intensive oscillations, and the duration of the periods of its very strong development, and the periods of its almost complete absence, can last for several years. However, only quite recently has it been found (Pokudov and Vel'yaots, 1979; Nishiyama *et al.*, 1980) that at least in some cases the evolution of the meander can be related to processes of generation or absorption of cyclonic eddies by the meander; the first of these processes is demonstrated in Figure 4.2.4 and the other in Figure 4.2.5. The observations showed that the eddy that merged with the Kuroshio (Figure 4.2.5(b)) was generated by the above-mentioned Kuroshio meander in May 1977 (Pokudov and Vel'yaots, 1979). Hence, the period from the formation of the eddy until its reverse absorption by the current (Figure 4.2.5(b)) lasted for three months. The eddy whose formation is demonstrated in Figure 4.2.4 began to move to the west-south-west with a velocity of 5–6 km/day after it had been completely separated from the Kuroshio (Nishiyama *et al.*, 1980).

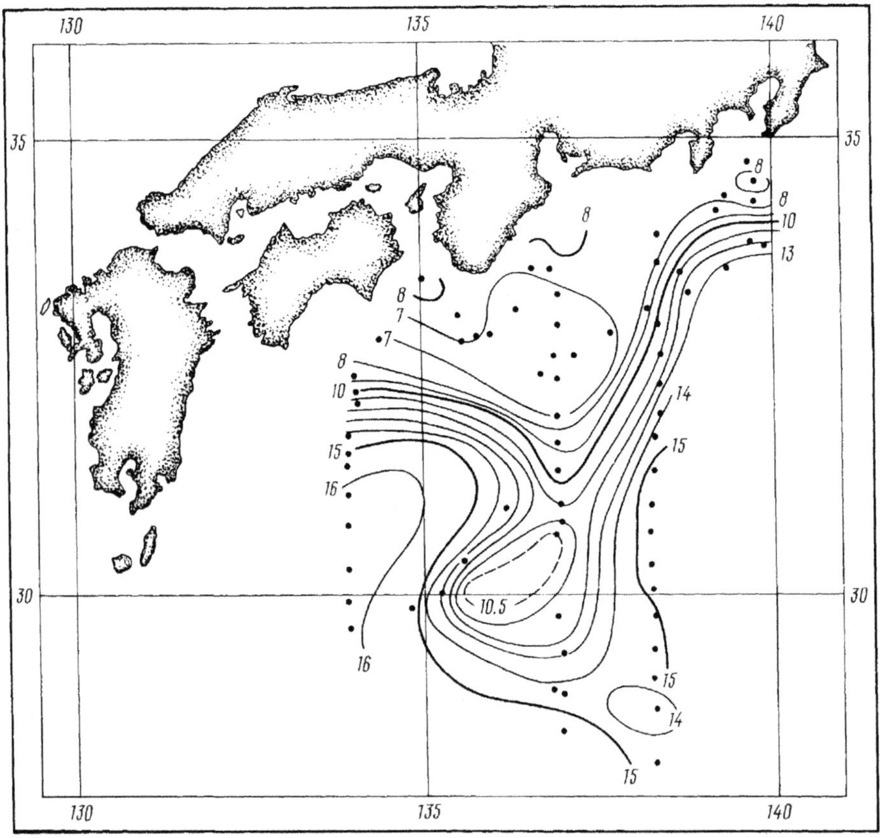

Fig. 4.2.4. Temperature distribution (°C) at a depth of 400 m in a region south of Japan in August 1979 (after Nishiyama *et al.*, 1980). The points show the sites of XBT soundings of the ocean.

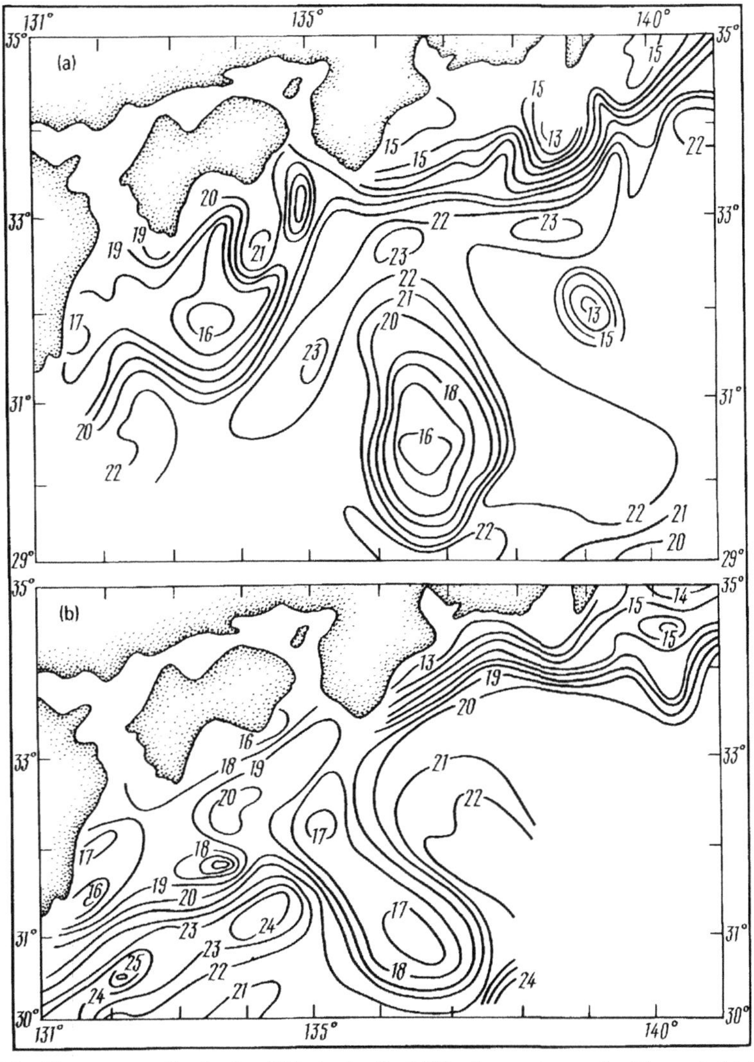

Fig. 4.2.5. Temperature distribution (°C) at a depth of 100 m in a region south of Japan in (a) July and (b) August 1977. (After Pokudov and Vel'yaots, 1979.)

The positions of the synoptic experiment areas of the Soviet expedition KISZ-80 (Figure 4.2.6) were chosen on the basis of two uniquely extensive multiship hydrographic surveys of the North-West Pacific carried out in July and August–September 1980 (Pokudov and Tunegolovets, 1983). The *T*, *S* analysis clearly showed that the dynamic formations seen in the north-west and north-east corners of the chart in Figure 4.2.7 were cyclonic Kuroshio meanders; the north-west meander was probably at a stage of its transformation into an eddy. The formation C_1 at 31°N, 143°E in Figure 4.2.7 was a cut-off cyclonic eddy of the Kuroshio. As can be concluded from Figure 4.2.7, the transverse dimension of the eddy attained a very large value of the order of 300 km and the current velocity in the eddy field

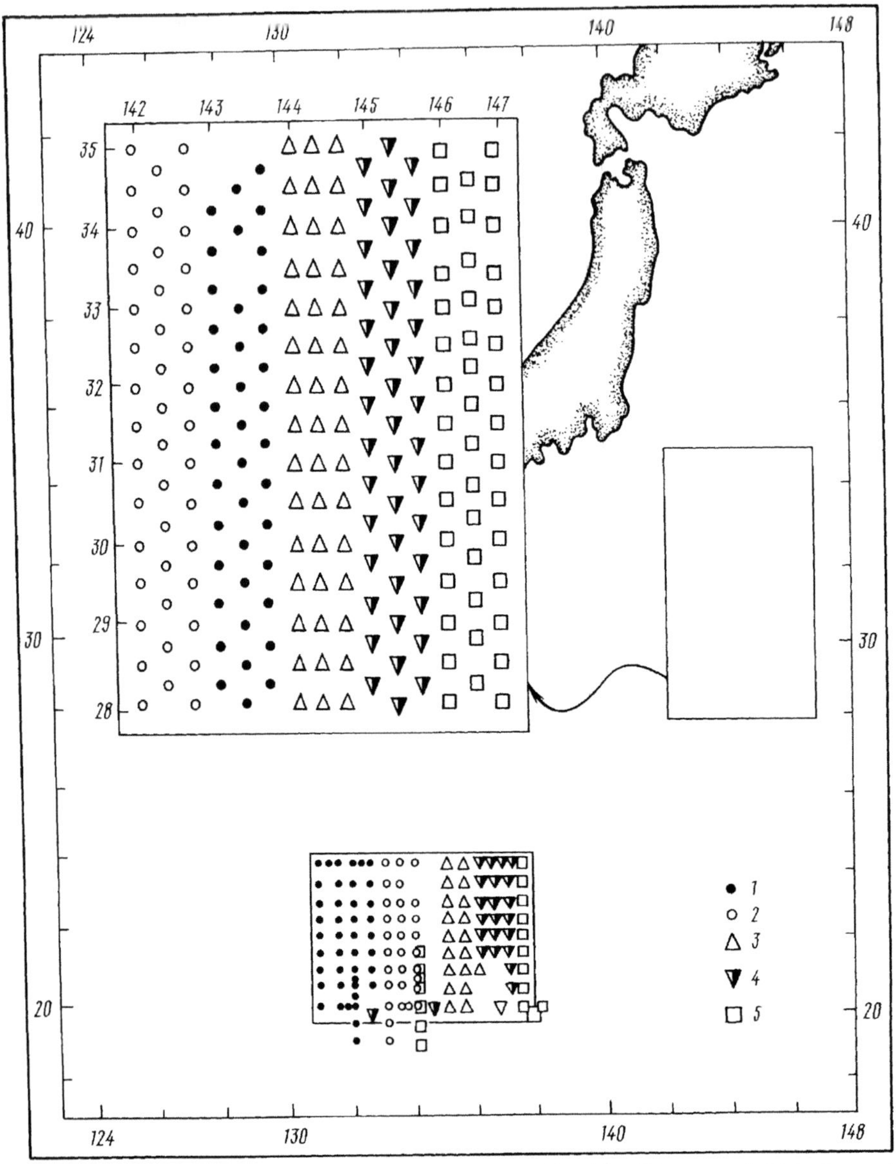

Fig. 4.2.6. Positions of hydrographic stations of different ships (the different symbols) in the south-west (October 4–10, 1980; observations above 1000 m) and north-east (October 21–30, 1980; observations above 1500 m) synoptic experiment areas of the expedition KISZ-80 (after Pokudov and Tunegolovets, 1983).

was about 1 m/s. The greatest velocities in the upper ocean and at depths of 600–1000 m were observed on the western (Figure 4.2.7) and eastern peripheries of the eddy, respectively. Consecutive hydrographic surveys of the region showed that from the middle of July to the end of October 1980 the eddy travelled approximately 100 km to the north and was probably at the initial stage of merging with

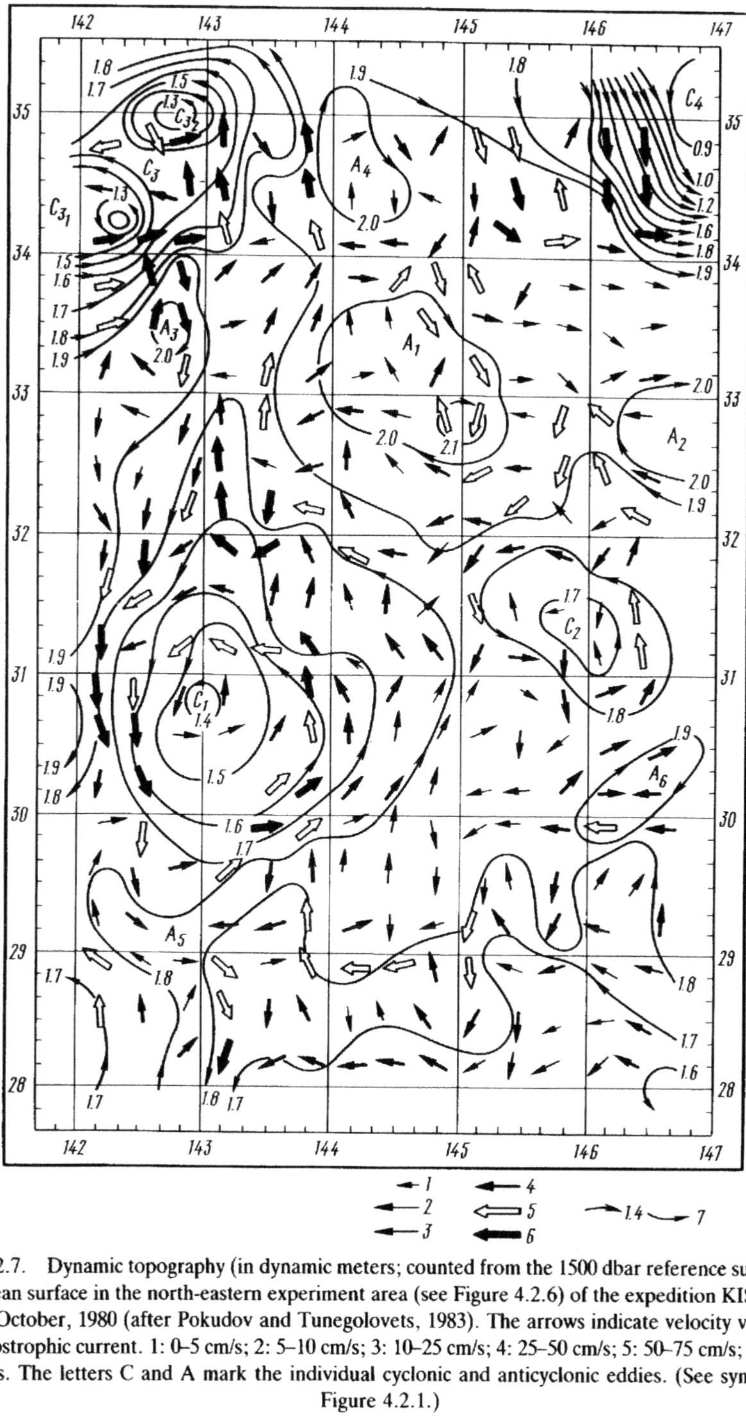

Fig. 4.2.7. Dynamic topography (in dynamic meters; counted from the 1500 dbar reference surface) of the ocean surface in the north-eastern experiment area (see Figure 4.2.6) of the expedition KISZ-80 on 21–30 October, 1980 (after Pokudov and Tunegolovets, 1983). The arrows indicate velocity vectors of the geostrophic current. 1: 0–5 cm/s; 2: 5–10 cm/s; 3: 10–25 cm/s; 4: 25–50 cm/s; 5: 50–75 cm/s; and 6: > 75 cm/s. The letters C and A mark the individual cyclonic and anticyclonic eddies. (See symbol 4 in Figure 4.2.1.)

another Kuroshio eddy-meander located north of the former (Figure 4.2.7).

Although the chart in Figure 4.2.8 does not demonstrate the positions of all the Kuroshio cyclones during the period from 1927 to 1977 and shows only those that could be registered by observations, it clearly indicates intensive generation of eddies in the region of the above-mentioned quasistationary cyclonic Kuroshio meander at 135–139°E and, particularly, in the region of rapidly repeating cyclonic Kuroshio meanders at 146–150°E. According to Figure 4.2.8, it can be assumed that after the formation of eddies in these regions they drift westward and most often undergo a reverse absorption by the current that generated them. Besides the above-mentioned route of Kuroshio cyclones (see Figure 4.2.3), Figure 4.2.8 allows us to indicate two more predominant routes of eddies – namely, from 35°N, 148°E to the passage through the Izu–Ogasawara Ridge at 29°N and from the region at 35°N, 162°E to the south-west. All these routes are likely to correspond to the Kuroshio quasistationary recirculation (counter current) branches. Further, Figure 4.2.8 shows a very good correlation between the intensity of the negative water-temperature anomaly in the central parts of Kuroshio cyclones (and hence their age) and the distance they travelled from the Kuroshio. Following earlier works (Ichiye, 1955; Koshlyakov, 1961), Kawai (1980) also assumes that the process of generation of Kuroshio cyclones is the main mechanism of the north-to-south transport of low-salinity water of subarctic origin across the Kuroshio and the North Pacific Current, which, in turn, results in the creation of the well-known interlayer of the above-mentioned low-salinity water throughout almost the whole North Pacific (owing to the process of the further eddies drawing in the

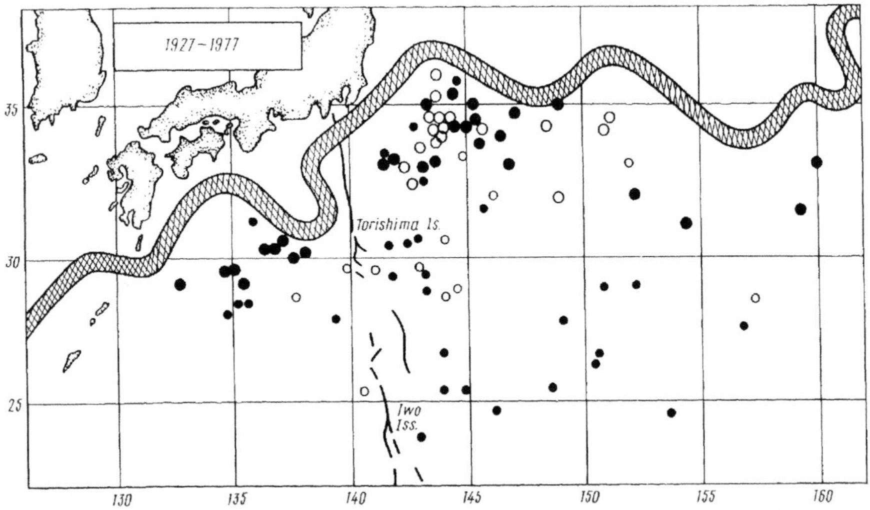

Fig. 4.2.8. Geographical distribution of the centers of the Kuroshio cyclones registered during the period from 1927 to 1977 (after Kawai, 1980). The large circles correspond to the eddies at whose centers the water temperature at a depth of 200 m was lower than 15 °C and the small circles correspond to temperatures between 15 and 16 °C. The light circles mark the eddies in which the salinity minimum was 0.1‰ or still lower than in the surrounding water. The shaded strip shows the typical position of the Kuroshio. Shown in the region 139–143°E are the positions of the parts of the Izu–Ogasawara Ridge at depths less than 1500 m.

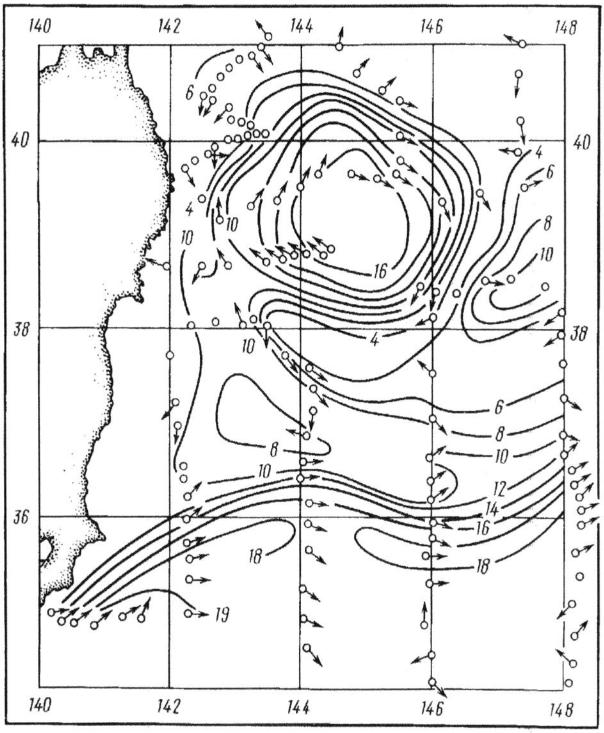

Fig. 4.2.9. Temperature distribution (°C) at a depth of 200 m in a region east of Japan on August 12–21, 1960 (after Kawai, 1972). The circles show the positions of the hydrographic stations. The arrows indicate the directions of surface currents measured by a geoelectromagnetic kinetograph.

anticyclonic subtropical water gyre in the North Pacific and 'spreading' the subarctic water at an intermediate depth in the ocean in accordance with its density).

Among all the eddies of the Kuroshio–Oyashio system, Kuroshio anticyclones east of Honshu underwent particularly thorough investigations because of their effect on the productivity of the fishing industry in this region. An example of a month-old Kuroshio anticyclone is demonstrated in Figure 4.2.9. The separation process developed rapidly and, as in the case of the formation of Gulf Stream anticyclones, was accompanied by an intensive inflow of the cold subpolar water from the north-east to the zone of separation (Figure 4.2.9). The power of this eddy is impressive. The total zonal and meridional dimensions of the eddy immediately after its formation were, respectively, 420 and 320 km, the temperature drop at a depth of 200 m was 14 °C (the greatest horizontal temperature drop ever registered in an oceanic eddy!), and the surface current velocity attained 2 m/s; and 13 months later the enumerated parameters had values of 320 and 180 km, 4 °C, and 50 cm/s, respectively. During this time the center of the eddy travelled 140 km to the south-west.

The process of formation of strong Kuroshio anticyclones in the region 143–146°E has a quasiperiodic character (Kawai, 1972; Kitano, 1975; Burkov and Pavlova, 1979). One more example of this kind of formation and further evolution of an anticyclone in February–November 1954 was described by Masuzawa (1955) and Ichiye (1956). In this case, in contrast to the above, two months after the generation of the eddy (in February 1954) it came into secondary contact with the Kuroshio and turned into an eddy-meander. In this form it existed until November 1954; during this time it travelled about 200 km to the east. In November a second and final separation of the eddy from the main Kuroshio jet took place. The authors of the above-mentioned works came to the conclusion that the phenomenon of multiple separations and reverse mergings of eddies with current jets generating them must be typical of the Kuroshio–Oyashio frontal zone and that the direction of motion of the centers of the eddies to the west or to the east is determined by the degree of contact of the eddies with these jets. We remind the reader (Section 1) that the same property also takes place in the case of Gulf Stream eddies. Kitano (1975) analyzed the observation data on Kuroshio anticyclones obtained from 1957 to 1973. It was found that eddy diameters vary mainly from 50 to 300 km and their average value is 180–200 km. The maximum water temperature at the centers of eddies predominantly varies from 16–17 °C at 37°N to 11–12 °C at 40°N, which can naturally be accounted for by the damping of eddies as they move away from the main Kuroshio jet. Of interest are the results of satellite surveys of the ocean (Kaz'min and Sklyarov, 1981) demonstrating, by analogy with the Gulf Stream region (Section 1), the development of secondary small cyclonic eddies (with a total transverse dimension about 100 km) of frictional origin on the periphery of particularly large Kuroshio anticyclones.

In the ocean region east of Japan, cold subpolar waters are located stationarily in the zone north of the main Oyashio jet (Figure 4.2.2). Spontaneous propagation of these waters to the south primarily takes place in two ocean regions (Figure 4.2.10); namely, in a region adjoining Honshu and a region approximately between 146 and 149°E. Multiple observations (e.g. see Sugiura, 1957; Kawai, 1972; Burkov and Pavlova, 1979) showed that this propagation relates to the quasiperiodic appearance of deep cyclonic Oyashio meanders in these regions and the transformation of these meanders into cold cyclonic eddies predominantly moving to the south and the south-west after separation. According to satellite temperature surveys of the sea surface, the transverse dimensions of these eddies vary mainly from 100 to 200 km, and the duration of the periods of their passage through a fixed point in the ocean ranges from 15 to 30 days (Bulatov, 1980). The temperature drop at a depth of 100 m in Oyashio cyclones can attain 12 °C and even more (Sugiura, 1957). Besides the absorption of Kuroshio cyclones in the open part of the Pacific (Figure 4.2.5) and the upwelling of cold deep waters, the motion of Oyashio cyclones to the south and then to the south-west along the east and south-east coasts of Honshu seems to be one of the mechanisms of filling the quasistationary Kuroshio meander in the region south of Japan (Figure 4.2.4) in the periods of its strongest development (Shoji, 1972; Burkov and Pavlova, 1979). Of interest is the work by Ichiye and Ichiye (1956) where it was shown that the fast southward motion of a cyclonic meander which then became an Oyashio eddy in the region east of the northern coast of Honshu in July–October 1950 was associated with a sharp local increase of intrusion of relatively low-salinity subpolar waters into the more saline waters of the interfrontal Kuroshio–Oyashio zone at

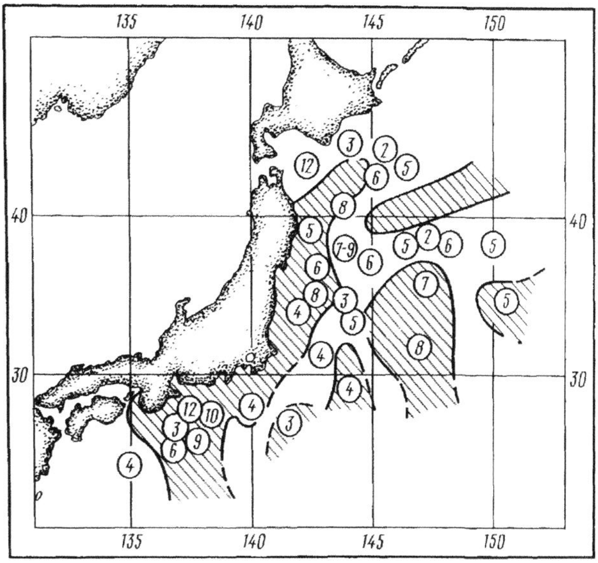

Fig. 4.2.10. Regions of predominant distribution of cold (shaded parts of the area) and warm (nonshaded parts) eddies in the ocean east and south of Japan during the period from 1955 to 1964 (after Burkov and Pavlova, 1979). The circles and the figures in them show the positions of cold and warm eddies at a depth of 200 m and their numbers in the specific regions during the indicated period.

intermediate depths of 150–500 m. This particular example once again confirms the general idea that the formation of the intermediate waters of the world ocean is closely related to processes of synoptic formation of eddies in frontal ocean regions (Ichiye, 1955; Koshlyakov, 1961; Kawai, 1980).

Examples of formation of very strong warm Oyashio anticyclones in a region east of the northern extremity of Honshu can be found, for instance, in the works of Sugiura (1957) and Kitano (1975). In the first of the above-mentioned cases (symbol 5 in Figure 4.2.1), the length of the major (zonal) axis of the eddy attained 300 km and the temperature drop at a depth of 100 m attained 11 °C. In the second case the transverse eddy dimension and the temperature drop at the 100 m depth were 250 km and 8 °C, respectively. Like the other eddies of the Kuroshio–Oyashio frontal region, Oyashio anticyclones often come into repeated contact with the current that generated them, which seems to specify the east-north-east direction of their mean drift in ocean regions east of the Tsugara Strait (Kitano, 1975).

The chart in Figure 4.2.10 was constructed by its authors from the data of hydrographic surveys of the part of the Pacific adjoining Japan, carried out from 1955 to 1964. As can be concluded from the chart, altogether about 80 cold eddies and 60 warm eddies were observed in this region during the indicated period. (It should be noted that the authors of this work regarded as eddies not only the cyclones and anticyclones formed owing to the separation of Kuroshio and Oyashio meanders but also the recirculation formations on the inner sides of the meanders.) The location of the regions of predominant propagation of cold and warm eddies in Figure 4.2.10 is in complete agreement with the descriptions presented earlier in this section. The analysis of hydrographic observation data made Burkov and

Pavlova (1979) conclude that throughout the ocean region under consideration, seven or eight cold cyclonic eddies and the same number of warm anticyclonic eddies are generated every year.

During the period from July 1980 to May 1981 American oceanographers (Schmitz *et al.*, 1982) performed direct current and water-temperature measurements at depths of 600, 1200, and 4000 m at ten moorings located along the 152°E meridian from 28 to 41°N (symbol 6 in Figure 4.2.1) which thus intersected the main Kuroshio jet (with the average position at 36°N) and the ocean regions north and south of the jet. They observed a rapid decrease of the average kinetic energy of synoptic disturbances of the current in the northward and southward directions from the region of the main jet; for instance, at a depth of 600 m, it decreased from 360 cm^2/s^2 at 34°N to 30–40 cm^2/s^2 at 28 and 41°N. The majority of the energy of synoptic oscillations of currents at all the depths was contained within the range of periods from 20 to 120 days. The values of the quantity $\overline{uv}$ (where u and v are the zonal and meridional components of the synoptic disturbance of the current velocity which are regarded as positive in the eastward and northward directions and the bar denotes averaging over the whole observation period) turned out to be positive in the region south of the Kuroshio axis and negative north of the axis. For a purely zonal Kuroshio model, this indicates that the kinetic energy of synoptic eddies is transformed into kinetic energy of the mean current in the regions north and south of the Kuroshio axis, i.e. the so-called 'negative viscosity' effect. We can say that, on the whole, synoptically, the region east of Japan is perhaps the most complicated in the ocean. Many details of the synoptic formation of eddies and the interaction between eddies and large-scale currents in this region have not yet been completely elucidated and require further purposeful investigations.

In this section we have by now considered the synoptic variability of the Kuroshio system mainly west of 150°E. As to the region east of this meridian, as early as the 1930s–1960s, hydrographic surveys of the Kuroshio carried out by Japanese oceanographers gave first indications of the fact that at least in the region between 150 and 165°E the main Kuroshio jet retains a rather concentrated character, undergoes intensive meandering, and generates cyclonic and anticyclonic eddies. For instance, a survey in 1939(!) revealed a clearly marked warm anticyclone immediately north of the main Kuroshio jet with center at 38°N, 156°E, transverse dimension around 300 km, and temperature drop at a depth of 200 m equal to 7 °C (Kawai, 1972). In June–August 1974, Soviet oceanographers carried out a thorough investigation of synoptic variability in this ocean region by current measurements at moorings and hydrographic surveys of the region (Byshev *et al.*, 1976; see also Figure 4.2.11 and symbol 7 in Figure 4.2.1). The velocity vectors in Figure 4.2.11(b) are obtained after filtering out the inertial and tidal oscillations of the currents. In the southern part of the chart in Figure 4.2.11(a) an anticyclonic meander of the main Kuroshio jet is clearly seen, and in the north-eastern part an anticyclonic Kuroshio eddy with transverse dimension about 300 km is observed. Attention should be paid to the good agreement between the measured and calculated currents. The current velocity in the field of this eddy attained 50 cm/s in the upper ocean. Comparing the pattern of currents in Figure 4.2.11(a) with the character of time variation of the velocity vectors at point 2 in

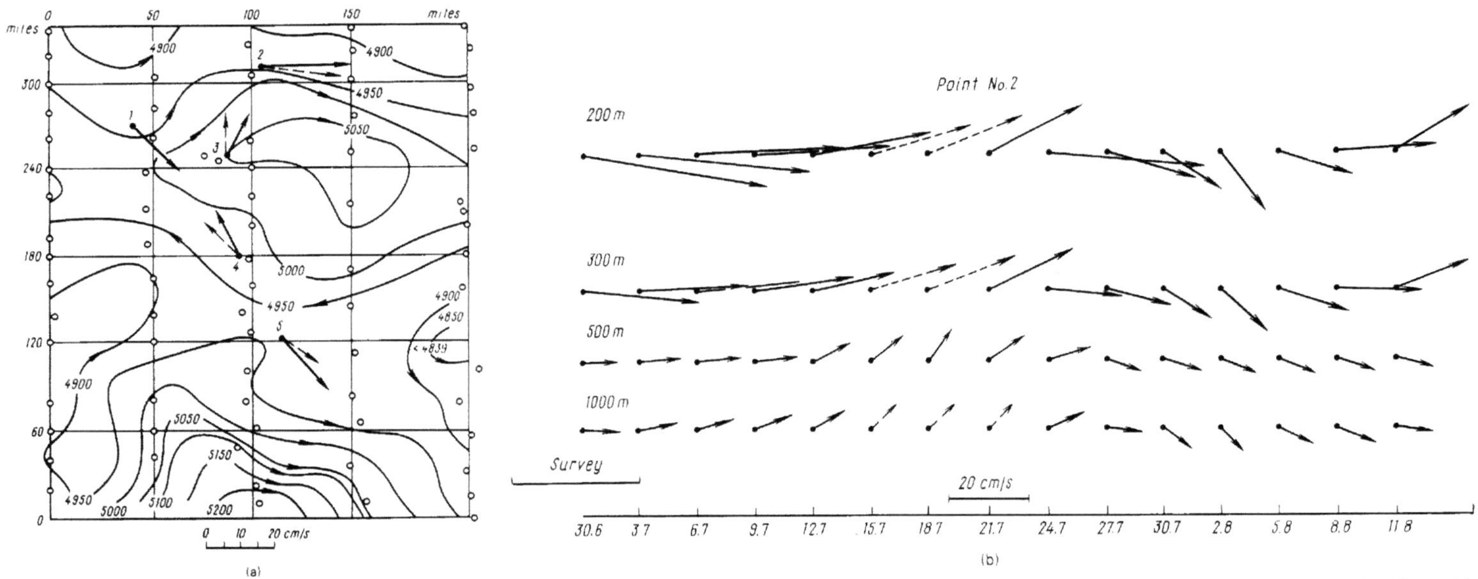

Fig. 4.2.11. (a) The dynamic topography of the 500 dbar isobaric surface relative to the 2000 dbar reference surface in a region north of the Kuroshio according to the data of the hydrographic survey at the end of June to the beginning of July 1974 and (b) the time variation of the velocity vectors of synoptic currents at several depths at point no. 2 (the position of the point is shown in (a)) (after Byshev *et al.*, 1976). The survey period is shown in (b). The dynamic height in (a) is measured in dynamic millimeters. The light and dark circles in (a) show the positions of the hydrographic and buoy stations. The solid arrows in (a) represent the average velocity vectors of the measured currents during the survey period at a depth of 500 m. The dashed arrows show the velocity vectors of geostrophic currents (See symbol 7 in Figure 4.2.1.)

Figure 4.2.11(b), the authors of the work (Byshev *et al.*, 1976) came to the conclusion that the system of synoptic disturbances seen in Figure 4.2.11(a) moved to the east with a velocity about 7 cm/s, which is close to the phase velocity of the eastward motion of Gulf Stream meanders. According to Figure 4.2.11(b), the velocity oscillation period relating to the passage of synoptic disturbances through a fixed point was 30–35 days.

Valuable information about the structure and variability of the Kuroshio and the North Pacific Current, which is a continuation of the Kuroshio, was given by XBT temperature measurements on ships plying between North America and Japan (Bernstein and White, 1977, 1981). Shown in Figure 4.2.12 is the temperature distribution in one of these transoceanic sections (symbol 8 in Figure 4.2.1) passing approximately two degrees north of the axis of the North Pacific Current. The basic specific property of this distribution is a sharp decrease of oscillation amplitudes of isotherm depths in the west-to-east passage across the 175°W meridian. Analogous results were obtained for another 47 transoceanic sections performed in 1975 in the zone from 35 to 40°N (Bernstein and White, 1977) where the longitude of the sharp damping of temperature field disturbances varied from 170 to 180°W. Obviously this damping relates to the general expansion and weakening of the North Pacific Current in the west-to-east direction. The temperature waves in the western part of the section in Figure 4.2.12 seem basically to be a consequence of the intersection of meanders of the main jet of the North Pacific Current by the section; however, it is not excluded that some of these waves relate to cut-off eddies. The data of *en route* shipboard water-temperature measurements in 1977–1978 confirmed the intensive character of the meandering of the Kuroshio–North Pacific Current up to 180° of longitude and demonstrated a number of specific cases of transformation of meanders of the current into cold cyclonic and warm anticyclonic eddies in the region 140–180°E (Bernstein and White, 1981). In qualitative accordance with the above results, the processing of data from hydrographic surveys by Japanese oceanographers in 1955–1977 of the region originating the North Pacific Current

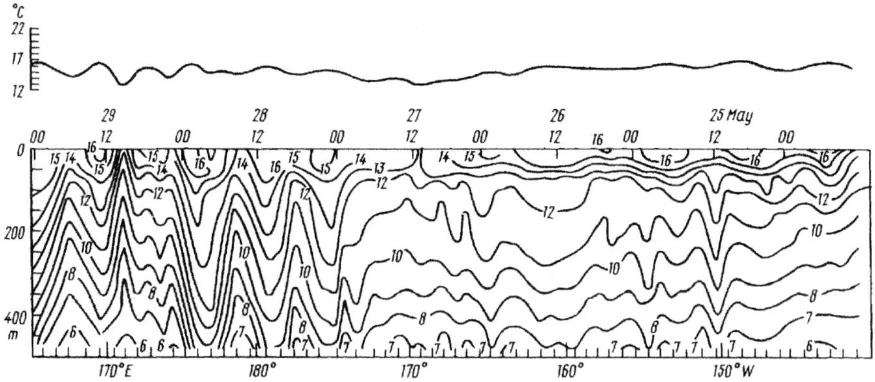

Fig. 4.2.12. Temperature distribution (°C) in the section along 38°N in the region of the North Pacific Current in May 1975 (after Bernstein and White, 1977). The marks below show the sites of the thermal soundings of the ocean. The curve above represents the sea surface temperature. (See symbol 8 in Figure 4.2.1.)

showed that the amplitudes of synoptic disturbances of the temperature, density, and geostrophic velocity fields in the main thermocline layer in the region of this current west of the Emperor Seamount Chain (approximately at 170°E) are, on average, 3–5 times higher than east of it (Roden *et al.*, 1982).

A synchronous temperature survey of a zonal ocean 'swath', 2500 km long and 270 km wide, extending from 156°E to 175°W and from 35°45′N to 38°15′N (symbol 9 in Figure 4.2.1) and adjoining the axis of the North Pacific Current on the north was carried out by six research vessels on parallel courses in May 1976 (Wilson and Dugan, 1978). The survey showed the presence of several warm anticyclonic eddies with diameters about 200 km which were formed owing to the separation of meanders of the main jet of the North Pacific Current. An analogous survey of a swath, 1800 km long and 220 km wide, extending from 160 to 180°E and from 29°45′ to 31°45′ (symbol 10 in Figure 4.2.1; the region south of the North Pacific Current) performed in November 1975 revealed two clearly marked cold cyclonic eddies whose diameters were also around 200 km (Wilson and Dugan, 1978).

The natural assumptions that the large-scale mean currents and their synoptic disturbances in the region of the Kuroshio–North Pacific Current have a geostrophic character, and that the absolute velocities of the mean and synoptic currents at a depth of 1000–2000 m are small in comparison with velocities in the upper ocean, made it possible to use the data of the above-mentioned *en route* shipboard temperature measurements in quasizonal sections across the Pacific to estimate the average meridional heat transfer across the North Pacific Current specified by the synoptic disturbances (Bernstein and White, 1982) and to study the kinetic energy exchange between the mean current and the synoptic disturbances (Nishida and White, 1982). The westward inclination of the 'vertical' axes of the temperature-field (water-density) synoptic disturbances which, for example, can be seen in Figure 4.2.12 turned out to be typical for almost all the above-mentioned transoceanic sections. In the light of the above-mentioned assumptions, this inclination specifies the presence of a south-to-north heat transfer across the North Pacific Current whose average magnitude during 1975 for the part of the vertical section of the ocean along 37°N bounded by the sea surface, the 2000 m depth, and the 165°E and 170°W meridians turned out to be equal to 3×10^{14} J/s. According to the estimates of Oort and Vonder Haar (1976), this amounts to 17% of the south-to-north heat flux transferred throughout the world ocean across the 40°N section, and indicates the substantial role played by eddies of the Kuroshio–North Pacific Current in the realization of this flux. The calculation of the average distribution of the specific kinetic energy of nonstationary synoptic disturbances of the Kuroshio–North Pacific Current during 1977–1978 at a depth of 100 m showed a sharp decrease in the distribution at 160°E; that is, from 150–200 cm^2/s^2 west of this meridian to 50 cm^2/s^2 east of it (Nishida and White, 1982). An investigation into the character of the balance of the current energy in the ocean region under consideration showed that, on average, the kinetic energy in synoptic eddies was transformed into kinetic energy in the mean current (the 'negative viscosity' effect) in a zone north of the axis of the Kuroshio–North Pacific Current and the reverse transformation occurred in the southern part of the region (see Figure 4.2.13). It was also shown that the transfer of kinetic energy in the mean current by synoptic

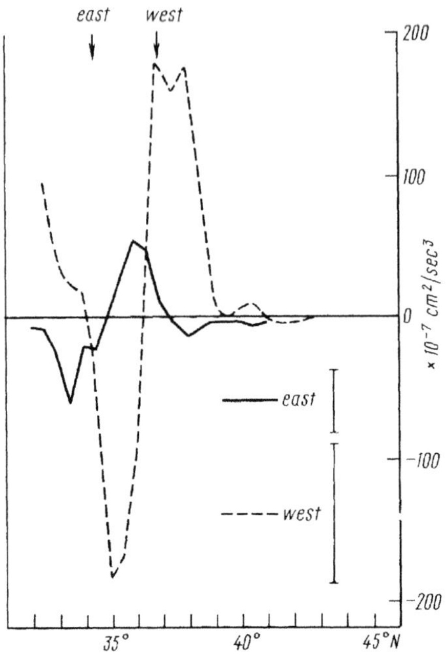

Fig. 4.2.13. Meridional distributions of the time- and longitude-averaged (during 1977–1978) rate of exchange of the specific kinetic energy between the mean current and the synoptic disturbances at a depth of 100 m in the Kuroshio regions from 140 to 160°E (dashed line) and from 160 to 180°E (solid line) (after Nishida and White, 1982). The positive values of the quantities represented correspond to the amount of disturbance energy that has been transformed into energy of the mean current. The arrows indicate the positions of the axis of the mean current west and east of 160°E. Also shown in the figure are the root-mean-square errors of the determination of the quantities for both parts of the region.

disturbances produces convergence of these energy fluxes in the axial zone of the current.

3. EDDIES OF OTHER WESTERN BOUNDARY CURRENTS

Eddies of the East Australian Current (EAC) began to be studied by Australian oceanographers at the end of the 1950s and the beginning of the 1960s, i.e. only a short time after the beginning of the investigation of Gulf Stream and Kuroshio eddies. Since the beginning of the 1960s up to the present time investigations of the structure and variability of the EAC have been carried out every year. They include density and temperature surveys of the region, current measurements by geoelectromagnetic kinetographs and some other measurements, and, in recent years, surface current measurements by satellite-tracked radio buoys (Hamon, 1965; Boland and Hamon, 1970; Nilsson, *et al.*, 1977; Godfrey *et al.*, 1980; Nilsson and Cresswell, 1981; Boland and Church, 1981). Despite the fact that the observations have revealed a complicated and nonstationary behavior of the EAC, nevertheless, as a rule, its main jet having a width of 80–100 km and flowing southward along the Australian continental slope can be identified. Usually at a latitude varying from 32 to 37°S this jet turns to the south-east, east, and north-east and forms an anticyclonic loop in the adjoining part of the ocean (Figures 4.3.1 and 4.3.2). The zone of the main current jet further to the east is known as the Tasman Front separating the warm Coral Sea water in the north from the colder Tasman Sea water in the south (Figure 4.3.1). The EAC penetrates the ocean to at least 2000 m and is characterized by highly time-varying velocities (probably with a period of half a year). In some periods the EAC transport in the 0–1300 m layer can attain 50×10^6 m^3/s, and the surface velocity in the part of the current adjoining Australia can attain 2 m/s (Boland and Church, 1981).

As is shown by numerous observations (e.g. see Godfrey *et al.*, 1980; Nilsson and Cresswell, 1981; Boland and Church, 1981), the EAC is characterized by a quasiperiodic process of formation of big anticyclonic eddies owing to the separation of anticyclonic meanders of the main current jet which are regularly formed in the region of the departure of the EAC from the Australian continent (Figures 4.3.1 and 4.3.2). On average, this separation occurs about twice a year, and it seems that there exists coherence between the eddy formation process and the velocity and transport oscillations of the current at the Australian coast. After separation the eddies have a tendency to move, on average, in the south-south-east direction with a typical velocity about 5 km/day which, however, attains 10 km/day and even more in some cases. However, in other cases eddies remain in the separation region for several months, which, for instance, was the case for eddy B in Figure 4.3.1 at the end of 1977 and the beginning of 1978. This can result in the reverse absorption of an eddy by the East Australian Current, i.e. a phenomenon which is analogous to the one observed in the Gulf Stream and Kuroshio regions and, as far as we can judge from Figure 4.3.1, is rather typical of the EAC as well. Owing to the alternating process of formation of eddies and their reverse absorption by the current, from one to three (Figure 4.3.1) and even more eddies of different age can be observed simultaneously south of the EAC. However, it can be assumed that some individual eddies (such as, probably, eddy A in Figure 4.3.1) avoid being absorbed by the current, and their life ends in the process of slowly mixing their water mass with the surrounding Tasman Sea water.

In September 1974 thorough investigations of an EAC anticyclonic eddy in the region 34–36°S, 151–154°E were carried out (Andrews and Scully-Power, 1976). The total diameter of the eddy, defined as the transverse dimension of the region of positive temperature anomaly at depths of 200–400 m, was 250 km; the isothermal surface depth difference in the main thermocline (100–500 m deep) attained 300 m; and the temperature drop at a depth of 300 m was 6 °C. We note that the internal Rossby scale in the ocean region under consideration is equal to 40 km, which corresponds to a disturbance wavelength of $2\pi \times 40$ km ≈ 250 km. The maximum orbital velocities of water motion in the eddy were observed at a distance of 60–80 km from the eddy center and they were equal to 60–180 cm/s in the surface layer. The strong current was concentrated in a ring only 50–70 km wide, which indicated that nonlinear effects were very clearly pronounced. The basic parameters of the eddy marked by the letter A in Figure 4.3.1 were close to those of the above-mentioned eddy of September 1974. According to the observations from October

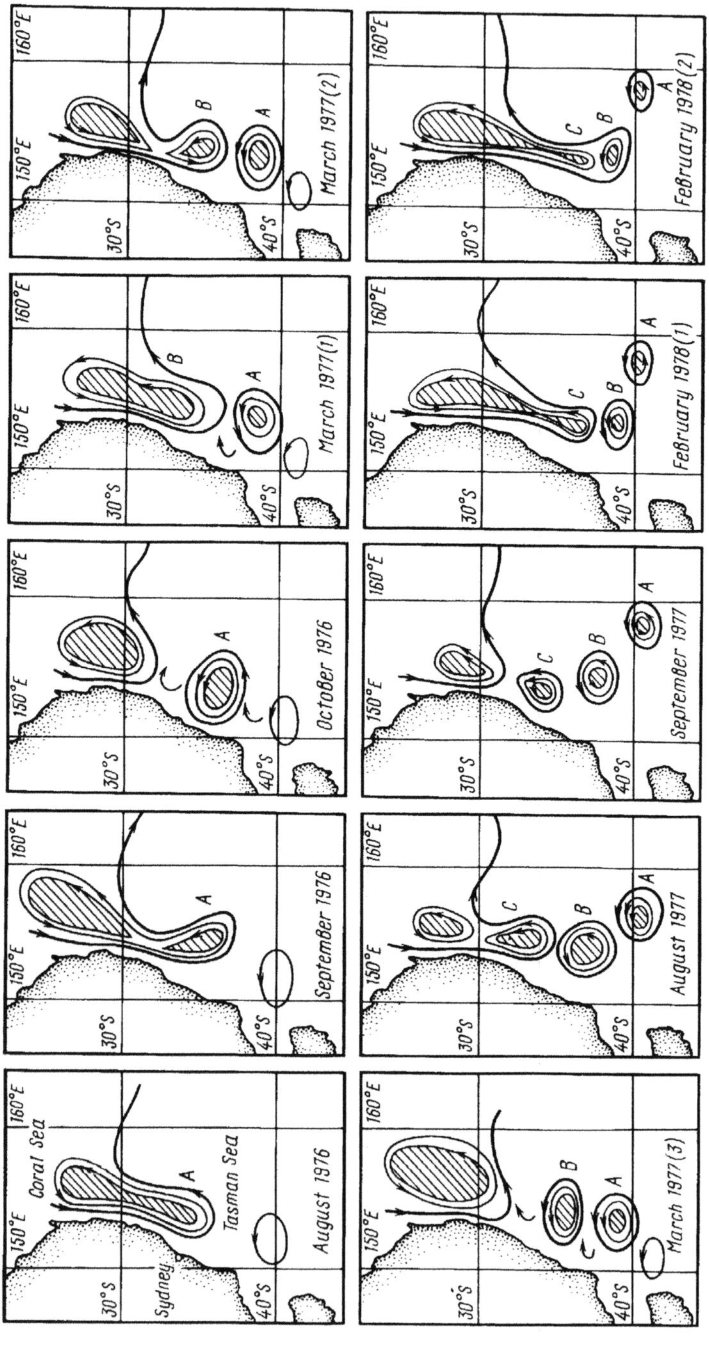

Fig. 4.3.1. Scheme of location of the main jet of the East Australian Current and several (A, B, and C) anticyclonic eddies in different months in 1976–1978 according to the data of hydrographic surveys and surface current measurements by satellite-tracked drifting buoys (after Nilsson and Cresswell, 1981). The thick streamlines separate the Coral Sea water (the temperature T_{250} at a depth of 250 m exceeds 15 °C) from the Tasman Sea water ($T_{250} < 15$ °C). The regions where $T_{250} > 17$ °C are shaded.

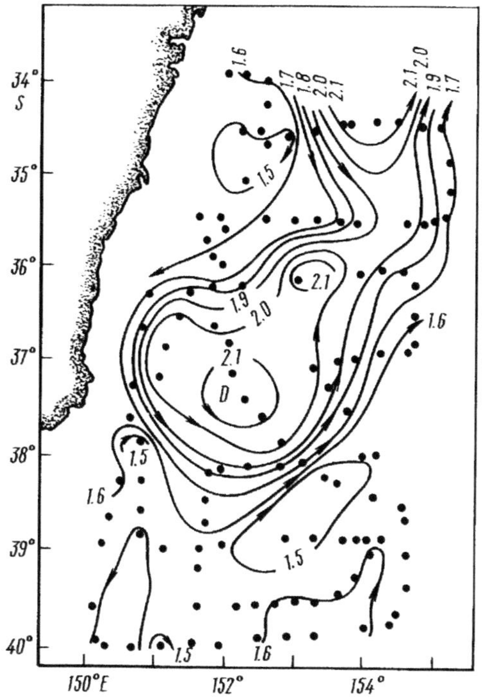

Fig. 4.3.2. Dynamic topography (in dynamic meters) of the sea surface relative to the 1300 dbar reference surface in a region east of Australia on May 22–June 6, 1978 (after Boland and Church, 1981). The points indicate the places of bathometer and XBT measurements.

1976 to March 1977, the former eddy was characterized by a diameter equal to 280 km, an isothermal surface depth difference and temperature drop equal to 350 m and 6 °C, respectively, and an orbital velocity of water motion in the upper ocean attaining 150 cm/s. The two satellite-tracked surface buoys which simultaneously drifted in eddy B (Figure 4.3.1) in April–May 1977 (one of them at an average distance of 40 km from the eddy center and the other at a distance about 80 km) showed the same period of orbital motion (about 7 days), which indicated that the motion of the central part of the eddy was close to solid rotation.

The permanent observations of eddy B (Figure 4.3.1) from March 1977 to February 1978 made it possible to study the time evolution of the eddy structure (Nilsson and Cresswell, 1981). During this period the eddy diameter decreased from 260 to 160 km and, simultaneously, the radial isothermal surface depth difference in the main thermocline decreased to a somewhat smaller degree (from 360 to 260 m). As a result, the orbital velocity of water motion in the eddy even increased slightly, which was confirmed by surface current measurements in the eddy by satellite-tracked drifting buoys showing velocities of 0.8 and 1.1 m/s for May 1977 and February 1978, respectively. The above-mentioned rate of smoothing of the isothermal surface relief in the eddy corresponds to the theoretical duration of the life of the eddy somewhat longer than two years.

One of the most interesting specific properties of EAC anticyclones is that many of them contain extensive intermediate isothermal water layers (cores) whose thickness in some cases attains 200 m (Figure 4.3.3) and even more. Australian oceanographers (Nilsson and Cresswell, 1981) explain the existence of the cores by the process of winter vertical convection and the further overlapping of the homogeneous water layer thus produced by the seasonal thermocline formed in the summer period. However, Fedorov and Belkin (1984) showed that this concept, by far, does not always account for the observed values of the core parameters, i.e. their thicknesses and temperatures, and suggested an alternative hypothesis according to which the cores are formed in the process of 'polhode circulation' (Monin and Fedorov, 1983) of waters in anticyclonic eddies, i.e. the radial-vertical circulation (produced by nonlinear effects and friction), one of whose branches is the downward isopycnic water motion from the periphery of an eddy to its center in the layer immediately above the main thermocline. This idea was confirmed by the authors by the T, S analysis of waters of the central and peripheral parts of eddy B (Figure 4.3.1).

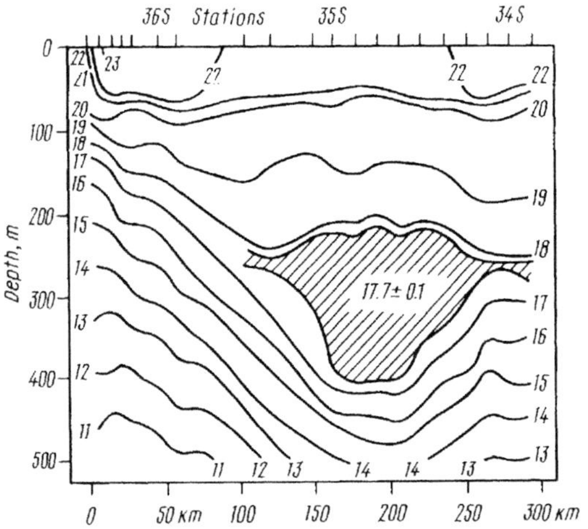

Fig. 4.3.3. Water temperature distribution (°C) in the meridional section across the center of an anticyclonic eddy in March 1975 in a region east of Australia (after Nilsson *et al.*, 1977). Shaded in the figure is the region of the isothermal core. The marks at the top of the figure show the places of XBT measurements.

The scheme in Figure 4.3.4 serves as an illustration of the hypothesis that the formation of EAC anticyclones is a consequence of the 'breaking' of baroclinic Rossby waves (propagating westward along the Tasman Front) owing to the incidence of these waves on the Australian continental slope. Observations (Nilsson and Cresswell, 1981) actually confirmed the existence of such waves with a wavelength about 350 km and a period about 170 days, which is in complete agreement with the average period of formation of eddies near the Australian coast.

The system of western boundary currents in the southern part of the Indian Ocean consists of the East Madagascar Current flowing southward along the eastern coast of Madagascar with a velocity attaining 150 cm/s at the sea surface

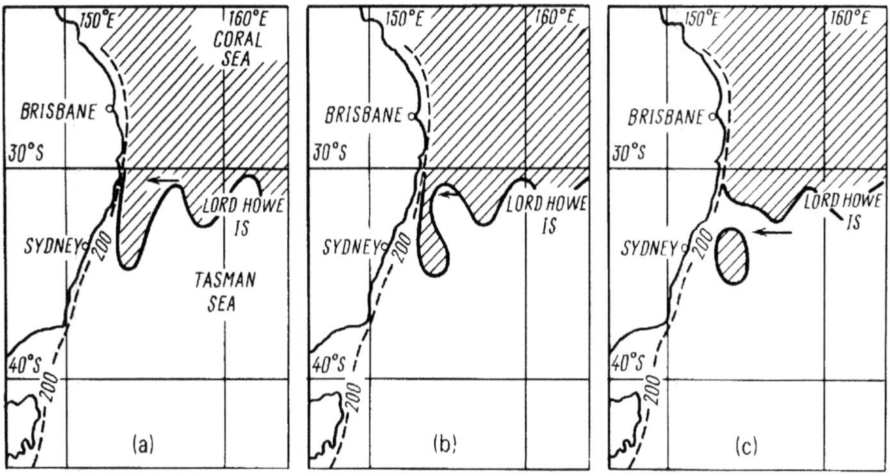

Fig. 4.3.4. Scheme of formation of anticyclonic eddies near the eastern coast of Australia (after Nilsson and Gresswell, 1981). The thick curve shows the axis of the East Australian Current (the Tasman Front). The arrows indicate the direction of propagation of waves on the front.

(Lutjeharms *et al.*, 1981), and the Mozambique and Agulhas Currents flowing to the south-west along the south-eastern and southern coasts of Africa. The surface current measurements by satellite-tracked drifting buoys performed in 1975–1979 in the ocean region south-east, south, and south-west of Madagascar revealed very well developed nonstationary eddying water-circulation disturbances throughout the zone (from 22 to 28°S) of the comparatively weak, on average, and rather wide western current joining the East Madagascar and Mozambique Currents in the zonal direction (Gründlingh, 1977; Lutjeharms *et al.*, 1981). In some cases the diameters of anticyclonic and cyclonic loops described by drifting buoys attained 200 km and the drift velocity of the buoys in the loops exceeded 1 m/s.

The Agulhas Current (Figures 4.3.5, 4.3.6, and 4.3.7), which is the most powerful component of the system of western boundary currents in the southern part of the Indian Ocean, comes close to the continental slope and has a particularly clearly pronounced jet character between 31 and 23°E; the width of the flow in this region does not exceed 80 km and the velocity in the axis of the current at the sea surface attains 250–300 cm/s (Gründlingh, 1977). At approximately 23°E the current goes away from the continental slope, and in the region 14–20°E (Gründlingh, 1978; Sarukhanyan, 1980a) it sharply turns to the east and generates the Agulhas Return Current. The axis of the latter current is located, on average, at 40°S and coincides with the axis of the Subtropical Convergence (Gründlingh, 1978), and the velocity in the axis at the sea surface decreases from 130 cm/s at the beginning of the return current to 80–90 cm/s at 40–50°E (Figure 4.3.5).

Shipboard hydrographic surveys (Harris and Van Forest, 1978; Sarukhanyan, 1980a), satellite infrared surveys of the sea surface (Harris *et al.*, 1978), and surface current measurements by satellite-tracked drifting buoys (Gründlingh, 1978) obviously showed that, after it leaves the continental slope, the Agulhas Current and, particularly, the Agulhas Return Current are nonstationary quasigeostrophic flows

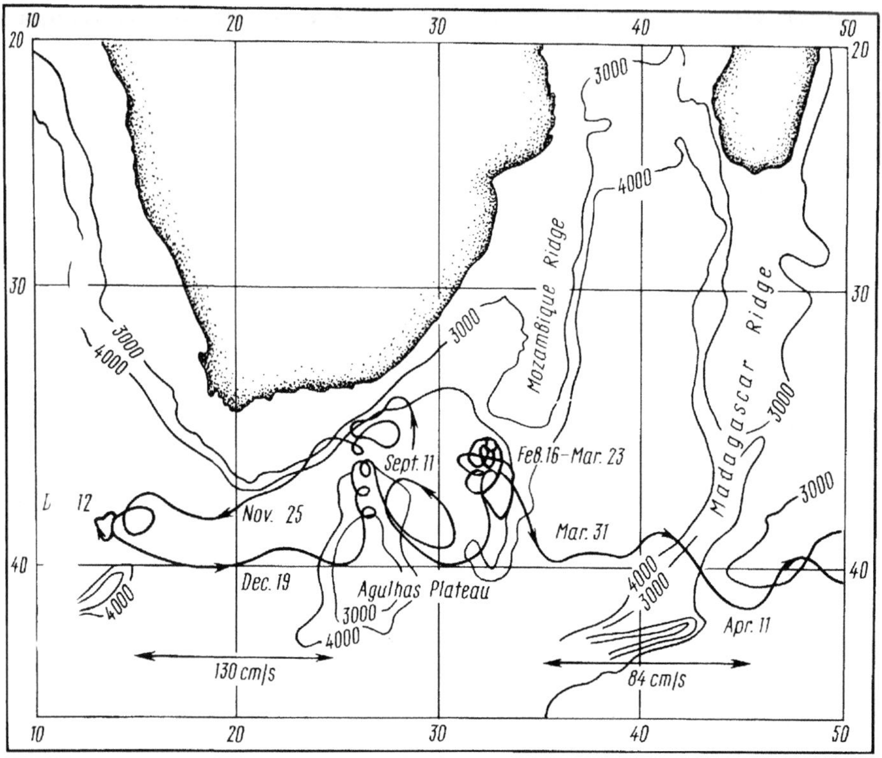

Fig. 4.3.5. Trajectory of motion of a satellite-tracked surface buoy in a region south of Africa in September 1975–April 1976 (after Gründlingh, 1978). The numbers below are the values of the average drift velocities of the buoy in the Agulhas Return Current. The 3000 and 4000 m isobaths are also shown.

which periodically generate synoptic-scale meanders and eddies (Figures 4.3.5, 4.3.6, and 4.3.7). It was found that there are three regions of predominant formation of particularly big meanders and eddies (Figures 4.3.5 and 4.3.7), namely the region of the Schmidt-Ott, Richardson, and other Seamounts at 11–16°E where the Agulhas Current turns to the east, the region of the Agulhas Plateau at 24–28°E, and the region of the southerly tip of the Mozambique Ridge at 31–35°E. In the first of these regions a spontaneous process of formation of anticyclonic eddies takes place owing to the periodic separation of the western end of the Agulhas Current–Agulhas Return Current loop. Figure 4.3.5 seems to correspond to the initial phase of transformation of an anticyclonic meander into an eddy. Figure 4.3.6 shows an eddy with center at 39°30′S, 14°E which has nearly been separated from the main current. The observations in March 1964 (Duncan, 1968) gave a result which is close to the one presented in Figure 4.3.6, and the hydrographic survey in March 1969 revealed a situation where the southward turn of the Agulhas Current took place at 20°E whereas the center of an anticyclonic eddy was located far away in the north-west at 35°30′S, 14°30′E (Harris and Van Forest, 1978). In March 1964 the transverse dimension of an anticyclonic eddy with center at 40°S, 15°E exceeded 300 km and the isothermal and isohaline surface depth drop

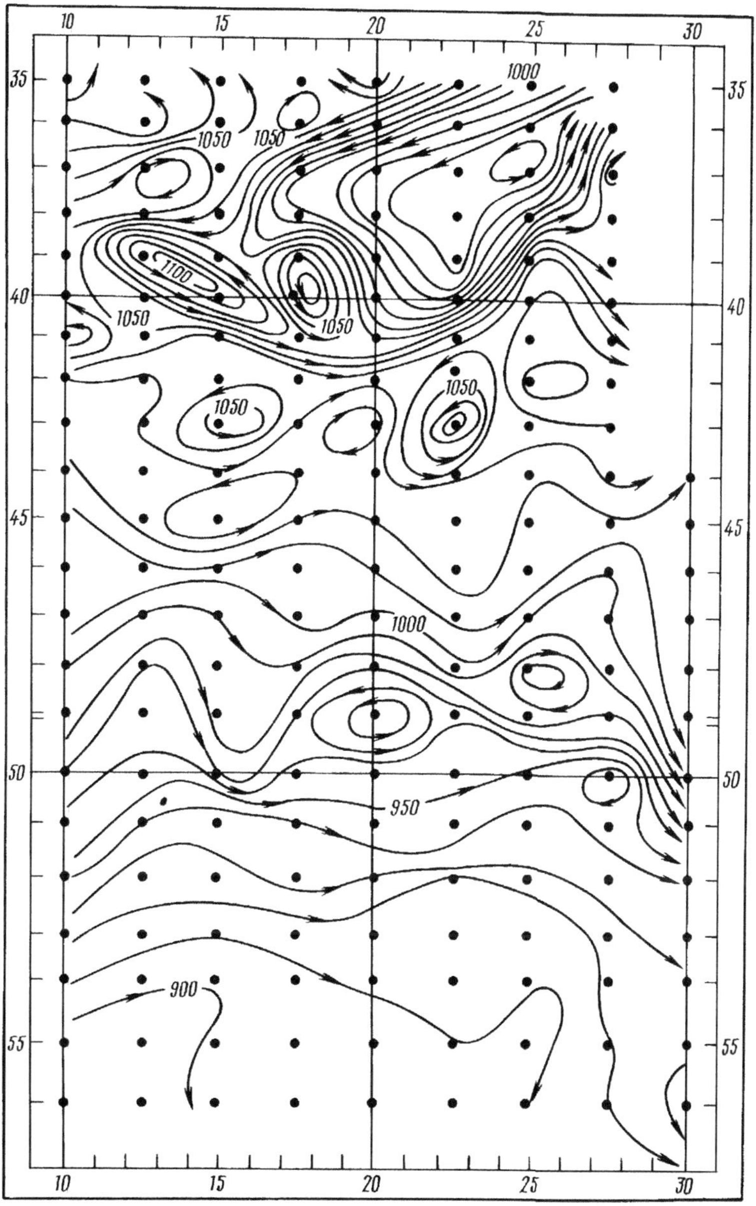

Fig. 4.3.6. Dynamic topography (in dynamic centimeters) of the sea surface relative to the 3000 dbar reference surface in a region south of Africa according to the data of the survey in December 1978–February 1979 (after Sarukhanyan, 1980a). The points indicate the positions of hydrographic stations.

in the main thermocline at the center of the eddy exceeded 500 m (Duncan, 1968). The values of salinity in the eddy core above 400 m clearly indicated the formation of the eddy from an Agulhas Current meander.

The permanent cyclonic meander of the Agulhas Return Current at 24–28°E is

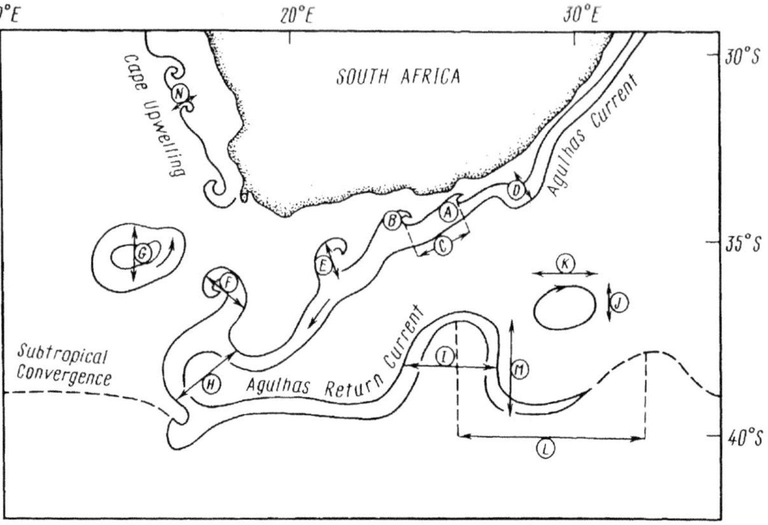

Fig. 4.3.7. Scheme of water circulation south of Africa (after Lutjeharms, 1981). The letters in the circles and the arrows indicate the linear parameters of synoptic circulation disturbances whose values are contained in Table 4.3.1.

undoubtedly related to the ocean bottom relief. It is interesting that as early as 1844 it was noted that Antarctic icebergs propagated far to the north in this region (Gründlingh, 1978). The cyclonic loops described by a surface float in the field of this meander (Figure 4.3.5) can be regarded as an indication of the existence of a cyclonic eddy in the meander.

The cyclonic meander of the Agulhas Return Current at 31–35°E (Figure 4.3.5) is also likely to relate to the bottom relief. The character of the drift trajectory of a surface buoy leaves hardly any doubt that the measurements corresponded to the completion phase of the formation of an intensive cyclonic eddy in the meander. In a four-week period the float trajectory described five loops with a typical 'diameter' of the order of 100 km; the velocity of the float motion was, on average, 60 cm/s and sometimes attained 120 cm/s. It seems that Figure 4.3.6 also confirms the process of formation of cyclonic and anticyclonic eddies from meanders of the Agulhas Return Current (see the eddies at 40°S, 14°E; 43°S, 22°E; and 37°S, 28°E).

Besides the above-mentioned large eddies formed by the separation of meanders of the Agulhas Current and the Agulhas Return Current, on the continental side of the Agulhas Current there occurs the formation and further south-westward drift of small cyclonic eddies of frictional origin (Figure 4.3.7) which are similar to those often observed in the Gulf Stream and the Kuroshio.

Several hundred satellite infrared surveys of the sea surface during 1978–1979 were used (Lutjeharms, 1981) to measure linear dimensions of the main synoptic disturbances of the Agulhas Current and the Agulhas Return Current (see Figure 4.3.7 and Table 4.3.1). In Table 4.3.1 particular attention should be paid to the large values of the parameters F, G, J, and K of cyclones and anticyclones of the Agulhas Current system.

TABLE 4.3.1
Linear dimensions of synoptic water-circulation disturbances south of Africa (after Lutjeharms, 1981)

Parameter (see Figure 4.3.7)	Average value, km	Root-mean-square deviation, km	The number of measurements
A. Transverse dimension of a frictional cyclone of the AC (Agulhas Current)	82	22	50
B. Dimension of an AC frictional cyclone	112	25	132
C Distance between AC frictional cyclones	252	70	34
D. Amplitude of an AC cyclonic meander	156	30	19
E. Dimension of an AC frictional cyclone	120	42	53
F. Dimension of an AC–ARC (Agulhas Return Current) anticyclone at the time of separation	307	89	56
G. Dimension of a separated AC–ARC anticyclone	324	97	29
H. Transverse dimension of an AC–ARC anticyclonic loop	341	72	32
I. Transverse dimension of an ARC cyclonic meander	291	65	35
J. Meridional dimension of an ARC cyclone	276	52	5
K. Zonal dimension of an ARC cyclone	228	24	5
L. Distance between ARC cyclonic meanders	450	110	23
M. Amplitude of an ARC cyclonic meander	323	60	51
N. Dimension of a cyclone in the upwelling region	120	31	175

The satellite infrared surveys of the sea surface in the region of convergence of the Falkland and Brazil Currents in the Southwestern Atlantic (Figure 4.3.8) carried out in September 1975–April 1976 and in January–July 1978 (Legeckis and Gordon, 1982) showed a very high level of dynamic activity and thermal contrasts in this ocean region. For example, in October 1975 in the region 40–42°S, 53–55°W the water temperature drop across the front separating the subtropical Brazil Current water from the subpolar Falkland Current water attained a record value in the world ocean (10 °C) for a synoptic front width of only several kilometers. The main mode of the very intensive quasiperiodic water-circulation disturbances in the region in question is the formation, approximately every two months, of an anticyclonic loop of the Brazil Current which extends far to the south, and the further one-fold or multiple separation of the end of the loop, thus forming a large,

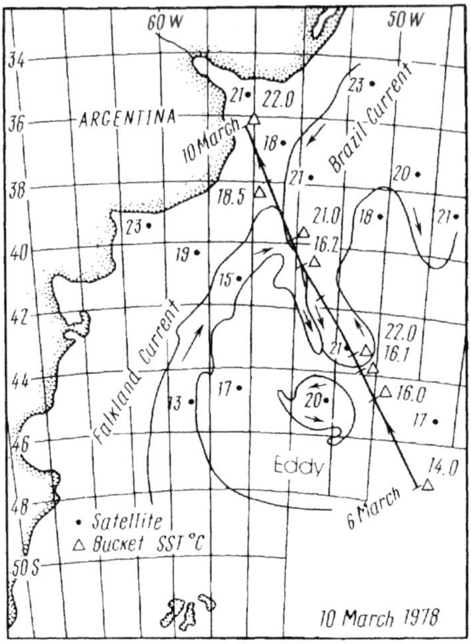

Fig. 4.3.8. Scheme of water circulation and location of the main temperature fronts in the region of convergence of the Falkland and Brazil Currents according to the data of the satellite infrared survey of the ocean on March 10, 1978 (after Legeckis and Gordon, 1982). Also shown is the position of the hydrographical section on March 6–10, 1978. The numbers near the points and the triangles are the water temperature values (°C) according to the data of satellite and shipboard measurements, respectively.

warm, anticyclonic eddy or a meridional chain of two or several eddies. In March–April 1978 a cycle of three consecutive separations of a loop of the Brazil Current, with intervals of only one week(!) between the separations, was observed. Figure 4.3.8 corresponds to the initial phase of the formation of the second eddy of the above-mentioned triple. Altogether at least 20 different Brazil Current anticyclones were observed in the region 40–48°S, 46–58°W during the observation time; these anticyclones seemed to make a substantial contribution to the interlatitudinal heat exchange in the ocean region under consideration. The total transverse dimensions of the eddies varied from 60 to 280 km; and the sea surface temperature drops between the central parts of the eddies and the surrounding water were 3–5 °C in the summer period for the horizontal temperature gradients on the fronts which bordered the eddies attaining 5 °C/2 km(!), and notably higher in the spring period. During the first days after separation, the velocity of the southward motion of the eddies attained very large values up to 35 km/day(!) and then, as a rule, decreased to several kilometers per day. It is remarkable that during the whole observation period not a single event of reverse absorption of eddies by the Brazil Current was registered. Episodical shipboard hydrological sections showed that water temperature disturbances in the field of Brazil Current anticyclones penetrate at least to a depth of 800–900 m. In 1975–1976 several cold cyclonic eddies were registered which were formed from the end of a cyclonic loop of the Falkland Current extending far to the north (Figure 4.3.8).

Satellite surveys of the sea surface performed in 1974–1976 in both the infrared and visible bands revealed very clearly marked eddies in the field of the Kamchatka Current and in the adjoining regions of the Bering Sea and the Pacific (Solomon and Ahlnäs, 1978) (symbol 11 in Figure 4.2.1). In particular, a series of infrared photographs taken in the summer–autumn period registered chains of eddies stretched along the eastern coast of Kamchatka and even groups of eddies with transverse dimensions from 100 to 150 km filling the south-western deep-water part of the Bering Sea. All these eddies possessed some very remarkable properties. First, almost without exception, they were all anticyclones. Second, despite the anticyclonic direction of rotation, the water temperature in their central parts in the surface ocean layer was lower than the temperature of the surrounding water, and the temperature contrast attained 5 °C, which indicated opposite deflections of the main and seasonal thermoclines. This property of oceanic eddies was, for instance, also observed during the experiment POLYMODE (see Section 2 of Chapter 5). Third, all the eddies had a clearly pronounced spiral-shaped character with evident indications of an intensive process of horizontal heat advection in their upper parts. Satellite surveys in the visible band performed in the winter–spring period also recorded chains of spiral-shaped anticyclonic eddies along the eastern coast of Kamchatka; in this case the role of an indicator of eddy motion was played by the sea ice drifting southward off Kamchatka coasts together with the Kamchatka Current. Successive photographs showed that these eddies were also drifting southward by the current. The authors (Solomon and Ahlnäs, 1978) assume that the appearance of these eddies is related to the irregularities of the eastern coast of Kamchatka.

Satellite surveys of the Greenland Sea in the visible and infrared bands showed an almost permanent presence in the sea of numerous synoptic water-circulation disturbances in the form of eddies, jets, sharp temperature fronts, etc. (Wadhams and Squire, 1983). In particular, in the region 79–79°30′N, 0–3°E at the eastern edge of the East Greenland Current synoptic-scale ice spirals were registered many times during the summer period. The special shipboard observations performed in this region in August 1980 elucidated, in general terms, the structure of this synoptic disturbance (see Figures 4.3.9 and 4.3.10). The thermocline in Figure 4.3.10 separates the cold and very freshened (and hence lighter) water of arctic origin in the upper ocean from the warm and saline Atlantic water in deeper layers. The emergence of the thermocline on the sea surface (Figure 4.3.9) results in the formation of the Polar Front, going parallel to and somewhat east of the edge of dense ice in the summer period in the ocean region under consideration (Figure 4.3.9). Figures 4.3.9 and 4.3.10 seem to make it possible to interpret the above-mentioned disturbance as a quasistationary eddy-meander of the East Greenland Current whose existence is probably due to the combined effect of the baroclinic instability of this current and the presence in this region (79°10′N, 2°50′E) of a deep ocean bottom trough with a transverse dimension about 30 km and a depth difference between the center of the trough and the surrounding ocean equal to 2000 m (Wadhams and Squire, 1983). The typical dimension (the diameter divided by 2π) of the anticyclonic eddy in Figure 4.3.9 is approximately 8 km, which coincides exactly with the estimate of the Rossby scale in the ocean region in

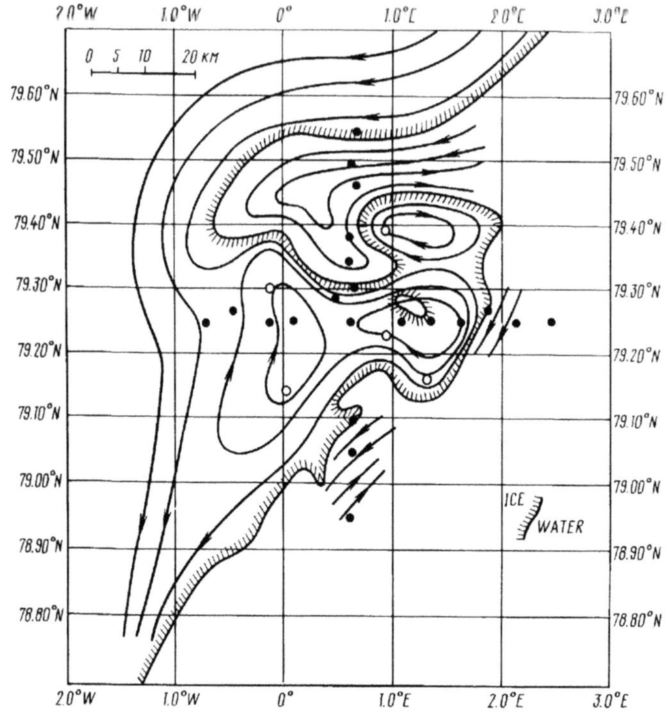

Fig. 4.3.9. Position of the ice edge with density of 80–90% and the scheme of water circulation in the upper ocean in the region of the East Greenland Current on August 23–27, 1980 according to the data of satellite observations, helicopter photography of the ice, CTD measurements (black points), and radar measurements of the ice drift (the initial positions of the radar transponders are shown by the light circles) (after Wadhams and Squire, 1983).

question. The results of speed-of-sound measurements (i.e., in fact, water-temperature measurements) on seven submarine tracks transverse to the East Greenland Current performed in October 1976 at depths from 67 to 125 m under the ice drifting together with the current (Wadhams *et al.*, 1979) revealed the presence of cold and warm temperature-field disturbances with dimensions from 5 to 30 km and temperature gradients at the edges of the disturbances attaining 1–4 °C km^{-1}. The authors relate these disturbances to the existence of warm and cold synoptic eddies in the current. Eddies, similar to the one described above, which are formed on the Polar Front were frequently registered by satellite surveys in the region of the Labrador Current as well (Legeckis, 1978).

The Somali Current, in contrast to the other western boundary currents of the world ocean, has a seasonal character. In the strong summer monsoon period it possesses very high velocities (up to 3 m/s(!); see Swallow and Fieux, 1982) and carries its waters from south to north along the Somali coast of Africa; in the weak winter monsoon period its velocities are small and the direction changes to the opposite. Multiple observations (e.g. Bruce, 1979; Swallow and Fieux, 1982) showed that in the summer period two anticyclonic gyres are formed in the field of the Somali Current whose major axes are stretched along the Somali coast and

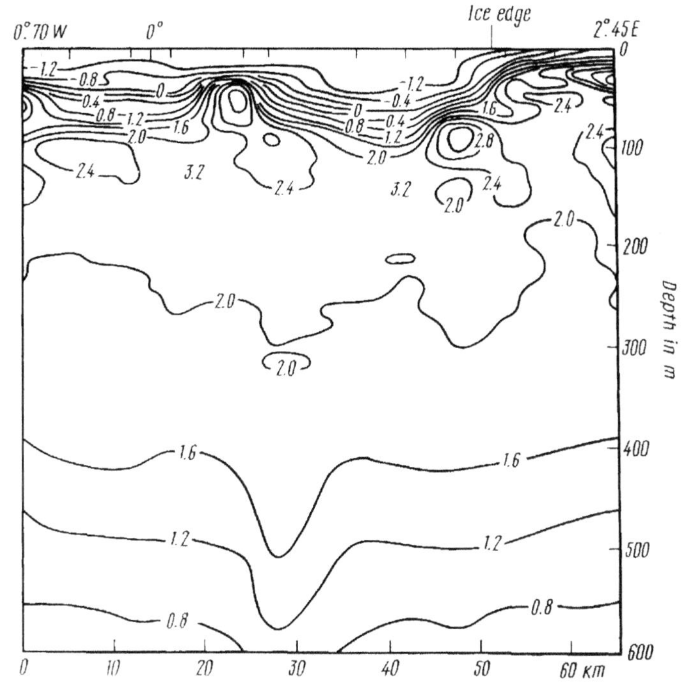

Fig. 4.3.10 Temperature distribution (°C) on August 25–26, 1980 in the zonal section along 79°15′N in the region of the East Greenland Current (see Figure 4.3.9) (after Wadhams and Squire, 1983). Shown above are the positions of CTD stations.

attain 700–800 km. However, owing to the quasistationary character of the position of these gyres and their vast dimensions, they should be regarded not as synoptic eddies but rather as elements of the large-scale oceanic circulation.

The sections performed at different times and in somewhat different hydrographic survey sites in the region of the Guiana Current and the adjoining part of the Atlantic (Cochrane, 1969; Mazeika, 1973; Cochrane *et al.*, 1979; Mazeika *et al.*, 1980; Sokolov and Tikhonov, 1984) were generally located insufficiently close to one another to reveal in detail the actual structure of the synoptic disturbances of the currents. However, the best of these surveys (e.g. see Figure 4.3.11) clearly demonstrated the complicated character of the water circulation in this region and the obvious presence of synoptic-scale meanders and eddies of the circulation field. This is probably one of the most interesting ocean regions from the viewpoint of further investigations of oceanic synoptic processes.

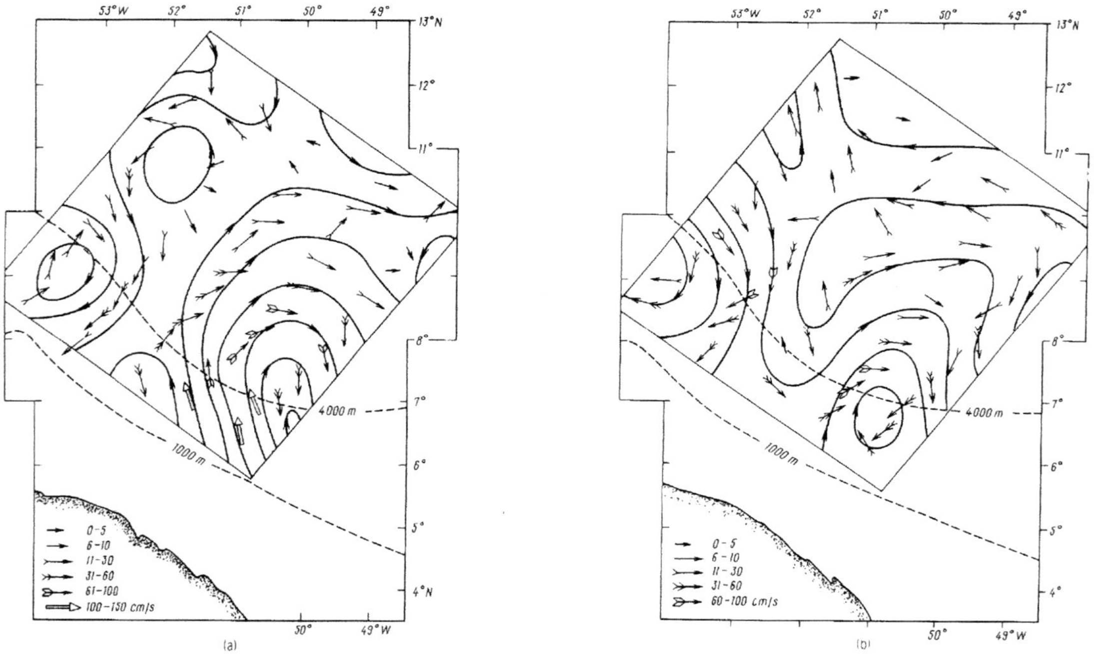

Fig. 4.3.11. Velocity vectors of the geostrophic current at a depth of 100 m calculated relative to a depth of 2000 m from the data of the hydrographic surveys of a region in the Atlantic north of the Guiana Current on (a) February 2–16 and (b) March 15–17, 1981 (after Sokolov and Tikhonov, 1984). The distances between the stations in the NE–SW and NW–SE directions were 30 and 60 miles, respectively. The streamlines are drawn by eye using velocity vectors. See symbol 10 Figure 4.1.1.

CHAPTER 5

Eddies in the Open Ocean

1. FIRST INDICATIONS. 'POLYGON-70' AND MODE

Eddies of western boundary currents described in the foregoing chapter are observed in comparatively limited ocean areas. What is the situation like in the rest of the ocean? Do formations which in some way resemble eddies of western boundary currents exist in the open ocean? The final answer to this question was given in 1970 by the Soviet expedition 'Polygon-70'. However, convincing indications of the existence of strong large-scale and long-period disturbances of oceanic currents in the ocean depth had been obtained earlier. First, we should mention the multiyear series of Soviet experiments which were specially intended for studying the variability of marine and ocean currents based on long-term current measurements by autonomous buoy stations (Koshlyakov and Monin, 1978). These experimental studies were carried out in the U.S.S.R. on the initiative and under the guidance of V. B. Shtokman who, as far back as 1935, conducted a series of long-term current measurements in the Caspian Sea (Shtokman and Ivanovskiĭ, 1937). After that the experiments in the Black Sea in 1956 (Ozmidov, 1962), in the North Atlantic in 1958 (Ozmidov and Yampol'skiĭ, 1965), and in the Arabian Sea in 1967 (Shtokman *et al.*, 1969) were developed. Among them 'Polygon-67' (Figure 5.1.1) was the first expedition where synoptic eddies were found (see Figure 5.1.2). The first of the hydrographic surveys, whose results are demonstrated in Figure 5.1.2, was carried out in the period of a comparatively weak but highly stable north-east monsoon and the other was performed in the period of weak variable winds corresponding to the change of monsoons. Figure 5.1.2 shows very clearly marked baroclinic disturbances of the geostrophic current characterized by a horizontal scale ('diameter') about 250 km, which is in good agreement with the Rossby scale value typical of this ocean region. The long interval between the surveys makes it impossible to perform mutual identification of eddies in the charts shown in Figure 5.1.2; however, the strong nonstationary quality of the process is obvious. According to the climatic data (Figure 5.1.1), the ocean region and the season under consideration are characterized by an eastward large-scale geostrophic current whose velocity varies from approximately 10 cm/s in the upper ocean to zero at a depth of the order of 1500 m. The dominant transfer in direction towards the east (or rather the north-east) is also clearly observed in Figure 5.1.2. The baroclinic character of the geostrophic current disturbances (Figure 5.1.2), their characteristic horizontal scale, and their strong nonstationary quality allowed

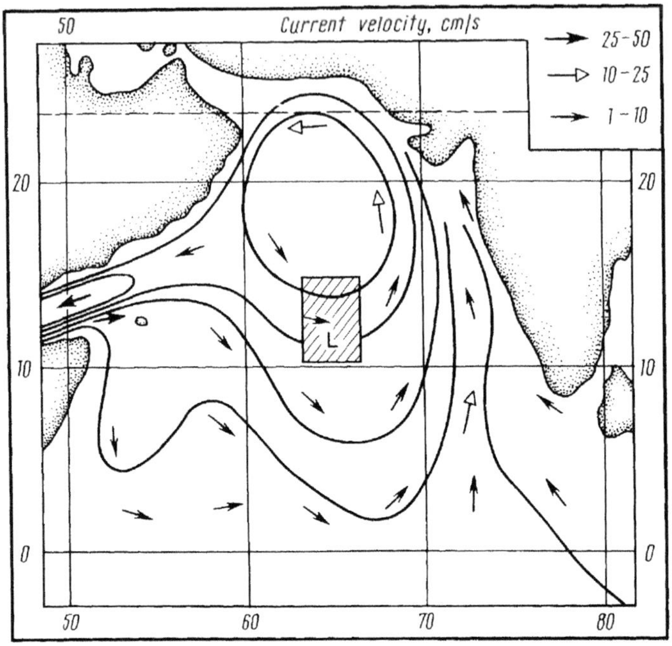

Fig. 5.1.1. Scheme of the large-scale geostrophic circulation of waters in the upper layer of the Arabian Sea during the winter monsoon period (after Neiman, 1970). Also shown in the figure are the region of 'Polygon-67' hydrographic surveys and the broken line along which the buoy stations with current meters were emplaced.

us to assume that they were caused by the baroclinic instability of large-scale currents.

The current measurement program of the 'Polygon-67' expedition was insufficient for revealing directly the spatial structure of the current field. However, for the period of the second hydrographic survey a comparison of the measured currents and those calculated by the dynamical method was made for several mooring sites (Koshlyakov *et al.*, 1972). After inertial and tidal oscillations were filtered out from the velocity time-series, good agreement was attained, on the whole, between the directions of the measured and calculated currents in the layer from 200 to 1000 m. The calculated current velocities turned out, on average, to be 30% smaller than the measured current velocities. This can be accounted for by the density field being smoothed in the horizontal plane before the geostrophic calculations.

Much interest was shown by oceanographers in deep current measurements carried out by J. C. Swallow in 1958 west of Portugal (Swallow and Hamon, 1960) and in 1959–1960 in a region south-west of Bermuda (Swallow, 1971) by means of neutrally buoyant floats (drifting together with the current). The observations near Bermuda (symbol 12 in Figure 4.1.1) – results that turned out to be particularly interesting – were performed mainly at depths of 2 and 4 km. Altogether about 80 float launchings, with typical durations from 4 to 10 days, were made over a 14-month period. Contrary to the then widely held belief that there was very weak

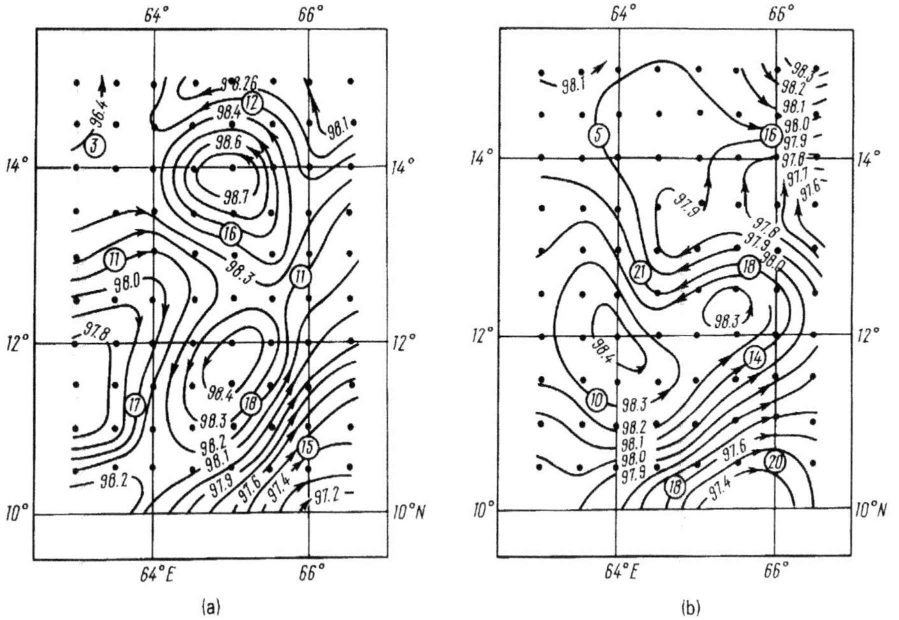

Fig. 5.1.2. Geostrophic currents at a depth of 150 m calculated from the data of the first ((a) January 21–February 7, 1967) and the second ((b) March 20–April 6, 1967) 'Polygon-67' hydrographic surveys (after Koshlyakov *et al.*, 1970). The numbers at the streamlines are the values of $Q \equiv -\int_{z_0}^{z} \bar{\rho}(x, y, z,)\, dz$, in units corresponding to the density measurement in $(\sigma_t - 20)$ and the depth measurement in hundreds of meters; $z_0 = 1500$ m is the depth of the reference surface; $\bar{\rho}\,(x, y, z,)$ is the distribution of density smoothed in the horizontal plane to remove traces of internal waves. The numbers in the circles give the velocity (cm/s). The small black dots mark the positions of hydrographic stations.

and constant southward deep-water transfer in this ocean region, the floats registered nonstationary currents with typical velocities of 5–15 cm/s, the current velocities at a depth of 4 km exceeding, on average, the velocities at 2 km and attaining a maximum value of 40 cm/s. Although the observation data were not sufficiently extensive for providing information about the actual structure of the current, nevertheless it became possible to crudely estimate the horizontal and temporal scales of the velocity field (the quarter wavelength and quarter period), which turned out to be equal to 100 km and 20 days, respectively. The resulting relationship between the spatial and temporal scales made it possible to interpret the measured nonstationary currents as planetary quasigeostrophic waves (Phillips, 1966). Phillips also considered the hypothesis that these waves were directly excited by the nonstationary wind field over the North Atlantic and found that the current velocities were too large for the hypothesis to be true. As another possible process generating the oscillations observed by Swallow, Stommel (1965) considered their excitation by Gulf Stream meandering. The deep-float measurements west of Portugal (see Swallow and Hamon, 1960; symbol 11 in Figure 4.1.1) also indicated the presence of strongly space- and time-varying deep currents whose velocities, however, were 4–5 times as small as those of the currents near Bermuda.

The long-term series of hydrographic observations which were performed on the

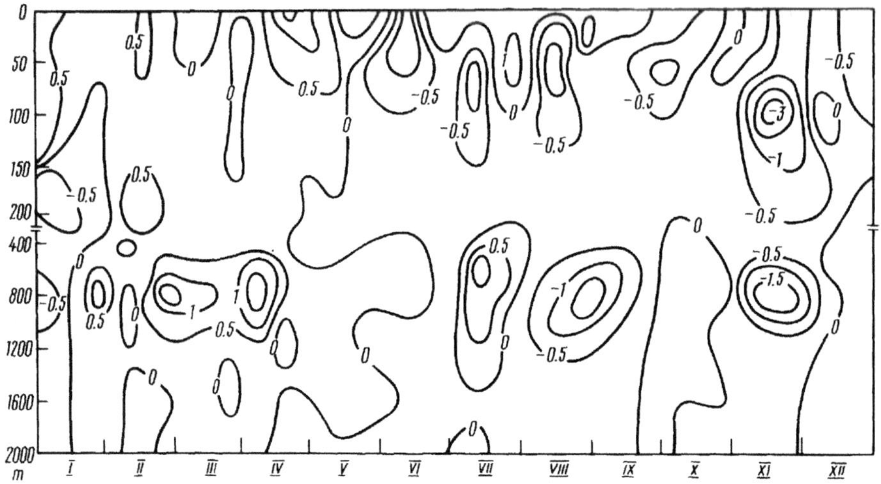

Fig. 5.1.3. Isopleths of the deviation of the water temperature (°C) from the average seasonal variation (the seasonal oscillations did not penetrate to depths below 350 m) constructed from the data of temperature measurements in 1960 at 32°N, 65°W (after Schroeder and Stommel, 1969). The vertical scale changes at 200 m. (See symbol 13 in Figure 4.1.1.)

initiative of Stommel from 1954 to 1969 at intervals of half a month at a site in the ocean with a depth around 3000 m located 25 km south-east of Bermuda (symbol 13 in Figure 4.1.1) is geographically relevant to the float current measurements in the North-West Atlantic described above. As an example, Figure 5.1.3 demonstrates the results of these measurements for 1960. They clearly show temperature oscillations with a period about 50 days and total amplitude up to 2 °C in the seasonal thermocline (depths down to 150 m) and with a period about 100 days and amplitude up to 2.5 °C in the main thermocline (depths from 400 to 1200 m). The oscillations of the sea surface level, calculated from the data of the above-mentioned hydrographic observations on the basis of the hydrostatic relation and under the assumption that there were no pressure oscillations at a depth of 2000 m, turned out to be coherent with the sea-level oscillations measured near Bermuda. It is obvious that the temperature anomalies shown in Figure 5.1.3 must be related to the propagation of some current field disturbances, having predominantly a geostrophic and baroclinic character, through the observation region. It is also clear that these disturbances were generated separately in the main and seasonal thermoclines.

The data resulting from the multiyear hydrographical observations at station 'Tango' (29°N, 135°E; symbol 12 in Figure 4.2.1) in the Pacific Ocean (Figure 5.1.4) were actually very close to those described above. In relation to hydrology, the 'Tango' region is approximately similar to the central part of the Sargasso Sea. The periodogram analysis of oscillations of depths of isopycnic surfaces (see Figure 5.1.4) in the 500–1100 m layer carried out for the measurement data during the period from 1950 to 1953 again showed the predominance of oscillations with a period of 90–100 days; the amplitude of oscillations attained 200 m. The temporal

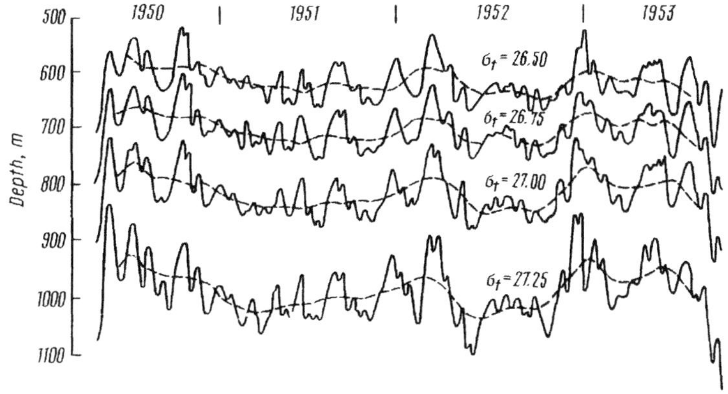

Fig. 5.1.4. Low-frequency oscillations (solid lines) of depths of isopycnic surfaces obtained from the data of standard hydrographic measurements at the station 'Tango' (24°N, 135°E; see symbol 12 in Figure 4.2.1) during 1950–1953 (after Yasui, 1961). The dashed lines represent the result of smoothing the solid lines with a period of 100 days

intermittency of these oscillations, which is clearly seen in Figure 5.1.4, was a very interesting fact making one think that separate groups of long-period waves with a typical amplitude-variation period of about one and a half years propagated through the observation point.

Wyrtki (1967) subjected to statistical processing the results of temperature measurements in the upper ocean layer (to a depth of 1200 m) at 44 meridional sections across the North Equatorial Current carried out in 1964–1965 in a region south-west of the Hawaiian Islands, and found a dominant scale in the meridional temperature oscillations corresponding to a wavelength of the order of 500 km and assumed, quite correctly, that it related to the expected 'geostrophic eddies' whose traces were already revealed earlier, as a result of hydrographic surveys, in the Hawaii region. Later more thorough processing of the above-mentioned section data (Bernstein and White, 1974) confirmed this assumption.

The observations described above do not exhaust all the works (although they seem to be most important) whose results made it possible, at the end of the 1960s, to find long-period ocean circulation disturbances, probably of eddy character, in various ocean regions. It became quite evident that it was necessary to organize an expedition whose primary purpose would be to carry out direct current measurements in a typical region of the open ocean far from continents and frontal zones, the extent, the structure, and the duration of the measurements being such that it was possible to reveal the actual spatial structure of ocean circulation disturbances and the character of evolution of the disturbances with time. The experiment 'Polygon-70' became such an expedition.

'Polygon-70' (Brekhovskikh, 1971; Brekhovskikh *et al.*, 1971a, b) was performed in February–September 1970 on the southern periphery of the North Equatorial Current in the Atlantic Ocean (symbol 14 in Figure 4.1.1). The ocean depth in the experiment area varied from 5000 to 5500 m. The vertical distribution of water density is characterized by a very sharply marked pycnocline occupying a layer with depths approximately from 80 to 200 m. The large-scale geostrophic current in the upper ocean is directed towards west-south-west and its velocity is around 4 cm/s; at depths below 400 m it is replaced by a deep counter current with velocities not higher than 1 cm/s (Enikeev and Koshlyakov, 1973). During almost the whole expedition the experiment area was in the zone affected by stable north-east trade wind. For instance, from the middle of March to the end of July 1970 the wind azimuth varied only from 45 to 80°. The wind velocity changed quasiperiodically with a period about a week and typical variation from 5 to 10 m/s (Kool' *et al.*, 1974).

The major observations during the experiment 'Polygon-70' were the current and temperature measurements carried out by the Institute of Oceanology of the U.S.S.R. Academy of Sciences at 17 buoy stations emplaced along the arms of a rectangular cross-shaped pattern with center at 16°30′N, 33°30′W (Figures 5.1.5 and 5.1.6); the length of each of the arms was 100 km. The current meters registering the current velocity and direction at 30-min intervals were located at ten depths at each of the stations in a layer from 25 to 1500 m. This system of stations was permanently maintained from the end of February to the beginning of September 1970. Several hydrographical surveys of the region of emplacement of buoy stations were also carried out.

The spectrum analysis of the velocity time-series obtained from the measurements at the 'Polygon-70' buoy stations showed that inertial and tidal oscillations of the velocity (with periods of 42 h and 12.5 h, respectively) were very clearly marked and also revealed a deep energy minimum at periods of the order of 3–4 days (Vasilenko *et al.*, 1976), which can be regarded as justification of the low-frequency filtration of the above-mentioned series aimed at separating out the synoptic component of the currents. This filtration was performed with an effective period of 3.5 days. The resulting smoothed velocity series were laid in the foundation of various synoptic current charts and plots, whose examples are shown in Figures 5.1.5–5.1.7. The analysis of all these data made it possible to draw the following basic conclusions (Koshlyakov and Grachev, 1973, 1974; Koshlyakov and Enikeev, 1977).

1. During the experiment 'Polygon-70' the direct current measurements reliably registered a perfectly clearly marked anticyclonic eddy which travelled across the experiment area in the WSW direction from the beginning of April to the beginning of July 1970 (Figures 5.1.5 and 5.1.6). At the end of May the eddy center almost coincided with the center of the experiment area (the patterns for May 24 in Figures 5.1.5 and 5.1.6). Moreover, the measurements registered the rear part of another anticyclonic eddy which moved ahead of the main anticyclone (the patterns for March in Figures 5.1.5 and 5.1.6) and the frontal part of a cyclonic eddy which moved across the experiment area behind the main eddy (the patterns for August). The circular currents in these eddy fields entirely overlapped the weak North Equatorial Current. As is seen from Figures 5.1.5 and 5.1.6, the current patterns in the 200–1500 m layer were highly depth-coherent.

2. The eddies moved close, one after another, which made an impression that the observed vortical disturbances were 'close-packed' or that their nature was at least partly of wave character. The periods of very weak currents at the central part of the experiment site (e.g. see the patterns for the second half of July in Figures

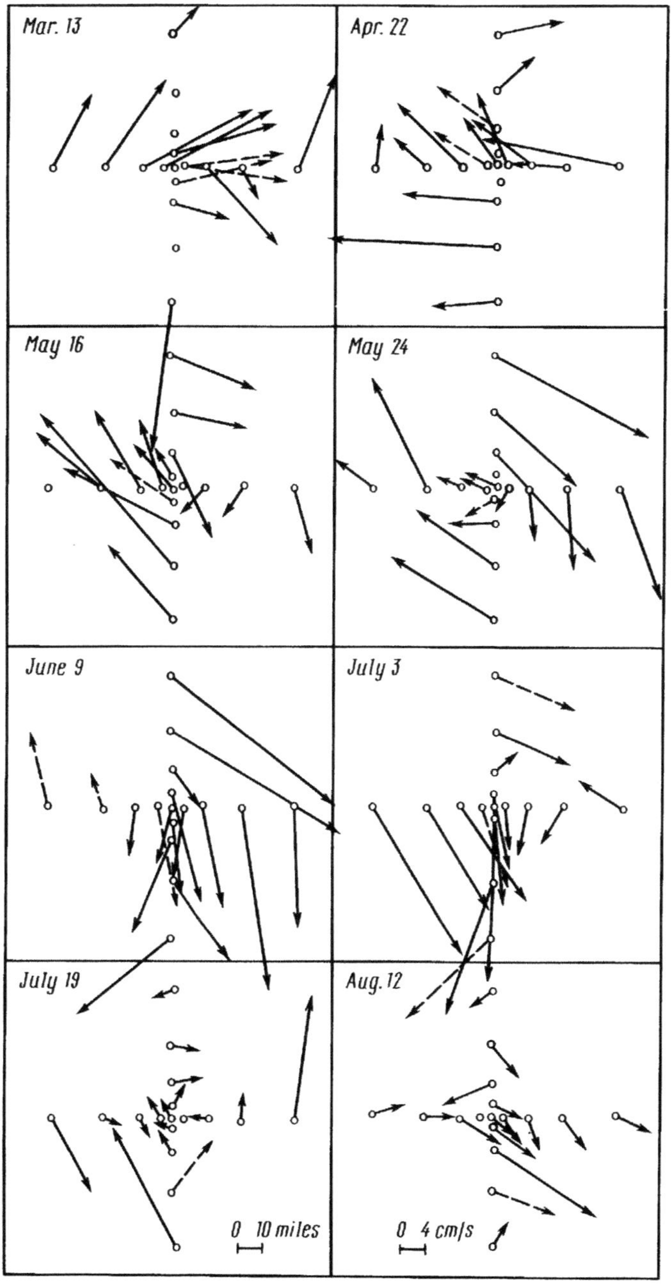

Fig. 5.1.5. Distributions of the velocity vector of the synoptic currents for different dates in 1970 at a depth of 400 m determined from 'Polygon-70' data (after Koshlyakov and Grachev, 1974). The center of the area is at 16°30'N, 33°30'W (symbol 14 in Figure 4.1.1). The circles indicate the positions of buoy stations. The dashed arrows represent the vectors obtained by interpolation over depth. Lack of arrows indicate a gap in the measurement data. The length and velocity scales are indicated at the bottom.

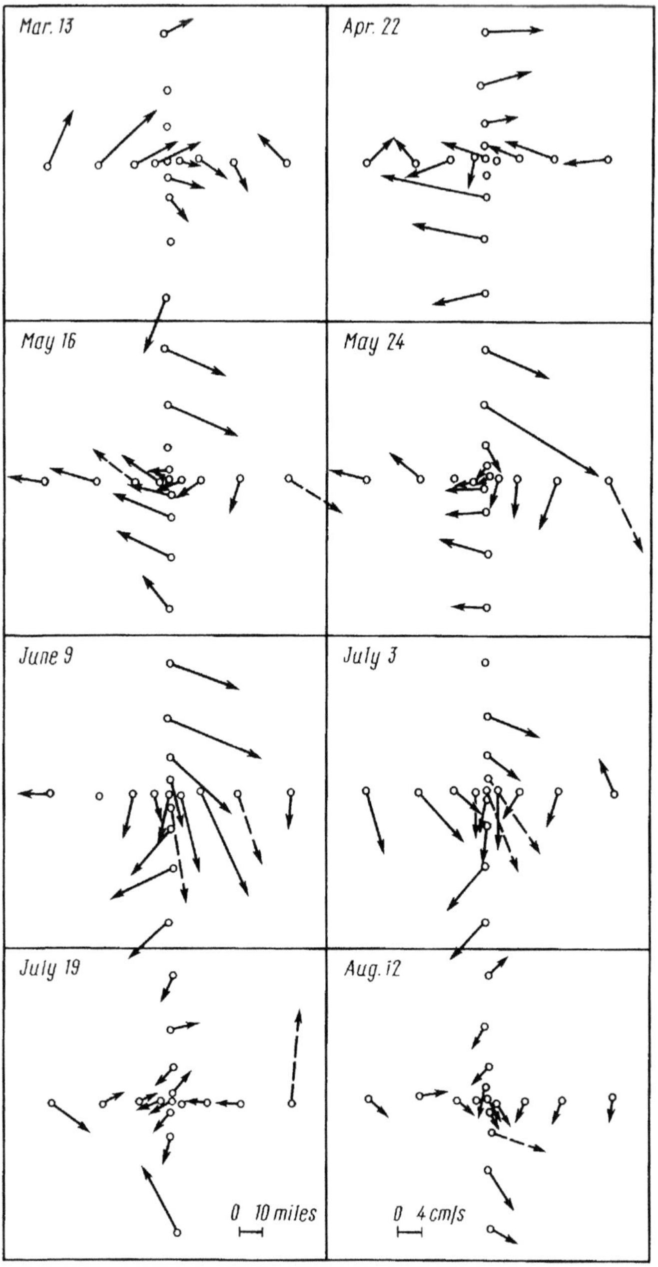

Fig. 5.1.6. Distributions of the velocity vectors of synoptic currents for different dates in 1970 at a depth of 1000 m determined from 'Polygon-70' data (after Koshlyakov and Grachev, 1974). Other details as on Figure 5.1.5.

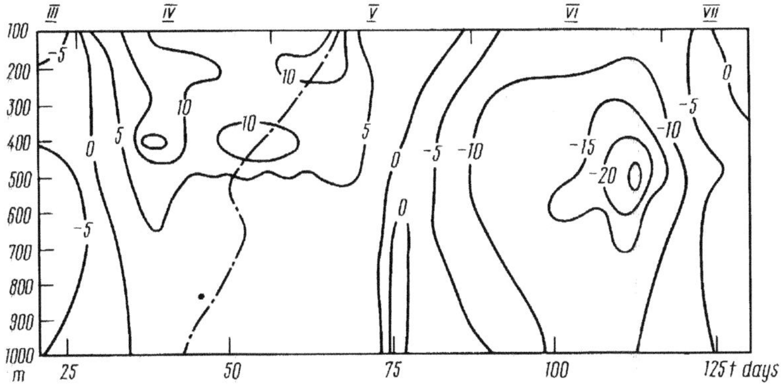

Fig. 5.1.7. Isopleths of the meridional component of the velocity of synoptic currents (cm/s) as functions of time (the number of days from the beginning of the work and months of 1970) and depths (meters) constructed from the data of the buoy station which was emplaced 5 miles east of the center of the cross-shaped pattern of the 'Polygon-70' buoy stations (after Koshlyakov and Enikeev, 1977).

5.1.5 and 5.1.6) can be interpreted as the periods during which saddle-shaped formations between four eddies travelled across the experiment area.

3. The main 'Polygon-70' eddy was of an elliptic shape with the ratio of the minor semi-axis to the major semi-axis approximately equal to 1 : 2 (the patterns for May in Figures 5.1.5 and 5.1.6). As the eddy moved westward, it rotated clockwise. The minor semi-axis of the eddy, defined as the distance from the eddy center to the point with maximum orbital velocity (the distance a in Figure 5.1.8), was around 100 km. The estimation of the Rossby scale (see Section 1, Chapter 4) for the 'Polygon-70' experiment site yields a value of 70 km. Interpreting the minor semi-axis of the eddy as the quarter wavelength of dominant oscillations in the velocity field (see Figure 5.1.8) and the Rossby scale as the wavelength divided by 2π, we obtain good agreement between experimental data and theory, suggesting that eddies are generated owing to the baroclinic instability of large-scale currents. Namely, according to this theory (see Chapter 3), disturbances within the Rossby scale are most unstable.

4. All the 'Polygon-70' eddies moved to the west with a small southward component. The average drift direction of the main 'Polygon-70' anticyclone during April–July 1970 formed an angle of 260° (counted clockwise) with direction towards the north; the average drift velocity of the anticyclone during this period was 5.5 cm/s, and the velocity increased from April–May to June–July. The comparison of the velocity of motion of the main eddy with its horizontal dimensions within the framework of the theory of Rossby waves will be discussed later.

5. The 'vertical' axis of the main 'Polygon-70' anticyclone was inclined approximately opposite to the direction of its motion. For instance, according to Figure 5.1.7, the zero value of the meridional velocity at a depth of 100 m at a point 5 miles east of the center of the experiment area was attained 13 days later than at a depth of 600 m at the same point, which, for the velocity of eddy motion of 5.5 cm/s, corresponded to a spatial shift of 62 km. Approximating the eddy by a plane quasigeostrophic wave with wavelength of 400 km propagating in the 215° direction

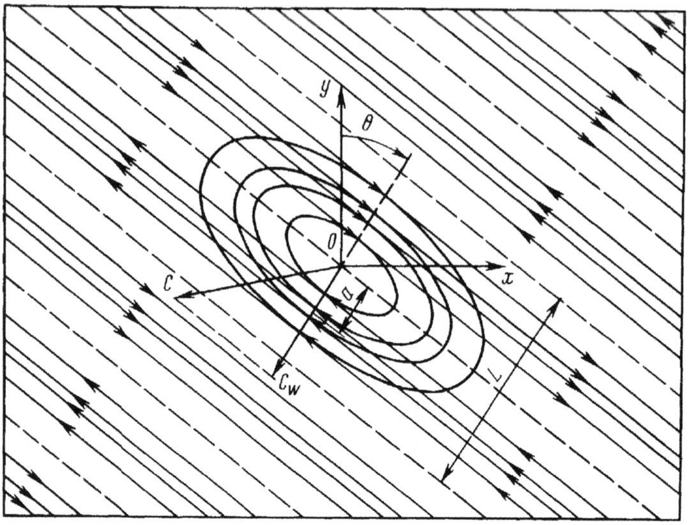

Fig. 5.1.8. Model of the main 'Polygon-70' anticyclone and a plane quasigeostrophic wave (after Koshlyakov, 1973). The ellipses and straight lines are the eddy field and wave field streamlines, respectively; a is the distance from the center of the eddy to the point of maximum velocity; L is the wavelength; c is the velocity of the translational motion of the eddy; c_W is the phase velocity of the wave.

(i.e. the average direction of the minor semi-axis of the eddy determined from observations; see Figure 5.1.8) we conclude that the indicated spatial shift corresponded to a phase shift of 40°. As is well known (see Chapter 3), the phase shift of a synoptic disturbance with increasing depth indicates that the disturbance was formed owing to the baroclinic instability of a large-scale current. In the case under consideration, the sign of the phase shift coincides with the one (e.g. see Kamenkovich and Reznic, 1978) corresponding to the increase of a disturbance when the sign of the vertical shear of the velocity of a large-scale motion is characteristic of the 'Polygon-70' area. The fact that this increase did take place is demonstrated in the next subsection.

6. The distribution of the average velocity of synoptic currents over the depth is seen from the data in Table 5.1.1. In certain periods the velocity attained 25 cm/s at depths of 200–300, 35 cm/s at 400–600 m, 20 cm/s at 1000 m, and 10 cm/s at 1500 m. The intermediate velocity maximum at depths of 400–500 m (Table 5.1.1) attracts attention. The increase in velocity at depths of 200–1000 m in the rear part of the main anticyclone during the period from May to June 1970 was determined rather reliably (Figures 5.1.5 and 5.1.6); e.g. the velocity increased from 20 to 35 cm/s at a depth of 400 m and from 12 to 20 cm/s at a depth of 1000 m.

The first interpretation of 'Polygon-70' results within the framework of the theory of Rossby waves was performed by Koshlyakov and Grachev (1973) and Koshlyakov (1973, 1974). Figure 5.1.8 demonstrates this interpretation in which the minor semi-axis of the main 'Polygon-70' eddy was identified with the quarter wavelength of a quasigeostrophic wave and the projection of the velocity of the eddy motion on the direction of the minor semi-axis was identified with the phase velocity of the wave. In the work by Koshlyakov (1974), a quasilinear model of

TABLE 5.1.1
Average moduli of the synoptic components of the velocity vectors calculated from 'Polygon-70' data (after Koshlyakov and Grachev, 1974)

Depth, m	Velocity, cm/s	Depth, m	Velocity, cm/s
25	23	400	11
50	17	500	11
100	9	600	10
200	9	1000	7.2
300	10	1500	3.5

Rossby waves in a two-layer ocean with a large-scale current (McWilliams and Robinson, 1974) was chosen as a specific theoretical model. In the case of stable waves, a small upper layer to lower layer thickness ratio, and the absence of any large-scale motion in the lower layer, this model gives the following dispersion relation for a baroclinic wave:

$$\omega = L_R^2(k^2 L_R^2 + 1)^{-1}\left[-\beta m + k^2(mU + nV)\right]. \tag{1.1}$$

Here m, n, and k are the zonal and meridional components of the wave vector and its absolute value, respectively, ω is the circular frequency, f and β are the Coriolis parameter and its latitudinal derivative, respectively, U and V are components of the large-scale current velocity in the upper layer, and L_R is the Rossby scale. For the experiment site we have $f = 4 \times 10^{-5}\,\mathrm{s}^{-1}$, $\beta = 2.2 \times 10^{-13}\,\mathrm{cm}^{-1}\,\mathrm{s}^{-1}$, $L_R = 70$ km, $U = -2$ cm/s, and $V = -1$ cm/s. Determining the quantities m, n, and k (according to the observation data) from the conditions $L = 4a = 4 \times 90$ km $= 360$ km and $\theta = 35°$ (Figure 5.1.8), and substituting the values of the variables and parameters written above into (1.1), we obtain the estimate $c_w = 3.7$ cm/s for the phase velocity $c_w = 2\pi/L\omega$, whereas the experiment, as follows from the estimates for velocity and direction of motion of the main 'Polygon-70' anticyclone presented earlier, yields $c_w = (5.5\ \mathrm{cm/s}) \times \cos 45° = 3.9$ cm/s. Thus, practically, the theoretical and experimental results completely coincide, which allows us to assert that during the 'Polygon-70' expedition baroclinic Rossby waves were for the first time measured in the open ocean directly and exhaustively.

The above interpretation of the 'Polygon-70' results in the framework of the theory of Rossby waves was successfully generalized by McWilliams and Robinson (1974) to the case of superposition of two baroclinic waves. Fomin and Yampol'skiĭ (1978) approximated the evolution of the pattern of synoptic currents at a depth of 300 m by a sum of four plane quasigeostrophic waves (Figure 5.1.9); in contrast to the above-mentioned attempts, they chose the wavelengths and wave periods independently. Good agreement between the four-wave approximation and the experimental data for the period from April to July 1970 was achieved (cf. Figures 5.1.9, 5.1.5, and 5.1.6). In relation to these models, of interest also are the results of the work by Vasilenko and Mirabel' (1977) who studied the vertical structure of synoptic currents by expanding the 'Polygon-70' current measurement data with respect to the system of vertical empirical orthogonal functions. It was found that

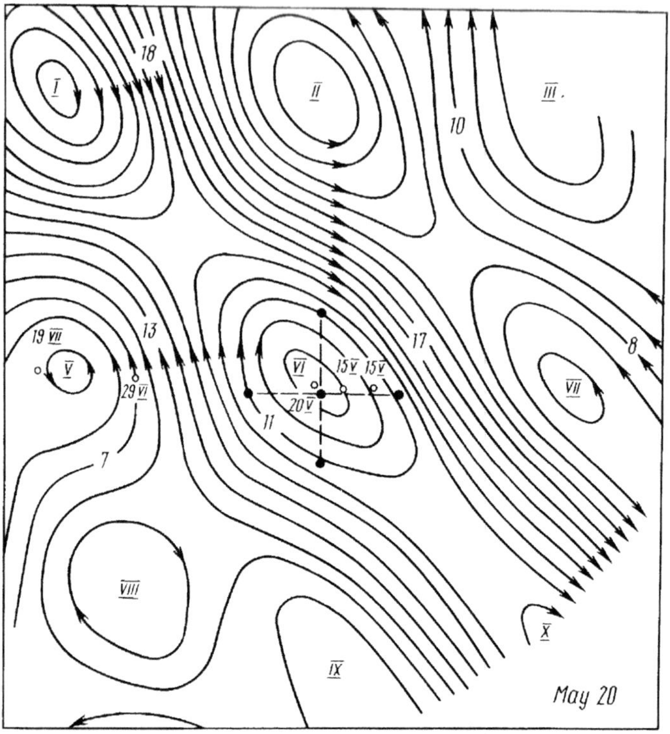

Fig. 5.1.9. Pattern of streamlines for May 20, 1970 in the field of the sum of four plane quasigeostrophic waves simulating the evolution of the field of synoptic currents at depth of 300 m in the 'Polygon-70' area (after Fomin and Yampol'skiĭ, 1978). The numbers at the isolines indicate the current velocity in cm/s. The Roman numerals indicate individual eddy centers. Also shown in the figure are the position of the cross-shaped pattern of the 'Polygon-70' buoy stations (Figures 5.1.5 and 5.1.6) and the positions of the center of the anticyclone VI (corresponding to the four-wave model under consideration) relative to the cross for different dates in 1970. Each of the four arms of the cross is 100 km.

the first three modes of this expansion almost exhaust the vertical variability of the currents.

To study more thoroughly the possibility of generation of 'Polygon-70' eddies owing to the baroclinic instability of large-scale currents (BILC), Koshlyakov and Yaremchuk (1984) subjected the time-series of the velocity of synoptic currents at all measurement depths to spectrum analysis. Some of the results of the analysis are demonstrated in Figure 5.1.10. Here oscillations with a period about 80 days propagating close to the zonal direction (under the natural assumption that the oscillations are transverse to this direction) are clearly marked. The joint spectrum analysis of synoptic oscillations of velocity at different depths showed that oscillations with a period of 80 days were highly depth-coherent and they reliably confirmed the presence of the average phase shift of these oscillations along the vertical, at least in the 200–2000 m layer, which, in particular, resulted in approximately a 30° phase lag of oscillations at a depth of 200 m in comparison with oscillations at a depth of 1000 m.

The BILC model (see Chapter 3) leads to the following formula for the rate of

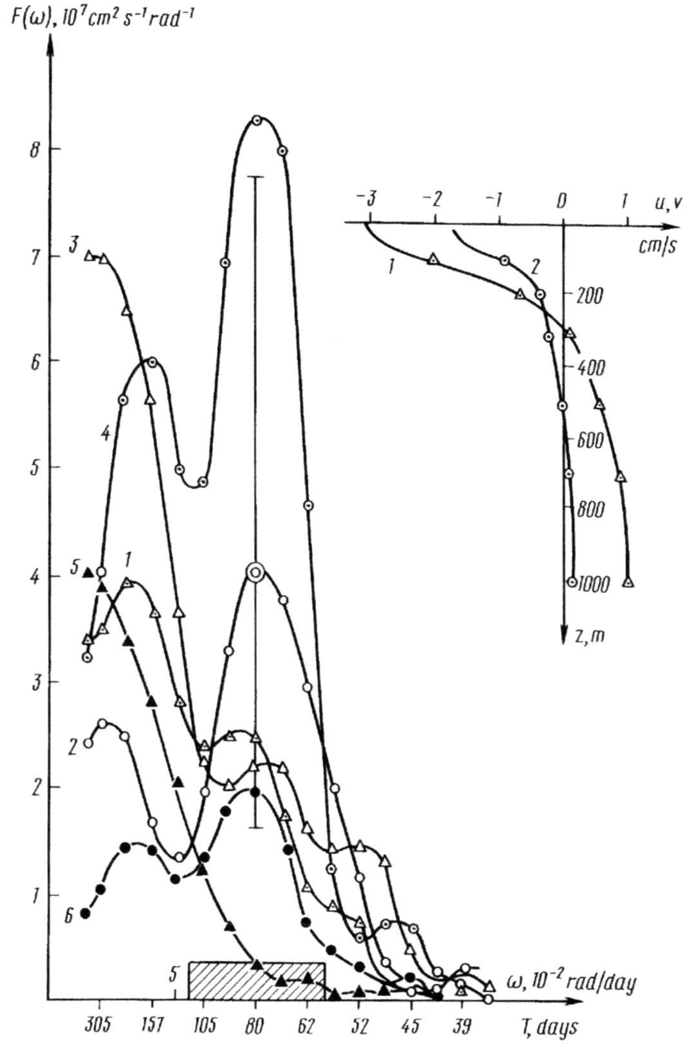

Fig. 5.1.10. Frequency spectra of the oscillations of the zonal (1, 3, and 5) and meridional (2, 4, and 6) components of the velocity of the synoptic current at depths of 200 (1, 2), 400 (3, 4), and 1000 m (5, 6) constructed from 'Polygon-70' data (after Koshlyakov and Yaremchuk, 1984). Also shown in the figure are the effective width of the spectral window (on the axis of abscissas) and the 90% confidence interval for the maximum of one of the spectra. The plot in the right upper corner demonstrates the vertical distributions of the (1) zonal and (2) meridional components of the large-scale geostrophic current in the 'Polygon-70' region for the summer period (after Enikeev and Koshlyakov, 1973).

increase of the integrated (over a layer (z_1, z_2)) available potential energy $\bar{P}$ (erg/cm^2) of a synoptic oscillation of velocity characterized by the wave vector (m, n), the average volume density $\bar{K}\,(z)$ of kinetic energy, and the phase $\theta(z)$:

$$\int_{z_1}^{z_2} \bar{P}_t \, dz = -\, 2\varrho_0 f^2 g^{-1} k^{-2} \int_{z_1}^{z_2} \theta_z \bar{\varrho}_z^{-1}(mU_z + nV_z)\bar{K} \, dz. \qquad (1.2)$$

The subscripts z and t in (1.2) symbolize differentiation with respect to the corresponding variables, and the bar denotes averaging over a time interval multiple of the oscillation period. Choosing the depths of 200 and 1000 m as z_1 and z_2 we find from the results of the spectrum analysis of the 'Polygon-70' data that, within the limits of the (z_1, z_2) layer, the ratio $\theta_z/\bar{\varrho}_z$ can be approximately regarded as constant over the depth and equal to 1.1. $\times 10^3$ rad/g·cm^{-3}. Further, estimating the distributions of $U(z)$, $V(z)$, and $\bar{K}(z)$ within this layer from the results presented in Figure 5.1.10 (to estimate $\bar{K}(z)$ the energy spectra are integrated with respect to the frequency exactly within the limits of the spectral slot shown in the figure) and putting, in accordance with the results obtained by Konyaev and Sabinin (1981) from the space-time spectrum analysis of current measurement data during the 'Polygon-70' experiment, $m = 1.35 \times 10^{-2}$ rad/km and $n = 0.36 \times 10^{-2}$ rad/km, we obtain for the left-hand side of (1.2) the estimate 1.2 erg/cm^2·s for the total energy transmitted to the 80-day synoptic oscillation of the velocity from the large-scale current. If we assume conditionally that there was no average integral transfer of energy across the exterior boundary of the 'Polygon-70' area, then the above result corresponds to the average energy dissipation rate equal to (1.2 erg/cm^2·s)/(5.5 km) = 0.22×10^{-5} erg/cm^3·s

The form of the synoptic-current patterns constructed from the 'Polygon-70' data relating to upper depths (of 25, 50, and 100 m) is much less regular in comparison with the form of the patterns for lower depths (Figures, 5.1.5 and 5.1.6) (Koshlyakov and Grachev, 1974). This can probably be accounted for, at least partly, by the fact that, besides the above-mentioned eddies with a scale of 100 km, the upper ocean also contains smaller vortical formations probably analogous to those specifying the temperature oscillations in the seasonal thermocline in the diagram in Figure 5.1.3. Traces of these small surface eddies were in fact found in the current measurement data during the 'Polygon-70' expedition (Kort *et al.*, 1974). The mesoscale thermal fronts discovered in the upper ocean during the 'Polygon-70' experiment, one of which, for instance, was located in the rear part of the main 'Polygon-70' anticyclone (Fedorov *et al.*, 1974), are indirect evidence of the existence of disturbances of the current field in the surface ocean layer whose scales are substantially smaller than those of the main synoptic eddies. The fact that this front was clearly marked not only in the horizontal surface but also in the isopycnic surface $\sigma_t = 25.00$ (with depths of 55–75 m), and produced a temperature difference of 0.4 °C at a distance of five miles in the latter surface, indicated the partly advective nature of the eddy motion in the uppermost ocean layer.

What has been said allows us to draw the conclusion that during the 'Polygon-70' expedition the Soviet oceanographers discovered an oceanic phenomenon of paramount importance which afterwards was called 'synoptic eddies of the open ocean'. The direct measurement method was used to determine the parameters and scales of these vortical formations and to reveal their energy level, actual spatial structure, and character of time evolution. Theoretical interpretation of the results was also given and experimental evidence on the eddy generation mechanism was obtained. All these facts specified the great importance of the 'Polygon-70' experiment for the further development of physical oceanology.

The American experiment MODE (Mid-Ocean Dynamics Experiment) (The

MODE Group, 1978) was an analog of the 'Polygon-70' experiment in another ocean region using somewhat different structure and equipment. In 1971–1972 a preliminary test experiment MODE-0 was carried out in the region chosen for the MODE experiment (symbol 15 in Figure 4.1.1). Current and water-temperature measurements with durations from 2 to 5 months were performed at depths of 1500 and 4000 m at about 20 buoy stations (Gould *et al.*, 1974). The velocity of synoptic currents turned out to be 6–7 cm/s at both depths. A spectrum analysis of oscillations of the synoptic currents at 4000 m revealed a dominant quasigeostrophic wave with a wavelength around 400 km which travelled to the north-west (in the Gulf Stream direction) with a phase velocity of about 5 cm/s. The MODE-0 results were in good agreement with the float measurement results for deep currents in the Bermuda region which were described in this section (Swallow, 1971) and with the results of two-year deep-current measurements at a buoy station located on the continental slope north of the Gulf Stream (symbol 4 in Figure 4.1.1; see Thompson, 1971) in the sense that all these measurements revealed a substantial barotropic component of oscillations of synoptic currents.

The main phase of the MODE experiment was carried out in the three-month period March–June 1973 in a Sargasso Sea region with center at 28°N, 69°40′W and radius of approximately 200 km (symbol 15 in Figure 4.1.1). The ocean depth in this region is 5300–5400 m. The main thermocline is very well developed; it occupies a layer from 500 to 1100 m and is characterized by a temperature drop from 17 to 6 °C. The seasonal thermocline is particularly well developed in the summer–autumn period when it is concentrated in a layer above 100 m and has a temperature drop from approximately 26 to 19 °C. The main observations during the MODE experiment were temperature and current measurements at more than 20 buoy stations in a layer from 400 m[1] to the bottom for, on average, seven temperature measurement depths and four current measurement depths at each of the stations. The current measurements by SOFAR floats (deep free-drifting neutrally buoyant floats which are tracked using an acoustic signal received by coastal stations) was a very interesting innovation. Altogether about 25 float launchings were carried out at a depth of 1500 m with durations from one to four months in the main MODE phase. Eight hydrographical surveys of the region and some other measurements were also carried out (The MODE Group, 1978).

Synoptic temperature and current-velocity charts for different depths and time periods were constructed for the MODE region from the data of buoy-station observations (Freeland and Gould, 1976; Richman *et al.*, 1977) and from the data of density surveys and float measurements (McWilliams, 1976a). Examples of charts of the first type are demonstrated in Figure 5.1.11; the temperature distributions shown in these charts were calculated from the data of temperature measurements at buoy-station sites using the optimal interpolation method worked out by Gandin (1963). As is seen from Figure 5.1.11, during the MODE experiment a clearly marked anticyclonic baroclinic eddy was registered; it was concentrated in

[1] American oceanographers use subsurface buoys for current measurements, which increases the durability of the system and reduces the short-period noise. However, in this case the upper ocean is not included in the observations. Moreover, a subsurface buoy system undergoes substantial vertical oscillation.

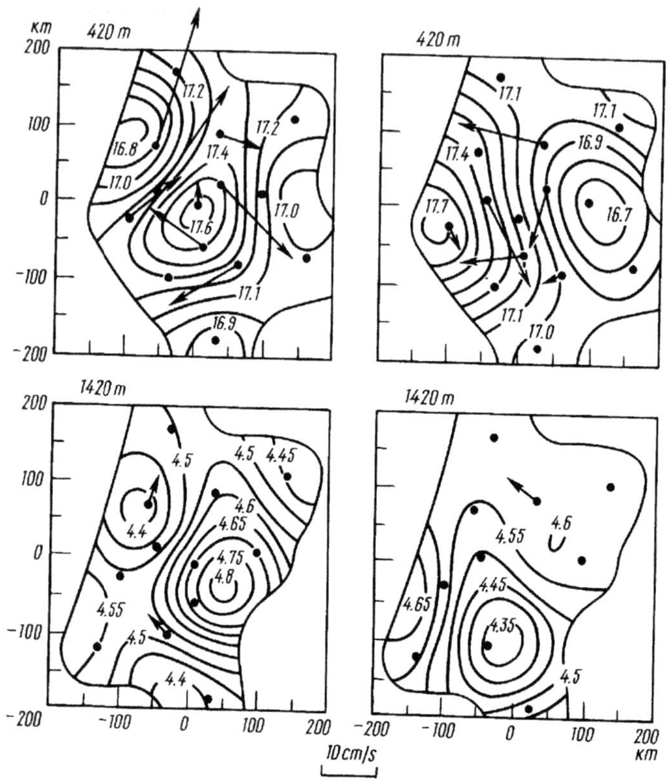

Fig. 5.1.11. Average distributions of water temperature (in °C) during the periods from April 27 to May 8, 1973 (the left patterns) and from June 14 to June 25, 1973 (the right patterns) at depths of 420 and 1420 m calculated from data obtained at the MODE buoy stations (after Richman *et al.*, 1977). The points indicate the positions of buoy stations. The arrows represent velocity vectors of the current averaged during the same periods. The center of the area is at 28°N, 69°40′W (symbol 15 in Figure 4.1.1). The isotherms (drawn at different temperature intervals for depths of 420 and 1420 m) are shown for those parts of the area where the average relative error of the reconstruction of the temperature field does not exceed 50%.

and above the main thermocline and resembled the main 'Polygon-70' eddy in a number of parameters. For instance, the horizontal scale of the MODE eddy (the distance from the eddy center to the point of maximum velocity) was about 80 km, i.e. somewhat less than that of the 'Polygon-70' eddy, which is completely accounted for by the difference in the local Rossby scale values (70 and 50 km for the 'Polygon-70' and MODE regions, respectively). The MODE eddy drifted westward with an average velocity of 3 cm/s, i.e. approximately twice as slow as the 'Polygon-70' eddy, which can be readily explained by both the smaller value of Rossby scale and the fact that the large-scale westward current in the upper ocean in the MODE area was approximately twice as weak as that in the 'Polygon-70' area (see formula (1.1)). The orbital velocities in the MODE eddy field were approximately one and a half times as high as those in the field of the main 'Polygon-70' eddy.

Figure 5.1.11 clearly shows that the disturbances of the baroclinic current in the MODE area penetrated to great depths. At the same time, the absence of similarity between the charts constructed for the 400 and 1400 m depths indicates that the vertical structure of the current was not very simple. McWilliams and Flierl (1976) managed to attain general qualitative agreement with the experimental data by approximating the evolution of the current field in the MODE region with a sum of two barotropic and two (first-mode) baroclinic linear Rossby waves propagating across the area in different directions. However, as was quite correctly noted in the work of Richman *et al.* (1977), this approximation is rather formal and does not exhaust the physical nature of the phenomenon, and hence it lacks predictability properties. Indeed, as will be indicated later, nonlinear effects were clearly marked in the MODE region.

In the course of the MODE experiment some interesting results were obtained by current measurements using deep floats. Figure 5.1.12 sums up the float trajectories at a depth of 1500 m in the period exceeding several times the duration of the MODE main phase. The root-mean-square current velocity for the interior square in Figure 5.1.12 was 3 cm/s. The complexity and variety of the float trajectories attracts attention. While some of the floats remained approximately in the same place for almost a year, others rapidly left the launching region and moved mainly to the west and south-west hundreds of kilometers away from this region for several months, the velocity of this motion exceeding several times that of the large-scale geostrophic current at the depth under consideration (Rossby and Freeland, 1977). This probably indicates the substantial role played by nonlinear (advective) effects in the formation of the synoptic component of the velocity field and a notable 'turbulization' of the motion. An estimate of the coefficient of horizontal large-scale turbulent diffusion gave a large value of 8×10^6 cm^2/s (Freeland *et al.*, 1975).

The float trajectories in the left part of Figure 5.1.12 are particularly interesting. They show that two floats were trapped by a single eddy with a diameter about 80 km and then travelled about 300 km westward together with the eddy. This motion lasted almost a year, and the orbital velocity attained 40 cm/s. The two floats were then drawn in a narrow (about 60 km wide) jet flowing southward along the continental slope (from 1500 to 5000 m) separating the Blake Plateau from the Hatteras Basin (the extreme western trajectories in Figure 5.1.12). Two months later a third float drifted in the same place and in the same direction with a velocity up to 52 cm/s. Besides the above-mentioned floats, several more floats revealed a clearly pronounced tendency to drift along sharply marked bottom slopes although, in some cases, the floats were at a height of 3 km above the sea bottom.

On analyzing the results of the current and temperature measurements obtained on the MODE buoy system, Bryden (1976) showed that the major contribution to the local synoptic oscillations of temperature in the 700–2000 m layer was made by the horizontal advection of heat. This indicates a high degree of deviation of synoptic processes from a linear wave regime in the thermocline and even below it. An analogous conclusion was drawn by Voorhis *et al.* (1976) for the upper ocean from the CTD survey data of the MODE region. Namely, it turned out that the temperature field in the upper layer is mainly formed owing to the advection of

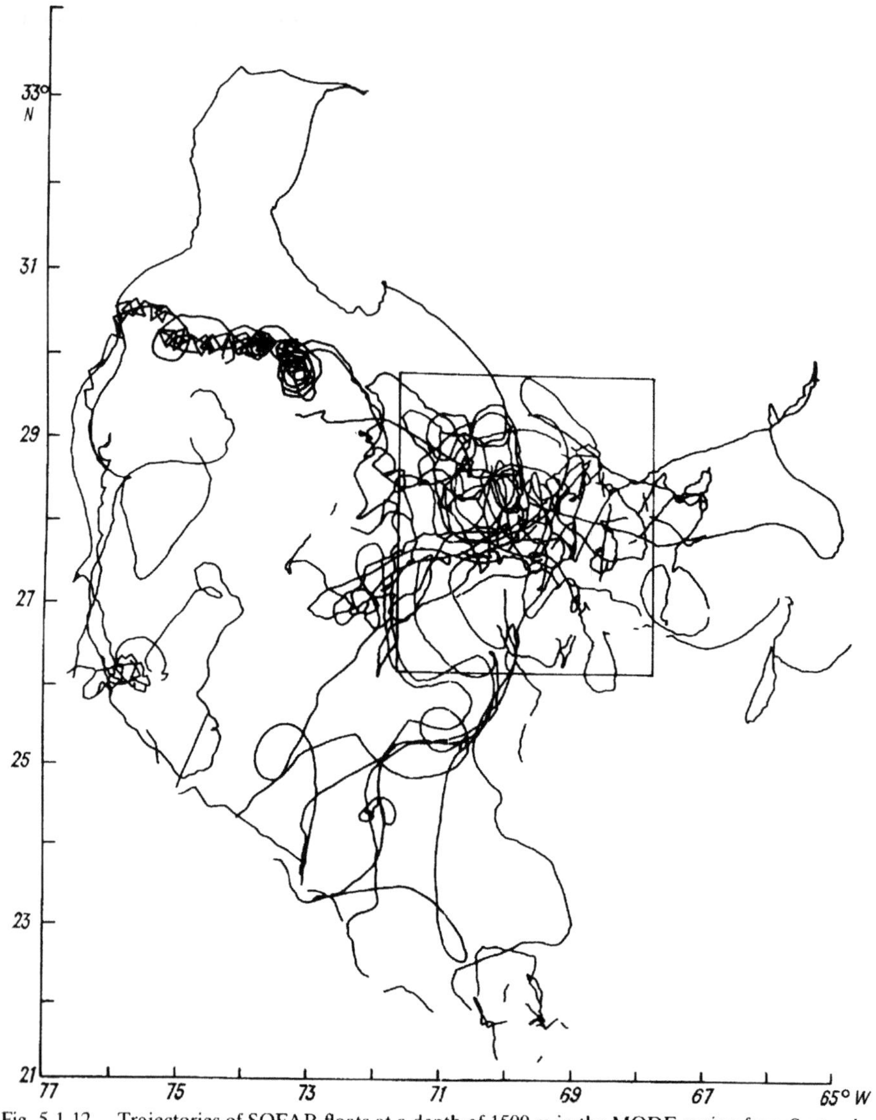

Fig. 5.1.12. Trajectories of SOFAR floats at a depth of 1500 m in the MODE region from September 1972 to June 1976 (after Rossby and Freeland, 1977). Most of the floats were launched within the limits of the interior square. The duration of the drift of the floats varied mainly from a month to two years.

heat by synoptic-scale currents (Figure 5.1.13); the appearance of synoptic temperature fronts in the upper layer of the Sargasso Sea also relates to this fact (Voorhis, 1969).

Using objective analysis of the data obtained from the hydrographic surveys and float measurements during the MODE experiment, McWilliams (1976b) carried out a comparative estimation of the terms of the vorticity equation for synoptic-scale currents (see Chapter 2). The estimation revealed a substantially nonlinear regime at least for the main thermocline and above it; that is, beside the local variation of vorticity, the vorticity advection and stretching of the vortex lines along

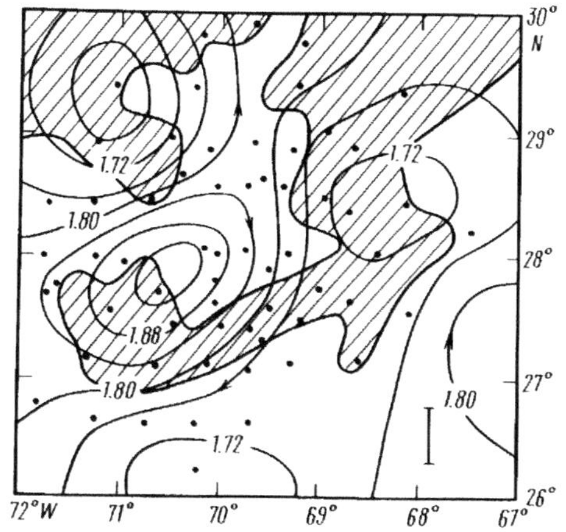

Fig. 5.1.13. Topography of the ocean surface (in meters) relative to the 1500 dbar reference surface in the MODE region, May 30–June 13, 1973 (after Voorhis *et al.*, 1976). The shaded part corresponds to the surface temperature above 25 °C; the unshaded part corresponds to the surface temperature below 25 °C. The points indicate the places of CTD soundings of the ocean.

the vertical turned out to be the major effects; the β-effect was comparatively weak.

Among the results of the statistical processing of the MODE data, we shall discuss briefly the calculations relating to frequency spectra of oscillations of temperature and velocity (an example of which is demonstrated in Figure 5.1.14). The spectrum in Figure 5.1.14 has a maximum at a period of about 140 days, which is in complete agreement with the above-mentioned estimates of dimension and westward drift velocity of the main MODE eddy. A spectrum analysis of the data of the current measurements by SOFAR floats at a depth of 1500 m showed a sharp predominance of meridional oscillations of velocity over zonal oscillations in the synoptic subrange of the spectrum, and revealed a kinetic energy spectrum maximum at a period around 50–70 days (Freeland *et al.*, 1975), which is approximately twice as small as the dominant period of temperature oscillations at a depth of 500 m (Figure 5.1.14). The results of calculating spatial correlation functions of the temperature field from the MODE data agree, in a certain sense, with the results of the above spectrum analysis. Namely, the horizontal scale of the temperature field (the value of the spatial shift at which the correlation curve intersects the axis of abscissas) at a depth of 500 m turned out to be twice as large as that at depths of 1400 and 4000 m (140 and 70 km, respectively; see Richman *et al.*, 1977). One possible explanation for this fact is the effect of mesoscale features of the bottom relief on the structure of the currents in deep ocean layers.

2. POLYMODE

The international experiment POLYMODE, designed and realized as development of the works performed during 'Polygon-70' and MODE (whence the name POLYMODE originates), was intended to investigate in more detail the structure and dynamics of synoptic eddies in the open ocean. The general program of

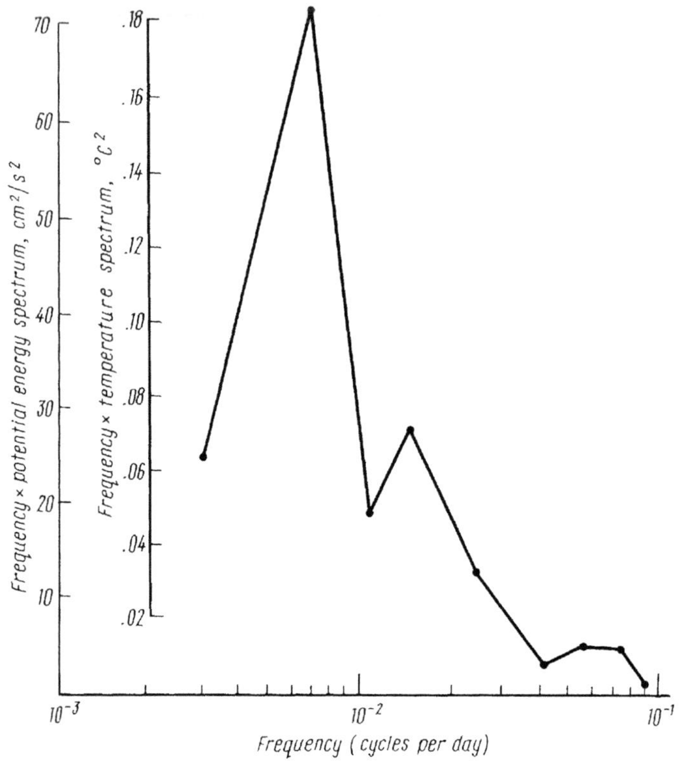

Fig. 5.1.14. Spectrum of temperature oscillations at a depth of 500 m constructed from the data of temperature measurements with a total duration of about 21 months at the MODE central buoy station (28°N, 69°40′W) (after Richman *et al.*, 1977).

POLYMODE was very extensive and formally it included more than ten individual local experiments vastly differing in their range which were carried out by Soviet, American, Canadian, and West-European oceanographers in the northern half of the Atlantic in 1974–1979. In this section we describe the basic results of the most extensive part of POLYMODE, the Soviet Synoptic Experiment carried out in the Sargasso Sea on the southern periphery of the Gulf Stream recirculation zone (Figure 5.2.1; symbol 16 in Figure 4.1.1) during the period from July 1977 to September 1978 and also of the American Local Dynamics Experiment performed in May 1978–July 1979 in a region adjoining the region of the Soviet experiment immediately to the north (symbol 17 in Figure 4.1.1). Some of the results of other components of the POLYMODE program will be described in Section 3.

The Soviet POLYMODE Synoptic Experiment was unique in its range. Ten research vessels took part and some of them made two or three cruises to the experiment region. The experiment was based on a system of 19 buoy stations of the P.P. Shirshov Institute of Oceanology of the U.S.S.R. Academy of Sciences located at the points of a grid of equilateral triangles with center at 29°N, 70°W and distances between the stations of 39 miles or 72.2 km (see below the figures demonstrating synoptic charts of currents). The vertical distributions of tempera-

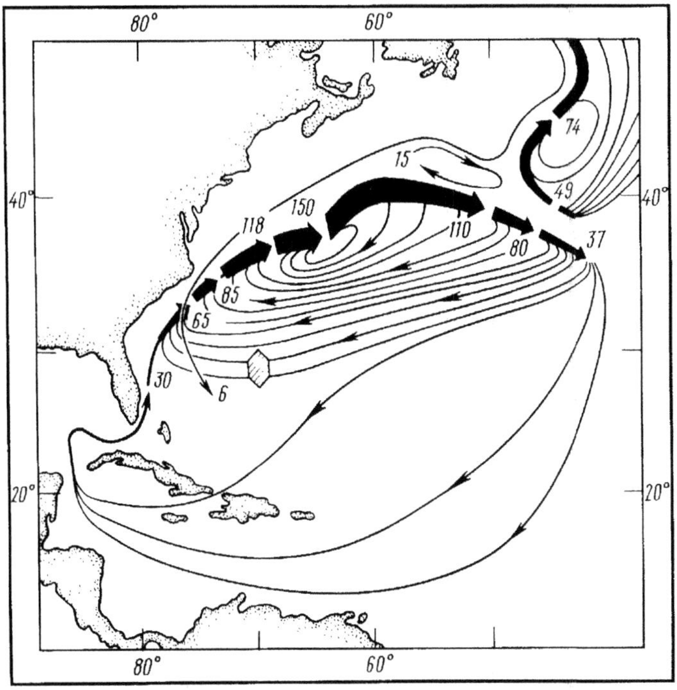

Fig. 5.2.1. Water circulation in the western part of the North Atlantic integrated over the ocean depth (after Worthington, 1976). The numbers express the water transport in 10^6 m^3/s. The hexagon represents the position of the Soviet POLYMODE buoy array.

ture, salinity, and density in the POLYMODE region are shown in Figure 5.2.2. The main thermocline is well developed, is located between 400 and 1100 m, and is characterized by a temperature drop from 17 to 5 °C. The seasonal thermocline is very well developed in summer when it lies above 100 m and has a temperature drop from approximately 26 to 19 °C; in winter the seasonal thermocline goes down below 100 m and the temperature drop across it is one or two degrees. Between the seasonal and main thermoclines there is a quasihomogeneous layer of '18-degree' Sargasso Sea water. The vertical density distribution reflects in a mirror-like manner the temperature distribution, but in the main thermocline it has a smoothed form since '18-degree' Sargasso Sea water is more saline than deep water (Figure 5.2.2).

The Soviet system of buoy stations was maintained in the POLYMODE experiment region from July 1977 to August 1978. At each point measurements at 100, 400, 700, and 1400 m were performed; they registered the current velocity and temperature at depths of 100, 400, and 700 m at 10 min intervals and the current velocity at 1400 m at hour-long intervals. The buoy station array effectively covered an ocean area equivalent to the area of three eddies if we take the model of round close-packed eddies and assume that the distance from the center of an eddy to the point of maximum velocity is equal to 80 km (see below).

Beside measurements at the above-mentioned buoy station system, an extensive

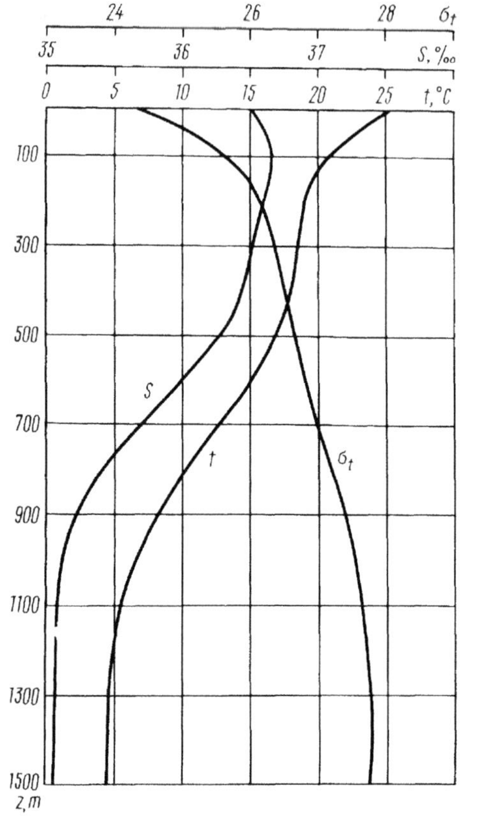

Fig. 5.2.2. Typical summer vertical distributions of temperature t, salinity S, and density σ_t in the region of the POLYMODE Synoptic Experiment (after Grishin *et al.*, 1979).

program of temperature (XBT), density (primarily CTD), and mixed temperature –density surveys of the ocean were performed during the POLYMODE Synoptic Experiment; the XBT and CTD measurements reached the depths 750 and 2000 m, respectively. Altogether, the measurements during a one-year period included 19 surveys of a region substantially exceeding the region of the main buoy array, 25 surveys of the region of the buoy station array itself, and 14 small-scale experiments. These small-scale experiments were carried out primarily by ships of the Marine Hydrophysical Institute of the Academy of Sciences of the Ukrainian S.S.R. and they also included current measurements at additional buoy stations over a period from several days to a month.

It is advisable to begin the review of the POLYMODE results with the direct current measurements at the main buoy station system. The kinetic energy spectra shown in Figure 5.2.3 were obtained by averaging the spectra calculated separately for each of the 19 buoy stations. Both in Figure 5.2.3 and in all the individual spectra there are clearly marked maxima at periods of 12.4 h (the semidiurnal tide), about 24 h (inertial oscillations and the diurnal tide), and in a broad low-frequency interval in the range of periods from 30 to 170 days (the synoptic scale). Of interest

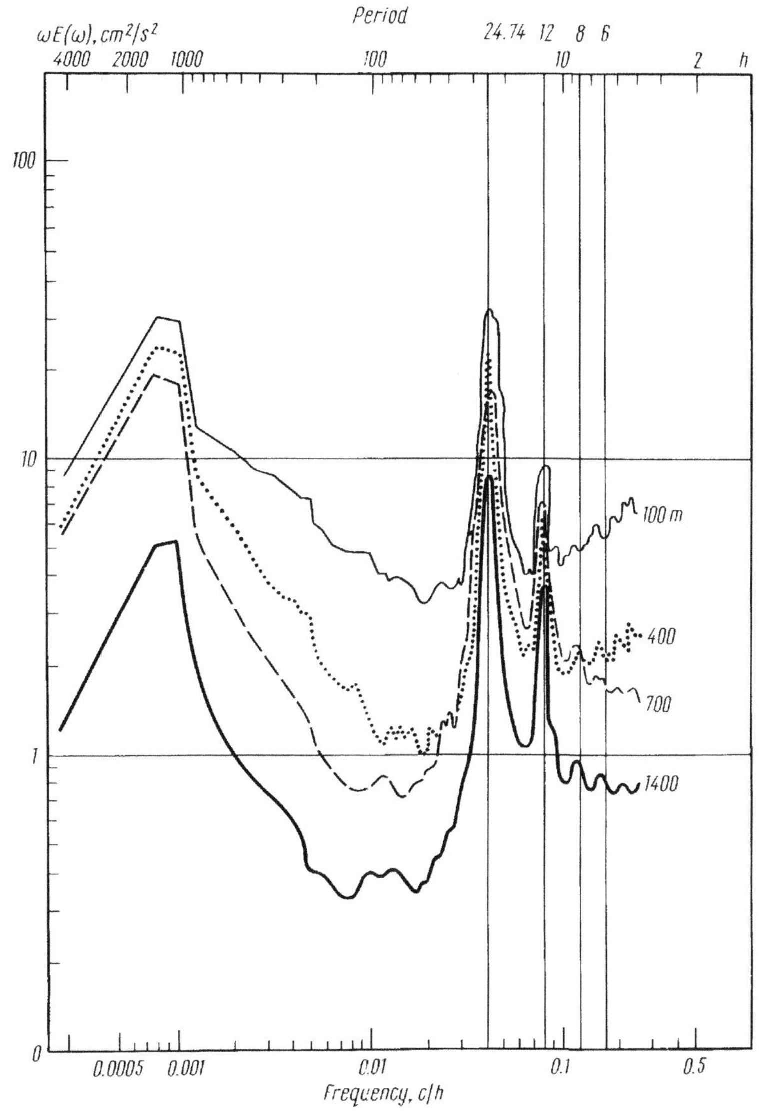

Fig. 5.2.3. Frequency spectra of the kinetic energy of currents at depths 100, 400, 700, and 1400 m according to the data of current measurements at the Soviet POLYMODE buoy system (after Drozdov *et al.*, 1981). Set off along the horizontal and vertical axes are the logarithm of the circular frequency and the logarithm of the product of the spectral density and frequency, respectively.

are the deep spectral minima at periods of 2–3 days at upper levels and 2–10 days at lower levels.

The form of spectra in Figure 5.2.3 can serve as a basis for the low-frequency filtration of a velocity series (obtained by current measurements at the POLYMODE buoy station system) to separate out synoptic currents intended for further investigation. This was done by a number of investigators, whose results are presented below. Prior to the calculation of the charts of synoptic currents presented in Figures 5.2.4–5.2.6 (in addition to the above-mentioned low-frequency filtration with an effective period of about 8 days) the gaps in the resulting synoptic velocity series were filled with the aid of consecutive interpolation of the velocity field with respect to time and depth and in the horizontal planes, and the subtraction of the velocities corresponding to the current averaged over the whole POLYMODE period. The total duration of the gaps amounted to 17.5% of the total duration of observations at all autonomous buoy stations (ABS) and at all levels. The mean current turned out to have a direction towards the south-west and a velocity of about 1.5 cm/s at the three upper levels and about 1.0 cm/s at 1400 m. The calculation of the charts (see the examples in Figures 5.2.4–5.2.6) was based on the synoptic current velocity series obtained at the observation points and was carried out by the optimal interpolation method (in the horizontal plane) for a two-dimensional random vector field assumed to be homogeneous, isotropic, and solenoidal (Gandin, 1964). The spatial correlation functions of the velocity field and its noise level, which were required for the calculation, were determined from the data subject to interpolation. The arbitrary constants in the resulting distributions of the stream function of synoptic currents for different levels and dates were found with the aid of a special algorithm based on the fact that the synoptic-scale currents are quasigeostrophic (Grachev *et al.*, 1982a).

The charts indicated well-developed synoptic-scale currents in the POLYMODE region. Altogether, during a one-year period direct current measurements reliably registered 13 eddy disturbances whose centers were located within the limits of the ABS array. Among the six anticyclones observed during this period there were four particularly clearly marked eddies that were registered from January to May 1978 (see the patterns for January, February, April, and May in Figure 5.2.4). They were all characterized by the Rossby scale (equal to 50 km for the POLYMODE region, which corresponded to the average distance from eddy centers to points of maximum velocity equal to $(\pi/2) \times 50$ km ≈ 80 km); they were very strong and sharply baroclinic (Figure 5.2.6); and they adjoined regions of weak currents. In some cases the current velocity in winter–spring anticyclones attained 40–45 cm/s at depths of 100 and 400 m and it decreased down to a magnitude of the order of 5–10 cm/s at 1400 m.

The horizontal scales of all the seven cyclonic disturbances reliably registered by the buoy array were one and a half times as great as the Rossby scale, or greater. Most of them were comparatively weak and were characterized by either a purely barotropic structure (e.g. the cyclones in the patterns for September and November 1977 and August 1978 in Figure 5.2.4) or a moderately baroclinic structure. The only exception was the remarkably powerful and sharply baroclinic cyclone in October 1977 (Figure 5.2.4). The current velocity in its field at the three upper levels attained a very high value of 60 cm/s. All POLYMODE eddies travelled in the general direction of the west with average velocities ranging from 2 to 6 km/day. The calculations also showed that the annual-average kinetic energy of synoptic currents at all the four measurement levels increased approximately twice from the south-east half of the POLYMODE buoy array region to its north-west edge (Grachev *et al.*, 1982a).

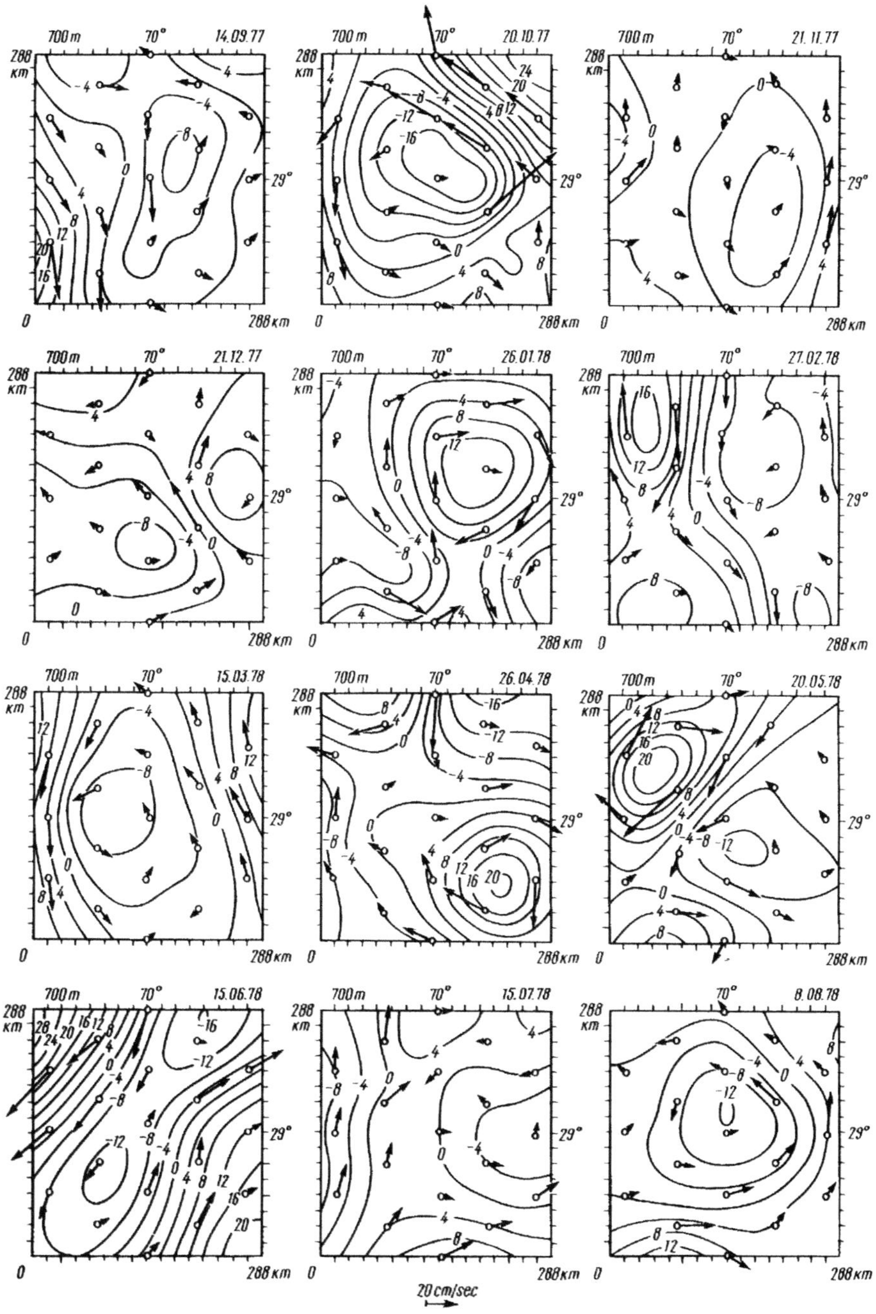

Fig. 5.2.4. Velocity vectors and streamlines of synoptic currents at depth 700 m in the region of the POLYMODE Synoptic Experiment for various dates in 1977–1978. The stream function is expressed in 10^7 cm^2/s. The circles show the positions of the ABS. The marks on the sides of the squares show the lines of the interpolation grid for the velocity field.

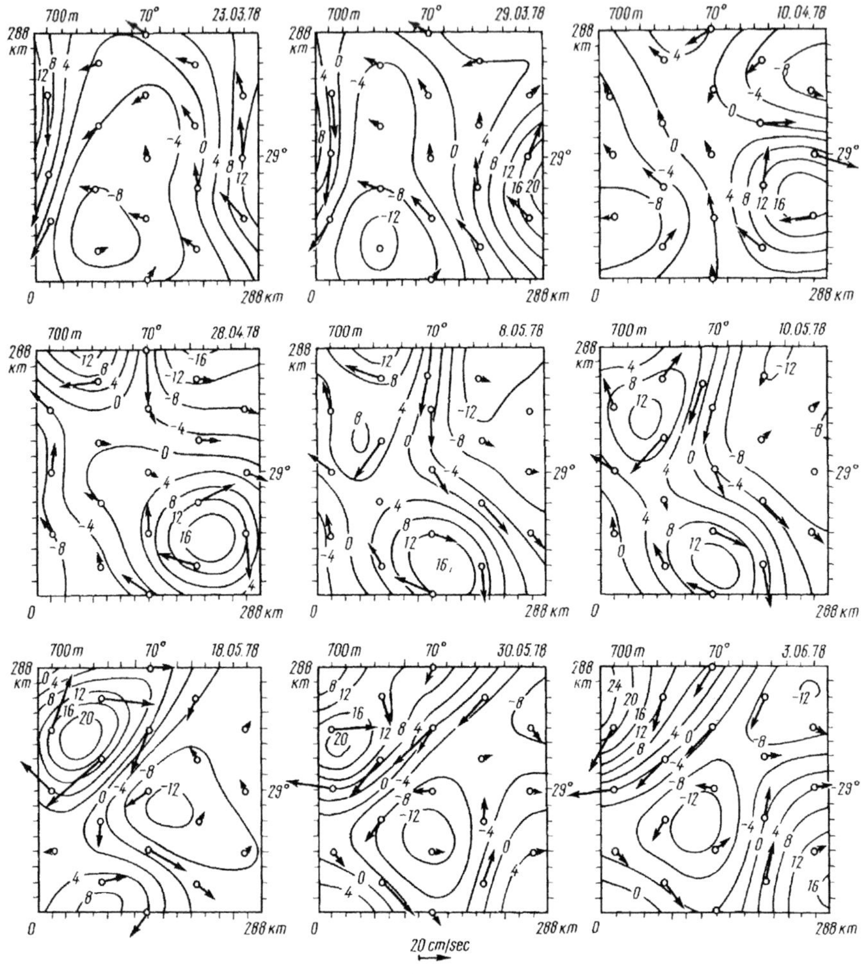

Fig. 5.2.5. Velocity vectors and streamlines of synoptic currents at depth 700 m in the region of the POLYMODE Synoptic Experiment in March–June 1978 (after Koshlyakov *et al*. 1980b). Other details as on Figure 5.2.4.

In addition to the charts in Figures 5.2.4–5.2.6, the kinetic energy curves in Figure 5.2.7 confirm the very strong nonstationarity of the synoptic eddy field in the POLYMODE region which, in particular, manifested itself in the sharply baroclinic character of eddies in January–May 1978 and in their partial or complete barotropization in November–December 1977 and in July–August 1978. The latter period, when the currents were almost purely barotropic, is particularly remarkable. The barotropic character of the currents was confirmed by the results of direct current measurements at depths of 2, 3, and 4 km performed in the POLYMODE region during this period (Vasilenko and Mirabel', 1980).

The synoptic charts of temperature and salinity constructed from the data of temperature and density surveys of the region of the POLYMODE synoptic experiment were, on the whole, in good agreement with the above-mentioned

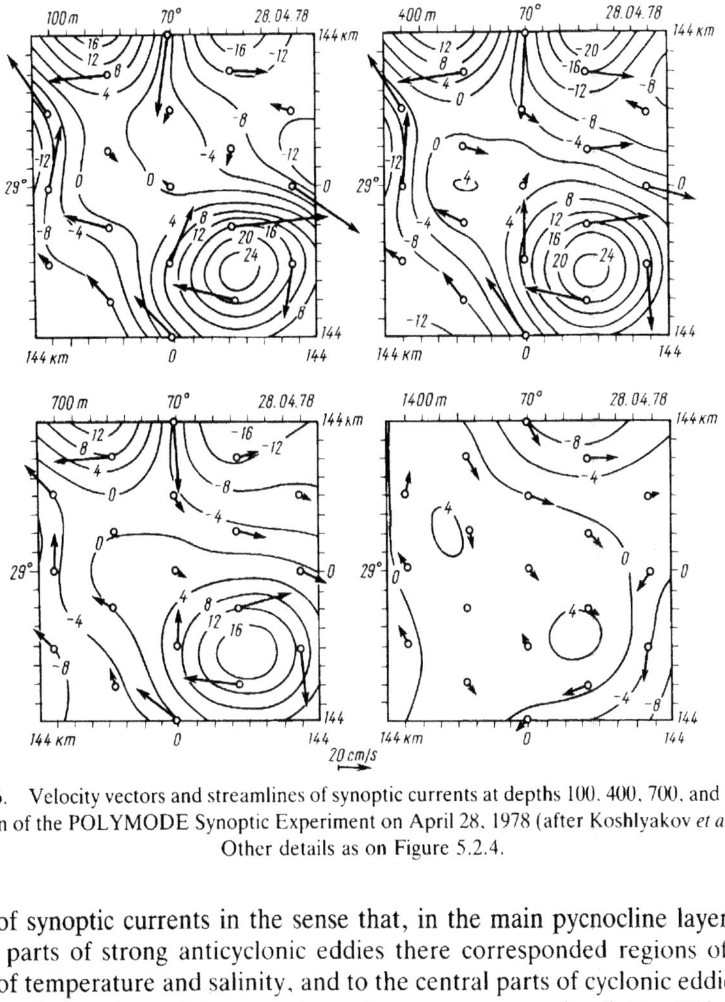

Fig. 5.2.6. Velocity vectors and streamlines of synoptic currents at depths 100, 400, 700, and 1400 m in the region of the POLYMODE Synoptic Experiment on April 28, 1978 (after Koshlyakov *et al.* 1980b). Other details as on Figure 5.2.4.

charts of synoptic currents in the sense that, in the main pycnocline layer to the central parts of strong anticyclonic eddies there corresponded regions of higher values of temperature and salinity, and to the central parts of cyclonic eddies there corresponded regions of lower values of temperature and salinity. This can be demonstrated by the charts in Figure 5.2.8 which must be compared with the chart for April 1978 in Figure 5.2.4. However, since a part of the CTD and XBT surveys covered a region whose area substantially exceeded that of the main ABS system, the surveys made it possible to register a considerably greater number of eddies than were recorded by the buoy stations (Kort *et al.*, 1980; Bulgakov and Paranichev, 1980). For example, Kort *et al.* (1980) registered 22 eddy formations, which consisted of 11 cyclones and 11 anticyclones. Data on 17 of these eddies are presented in Table 5.2.1 confirming that all these eddies travelled across the observation region in the general direction of the west. On average, the velocity of this motion was notably higher than the one that could be obtained from the model of linear baroclinic Rossby waves for the direction and velocity of the large-scale current which are typical of the POLYMODE region, which indicated the nonlinear character of synoptic disturbances in the current. The average depth of the 15 °C isothermal surface in the POLYMODE region is 600 m. Hence, according to

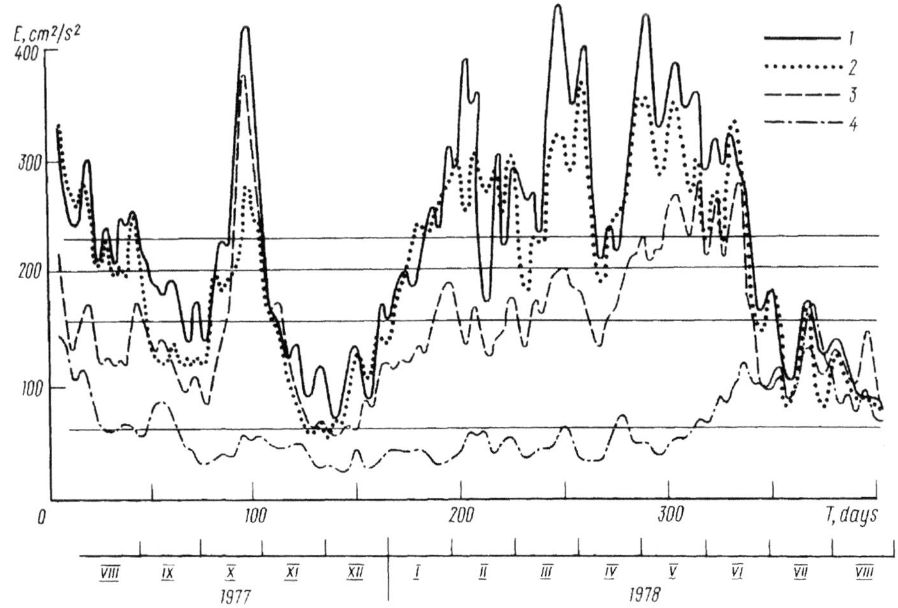

Fig. 5.2.7. Time variations of the area-averaged kinetic energy of synoptic currents at levels (1) 100, (2) 400, (3) 700, and (4) 1400 m for the POLYMODE experiment area (after Enikeev *et al.*, 1982). Also shown in the figure are the energy levels at each of the horizons averaged over the POLYMODE period.

Table 5.2.1, maximum vertical displacements of this surface in individual eddies attained a magnitude of the order of 100 m. In this sense only the sharply marked single cyclone C_5 is anomalous; this will be described in detail later.

The XBT surveys also showed (Kort *et al.*, 1980) that, in good agreement with the above-mentioned data of direct current measurements in the POLYMODE region, the winter–spring period of 1978 was characterized by a particularly high eddy energy, predominance of strongly concentrated anticyclones separated by vast regions of weak cyclonic vorticity, and intensive interaction between individual eddies. Moreover, a detailed analysis of the results of XBT, CTD, and hydrographical measurements during the whole POLYMODE period led Korotaev (1980) and then Harrison and Heinmiller (1983) to the conclusion that, on the whole, the synoptic current field in the POLYMODE region can be divided into two components (modes) substantially differing in their kinematic characteristics and, probably, in their physical nature. Namely, these are strong quasi-isolated and concentrated eddies (primarily but not exceptionally anticyclones) with a diameter of 100–200 km carrying away their water in their translatory motion, and a more large-scale field of weak eddy-like current disturbances, probably of a dominant wave nature, separating the eddies. According to Korotaev (1980), only six of the eddy disturbances observed during POLYMODE can be reliably placed in the first category; namely, the cyclones in October and December 1977 and the anticyclones in January, February, April, and May 1978 (see the charts for the corresponding months in Figure 5.2.4; in Table 5.2.1 these eddies are marked as C_3, C_5, A_4, A_6, A_7, and A_8). The relative part of the ocean area occupied by the strong

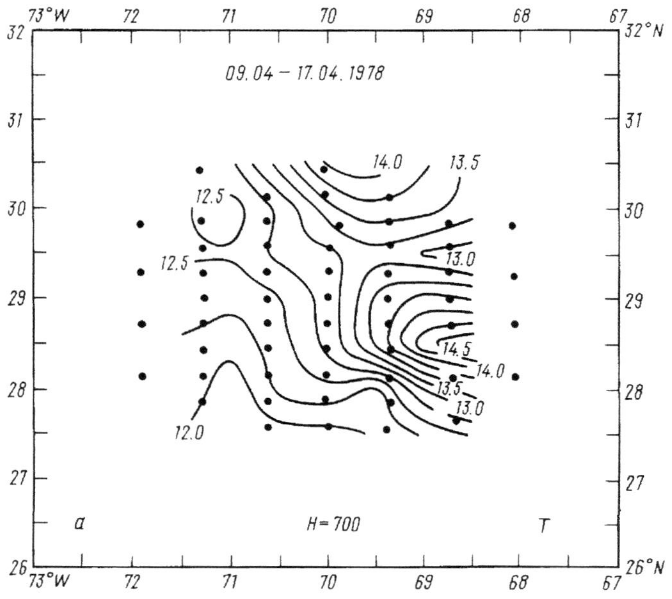

(a)

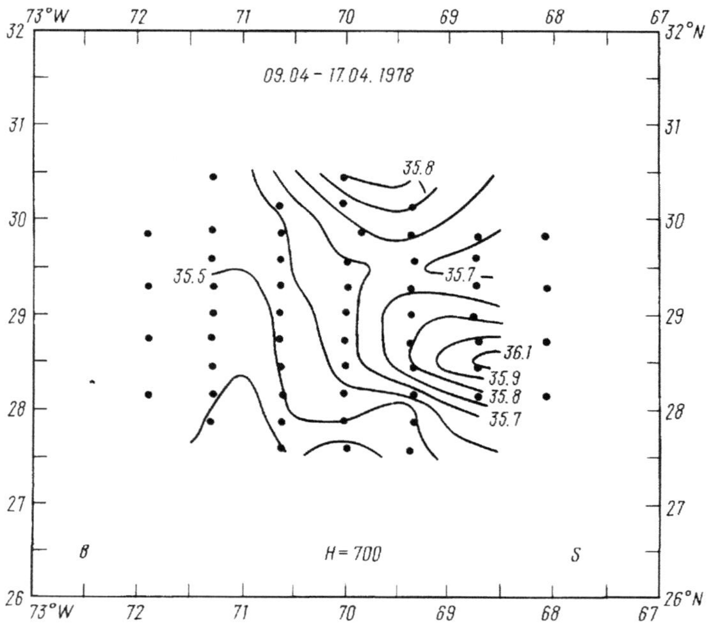

(b)

Fig. 5.2.8. Distributions of (a) temperature T (°C) and (b) salinity S (‰) at a depth of 700 m according to the data of the CTD survey of the region of the POLYMODE Synoptic Experiment on April 9–17, 1978 (after Tsybaneva, 1984). The circles show the sites of the CTD soundings of the ocean.

TABLE 5.2.1

Parameters of synoptic ocean eddies according to the data of XBT surveys of the region of the POLYMODE Synoptic Experiment (after Kort *et al.*, 1980). A and B mean anticyclones and cyclones, respectively.

(1) Eddy index	(2) Time of location in the experiment area		(3) Direction of motion	(4) Velocity of motion, km/day	(5) Extremal depth of the 15 °C surface, m
	Months of 1977 and 1978	Days			
A_2	VII–X(77)	76	WNW	4.4	700
A_3	X–XII(77)	73	W	7.6	720
A_4	I–III(78)	55	W	7.2	700
A_5	XI(77)–II(78)	66	WNW	6.0	680
A_6	I–III(78)	55	W	7.2	700
A_7	II–VI(78)	94	WSW	4.0	700
A_8	II–VI(78)	94	WSW	4.6	700
A_9	V–VII(78)	34	WSW	7.7	680
A_{10}	V–VIII(78)	56	W	5.0	660
C_2	VII–X(77)	63	W	4.3	540
C_3	VII(77)–I(78)	102	WNW	5.7	500
C_5	XII(77)–II(78)	75	NNW	4.8	380
C_6	I–III(78)	55	W	4.6	520
C_7	II–V(78)	73	WNW	7.6	520
C_8	IV–VII(78)	86	WSW	4.5	500
C_9	V–VII(78)	65	W	3.3	520
C_{10}	VI–VII(78)	56	W	3.0	520

eddies varied with time in the POLYMODE region and was proportional to the average energy level of synoptic currents. On average, according to G. K. Korotaev's estimates, during the POLYMODE period this part amounted to 30% of the ocean area but it contained 80% of the entire available potential energy (APE) of synoptic disturbances in the water circulation. The average density of the APE of intensive eddies integrated across the ocean depth (within the 0–1500 m layer) amounted to 6×10^4 J/m^2 and that of the wave background between the intensive eddies was equal to 7×10^3 J/m^2. Assuming that the radius of an intensive eddy was equal to 100 km, Korotaev (1980) estimated the total APE of an eddy to be 2×10^{15} J, which is approximately 50 times as small as the average total APE of a Gulf Stream cyclonic ring.

Very strong inhomogeneity in the energy distribution of synoptic currents over the area of the POLYMODE region is also clearly seen in the example demonstrated by Figure 5.2.9, the APE distribution of currents in the main thermocline (the chart *c* in Figure 5.2.9) being particularly contrasting. According to the data of temperature–density surveys (Sazhina, 1982), in the 0–1400 m layer the depth-integrated KE and APE densities of synoptic currents during the winter–spring period of 1978 in the POLYMODE region were of the same order (of the order of 1 J/cm^2), the APE dominating in the main pycnocline and the KE dominating in the layer of the '18-degree' Sargasso Sea water and below 1200 m.

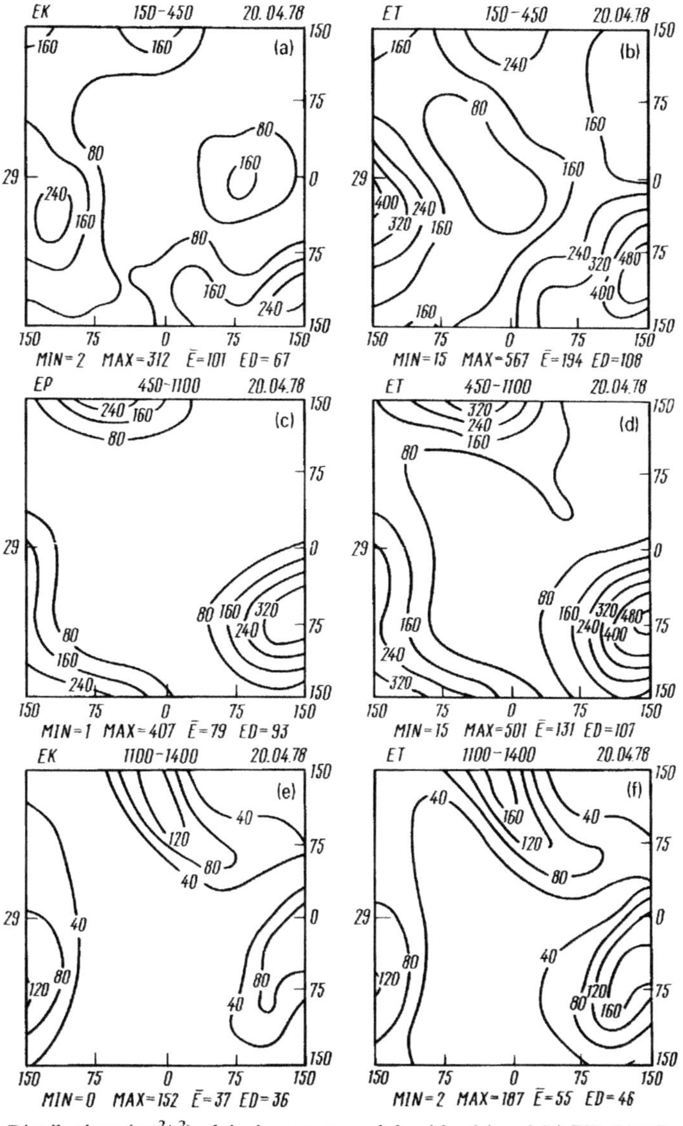

Fig. 5.2.9. Distributions (cm^2/s^2) of the layer-averaged densities (a) and (e) EK, (c) EP, and (b), (d) and (f) ET of the kinetic, available potential, and total energies of synoptic currents for (a) and (b) 150–450 m, (c) and (d) 450–1100 m, and (e) and (f) 1100–1400 m layers in the POLYMODE experiment area according to the data of the temperature-density survey of the area in April 1978 (after Sazhina, 1982). The positions of the CTD and XBT stations are shown in Figure 5.2.14; ED is the root-mean-square deviation of the energy density from its average value $\bar{E}$.

The fact that the orbital velocity of water motion in the field of strong POLYMODE eddies substantially exceeded the velocity of their westward translatory motion obviously shows that, in the process of this motion, the eddies carried away their water in their central parts. At the same time, the cores of all eddies, with the exception of one anomalously concentrated cyclone (eddy C_5 in Table 5.2.1), were characterized by the same *T, S* relation which is climatically typical of the western part of the Sargasso Sea (Bulgakov and Paranichev, 1980). This fact is accounted for by the short duration of the life of each of the eddies (with the exception of cyclone C_5) as an individual formation, which was revealed by both direct current measurements (Figure 5.2.5) and XBT–CTD surveys and, in turn, was due to the intensive interaction between individual eddies when they came closer together. For instance, these interactions are clearly seen in Figure 5.2.5, demonstrating an active water transfer in the first ten-day period of May 1978 from an anticyclonic eddy in the southern part of the experiment area to a clearly marked jet of a quasimeridional direction formed on the north-east periphery of that eddy, and then from this jet to another strong anticyclone in the north-west part of the area. Owing to these interactions, at least during the winter–spring period of 1978 none of the eddies travelled across the observation region without changing substantially its size, shape, and energy. In a sense, these properties of intensive POLYMODE eddies admit of their interpretation as structural elements of a large-scale ocean turbulence. This conclusion is also in complete agreement with the results of an analysis of the character of the synoptic heat and vorticity balances in the field of the most powerful POLYMODE eddies (Korotaev and Shapiro, 1980) carried out using the data of CTD–XBT surveys of the POLYMODE region. The analysis revealed the leading role of advective (nonlinear) effects in the formation of these balances.

As to the above-mentioned background field of weak eddy-like disturbances located between intensive eddies in the POLYMODE region (Korotaev, 1980), the space–time spectrum analysis of synoptic disturbances in water temperature performed on the basis of the data of temperature and density surveys of the region (Korotaev, 1980) showed that this field could be interpreted as a result of superposition of about ten first-mode baroclinic Rossby waves propagating in various directions within the limits of the south-west quadrant and characterized by wavelengths of the order of 500–600 km. It turned out that the relationships between the wave vectors and the periods of these waves and the average velocity of the westward drift of intensive eddies did not contradict the hypothesis that the waves were radiated by the intensive eddies. Moreover, it was possible to estimate the average value of the depth-integrated energy flux in the 0–1500 m layer from eddies to waves, which turned out to amount to 2×10^{-3} W/m^2. The mechanism of generation of intensive eddies, according to the above concept (Korotaev, 1980), is the baroclinic instability of large-scale currents.

Both the results of direct current measurements in the POLYMODE experiment area and the results of density and temperature surveys make it possible to analyze some aspects of the eddy field dynamics. The predominance of Rossby-scale baroclinic eddies in the POLYMODE region during the winter–spring period of 1978, and the above-mentioned growth of the average energy of eddies towards the north-western edge of POLYMODE experiment area, indicate that the baroclinic instability of the Gulf Stream Counter Current (Figure 5.2.1) is the most probable eddy generation mechanism in this period. Kinematically this process must manifest itself in the spontaneous formation of narrow and strong local jets in the field of the large-scale current (LSC), their meandering, and the further separation of meanders. It is obvious that on the southern periphery of the westward LSC (Figure 5.2.1) *anticyclonic* meanders of the currents and *anticyclonic* Rossby-scale cut-off eddies must be primarily formed. This is what was observed in the POLYMODE region in January–May 1978 (Figures 5.2.4–5.2.6). Moreover, Figure 5.2.5 clearly reveals the process of fast southward motion of an anticyclonic meander of the westward current in the north-western part of the POLYMODE experiment area, and the further transformation of the meander into an anticyclonic eddy in the first ten-day period of May 1978.

To investigate quantitatively the eddy generation process in the POLYMODE region in the winter–spring period of 1977–1978, the time series of the velocity of synoptic currents during this period were subjected to a detailed spectrum analysis. An average of these series over the entire period from January 6 to May 20, 1978 and over all 19 measurement points showed that during the indicated period the LSC was well developed and was characterized by an azimuthal direction of 240° (the azimuth is counted clockwise from the north) and by the velocity ranging from 4.6 cm/s at a depth of 400 m to 0.7 cm/s at 1400 m (Koshlyakov *et al.*, 1984a). On the other hand, Figures 5.2.10 and 5.2.11 show that during this period of 1978 there were clearly marked sharply baroclinic oscillations of currents with a period of 80–90 days in the POLYMODE experiment area which had a propagation azimuth of about 240° and a wavelength of the order of 380 km. It is of interest to note that the direction of propagation of these oscillations completely coincided with the LSC direction, which probably indicates that they are genetically interrelated. However, the most important result is demonstrated by Figure 5.2.12 which shows that in the field of the '85-day' velocity oscillations there was a clearly marked vertical phase shift in the main pycnocline (see Figure 5.2.2), the sign of the phase shift being such that for the sign of the LSC velocity shear it corresponded to the transformation of the available potential energy (APE) of the LSC into the APE of eddies (formula (1.2) in Section 1 of the present chapter). Applying formula (1.2) to estimate the average rate of growth of the depth-integrated eddy energy in the 700–1400 m layer in the POLYMODE region in January–May 1978 and substituting the parameters of the LSC and eddies, we obtain an estimate of 4.7 erg/cm^2·s for the rate of growth (Koshlyakov *et al.*, 1984a). If we conditionally assume that the currents in the POLYMODE experiment area during the winter-spring period of 1978 were in a quasistationary regime and that there were no mean energy fluxes across the lateral boundaries of the experiment area, then the depth-averaged rate of dissipation of energy is estimated as (4.7 erg/cm^2·s)/(5.4 km) = 0.90 × 10^{-5} erg/cm^3·s.

As is known, barotropization of an eddy field of the type of the barotropization observed in the POLYMODE experiment area in November 1977 and, particularly clearly, in June–July 1978 (Figure 4.2.7) is a property of free quasi-two-dimensional turbulence (Rhines, 1977; Mirabel' and Monin, 1980) and indicates a sharp weakening or a complete suppression of the process of generation of baroclinic eddies. Indeed, the processing of the data of current measurements performed in the POLYMODE experiment area in June–August 1978 showed that during this period there were almost no LSC and no vertical phase shift of synoptic oscillations of velocity in this region (Yaremchuk, 1982). From the theory (Rhines,

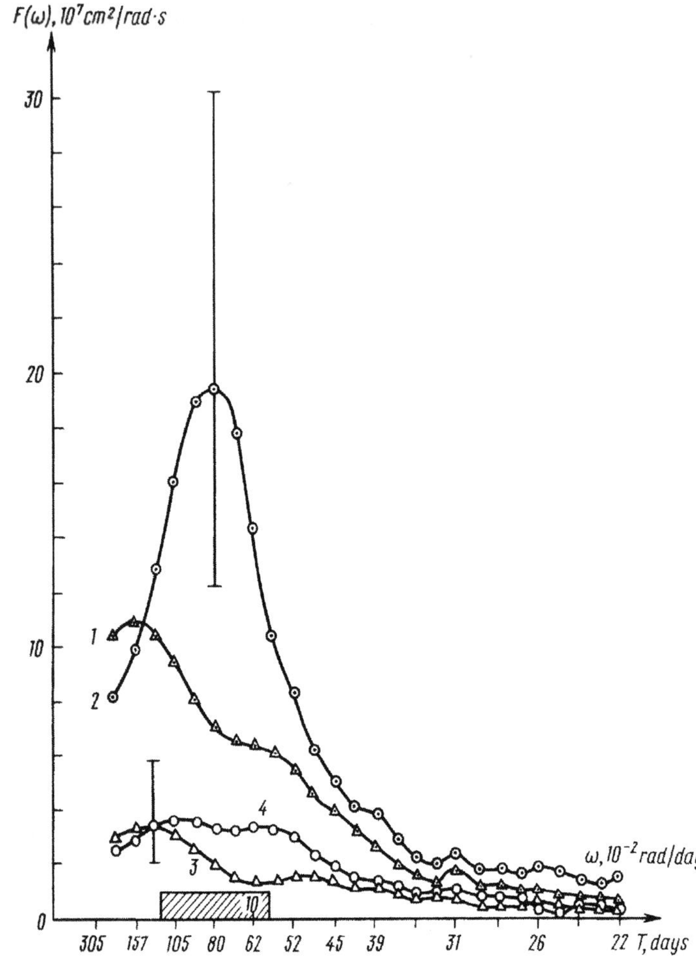

Fig. 5.2.10. Time spectra of oscillations of the (1 and 3) zonal and (2 and 4) meridional components of the current velocity at depths (1 and 2) 700 and (3 and 4) 1400 m in the POLYMODE experiment region for the period from January 6 to May 20, 1978 (after Koshlyakov *et al.*, 1984a). Shown in the figure are the length of the effective frequency-smoothing interval for the spectra and the 90% confidence intervals for the two maximum values of spectral density.

1977; Mirabel' and Monin, 1980) it follows that barotropization of eddies must be accompanied by growth of their horizontal dimensions from the scale L_R to the scale $L_\beta = (2U/\beta)^{1/2}$, where U is the typical velocity in the eddy field. In the case under consideration we can put $U = 15$ cm/s and then we have $L_\beta = 120$ km, which is more than twice as great as the estimate for L_R in the POLYMODE region. Figure 5.2.4 shows that in the course of barotropization the eddy dimensions did in fact increase approximately two-fold.

The barotropic character of the field of synoptic currents in July–August 1978 (Figure 5.2.7; Vasilenko and Mirabel', 1980) makes it possible to estimate the kinetic energy (KE) of the currents integrated across the ocean depth, which turns out to be 110 erg/cm^3 × (5.4 km) = 5.9 J/cm^2. Taking into account the dissipation

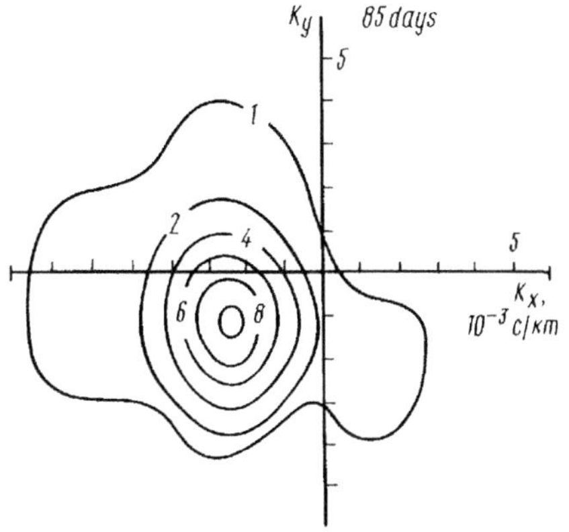

Fig. 5.2.11. Normalized 85-day cross-section of the space–time spectrum of the kinetic energy of synoptic currents at a depth of 700 m in the POLYMODE experiment region for January–May 1978 (after Konyaev and Sabinin, 1981). The isolines are drawn for the values of spectral density amounting to 10, 20, 40, 60, 80, and 95% of the maximum value.

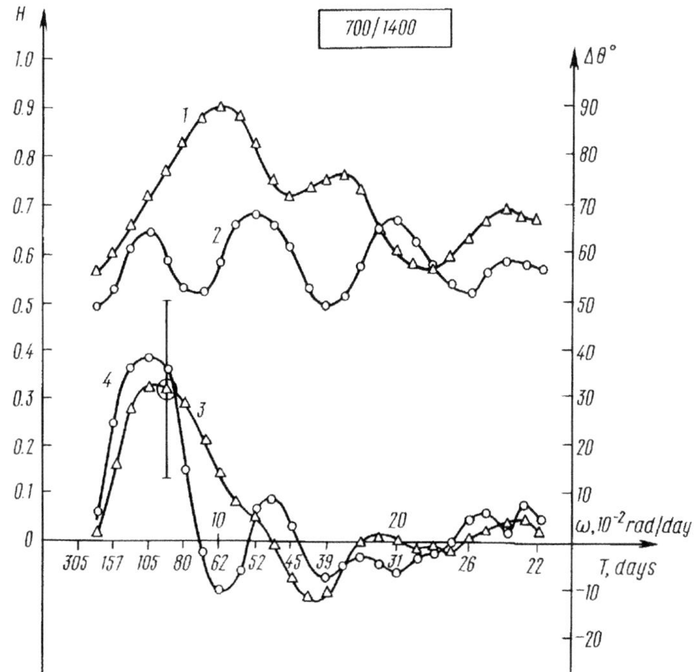

Fig. 5.2.12. Time spectra of coherence H (linear form: curves 1 and 2) and phase difference $\Delta\theta$ (curves 3 and 4) between oscillations of the current velocity at depths 700 and 1400 m in the POLYMODE experiment area for the period from January 6 to May 20, 1978; 1 and 3, the zonal component; 2 and 4, the meridional component (after Koshlyakov *et al.*, 1984a). $\Delta\theta > 0$ corresponds to a delay in the oscillations at 700 m relative to the oscillations at 1400 m. For the maximum of the curve $(\Delta\theta)_u$ the 90% confidence interval is shown.

of energy during the 20-day barotropization period (Figure 5.2.7), we estimate the depth-integrated total energy of eddies (APE + KE) before the barotropization as 5.9 + 0.9 = 6.8 J/cm^2. In the POLYMODE region, the level 1400 m corresponds to the depth of the zero amplitude function of the first baroclinic mode of velocity oscillations in a field of linear Rossby waves (Richman *et al.*, 1977). Assuming conditionally that the actual nonlinearity of synoptic currents in the POLYMODE region and their interaction with the LSC produce no fundamental changes in their vertical structure (Gill *et al.*, 1974; Rhines, 1977), and that this structure can crudely be described by the sum of the barotropic and first baroclinic modes, we estimate the depth-integrated KE of the barotropic component of currents in the spring of 1978 (Figure 5.2.7) as 50 erg/cm^3 × (5.4 km) = 2.7 J/cm^2. The rate of transformation of the integrated APE of synoptic currents and the KE of their baroclinic component into the integrated KE of the barotropic component during the barotropization period (with inclusion of the dissipation of the APE and the KE of the baroclinic component) is estimated as ((6.8 − 2.7 − 0.5) J/cm^2)/(20 days) = 21 erg/cm^2·s (Koshlyakov *et al.*, 1984a).

To study in more detail the variability of the energetic regime of eddies and of their interaction with LSC in the POLYMODE experiment area, it was first necessary to reconstruct in greater detail the LSC structure and its variability during the observation period from the observation data. In accordance with the results of the calculation of the mean geostrophic currents in the POLYMODE region from historical temperature and density data (Ebbesmeyer and Taft, 1979), it was assumed (Koshlyakov *et al.*, 1984b) that at any depth and at any time the LSC in the POLYMODE region is a plane-parallel flow with a constant 240° (or 60°) direction and a linear velocity distribution in the transverse direction (Figure 5.2.13(a)). The time variation of the LSC parameters at each measurement level (Figure 5.2.13(b)) was determined by the least-squares method from the data of current measurements at the POLYMODE ABS after the velocity time-series were smoothed with a period of 170 days at all the measurement points (at the ends of the POLYMODE time interval the smoothing had a one-sided character). Figure 5.2.13 shows that, according to the above-mentioned model, the LSC was directed to the west-south-west during almost the entire POLYMODE time interval and its velocity increased towards the north-north-west. The velocity maximum was attained at a depth of 400 m.

Taking into account the above LSC model we write the following expressions for the densities P' and K' of the APE and KE of synoptic currents (SC) and the densities $(\bar{P} \rightarrow P')$ and $(\bar{K} \rightarrow K')$ of the rates of transformation of the APE and KE of the large-scale current into those of synoptic currents: $P' = (g^2/2\,\varrho_0)(\varrho'^2/N^2)$, $K' = (\varrho_0/2)(u^2 + v^2)$, $(\bar{P} \rightarrow P') = (gf/2)\,[N^{-2}\varrho'(u - v\sqrt{3})(\partial U/\partial z)]$, and $(\bar{K} \rightarrow K') = (\varrho_0/2)\,\{[(\sqrt{3}/2)(u^2 - v^2) - uv]\,(\partial U/\partial n)\}$. Here N^2 is the square of the Brunt-Väisälä frequency, U is the LSC velocity (see Figure 5.2.13), $u = -\partial\psi'/\partial y$ and $v = \partial\psi'/\partial x$ are the zonal and meridional components of SC velocity (ψ' is the synoptic disturbance of the stream function), and $\varrho' = -\varrho_0 f g^{-1} \partial\psi'/\partial z$ is the synoptic disturbance of the water density. We use a local Cartesian coordinate system with origin at the undisturbed sea surface and the axes x, y, and z pointing east, north, and upward. The spatial distributions of the quantities u, v, and ϱ' in

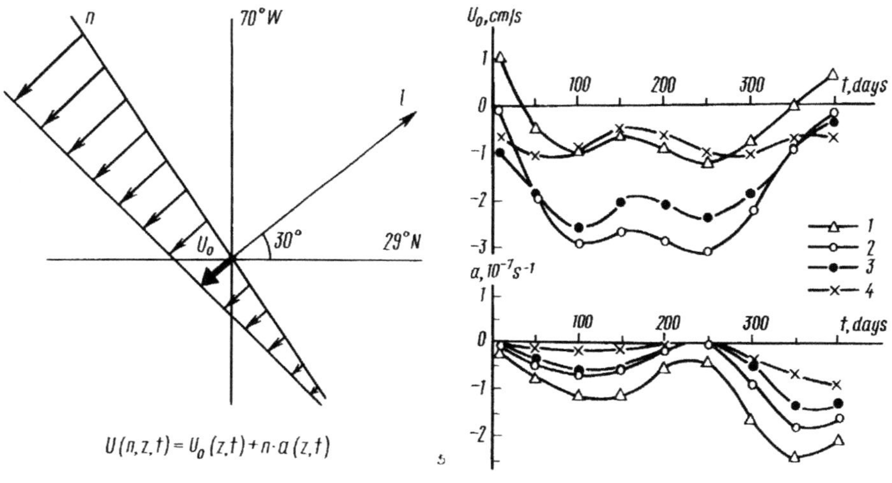

(a) (b)

Fig. 5.2.13. (a) Model of the large-scale current (LSC) in the POLYMODE region. (b) Time variations ($t = 0$ and $t = 400$ correspond to July 29, 1977 and September 3, 1978 respectively) of the parameters U_0 and a of the large-scale current at depths (1) 100, (2) 400, (3) 700, and (4) 1400 m according to the data of current measurements in the POLYMODE experiment area (after Koshlyakov *et al.*, 1984b). Simultaneous negative values of U_0 and a correspond to a current direction of 240° (the angle is counted clockwise from the direction to the north) and an increase in the modulus of velocity in the 330° direction.

the POLYMODE experiment area and their time variability during the whole POLYMODE period which are required to calculate the field P', K', $(\bar{P} \to P')$, and $(\bar{k} \to K')$ were found by means of depth interpolation of the distributions of $\psi'(x, y)$ obtained earlier for current measurement levels with the aid of optimal interpolation of the velocity field (examples of such distributions are presented in Figures 5.2.4–5.2.6) (Koshlyakov *et al.*, 1984b). The depth interpolation of the field ψ' on each day during the POLYMODE period was performed using cubic splines for the boundary conditions for $\partial\psi'/\partial z$ at depths 0 and 1600 m, corresponding to the relative depth variations of the amplitude function of the first baroclinic mode of current velocity oscillations in the Rossby wave field characteristic of the POLYMODE region (Richman *et al.*, 1977). Simultaneously, the spatial distributions of ϱ', ψ', u, and v were calculated from the data of five CTD–XBT surveys of the region of the buoy array which were carried out in February–May 1978. Besides some other procedures, this processing included the reduction of temperature data to density data using the T, S relation characteristic of the POLYMODE region, the calculation of the vertical distributions of ψ' at the sites of CTD and XBT stations by the dynamic method for the reference distributions of ψ' at depth 1400 m obtained from the data of direct current measurements, and the further optimal interpolation of the fields ϱ' and ψ' to the points of the interpolation grid (Figure 5.2.14).

The example in Figure 5.2.14 demonstrates good qualitative agreement between the distributions of ϱ' and ψ' calculated from the two above-mentioned different types of data. At the same time, the amplitudes of the disturbances ϱ' and ψ'

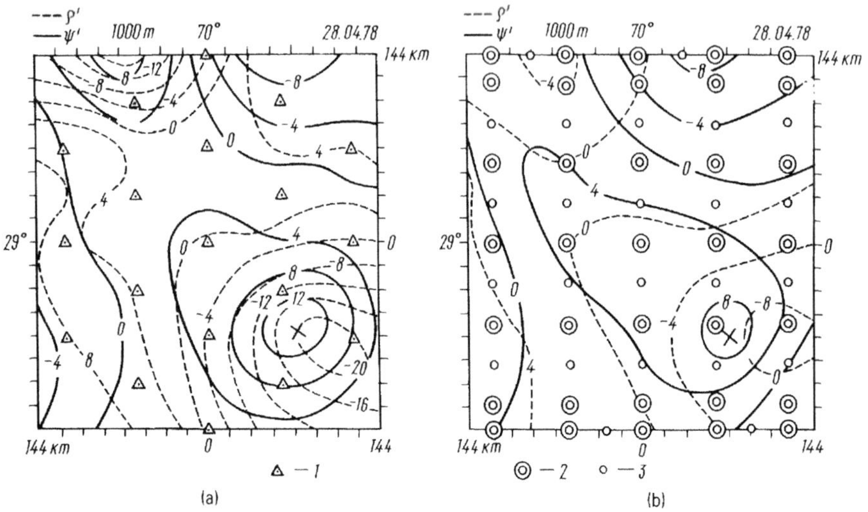

Fig. 5.2.14. Distributions of synoptic disturbances ψ' (10^7 cm^2/s) and ϱ' (10^{-5} g/cm^3) of the stream function and water density fields at a depth of 1000 m in the POLYMODE experiment area on April 28, 1978 calculated from the data of direct current measurements (on the left) and a temperature-density survey of the area (on the right) (after Koshlyakov *et al.*, 1984b). The symbols 1, 2, and 3 indicate the positions of the ABS, CTD, and XBT stations, respectively. The cross shows the center of an anticyclonic eddy. The marks on the sides of the square indicate the positions of the lines of the interpolation grid.

calculated from current measurement data are considerably greater than the 'density' amplitudes, which can be accounted for both by some instrumental overestimation of the current velocity produced by the current meters and by the fact that the distance between the sections of the CTD–XBT surveys was too large. Both charts in Figure 5.2.14 demonstrate clearly marked zonal phase shifts between the disturbances of the fields ϱ' and ψ'; the sign of the shifts corresponds (if the sign of the vertical shear of the LSC velocity during the period under study is taken into account; see Figure 5.2.13) to the transformation of the APE of the LSC into that of the SC. Analogous phase shifts of the same sign resulted from the processing of the two types of data for all clearly marked eddies during the period February–May 1978.

Figure 5.2.15 demonstrates the results of calculating the time variation of the quantities P', K', $(\bar{P} \to P')$, and $(\bar{K} \to K')$ integrated over the depth within the limits of the main pycnocline which is the most important ocean layer from the viewpoint of studying eddy dynamics. The basic result consists in calculating reliably the energy flux from the APE of the LSC to the APE of the SC (curve 3 in Figure 5.2.15) which attained particularly large values during the period of days 180–310 (from the middle of January to May 1978). The growth of the total eddy energy which attained its maximum approximately 50 days after the maximum of $(\bar{P} \to P')$ (Figure 5.2.15) is in agreement with this result. However, beginning with the end of May to the beginning of June 1978 (days 310–330), a sharp decrease of vertical gradients of the LSC and of the flux $(\bar{P} \to P')$ in the main pycnocline

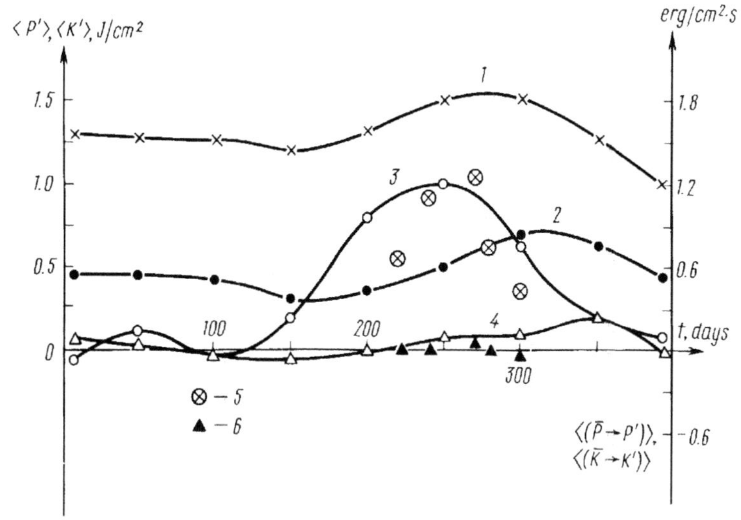

Fig. 5.2.15. Variations (according to the data of direct current measurements in the POLYMODE experiment area) of the densities P' and K' of the available potential and kinetic energies of synoptic currents (curves 1 and 2, respectively) and densities $(\bar{P} \rightarrow P')$ and $(\bar{K} \rightarrow K')$ of the rates of transformation of the APE and KE of the large-scale current into the APE and KE of synoptic currents (curves 3 and 4, respectively) subjected to the following three-fold operation (denoted $< >$): time smoothing with a period of 170 days, averaging over the experiment area, and integration with respect to depth within the limits from 550 to 1100 m (after Koshlyakov *et al.*, 1984b). Symbols 5 and 6 correspond to values of $<(\bar{P} \rightarrow P')>$ and $<(\bar{K} \rightarrow K')>$ (without time-smoothing) found from the data of temperature-density surveys of the experiment area.

took place (Figures 5.2.13 and 5.2.15) which was accompanied by barotropization of the eddy field (Figure 5.2.7). Simultaneously with the decrease of the vertical gradient of the LSC velocity in July–August 1978 (days 320–400) its horizontal gradient sharply increased (Figure 5.12.13), which resulted in notable values of the flux $(\bar{K} \rightarrow K')$ from the LSC to the eddies (Figure 5.2.15) during the final period of POLYMODE. For the deep ocean layer (1100–1600 m) the curves analogous to the curves shown in Figure 5.2.15 are qualitatively similar to the latter curves, the major distinction between them lying in the fact that, for the former curves, the maximum flux $(\bar{P} \rightarrow P')$ is attained considerably later (on day 300) and amounts to only 0.3 erg/cm^2·s. According to the data of direct current measurements, no notable fluxes $(P \rightarrow P')$ were observed in the ocean layer above 550 m. The annual average integrated rate of generation of the eddy energy in the 0–1600 m layer produced by the fluxes $(P \rightarrow P')$ and $(K \rightarrow K')$ turned out to be 0.72 erg/cm^2·s. Under the assumptions that this generation process was absent at depths below 1600 m and that the value of the annual average divergence of the horizontal fluxes of eddy energy in the POLYMODE region was equal to zero, the above result corresponds to the annual average value of the rate of dissipation of eddy energy averaged over the entire ocean depth equal to 0.13×10^{-5} erg/cm^3·s.

On the whole, Figures 5.2.13 and 5.2.15 allow us to assume that the processes of energy variation of the LSC and eddies in the POLYMODE region are cyclic (probably with a year-long period); for example, the curves in these figures behave similarly during days 0–50 and 350–400. It is likely that after a partial or total barotropization of currents has been completed there occurs reconstruction of the baroclinic LSC owing to the spatial inhomogeneity of the wind field above the ocean and the related vertical motions in the upper ocean, after which the whole cycle is repeated.

As is seen from Figure 5.2.15, the results of calculating the quantity $(\bar{P} \to P')$ for the main pycnocline from data of CTD–XBT surveys (in these calculations the same LSC model was used as in the calculations from current measurement data) are in good agreement with the corresponding curves constructed from current measurement data. This agreement is also confirmed very convincingly by the curves in Figure 5.2.16 which however, in contrast with the results of the calculations from buoy data, indicate the existence of very strong fluxes $(\bar{P} \to P')$ from the LSC to the eddies in the upper ocean.

The data of the CTD–XBT surveys of the region of the POLYMODE buoy array carried out in the winter–spring period of 1978 were also used to estimate the

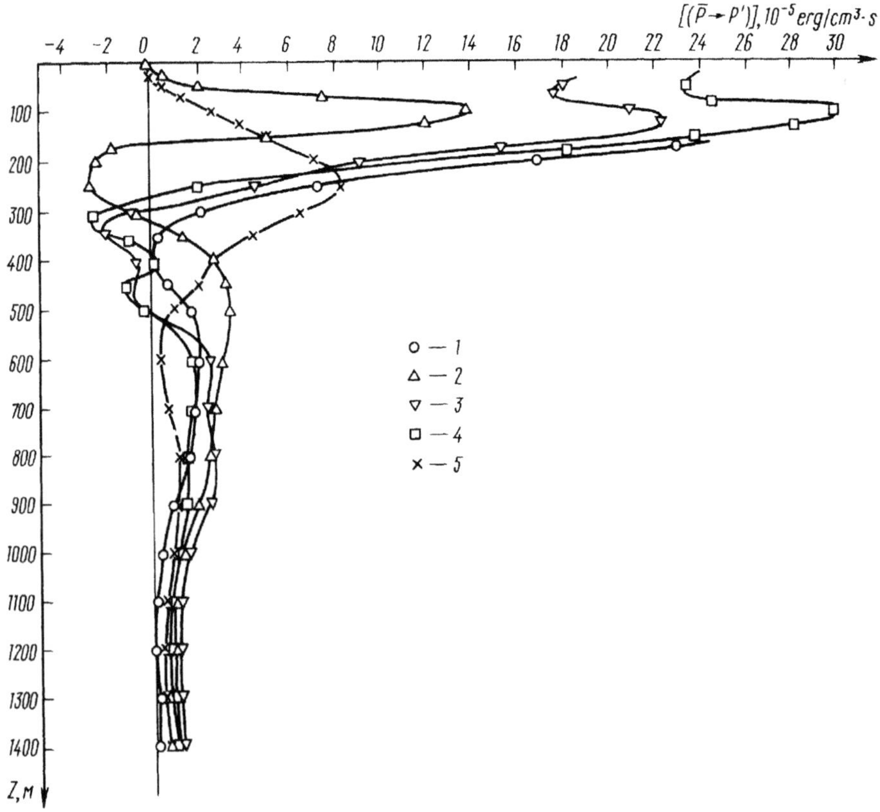

Fig. 5.2.16. Depth distributions of the density $(\bar{P} \to P')$ of the rate of transformation of the APE of the large-scale current into the APE of synoptic currents averaged over the area of the POLYMODE experiment region (this operation is denoted by square brackets) according to the data of five CTD-XBT surveys of the region of the POLYMODE buoy array on (1) February 17–March 5, (2) March 13–18, (3) April 20–26, (4) April 26–May 1, and (5) May 10–21, 1978 (after Grachev *et al.*, 1984).

directions and average values of heat fluxes produced by the synoptic disturbances of circulation (Figure 5.2.17). Figure 5.2.17 clearly shows that in both the upper ocean (layer 1 in Figure 5.2.17) and the main thermocline (layer 3), the integral effect of synoptic eddies tends to decrease the large-scale temperature gradient, which corresponds to the process of transformation of the APE of the LSC to the APE of the eddies (Figures 5.2.15 and 5.2.16). In accordance with the results of Bryden and Hall (1980), who estimated the average meridional transfer of heat by synoptic eddies in the subtropical Atlantic from the data of zonal transoceanic temperature sections, the average heat flux in the POLYMODE region during the period March–May 1978 integrated across the 0–1600 m layer was directed southward (Figure 5.2.17). The magnitude of this flux (3.5×10^5 W per horizontal distance of 1 cm in the tranverse direction) was approximately five times as small as the one obtained by Bryden and Hall.

Besides the above-mentioned Rossby-scale eddies consisting of water of local origin, a strong cyclonic eddy with smaller dimensions (eddy C_5 in Table 5.2.1) was detected and thoroughly investigated during POLYMODE. It penetrated the ocean depth at least to the lower boundary of the main thermocline, differed sharply in *T, S* characteristics of its core from the surrounding Sargasso Sea water, and moved across the southern part of the POLYMODE region in December 1977–February 1978 (Figure 5.2.18). In January 1978 the eddy travelled to the north with a velocity of 4–5 cm/day; at the beginning of February its direction changed to north-west, and its velocity of motion began to increase sharply, attaining 15 km/day at the end of February (Ivanov and Paramonov, 1980). Simultaneously, the eddy became more concentrated, which manifested itself in the contraction of its horizontal dimension from 130–150 to 80–100 km. During this period the isothermal surface rise in the main pycnocline, which was of the order of 200 m, remained almost unchanged. As a result, the orbital velocities in the eddy field began to increase, and this was registered by direct current measurements which showed that the velocity of the rotational motion of water in the eddy field at depths of 200–500 m increased from 50–60 to 80–90 cm/s during February (Zhidkov and Dzhiganshin, 1978). The eddy possessed a remarkable property: the deflections of the main and seasonal thermoclines had opposite direction, and therefore there was almost no '18-degree' Sargasso Sea water layer in the eddy. The *T, S* and *T*, O_2 characteristics of the water in the upper part of the eddy (approximately above 600 m) made it possible to assume that the eddy had been transported to the observation region from the south-east part of the North Atlantic subtropical gyre (Ivanov and Paramonov, 1980). The very intensive character of the cyclonic eddy under consideration is also demonstrated by Figure 5.2.19, where the maxima of synoptic disturbances of the water temperature in December 1977–January 1978 relate to this particular eddy. We also note the weakening of temperature disturbances in August–September 1978 (Figure 5.2.19) owing to the above-mentioned barotropization of the current field in the POLYMODE region during that period.

The phenomenon of the opposite deflections of the main and seasonal thermoclines was characteristic not only of the above-mentioned single cyclonic eddy but also of a number of other cyclonic and anticyclonic eddies observed during POLYMODE (Fedorov *et al.*, 1979b; Bulgakov and Paranichev, 1980; Bubnov and

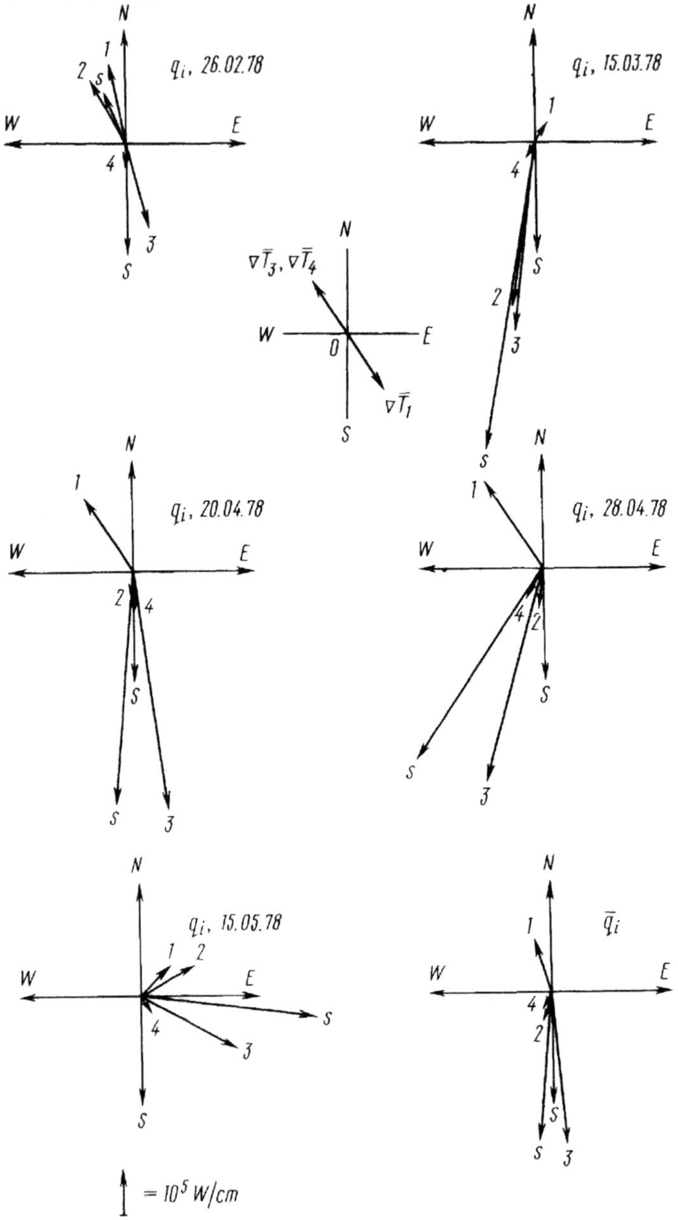

Fig. 5.2.17. Horizontal heat flux density vectors q_i (= 1, 2, 3, 4, s) integrated along the vertical within the limits of (1) 0–150 m, (2) 150–550 m, (3) 550–1100 m, and (4) 1100–1600 m layers and (s) over the entire 0–1600 m layer and averaged over the area of the region of the POLYMODE buoy array according to the data of five temperature–density surveys of the experiment region in February–May 1978 and the averaged values of the heat flux over these surveys. Also shown in the figure are the directions of the gradients $\nabla \bar{T}$ of the large-scale temperature distributions in layers 1, 3, and 4 in the POLYMODE region in accordance with the given model of the large-scale current (Fig. 5.2.13).

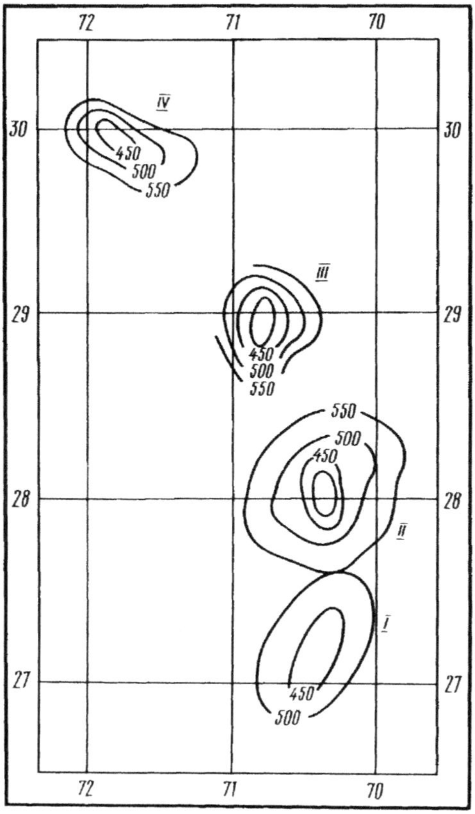

Fig. 5.2.18. Successive positions of a single cyclonic eddy in the POLYMODE region on (I) December 27–30, 1977, (II) January 17–19, 1978, (III) February 9–10, 1978, and (IV) February 19–21, 1978 (after Ivanov and Paramonov, 1980). The curves represent the depth isolines (in meters) of the 15 °C isothermal surface.

Fedorov, 1980). For example, Figure 5.2.20 demonstrates the motion of an anticyclone of this type where the main thermocline had a downward deflection of 60–70 m and the seasonal thermocline had an upward deflection of 30–40 m. The highest orbital geostrophic velocity was attained at a depth of 200 m and was equal to 25 cm/s. The eddy moved to the west-south-west with a velocity of about 7 km/day.

The curves in Figure 5.2.21 were calculated by the well-known method of expanding empirical distributions with respect to the system of vertical empirical orthogonal functions (Obukhov, 1960). As is clearly seen from Figure 5.2.21, the first and second modes of temperature oscillations corresponded, respectively, to the same signs and to the opposite signs of deflections of the main and seasonal thermoclines. As was shown by calculations, the sum of the first two temperature-oscillation modes presented in Figure 5.2.21 contained 80–90% of the total energy of these oscillations (it is proportional to the square of the temperature disturbance) in the POLYMODE region. The first mode was dominant in July–October 1977; the second mode was dominant in November 1977–April 1978; and the two

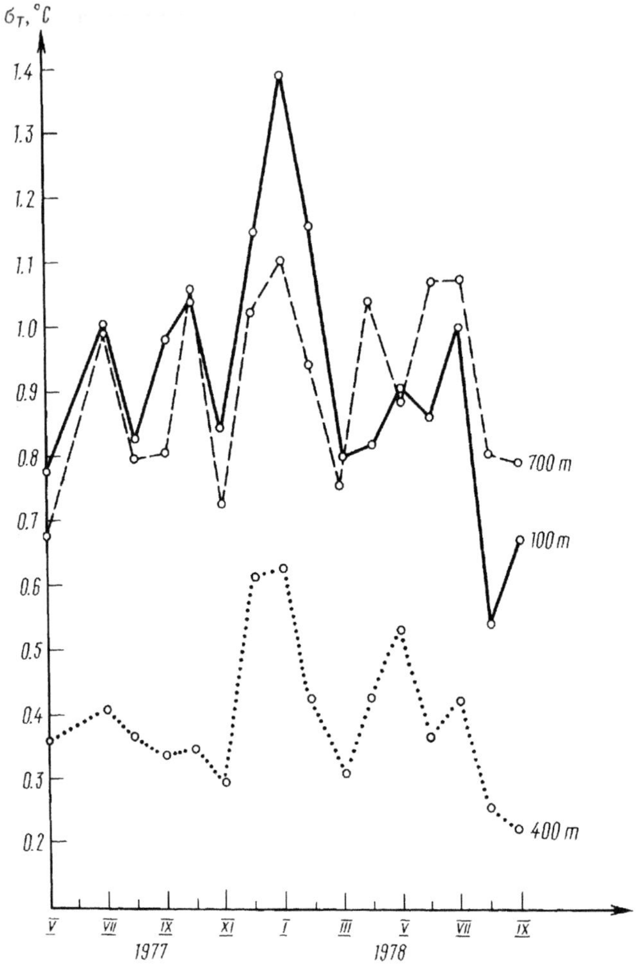

Fig. 5.2.19. Root-mean-square values of synoptic disturbances of the water temperature at depths 100, 400, and 700 m in the region of the POLYMODE Synoptic Experiment for different months of 1977 and 1978 according to the data of XBT surveys of the region (after Nikitin, 1985).

modes were marked approximately to the same degree beginning with May 1978 (Korotaev and Kosnyrev, 1980).

Rather sharp deviations of the observed *T, S* relations from the standard *T, S* relation for the western part of the Sargasso Sea were detected in the POLYMODE region in summer 1978 in the axial parts of two more eddies, one of which was cyclonic (Figure 5.2.22) and the other anticyclonic (Fedorov *et al.*, 1980). However, in contrast to the above-mentioned single cyclone in December 1977–January 1978, these deviations were observed not in the upper ocean but in the lower part of the main thermocline corresponding to the water temperature 8–10 °C (Figures 5.2.22 and 5.2.2). In both the cases of summer 1978 the sign of the isothermal salinity anomalies was positive (Figure 5.2.22). This means, as was noted by the authors (Fedorov *et al.*, 1980), that these anomalies could be formed

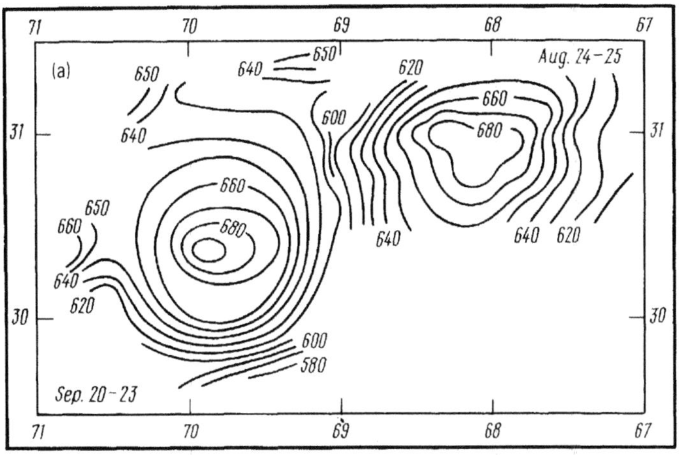

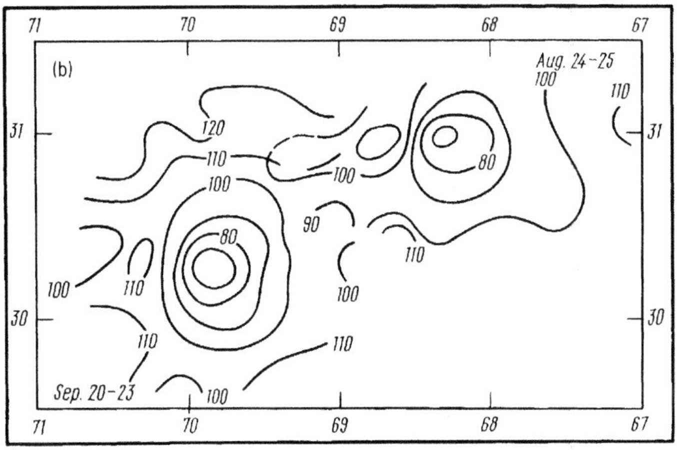

Fig. 5.2.20. Depth (in meters) of (a) 15 °C and (b) 20 °C isothermal surfaces in the northern part of the region of the POLYMODE Synoptic Experiment on August 24–25 (the right parts of the charts) and September 20–23, 1978 (the left parts of the charts) (after Bubnov and Fedorov, 1980).

owing to both the vertical mixing of waters in the corresponding ocean layer and the horizontal advection transporting water of some other origin to the POLYMODE region from another ocean region. Both positive and negative isothermal salinity anomalies were also found in the POLYMODE region in summer 1978 in upper parts of several cyclonic and anticyclonic eddies (Fedorov *et al.*, 1980).

The advective character of motion of particularly strong eddies in the POLYMODE region is also confirmed indirectly by the properties of temperature fronts in the upper vertical-homogeneous ocean layer and in the seasonal thermocline. These fronts were observed during the winter-spring period of POLYMODE (Koshlyakov *et al.*, 1980a). As a rule, the fronts were located on the periphery of strong eddies (Figure 5.2.23) and moved together with them. The front shown in Figure 5.2.23(a) was characterized by a width of 10–15 km and a temperature drop of 1.5–2 °C at a depth of 100 m and above. As was shown by consecutive XBT

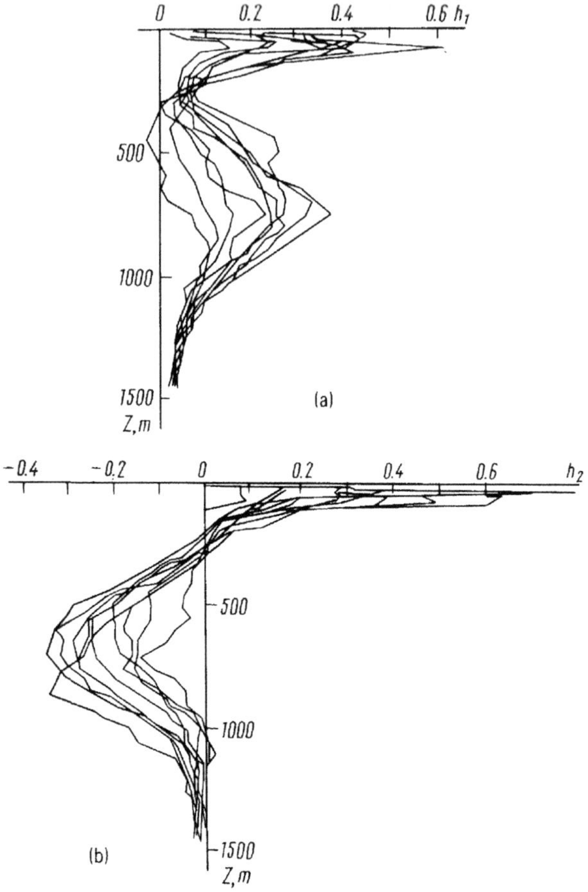

Fig. 5.2.21. Vertical distributions of the relative amplitudes of (a) the first and (b) the second empirical modes of synoptic disturbances of the temperature field according to the data of nine density surveys of the POLYMODE region from August 1977 to June 1978 (after Korotaev and Kosnyrev, 1980).

surveys, the front moved to the west together with an intensive anticyclonic eddy adjoining it with a velocity of 3–3.5 km/day. Besides long-period velocity oscillations with periods of 50–100 days, the current meters placed at a depth of 100 m at POLYMODE buoy stations registered clearly marked oscillations with typical periods of 4–12 days during the winter–spring period of 1978 which were obviously related to the passage of thermal fronts through the observation points (Koshlyakov *et al.*, 1980a). The velocity vectors of these 'frontal' currents obtained with the aid of high-frequency filtration of synoptic velocity series are shown in Figure 5.2.23. In some cases one can see that the current has a tendency to be oriented along the front so that the warm water remains on the right of the current. The observations also indicated an increase of the amplitude of internal gravitational waves and particularly strong development of the fine structure of the ocean on the periphery of strong cyclonic and anticyclonic eddies in the POLYMODE region (Kushnir, 1980; Aretinskiĭ *et al.*, 1980).

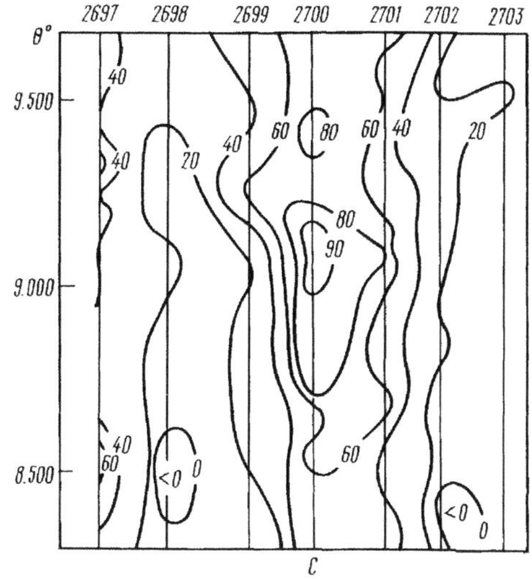

Fig. 5.2.22. Deviation of salinity (in 10^{-3}‰) from its values corresponding to the standard *T*, *S* relation for the POLYMODE region as a function of the horizontal distance and potential temperature (°C) according to the data of a CTD section from the north-west (the left edge of the figure) to the south-east across the central part of a cyclonic eddy in the region 28°N, 70°30′W on August 21–22, 1978 (after Fedorov *et al.*, 1980). The vertical lines and the figures at the top indicate the positions and numbers of the CTD stations. The station corresponding to the center of the eddy is shown by the line marked C.

The American Local Dynamics Experiment (LDE) of POLYMODE was carried out in May 1978–July 1979 in a region with center at 31°N, 70°W and a transverse dimension of about 200 km (symbol 17 in Figure 4.1.1). The experiment included the following basic observations: long-term current and temperature measurements at a rhombic system of nine ABS, the observations at the central ABS and at the other eight ABS covering the whole ocean depth and the central part of the main thermocline between 600 and 850 m, respectively; an extensive series of current and temperature measurements by SOFAR floats at depths of 700 and 1300 m; a series of CTD and XBT surveys of the region performed in May–July 1978; current measurements by free-falling velocity profilometers. The current measurements at ABS (Owens *et al.*, 1982) revealed a mean (large-scale) current directed to the south-west in the upper ocean (above 1800 m) and attaining 3–4 cm/s at depths 200–700 m. At greater depths the mean current was directed to the east-north-east and had a velocity of about 3 cm/s at depths of 5200–5300 m. As in the region of the POLYMODE Synoptic Experiment, the structure of synoptic-scale currents in the LDE region was essentially nonstationary, which manifested itself particularly clearly in the passage of two strong and narrow current jets across the observation region, one of them flowing in the south-west direction during June 1978 and the other flowing in the direction to the west-south-west from April 20 to May 20, 1979. The two jets penetrated to the whole ocean depth, and the current velocity in their field attained 40 and even 50 cm/s (in the second jet) in the upper ocean and decreased down to 5–10 cm/s at depths of 4–5 km. The two jets were displaced in directions transverse to them (north-west and north-north-west directions, re-

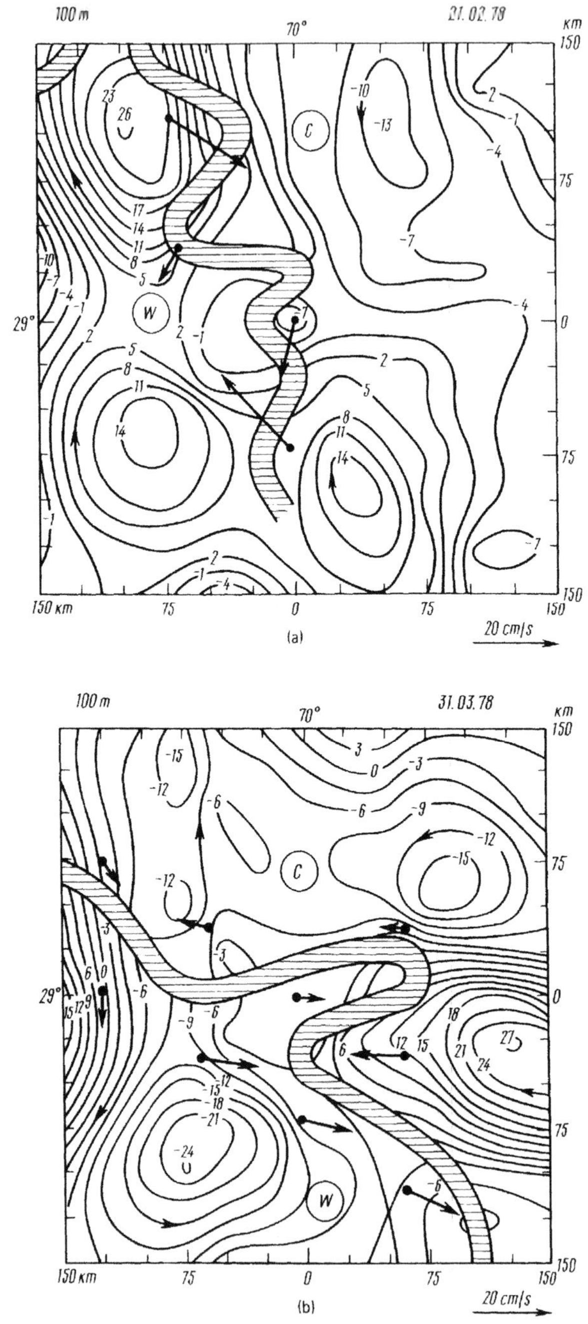

Fig. 5.2.23. Streamlines of synoptic currents and positions of thermal fronts at a depth of 100 m on (a) February 21 and (b) March 31, 1978 in the region of the POLYMODE Synoptic Experiment (after Koshlyakov *et al.*, 1980a). The stream function is measured in 10^7 cm^2/s. The arrows represent velocity vectors of the frontal currents at the same depth. The letters W and C mark the warm and cold waters.

spectively) with a velocity of 4 cm/s and their width was about 100 km. As far as it can be judged from the data under discussion, during the rest of the observation period the dominant role in the main thermocline was played by Rossby-scale baroclinic eddy formations.

Joint analysis of the data of current and temperature measurements at the LDE buoy system indicated reliably the presence of transformation of the APE of the large-scale current into the APE of synoptic eddies (Bryden, 1982). The average rate of this transformation in the 600–850 m layer turned out to be 3.3×10^{-5} erg/cm^3·s. The average synoptic heat flux in this layer was directed to the south-south-east and its density was about 3 cal/cm^2·s. Some special estimates (McWilliams, 1983) also revealed the presence of convergence of this flux in the LDE region which is required to compensate for the 'advection of cold' related to the average ascending water motion in the main thermocline.

Within the framework of the LDE during the period from May 1978 to October 1979 altogether 20 SOFAR floats were launched at each of the levels 700 and 1300 m (Spain *et al.*, 1980). This was accompanied by current measurements with five SOFAR floats (three floats at 700 m and two floats at 2000 m) with duration from 417 to 734 days which were performed in 1977–1979 in the region 26–31°N, 67–75°W (Riser and Rossby, 1983). At deep levels (1300 and 2000 m) the dominant part was played by reciprocating motions of the floats which indicated the presence of quasibarotropic planetary-topographic waves (Figure 5.2.24). Namely, as was shown by calculations (Price and Rossby, 1982), in the field of motion of a group of floats (demonstrated by Figure 5.2.24) the condition of conservation of the potential vorticity $d[(\zeta + f)/H]/dt = 0$ (where $\zeta = (\partial v/\partial x) - (\partial u/\partial y)$ is the relative vorticity, f is the planetary vorticity, and H is the ocean depth) held with a high accuracy. In the calculations the parameters f and H were estimated from the motion of the center of the group of floats and the values of ζ for the group as a whole at different instants of time were estimated taking into account the variations of direction and velocity of motion of each of the floats. The float motion shown in Figure 5.2.24, corrected for the Doppler shift due to the presence of the westward mean current with a velocity of 2 cm/s, corresponded to a barotropic planetary-topographic wave propagating in the direction with azimuth 300° and having a period of 61 days, a wavelength of 340 km, the root-mean-square amplitude equal to 12 cm/s, and an eastward group velocity of 5 cm/s (Price and Rossby, 1982).

The behavior of float trajectories at a depth of 700 m was mainly chaotic although, on some of their segments, reciprocal motions were seen. On the whole, this indicates a complex structure of water motion in the main thermocline, which probably is a consequence of superposition of actively interacting turbulent eddies and Rossby waves. The loops of the trajectory of one of the floats which drifted at a depth of 700 m covered a vast region from 22 to 29°N and from 68 to 75°W in 650 days, and at the end of the drift this float turned out to be only at a distance of 200 km from its launching site (25°N, 71°W) (Riser and Rossby, 1983). The values of the coefficients of horizontal large-scale turbulent diffusion in the zonal and meridional directions estimated from the above-mentioned float measurements at 700 m were 7×10^7 and 3×10^7 cm^2/s, respectively (Riser and Rossby, 1983).

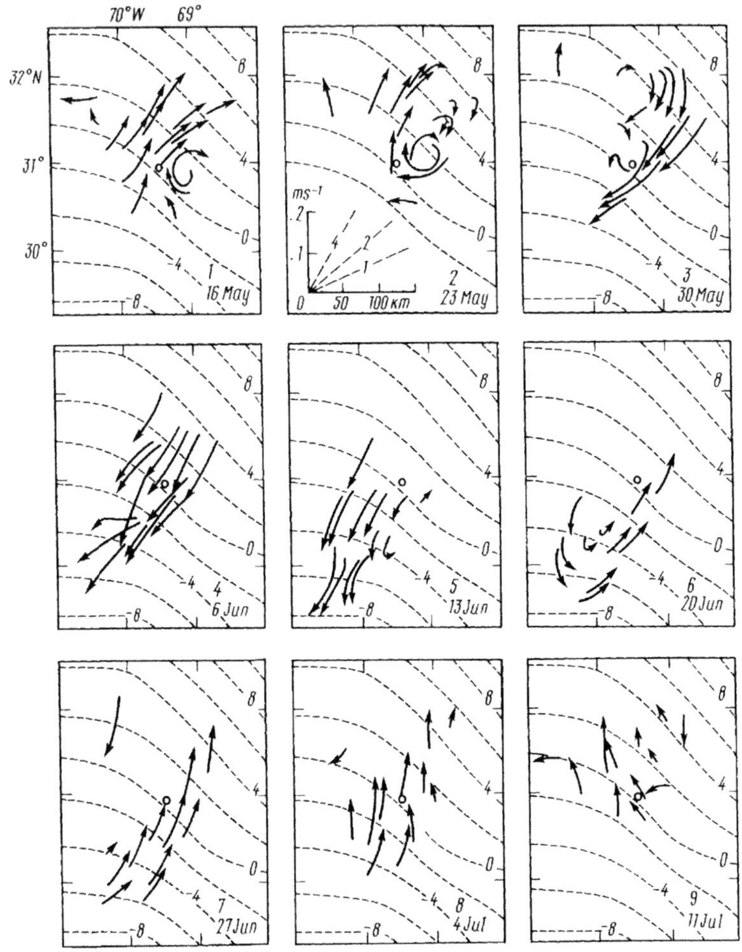

Fig. 5.2.24. Week-long drift trajectories of 17 SOFAR floats at a depth of 1300 m in the region of the Local Dynamics Experiment (LDE) of POLYMODE in May–July 1978 (after Price and Rossby, 1982). The dates indicate the beginnings of the week-long cycles; the circle at the center of the region marks the position of the central ABS of the LDE. The dashed curves and their figures indicate the isolines of the planetary-topographic vorticity and its values in the units $[(f/H)/(f_0/H_0) - 1] \times 10^2$ (f is the Coriolis parameter and H is the ocean depth; f_0 and H_0 are their values at the center of the region). Presented in the pattern for May 23 are the distance and velocity scales (relative to the week-long path of a float) and also the isolines of the relative vorticity in percentages of f_0.

The cross-sections of space–time spectra in Figure 5.2.25 are presented for the values of the frequency (0.01 and 0.02 c/day) corresponding to the maxima of the energy time-spectra calculated from data of LDE float measurements (Konyaev and Merinova, 1983). The maxima of energy spectra for a depth of 1300 m shown in Figure 5.2.25 correspond to a wave with a wavelength of about 300 km propagating along the azimuth about 300°, and they lie rather accurately within the dispersion corridor for a barotropic Rossby wave, which is in good agreement with the above-mentioned results of the analysis of float data for 1300 m (Figure 5.2.24).

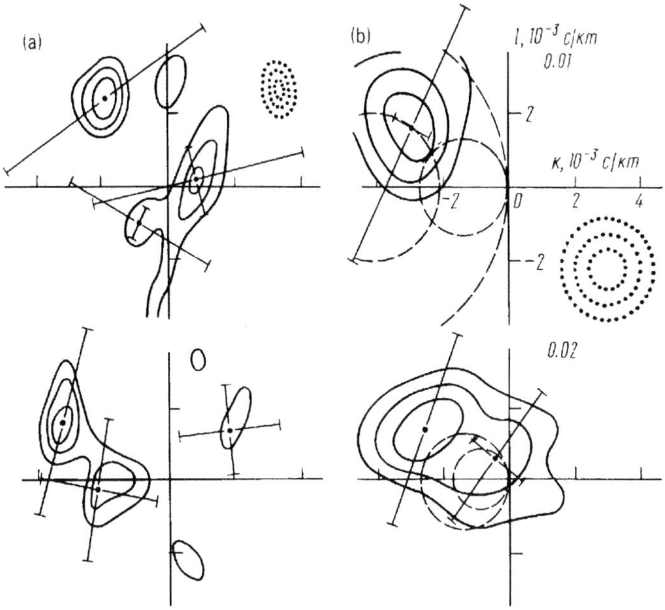

Fig. 5.2.25. Normalized cross-sections of the space–time spectra of the kinetic energy of synoptic currents at depths (a) 700 and (b) 1300 m corresponding to periods of 100 (at the top) and 50 days (at the bottom) according to the data of float current measurements in May–November 1978 during the Local Dynamics Experiment of POLYMODE (after Konyaev and Merinova, 1983). The isolines of the spectral density are drawn for the levels corresponding to 40, 60, and 80% of the main maxima of each of the spectra. The straight lines are the axes of the ellipses of the root-mean-square velocity for the maxima. The dotted curves represent the spectral windows corresponding to measurements at 700 and 1300 m. The dashed circles are the dispersion corridors for the barotropic and first-mode baroclinic Rossby waves.

The spectra for a depth of 700 m (Figure 5.2.25) can be interpreted as a result of the passage of elliptical-shaped eddies with major axes along the NE–SW direction which propagated across the experiment area to the north-west with a velocity of about 5 km/day (Konyaev and Merinova, 1983).

The drift of one of the floats launched at a depth of 700 m during the LDE (float 52; Figure 5.2.26) registered a small single anticyclonic eddy which travelled to the south-west with an average velocity of 15 km/day in June 1978. As was shown by a hydrographic survey performed at the end of July 1978, the eddy had a diameter of about 40 km, was entirely concentrated in the 500–1000 m layer, was characterized by the velocity of rotational water motion of the order of 15 cm/s, and had a well mixed core in the 600–850 m layer. The core had a very low oxygen content, which indicated that the eddy was formed in the eastern part of the tropical zone of the North Atlantic. This implies that the eddy was several years old and indicates that the process of eddy water mixing with the surrounding ocean water was very slow.

Two other clearly marked small single eddies were registered by density surveys of the LDE region and by current profilometers. One of the eddies which had a transverse dimension of 50–70 km (Figure 5.2.27) was remarkable. As was shown by current soundings (Sanford, 1978), it was completely concentrated in the

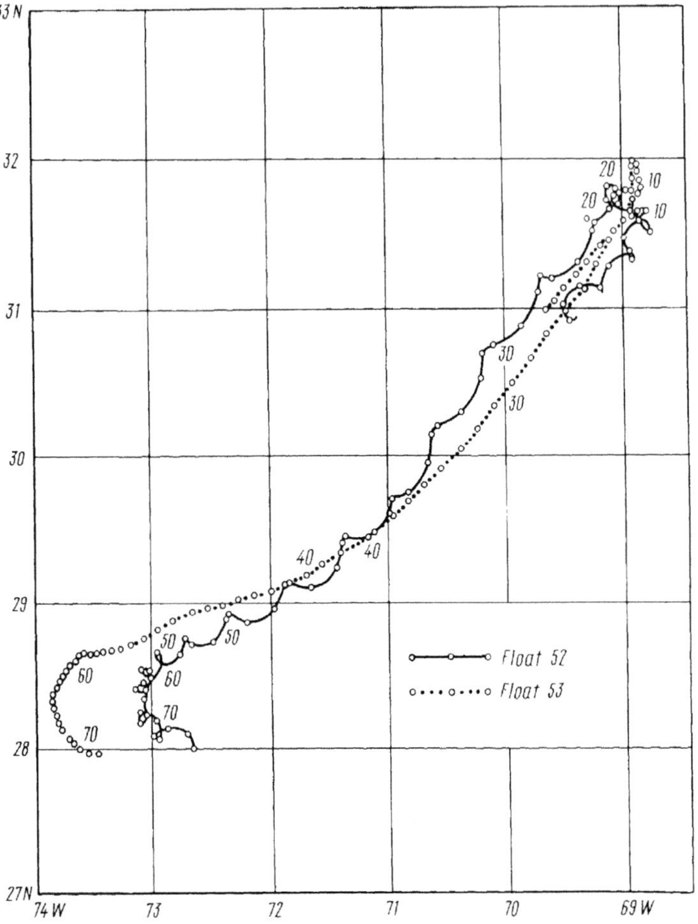

Fig. 5.2.26. Drift trajectories of two SOFAR floats at a depth of 700 m in the Sargasso Sea in May–July 1978 (after Rossby *et al.*, 1982). The circles and adjacent figures indicate the positions of the floats on each day and the numbers of the days; day 10 corresponds to May 21, 1978.

1000–2500 m layer, which is also confirmed by Figure 5.2.27, where no traces of the eddy are seen in the main thermocline. The velocity of the anticyclonic rotation of water in the field of this eddy attained its maximum at a depth of 1600 m and was equal to 15–20 cm/s. The eddy was transferred to the west-south-west with a velocity of 10–15 cm/s. As was shown by current soundings at surrounding points, this transfer had an advective character and was produced by a jet of a synoptic current with a larger scale whose manifestation in the temperature field is, for instance, seen in Figure 5.2.27(g). The T, S and T, O_2 relations for the eddy core made it possible to assume that the eddy was formed by the water transported to the observation region from a region north of the Gulf Stream (Taft *et al.*, 1978). Another single eddy with a transverse dimension of 70–90 km was observed in the LDE region at the end of June to the beginning of July 1978 (Ebbesmeyer *et al.*, 1978). However, in contrast to the above-mentioned eddies, the eddy under

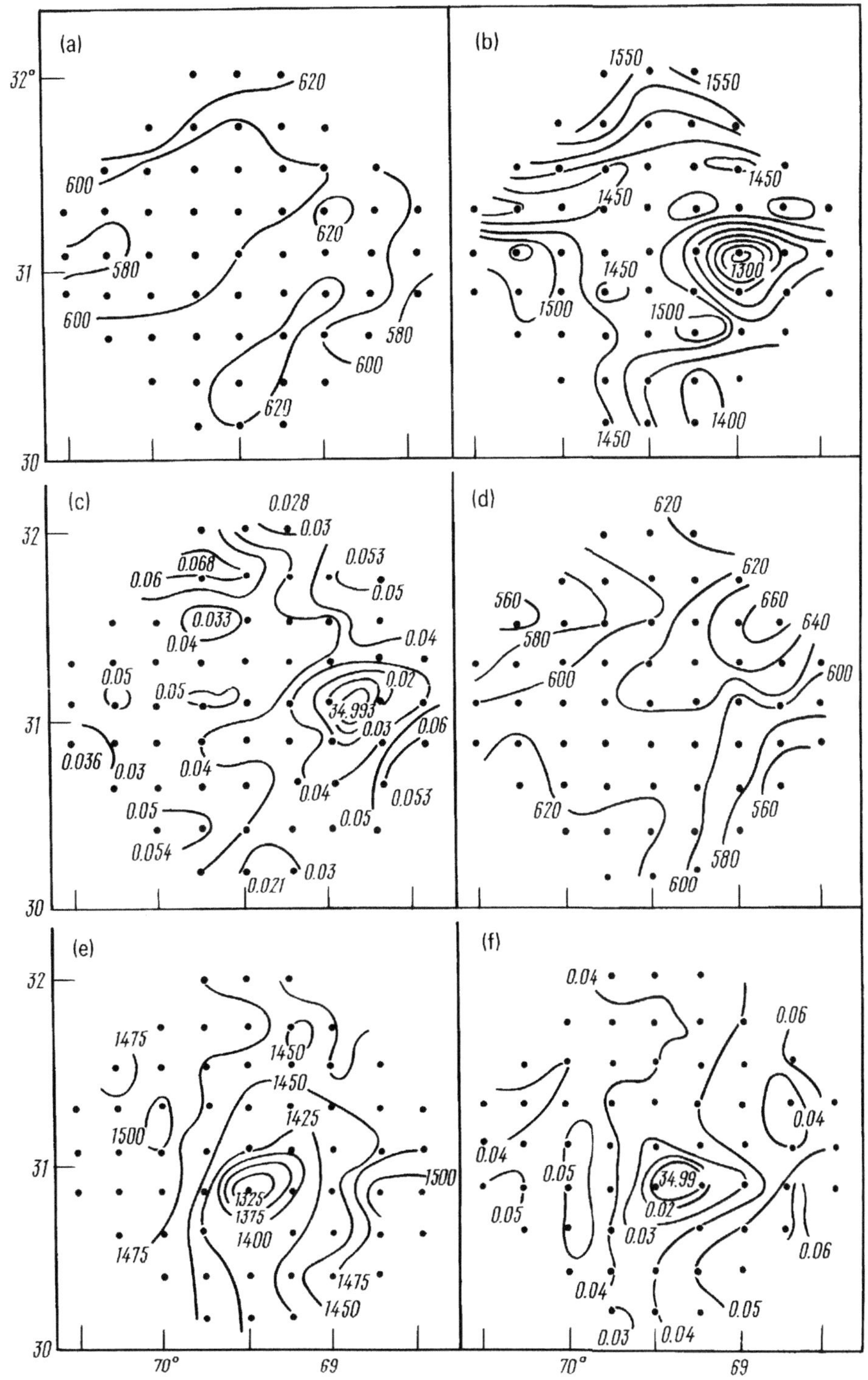
(a)
(b)
(c)
(d)
(e)
(f)
32°
31
30
70°
69
620
600
580
1550
1450
1500
1300
1400
0.028
0.03
0.053
0.068
0.06
0.05
0.033
0.04
0.02
34.993
0.06
0.036
0.054
0.021
660
640
560
1475
1425
1325
1375
34.99

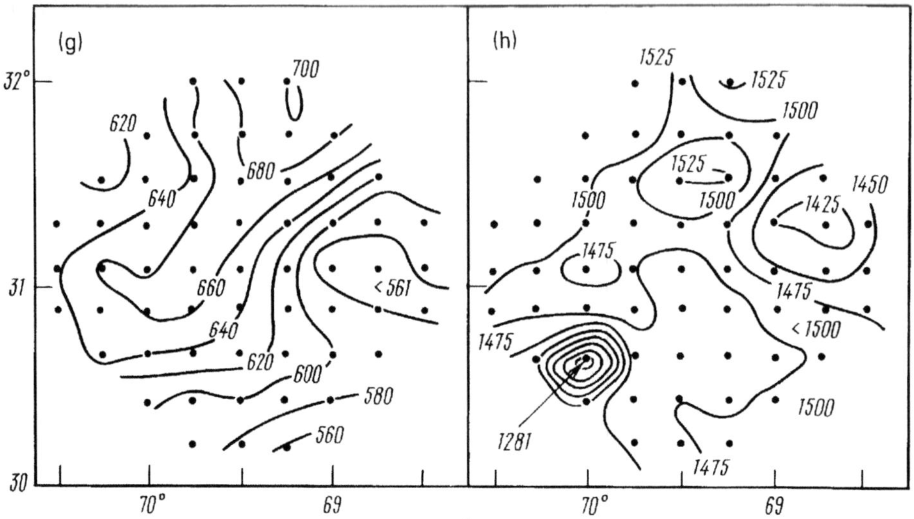

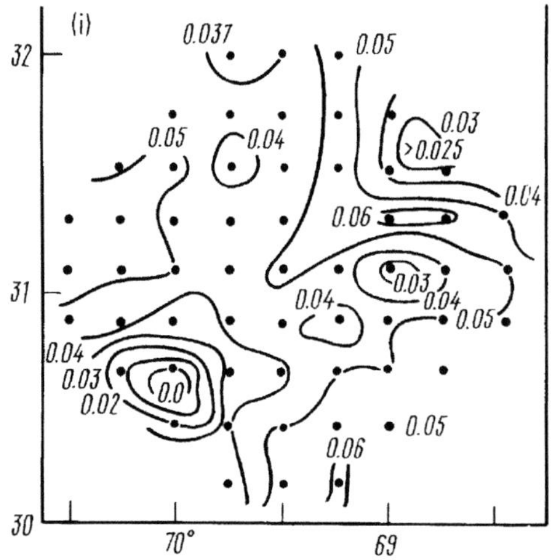

Fig. 5.2.27. Topography (in meters) of the (a, d, g) 15 °C and (b, e, h) 4.5 °C isothermal surfaces and (c, f, i) salinity distribution (the excess of salinity over 35 000 ‰ in ‰) on the isopycnic surface $\sigma_t = 27.765$ (its average depth is equal to 1325 m) in the region of the LDE of POLYMODE according to the density surveys on (a, b, c) May 20–25, (d, e, f) May 29–June 3, and (g, h, i) June 6–11, 1978 (after Taft *et al.*, 1978). The points indicate the sites of CTD soundings of the ocean.

discussion was concentrated in and above the main thermocline, and the water rotation in its field, whose velocity attained 40 cm/s at a depth of 600 m, had a cyclonic character. The eddy moved to the south-west with a velocity of about 18 cm/s and was located at the south-east periphery of a sharply marked south-west current jet with width of about 70 km; the velocity in the axis of the jet attained 60 cm/s at depth 600 m and the jet was shifted to the north-west with a velocity of about 3 km/day.

3. EDDIES AT LOW LATITUDES

The results of the experiments 'Polygon-70' and 'MODE' gave impetus to the rapid development of special experimental studies of synoptic ocean eddies by oceanographers in many countries. At the same time some data from previous observations were revised from the new viewpoint. As a result, at the present time (at the middle of 1983) a great number of very interesting results relating to the distribution and properties of eddies in various ocean regions have been accumulated. Here and in the following section we briefly discuss the most important of these results. First we consider eddies at low latitudes and then pass to eddies at high latitudes. We begin the review of the results concerning eddies at low latitudes with the Pacific Ocean.

Among all the observed eddies the nearest to the equator was an anticyclonic eddy whose traces were found in April 1958 in the western part of the Equatorial Counter Current of the Pacific Ocean in the region 3–5°N, 142–145°E (Burkov and Ovchinnikov, 1960). According to the data of a hydrographic survey, and surface current measurements by an geoelectromagnetic kinetograph, the total meridional dimension of the eddy was about 160 km, and the current velocity in the eddy field attained 75 cm/s at the sea surface and 50 cm/s(!) at a depth of 500 m. A repeated hydrographic survey made it possible to assume that the eddy which was located in the southern (the strongest) part of the Equatorial Counter Current moved eastward with a velocity of about 16 cm/s.

Very remarkable are the results of satellite observations of synoptic eddies in a part of the Pacific Ocean adjoining the southern coast of Central America (Stumpf and Legeckis, 1977). In this region, all the year round and, particularly frequently, during the period November to April there appears periodically (including 10–15 times in winter) a synoptic situation characterized by strong north and north-east winds blowing from the Gulf of Mexico and the Caribbean Sea. Owing to specific orographic properties of Central America, these winds are particularly strong in the regions of the Gulfs of Tehuantepec and Papagayo (Figure 5.3.1). This situation leads to the following consequences: (a) the departure of the Costa Rica Current from the continent and its replacement by a weak eastern coastal current flowing from the southern coast of Mexico, which results in the formation of an anticyclonic vorticity of currents in this ocean region (Figure 5.3.1); and (b) the upwelling of cold subsurface waters and intensive mixing of the upper ocean. As is shown by satellite temperature surveys of the sea surface (Stumpf and Legeckis, 1977), by the time the north wind weakens, the above-mentioned processes result in the appearance of one or two anticyclonic eddies in this ocean region; their diameters are 250–300 km and their central parts contain low-temperature water in the surface layer (which indicates that these eddies are characterized by the opposite signs of the deflection of isothermal surfaces in the main thermocline and in the upper layer). The pattern shown schematically in Figure 5.3.1 was formed by the end of February 1976 as a result of several successive strengthenings of the north-east wind. The extreme western eddy in Figure 5.3.1 was formed in the Gulf of Tehuantepec at the beginning of February; the others were probably formed in the region of the Gulf of Papagayo. All the eddies had a tendency to move westward with a high velocity of 10–15 cm/s. The surface water temperature in the central part of the western eddy in Figure 5.3.1 was 1.5–2 °C lower than the temperature of the surrounding water; for the other eddies the temperature contrast was several tenths of a degree. On the whole, eddies in this ocean region

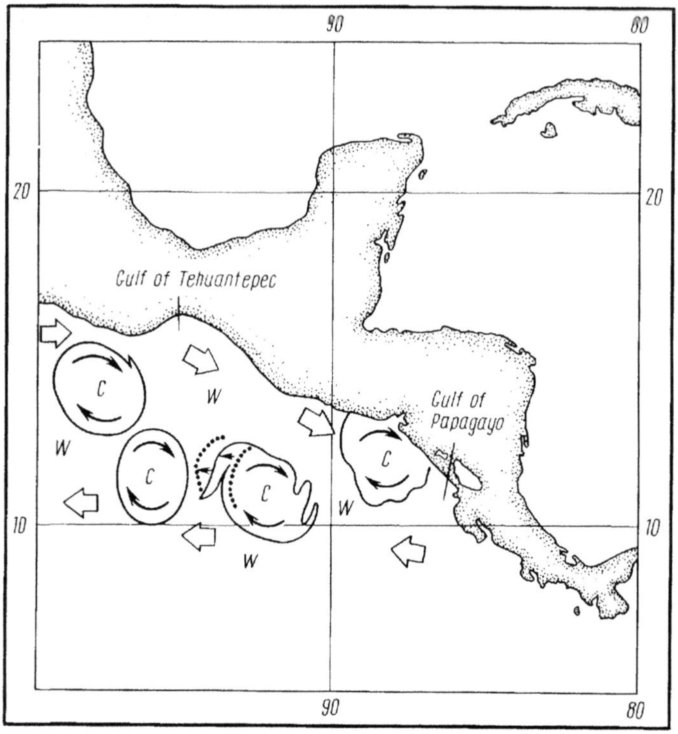

Fig. 5.3.1. Scheme of location of anticyclonic eddies and jets of large-scale currents in the part of the Pacific adjoining Central America at the end of February 1976 (after Stumpf and Legeckis, 1977). The letters C and W mark cold and warm surface waters.

are a peculiar example of quasisingle anticyclones formed under the direct action of atmospheric motion.

Every month, during the period from February 1964 to June 1965, American oceanographers carried out temperature measurements to a depth of 250 m in a quasizonal section about 1000 km long in a region at 25°N north of Hawaii (symbol 13 in Figure 4.2.1; Bernstein and White, 1977). The section was located on the northern periphery of the North Equatorial Current and the maximum velocity of the western geostrophic current, attained at a depth of 250 m, was 1.5 cm/s. The measurements registered clearly marked zonal nonstationary disturbances of the temperature field in the thermocline which were characterized by a predominant wavelength about 450 km, an average phase velocity of 4.5 cm/s directed towards the west, and an amplitude (isothermal surface depth difference) attaining 100 m. The relationship between the horizontal and temporal scales of these disturbances is in good agreement with the model of first-mode baroclinic Rossby waves. Of interest also is the result obtained by Seckel (1975), who showed that the amplitude of synoptic disturbances of density in the field of the North Equatorial Current in the Hawaii region increased when the velocity of the large-scale currents increased. This provides an indirect indication of the baroclinic instability of the North Equatorial Current as the most probable cause of the formation of synoptic disturbances in the ocean region under consideration.

Traces of baroclinic synoptic eddies in the central part (between 14 and 19°N) of the North Equatorial Current in the Pacific Ocean were found in 24 (of 51) aircraft and ship meridional XBT sections along 150, 153, and 158°W performed during the period from February 1979 to June 1980 in the equatorial–tropical part of the ocean (Wyrtki, 1982). Altogether, during the indicated period six individual quasisingle cyclonic eddies, and not a single anticyclonic eddy, were registered from the data on temperature field disturbances. The horizontal dimension of the eddies varied from 150 to 350 km and the isothermal surface height difference in the main thermocline in the eddy field varied from 25 to 65 m. The eddies drifted westward with a velocity around 10 km/day, which is in good agreement with the average velocity of the central part of the North Equatorial Current in the main thermocline. The disturbances of the temperature field extended to the whole thermocline and reached a depth of at least 800 m. The frequency of the appearance of traces of eddies in the sections, and the average velocity of their westward drift, were such that, on average, in the region bounded by the 10 and 20°N parallels and the 150 and 158°W meridians (whose area is about 900 000 km^2) one eddy was observed at a time; this corresponds to 7% of the entire territory occupied by the eddies.

A hydrographic survey, whose results are presented in Figure 5.3.2, was carried out in an ocean region where the mean (large-scale) current was almost indis-

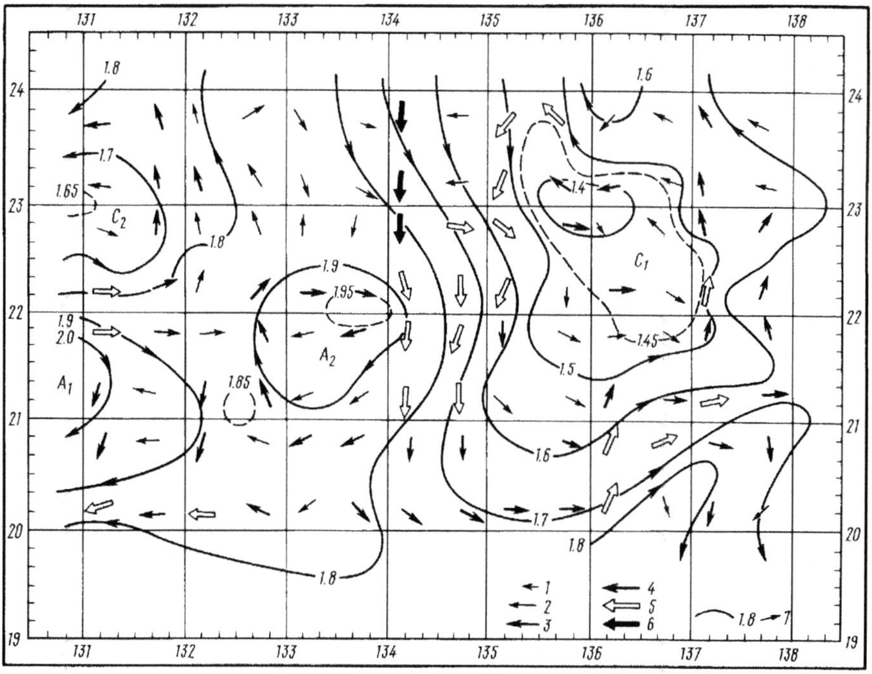

Fig. 5.3.2. Dynamic topography (in dynamic meters) of the sea surface relative to the 1000 dbar reference surface on the experiment area in the northern part of the Philippine Sea according to the data of a hydrographic survey of the area by five ships (Figure 4.2.6) on October 4–10, 1980 (after Pokudov and Tunegolovets, 1983). The arrows show velocity vectors of the geostrophic current. 1: 0–5 cm/s; 2: 5–10 cm/s; 3: 10–25 cm/s; 4: 25–50 cm/s; 5: 50–75 cm/s; 6: > 75 cm/s. C_1, A_1, etc., are different cyclonic and anticyclonic eddies. (See symbol 3 in Figure 4.2.1.)

tinguishable (symbol 3 in Figure 4.2.1). Therefore, still more impressive are the velocities of synoptic-scale geostrophic currents which attained 105 cm/s(!) at the sea surface in a region between two synoptic centers (Figure 5.3.2). The cold eddy C_1 (Figure 5.3.2), with a transverse dimension of 350–400 km, was rather rapidly damped with increasing depth and was indistinct in the temperature field even at a depth of 800 m; by contrast, the warm eddy A_1 was still clearly marked at a depth of 1000 m.

The California Current in the region between 23 and 40°N is one of the most thoroughly studied currents in the ocean. Beginning with 1950, several hydrographic surveys of the current have been performed every year; in recent years they have been accompanied by satellite temperature surveys of the sea surface (Bernstein *et al.*, 1977). Although the distances between the stations in Figure 5.3.3 were somewhat too large to allow an accurate description of synoptic disturbances of the current, the figure generally provides a correct indication of the character and variability of the California Current. Despite the fact that the current is somewhat diffuse and multijet, it is possible to separate out its main jet with a width of 100–150 km which is particularly clearly marked in the southern part of the region. The velocity in the axis of the jet attains 30–40 cm/s at the sea surface. The current is entirely concentrated in a layer above the depth of 500 m. The observations clearly show that the California Current is unstable and that meanders with a wavelength of 300–400 km (Figure 5.3.3) permanently develop in its field; in some cases they have a tendency to be transformed into cut-off eddies, and the duration of the process of complete development of a meander from the beginning of its formation to the separation of an eddy is usually several months. The development process of such a clearly marked cyclonic meander with an internal vortical motion is seen in the southern part of the region in Figure 5.3.3. Other parts of the charts in Figure 5.3.3 show several more comparatively weak and probably not very stable vortical formations mainly associated with the interior sides of the meanders. Satellite temperature surveys of the sea surface also registered the formation of spots of warm and cold water related to the development and separation of meanders of the California Current (Bernstein *et al.*, 1977). In some cases the dimensions of the spots along their major axes attained 400 km and the temperature drop was several degrees. An analysis of the data of observations of the California Current – in the light of the theory of the baroclinic instability of large-scale currents – clearly indicated the instability of the California Current, and the observed spatial and temporal scales of synoptic disturbances of the current turned out to be close to those predicted by theory (Bernstein *et al.*, 1977).

Burkov and Pavlova (1980) analyzed the data of the hydrographic surveys of the California Current performed during the period from 1949 to 1965 specially to separate out vortical formations. Altogether, at a depth of 200 m they separated out 26 strong cyclonic eddies (i.e. contoured by two or more closed dynamic horizontals in the chart of dynamic topography of the 200 dbar surface relative to the 500 dbar reference surface for the interval between the dynamic horizontals of 4 dynamic centimeters) and 30 strong anticyclonic eddies with transverse dimensions from 100 to 300 km. Approximately half of these eddies were located in the interior parts of meanders. Figure 5.3.4 indicates the existence of two regions of predomi-

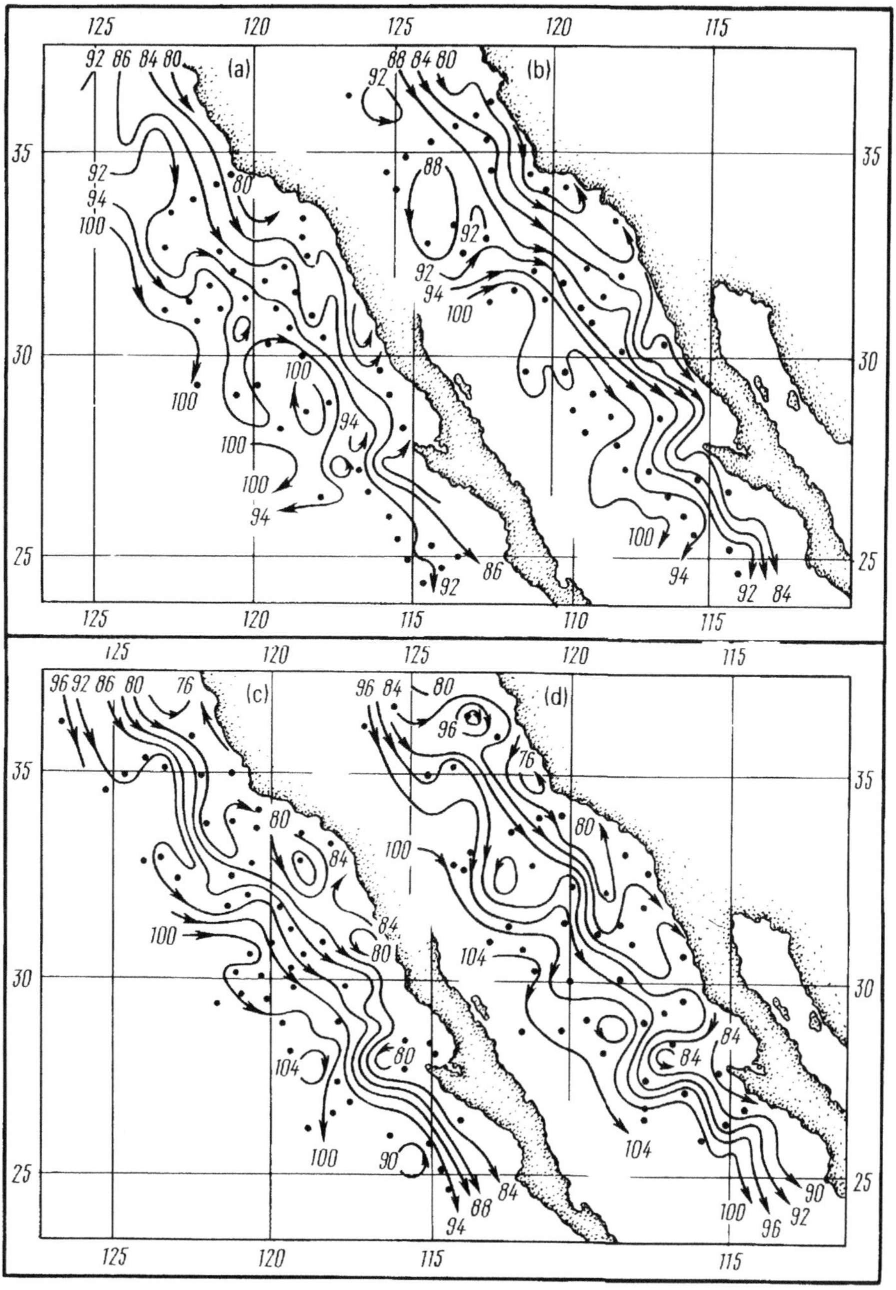

Fig. 5.3.3. Dynamic topography (in dynamic centimeters) of the sea surface relative to the 500 dbar reference surface on (a) April 2–15, (b) May 9–26, (c) June 3–21, and (d) July 2–23, 1952 in a region of the California Current (after Bernstein *et al.*, 1977). The points indicate the positions of the hydrographic stations.

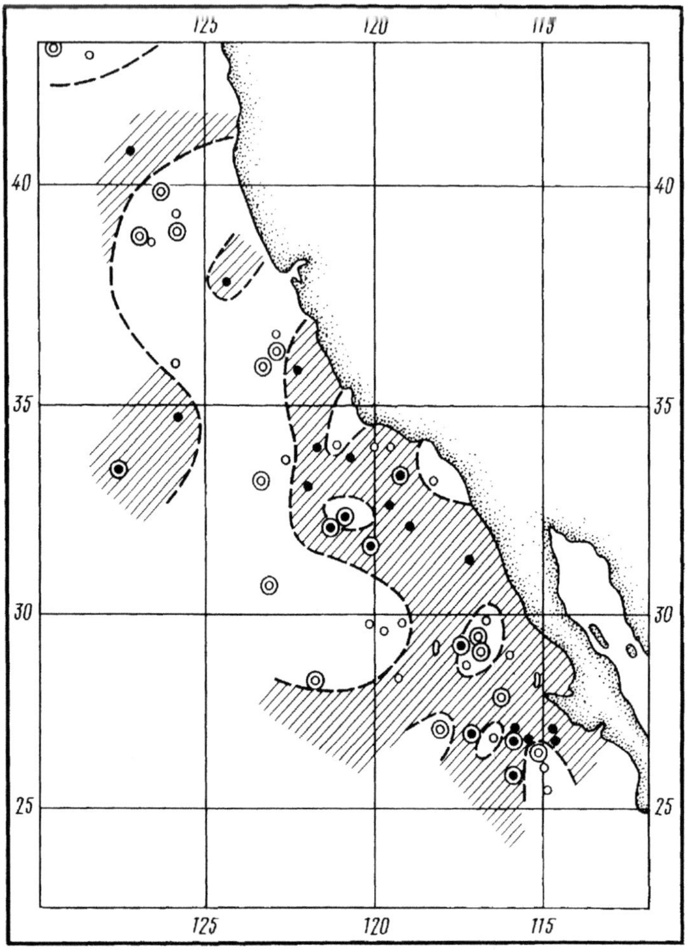

Fig. 5.3.4. Location of the centers of cyclonic and anticyclonic synoptic eddies (dark and light circles, respectively) at a depth of 200 m in the California Current during the period 1949–1965 (after Burkov and Pavlova, 1980). The double circles show the eddies located inside meanders. Shaded and nonshaded in the figure are the regions of predominant presence of cyclonic meanders and eddies and anticyclonic meanders and eddies, respectively.

nant formation of cyclonic eddies: (a) the region of a quasipermanent cyclonic meander of the California Current at 27–28°N (cf. Figure 5.3.3); (b) the region at 31–34°N where a northern geostrophic counter current (the Davidson Current) is formed at times in coastal waters (e.g. see Figure 5.3.3(c), (d)), which results in the appearance of a cyclonic situation in this ocean region. The anticyclonic eddies in the region from 27 to 40°N and from 122 to 128°W in Figure 5.3.4 are related to the predominant appearance of anticyclonic meanders of the current in this area, which can also be clearly seen in Figure 5.3.3. In the spring–summer period when, on the whole, the California Current is stronger than in autumn and winter the number and the intensity of the eddies generated in its field are somewhat greater (Burkov and Pavlova, 1980).

In September 1976–August 1977 measurements of surface oceanic currents were performed in the North-East Pacific (Figure 5.3.5) with the aid of 22 satellite free-drifting floats (Kirwan *et al.*, 1978). The floats were launched in the eastern part of the North Pacific Current and then, depending on their latitudinal location, were drawn into the initial parts of either the California Current or the Alaska Current. One of the main results of the measurements was that the kinetic energies of the mean (large-scale) and synoptic currents turned out to be of the same order of magnitude. For instance, for the floats drawn into the subtropical gyre the average velocity of the large-scale current was 20–25 cm/s and the root-mean-square velocity of synoptic fluctuations was about 10 cm/s. For the floats drawn into the subpolar gyre the corresponding quantities coincided and were equal to 10 cm/s. Clearly marked predominance of synoptic fluctuations over the mean current was observed only for the north-east subregion adjoining the coasts of Canada and

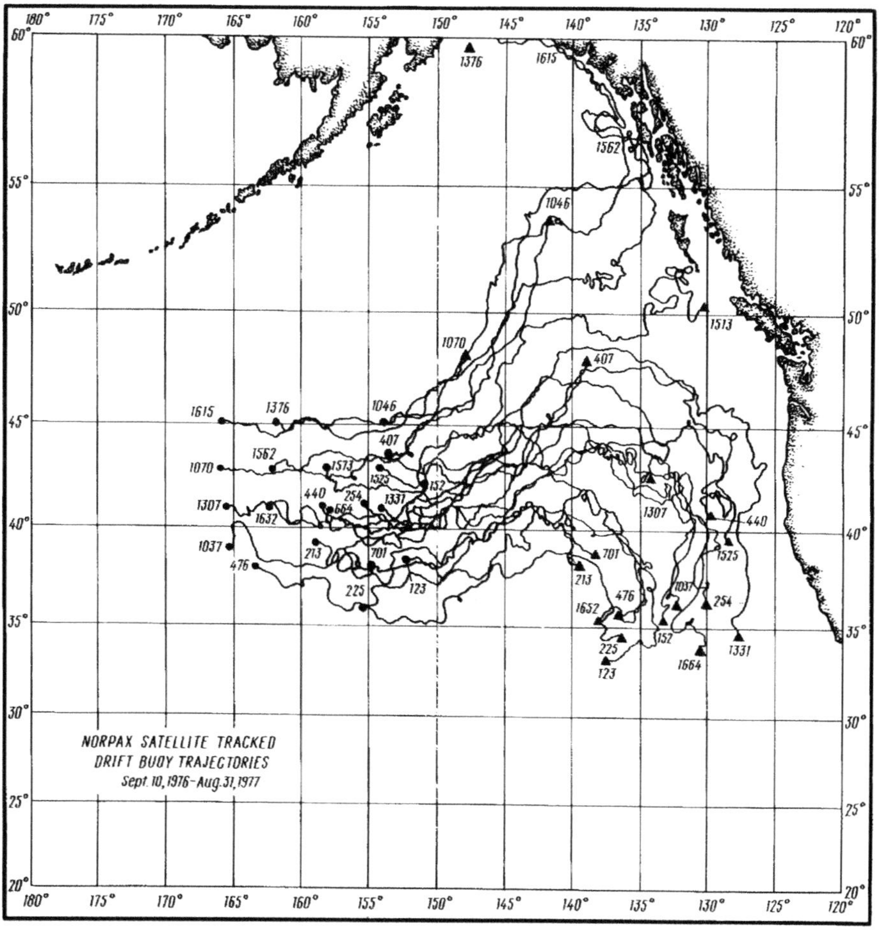

Fig. 5.3.5. Trajectories of motion of 22 satellite-tracked surface buoys during the period from September 10, 1976 to August 31, 1977 (after Kirwan *et al.*, 1978). The figures at the initial and final points of the trajectories are the numbers of the buoys.

south-east Alaska in the zone from 50 to 58°N (Figure 5.3.5). The dominant time scale (the half period in the case under consideration) of the synoptic oscillations of velocity was 20–30 days. One of the particular results of the measurements was the detection of a single cyclonic eddy south of North Pacific Current in the region 37–38°N, 161–163°W. The diameter of circular motion of the float in the eddy field was around 100 km and the average velocity of this motion was about 15 cm/s. The eddy moved westward with an average velocity of 2 cm/s.

From the end of the 1960s up to the present time Soviet oceanographers have performed more than 100 hydrographic surveys of regions of some individual seamounts in the Pacific (Darnitskiĭ, 1980a). In almost every case the surveys revealed single anticyclonic or cyclonic eddies with horizontal dimensions comparable with seamount scales (which, as a rule, are tens of kilometers) or systems of two or three or four such eddies located over seamount tops or shifted to 'lee' slopes of the seamounts relative to the large-scale current. Examples of such eddies are given in Figures 5.3.6 and 5.3.7; in the first case the dimension of the eddy completely corresponded to the total transverse dimension of the seamount and in the other case four eddies were located on the slopes of a seamount whose total transverse dimension was approximately 100 km. The Seamount Vostochnaya (Figure 5.3.7; symbol 14 in Figure 4.2.1) lies in the zone of a weak eastward large-scale current. As can be judged from Figures 5.3.6 and 5.3.7, all the eddies observed had a clearly marked baroclinic character, and, as a rule, the velocity of the geostrophic current in their field was several tens of centimeters per second but sometimes attained 130–140 cm/s (Kozlov *et al.*, 1982). In some cases one could observe correspondence between the direction of rotation of water in the field of a single eddy above a seamount top and the sign of the vorticity of the wind field over the corresponding part of the ocean during the observation period (Darnitskiĭ, 1980b), which perhaps determined the sign of the local vorticity of the large-scale current. The observations registered both cases of stationary behavior of eddies when eddies remained in seamount regions for one or two months (e.g. this was the case in the situation demonstrated by Figure 5.3.6) and cases of separation of eddies from seamounts and their drift in the direction of the large-scale current. Increased concentrations of dissolved silica and phosphates were frequently observed in the upper ocean in regions of topographic eddies (Darnitskiĭ, 1980a; Kozlov *et al.*, 1982), which can be accounted for by an intensive vertical mixing of waters or by vertical motions in the eddy field.

The result demonstrated by Figure 5.3.8 can be associated with those described above. For a period of 100 days a satellite surface float, a part of whose trajectory is shown in the figure, travelled a path about 2800 km drifting eastward in the southern branch of the Kuroshio–North Pacific Current before reaching the region of the Emperor Seamounts at 171–173°E (Cheney, 1978b). Then, as can be seen from Figure 5.3.8, the drift direction of the float changed from east to south-west, and the float made several anticyclonic rotations over the eastern slope of a seamount with a top at a depth of 2400 m, after which it resumed an eastward drift. The maximum 'diameter' of circular motions of the float was 60 km, which corresponded to the horizontal dimensions of the seamount (Figure 5.3.8); the maximum velocity of these motions was 70 cm/s, i.e. much greater than the velocity

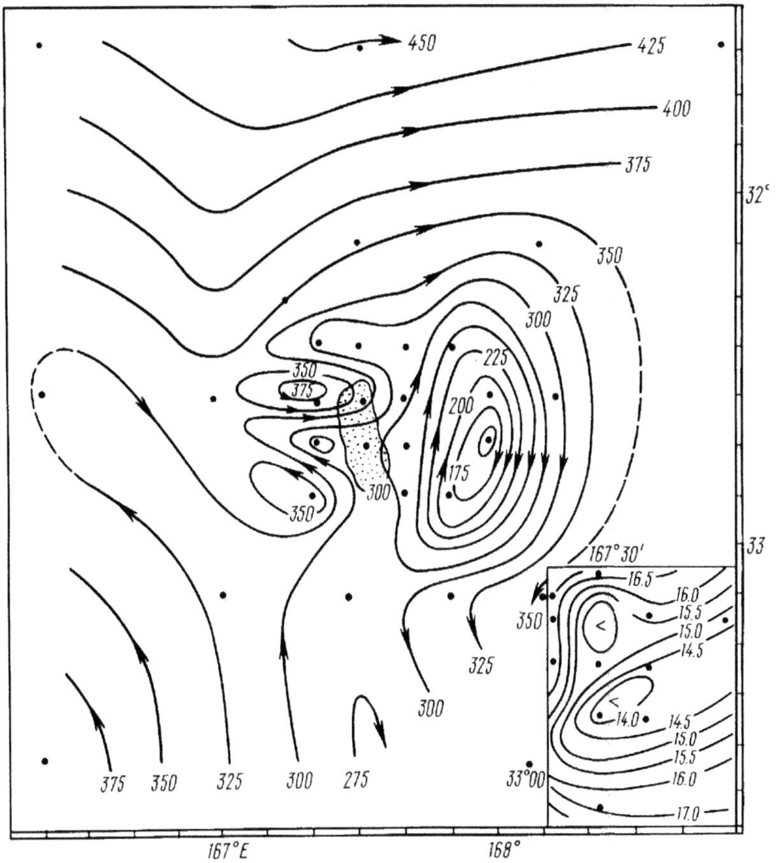

Fig. 5.3.6. Dynamic topography (in dynamic millimeters) of the 100 dbar surface relative to the 1000 dbar reference surface in the region of a seamount in the southern part of the Tasman Sea on November 14–30, 1977 (after Darnitskii, 1979a). The dotted area shows a region of depths less than 300 m. The points indicate the positions of hydrographic stations. The chart in the right lower corner shows the temperature distribution at a depth of 100 m in the central part of the region in January 1978.

of the preceding eastward drift of the float. The author (Cheney, 1978b) interpreted this result as the formation of a Taylor column in a stratified ocean, i.e. an anticyclonic eddy appearing above a seamount when a slow and wide large-scale current flows over it.

The most important results of the investigation into synoptic eddies of the tropical and subtropical Atlantic were obtained during the 'Polygon-70', MODE, and POLYMODE expeditions (see Sections 1 and 2). Among the other results we should first mention the spectrum of temperature oscillations (Wunsch, 1972) at a depth of 500–600 m at a point 25 km east of Bermuda (symbol 13 in Figure 4.1.1). The spectrum was calculated from the data of hydrographic observations carried out at half-monthly intervals from 1954 to 1969; this revealed an obvious predominance of temperature oscillations with periods from 50 days to one year, which can be interpreted as an effect produced by baroclinic synoptic eddies passing across the observation region.

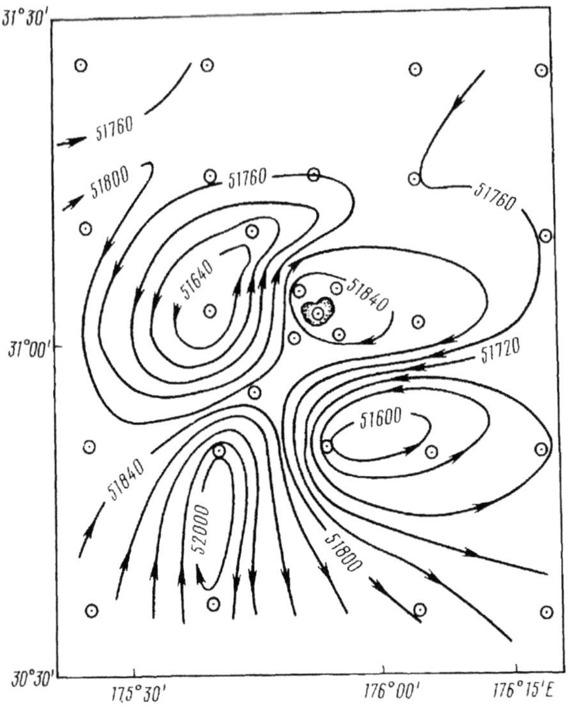

Fig. 5.3.7. Dynamic topography (in dynamic millimeters) of the 300 dbar surface relative to the 1000 dbar reference surface in the region of the Seamount Vostochnaya of the Hawaii Ridge on April 5–9, 1976 (after Darnitskiĭ, 1980a). The dotted area shows a region with depths less than 300 m. The circles with points indicate the positions of the hydrographic stations. (See symbol 14 in Figure 4.2.1.)

Within the framework of the POLYMODE program, whose main experiments were described in Section 2, American oceanographers carried out current and water temperature measurements between May 1977 and May 1978 with the aid of three clusters (systems) of autonomous buoy stations (ABS) (Fu *et al.*, 1982) two of which (A and B; accordingly, symbols 18 and 19 in Figure 4.1.1, with cluster centers at 27.5°N, 48.5°W and 27°N, 41°W) were located on the western and eastern slopes of the North Atlantic Ridge in the zone of a very weak south-west large-scale current, and the third cluster (C; symbol 20 in Figure 4.1.1, with its cluster center at 16°N, 54°W) was located in the central part of the western region of the North Equatorial Current. The region of cluster C is characterized by an ocean depth of approximately 5300 m and a comparatively flat bottom, and the regions of clusters A and B have depths of 4–5 km and a very rough bottom with a typical scale (half 'wavelength') of irregularities on the bottom relief equal to 10–30 km and depth differences up to 1 km. Each of clusters A and B involved five ABS and cluster C consisted of four ABS; the distances between neighboring stations ranged from 15 to 150 km. At each of the ABS, measurements were carried out at several (from three to seven) depths in the range 100–4000 m. One autonomous buoy station from each of clusters A and B was retained during the period from May 1978 to September 1979. All these measurements showed that synoptic

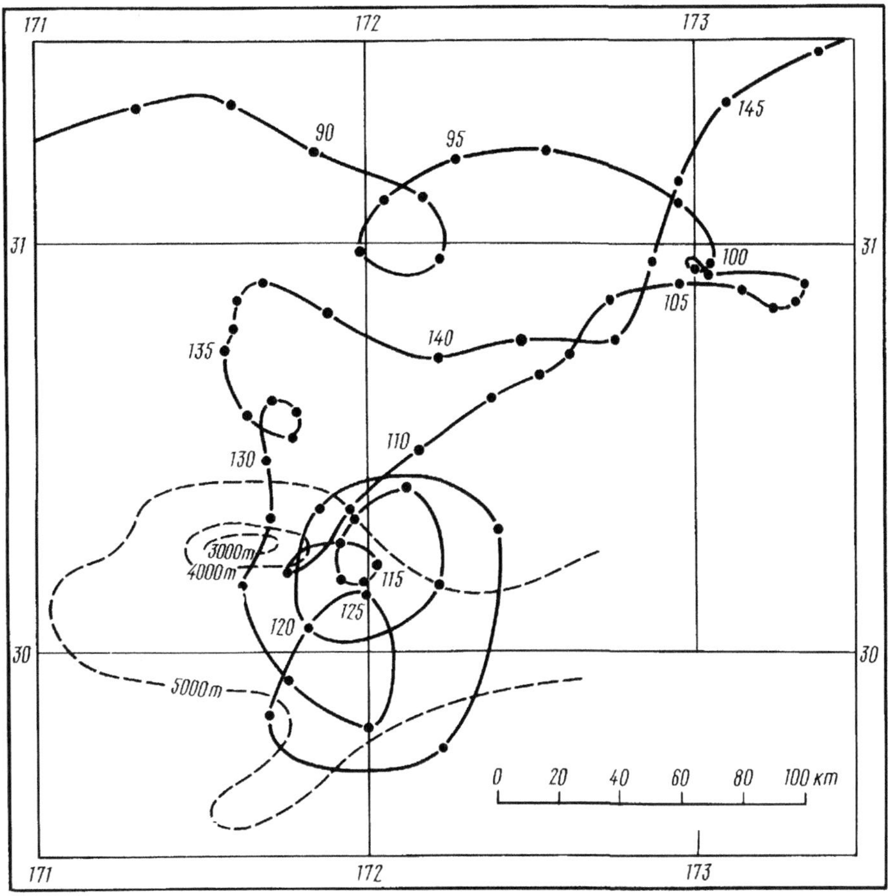

Fig. 5.3.8. Trajectory of motion of a satellite surface float in the region of the Emperor Seamounts in April–May 1977 (after Cheney, 1978b). The points and adjacent figures indicate the positions of the float on each of the days and the serial numbers of the days in 1977. The 3000, 4000, and 5000 m isobaths are shown. (See symbol 15 in Figure 4.2.1.)

oscillations of currents and temperatures were very clearly marked in the regions of the three clusters. In the layer from the sea surface to a depth of approximately 2000 m in the regions of clusters A and B and in the whole ocean depth in the region of cluster C, the greatest energy was contained in disturbances with periods of 50–100 days, characterized by rather high coherence over the depth and by the predominance of oscillations of the meridional component of the velocity of the current over the oscillations of the zonal component. By contrast, at lower frequencies the oscillations of the zonal component were dominant. At a depth of 4000 m in the regions of clusters A and B, oscillations with periods of only 10–30 days were dominant. The joint spectrum analysis of oscillations at the various ABS revealed a westward drift of the disturbances of velocity and temperature fields with a phase velocity of 3–5 km/day for all the three clusters. This can probably be accounted for by the combined effect of the advection of disturbances by large-scale currents and the westward propagation of baroclinic Rossby waves (Fu *et al.*, 1982; Keffer, 1983).

Among the basic purposes of the above-mentioned observations were the problems of studying the difference between the energy levels of synoptic oscillations of currents on different sides of the North Atlantic Ridge and finding evidence for the generation of eddies in the zone of the North Equatorial Current owing to its baroclinic instability. As to the first problem, no substantial difference was found; in the regions of both cluster A and cluster B the average specific kinetic energy of synoptic currents was 55–75 cm^2/s^2 at depths of 100–200 m, 2–4 cm^2/s^2 at depths of 1400–1500 m, and around 1 cm^2/s^2 at 3500–4000 m. Attention should be paid to the very sharp damping of the currents with increasing depth, which seems to be related to the dissipative effect of the mesoscale bottom relief.

The average synoptic heat flux vectors calculated from the data of stations 80 and 82 of cluster C (Figure 5.3.9) are directed so that, as is clearly seen from the figure, they must produce a reduced climatic temperature gradient in the field of the North Equatorial Current. This is in complete agreement with the concept of the generation of synoptic eddies of this current owing to its baroclinic instability. However, the directions of the heat flux vectors calculated from the data of stations 79 and 81 (Figure 5.3.9) contradict this concept. If only the data of stations 80 and 82 are taken into account, the average rate of transformation of the APE of the large-scale current into eddy energy (see Section 1 of Chapter 4) in the 300–500 m layer turns out to be 1.3×10^{-5} erg/g·s. For the average total energy (i.e. KE plus APE)

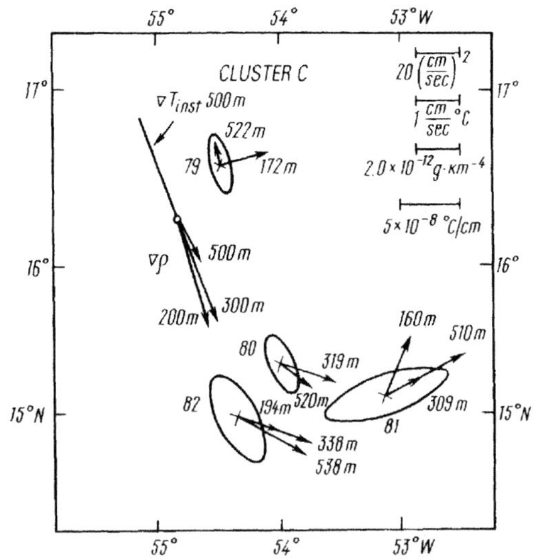

Fig. 5.3.9. Dispersion ellipses of the velocity of the synoptic current at a depth of 500 m and the average synoptic heat-flux vectors at depths of 160–540 m according to the observation data from the ABS of cluster C (after Fu *et al.*, 1982) in the North Equatorial Current of the Atlantic Ocean in May 1977–May 1978. The numbers near the vectors and the ellipses indicate the corresponding depths of measurements and numbers of the ABS respectively. Also presented in the figure are the gradients of the mean temperature distribution at a depth of 500 m and of the mean density distributions at depths of 200, 300, and 500 m. (See symbol 20 in Figure 4.1.1.)

of the eddies in this layer, which is 76 erg/g, this rate corresponds to the time of complete renewal of the eddy energy equal to 68 days.

The two temperature-density surveys of an Atlantic region north of the Bahamas and the Antilles performed in July–August 1972 and January–February 1973 showed that synoptic oceanic eddies were well developed in this region in the summer period and entirely disguised the large-scale Antilles Current (Gunn and Watts, 1982). In the period from March 1976 to October 1978, within the framework of the POLYMODE program, long-term current and water temperature measurements by SOFAR floats were carried out at depths of 700 and 2000 m in the vast region of the subtropical Atlantic between 18 and 27°N, 44 and 74°W (Figure 5.3.10). In complete agreement with the results of the spectral processing of the data of current measurements at the ABS of the 'Polygon-70', MODE, and POLYMODE expeditions (see Sections 1 and 2) and the conclusions of the theory of oceanic synoptic-scale currents (e.g. see Rhines, 1977), the float trajectories in Figure 5.3.10 involve zonal motions with very long periods (of the order of hundreds of days) including, possibly, the mean large-scale eastern current with a velocity of 3–4 cm/s and relatively short-period oscillations (with periods of 40–80 days) of water particles in the meridional direction superimposed on those zonal motions. According to the data from several of the launched floats, these oscillations were accompanied by coherent oscillations of water temperature along the float trajectories, as is expected in a field of baroclinic Rossby waves. The drift of the floats launched at a depth of 2000 m also indicated predominance of long-period zonal motions of water. The statistical processing of data from the above-mentioned measurements yielded estimates of the Lagrangian coefficients of turbulent diffusion in the zonal and meridional directions in the order of 3.7×10^7 cm^2/s and 1.3×10^7 cm^2/s for a depth of 700 m and 1.2×10^7 cm^2/s and 0.3×10^7 cm^2/s for a depth of 2000 m, respectively (Riser and Rossby, 1983). This is the diffusive contribution of synoptic eddies to the large-scale motion of ocean waters. The real spatial structure of the above-mentioned 'short-period' disturbances of the current field can be seen in Figure 5.3.11, which clearly indicates the presence of cyclonic and anticyclonic eddies in the observation region; they are characterized by transverse dimensions of the order of 150 km and a westward motion with a velocity of about 4 cm/s. In Figure 5.3.11 we can see, but cannot easily explain, the lack of coincidence among the simultaneous geographical positions of eddy centers in the upper ocean, in the layer of the '18-degree' Sargasso Sea water (at a depth of 300 m), and in the main thermocline (at a depth of 700 m).

Within the framework of the above-mentioned experiments according to the POLYMODE program in the region 24–25°N, 68.5–69.5°W (symbol 22 in Figure 4.1.1) in October 1976, temperature and density surveys of the ocean and current measurements by SOFAR floats at a depth of 700 m (McDowell and Rossby, 1978) revealed a deep single eddy in which a number of parameters somewhat resembled one of the eddies of the Local Dynamics Experiment (LDE) of POLYMODE (Section 2). The eddy of 1976 was an anticyclone, had a transverse dimension around 100 km, and was entirely concentrated in a layer from 200 to 1400 m. The velocity of water rotation in the field of this eddy attained 20–30 cm/s at depths of 500–600 m. The lower part of the eddy (from 700 to 1300 m in the σ_t range of 27.30

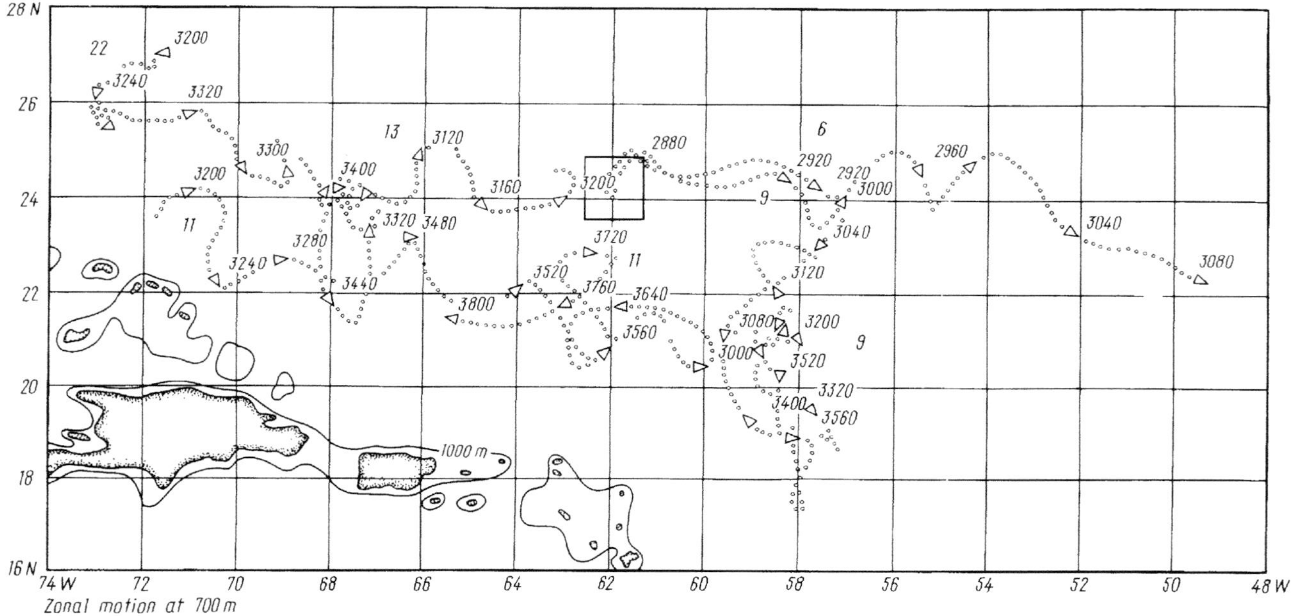

Fig. 5.3.10. Trajectories of motion of five SOFAR floats (nos 6, 9, 11, 13, and 22) at a depth of 700 m in the south-west part of the North Atlantic in March 1976–October 1978 (after Riser and Rossby, 1983). The figures at the trajectories are the serial numbers of days. No. 2750 corresponds to December 4, 1975; no. 3000, August 10, 1976; no. 3250, April 17, 1977; no. 3500, December 23, 1977; no. 3750, August 30, 1978. The square is the position of the XBT experiment area (see Figure 5.3.11).

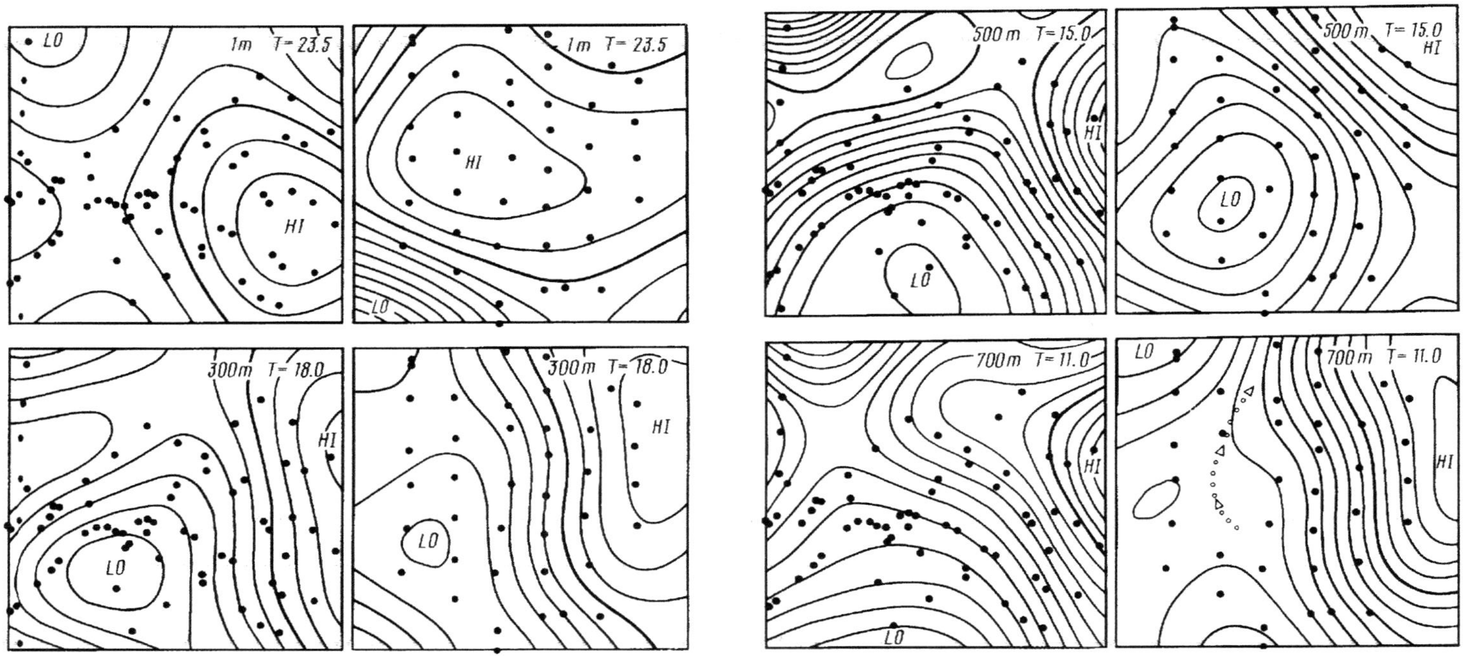

Fig. 5.3.11. Water temperature distribution at depths of 1, 300, 500, and 700 m in an experiment area with center at 24°12′N, 61°52′W in the southern part of the Sargasso Sea (see Figure 5.3.10) on March 1–3 (the left charts) and March 7–9, 1976 (the right charts) (after Riser and Rossby, 1983). The values of temperature corresponding to the thick isotherms are given in the right upper corners of the charts; the isotherms are drawn at 0.1 °C intervals. The temperature maxima and minima are marked by the symbols HI and LO. The points show the places of temperature soundings of the ocean. The chart for a depth of 700 m and the survey on March 7–9, 1976 shows the drift trajectory of a SOFAR float at the same depth from March 3 to 18, 1976. (See symbol 21 in Figure 4.1.1.)

to 27.55) was formed by water sharply differing in its *T*, *S* characteristics from the Sargasso Sea water surrounding the eddy. It was assumed by McDowell and Rossby (1978) that this eddy water had its origin in the Mediterranean Sea. This implies that the eddy core travelled a path of several thousand kilometers across the ocean without mixing with the surrounding waters. The measurements by floats – one of which had remained in the eddy core for 300 days – showed that the eddy moved to the south-west with an average velocity of about 6 cm/s.

In November 1980 in almost the same region of the ocean (24–27°N, 70–72°W) another deep single anticyclonic eddy was found when a temperature–density survey of the ocean was being performed (Zantopp and Leaman, 1982). This eddy had a transverse dimension of about 200 km and a completely mixed core located between depths of 650 and 730 m (σ_t = 26.82). The observed values of temperature, salinity, and oxygen content in the core led Zantopp and Leaman to the assumption that the water mass of the core was formed in the Gulf of Cadiz during the process of winter vertical convection and that the age of the core was 3–4 yr. The results of the above-mentioned observations of the two eddies in the region 25°N, 70°W and the results of the LDE of POLYMODE presented in Section 2 make us assume that the formation of deep single anticyclonic eddies with well-mixed cores in the eastern part of the subtropical zone of the North Atlantic, and the later transfer of these eddies to the western part of the ocean through the large-scale circulation of waters, is a widespread phenomenon.

Detailed investigations of a powerful cyclonic eddy in the subtropical region of the Atlantic west of the North Atlantic Ridge (symbol 23 in Figure 4.1.1) were carried out by Soviet oceanographers in the autumn–winter period of 1976. In Figure 5.3.12, constructed from the data of these observations, the immensity and irregular form of the eddy attract attention. The eddy moved westward and simultaneously rotated clockwise (Figure 5.3.12). According to the current measurements at buoy stations, the current velocity in the upper part of the eddy attained 120 cm/s (Bulgakov *et al.*, 1977). In the rear (eastern) part of the eddy the current velocity at depths of 1000 and 1500 m (and probably also at greater depths) attained 15 cm/s and was directed approximately opposite to the current at a depth of 500 m and in the layers above it. The presence of relatively low-salinity water in the upper part of the eddy makes it possible to assume that the eddy was formed in an ocean region lying further to the north. The permanent presence of synoptic eddies and their motion in the Atlantic region under consideration is also confirmed by the results of the spectrum analysis (Gill, 1975) of the data of temperature measurements over a period of three years (1966–1969) at a depth of 250 m aboard the weathership *Echo* anchored at 35°N, 48°W (symbol 24 in Figure 4.1.1). The spectrum analysis revealed clearly marked temperature oscillations with periods from 40 to 200 days, which probably should be related to synoptic ocean eddies passing through the observation point.

The results of Dantzler's calculations (1977) of the distribution of the APE of synoptic currents (see Section 5) in the North Atlantic made it possible to assume that the region of intersection of the North Atlantic Ridge by the southern branch of the Gulf Stream approximately at 35°N, 35°W (Figure 4.1.1) is a region of increased eddy activity. This was convincingly proved by the Soviet expedition of

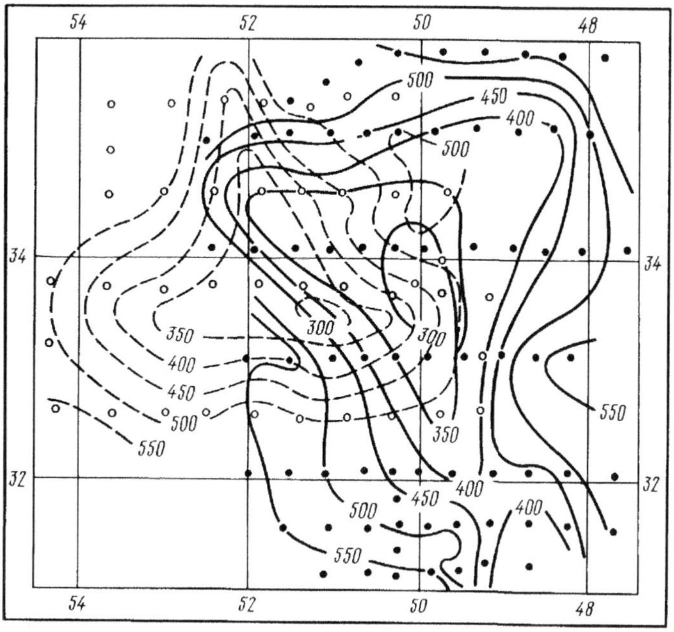

Fig. 5.3.12. Topography (depth in meters) of the 15 °C isothermal surface in the field of a cyclonic eddy in an Atlantic region west of the North Atlantic Ridge (symbol 23 in Figure 4.1.1) on November 9–23 (solid curves) and December 17–27, 1976 (dashed curves) (after Bulgakov *et al.*, 1977). The dark and light circles indicate the positions of the hydrographic stations of the first and second surveys of the region, respectively.

1980 which carried out a series of XBT sections of the ocean, several hydrographic surveys (the survey region is marked by symbol 25 in Figure 4.1.1), and direct current measurements at several buoy stations, each of which lasted for several days in the indicated ocean region (and in adjoining regions; see Figure 5.3.13) (Kort, 1982; Kort *et al.*, 1983). The positions of the cyclonic and anticyclonic eddies in the sections (Figure 5.3.13) were determined from the rising and the lowering of isotherms, respectively, and this was preceded by removing the traces of internal gravitational waves by smoothing the temperature distributions along the sections with a scale of 110 km. Since the sections passed at different distances from the centers of different eddies, the diameters of most of the eddies in Figure 5.3.13 are underestimated. However, this is not the case for eddies A8, C7, A6, A12, and C11 (Figure 5.3.13) which were located in the above-mentioned region of intersection of the North Atlantic Ridge by a branch of the Gulf Stream and were registered in a series of closer sections and surveys. The diameters of these eddies were 150–200 km and the isothermal surface height differences in their field attained 150–200 m. Repeated surveys made it possible to assume that eddy A8 (in which, according to the data of direct current measurements, the current velocity at a depth of 100 m attained 55 cm/s) moved to the south-west with a velocity about 10 km/day. Isothermal surface height differences attaining values up to 150 m were also registered in eddies A16, C17, and A26 (Figure 5.3.13). Of interest is the absence of traces of any strong eddies in all sections south of the Tropic of Cancer (Figure

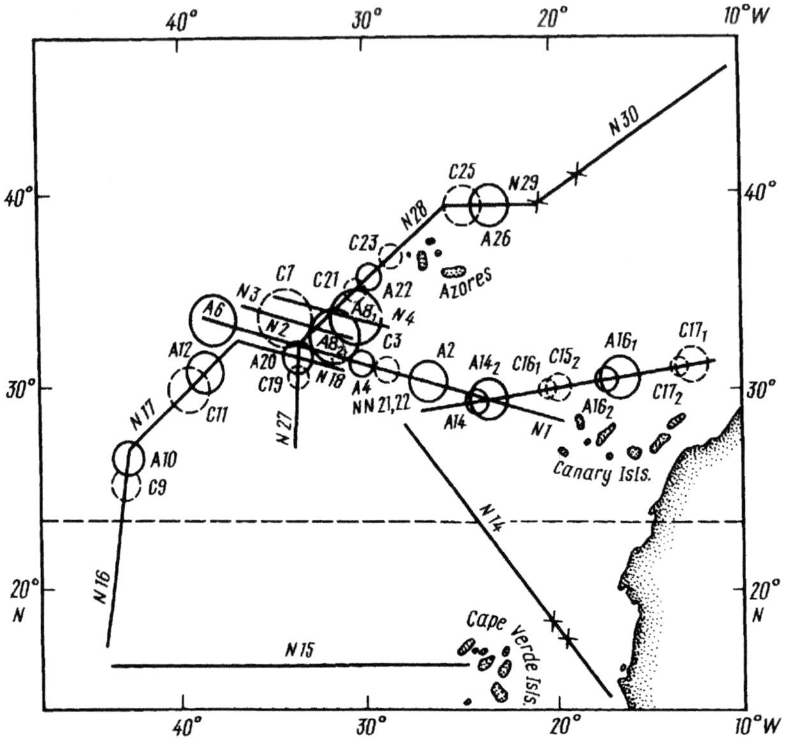

Fig. 5.3.13. Positions of the anticyclonic and cyclonic synoptic eddies (solid circles marked by the letters A and dashed circles marked by the letters C, respectively) in XBT sections to a depth of 500 m in the eastern part of the North Atlantic in June–August 1980 (after Kort, 1982). Given in the figure are the numbers of sections and of individual eddies. The crosses in sections 14 and 30 show the boundaries of the temperature frontal zones.

5.3.13). In the frontal zones at sections 14 and 30 (Figure 5.3.13), the temperature gradients directed to the south-east and south-west, respectively, were 4 °C per 100 km and 3 °C per 150 km at a depth of 100 m.

The small seamount in whose region (mainly on its western slope) direct deep current measurements were performed in October 1976–May 1977 (Figure 5.3.14; symbol 26 in Figure 4.1.1) is located on the northern periphery of a very weak westward large-scale current and its elevation over the surrounding ocean bottom is 400 m. As can be seen from Figure 5.3.14, the measurements show the almost permanent presence of an anticyclonic eddy above the seamount, the horizontal dimension of the eddy being comparable with the seamount scale. The current velocity in the eddy varied over a period of 100–120 days and attained 25–30 cm/s at a height of 1 km above the seamount during the phases of the eddy's maximum development (Figure 5.3.14). The CTD sections along 55°W performed in October 1976 and July 1977 revealed a rise of isopycnic surfaces over the seamount which was damped towards depths of 1200–1600 m; this indicated a general intensification of the vortical motion towards the ocean bottom. The analysis of the data of the direct current measurements also showed that, during the whole observation time,

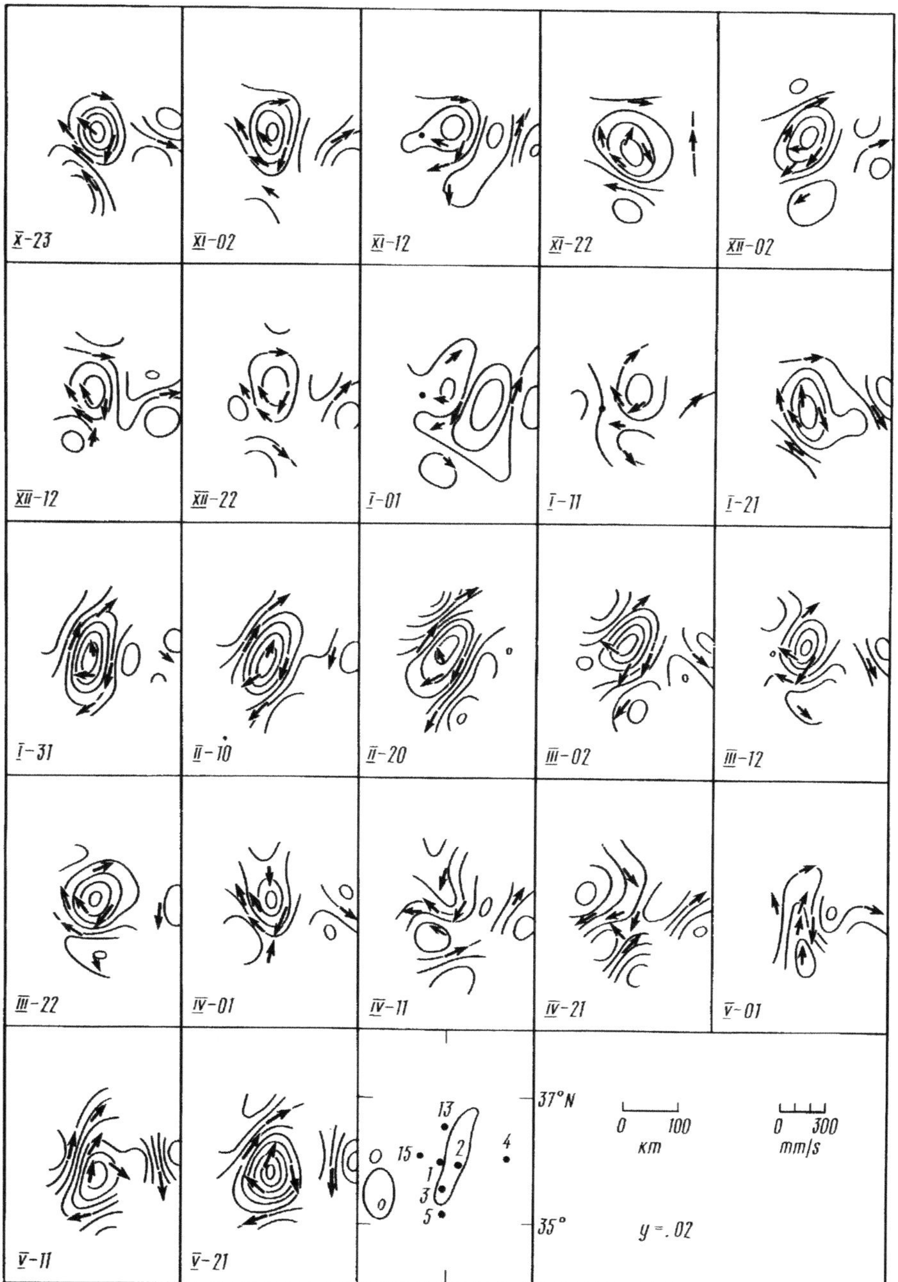

Fig. 5.3.14. Velocity vectors and corresponding streamlines of mesoscale currents (averaged with a period of 10 days) at a depth of 4000 m in the region of a seamount in the north-east part of the Sargasso Sea in October 1976–May 1977 (after Owens and Hogg, 1980). The streamlines are drawn at intervals of 1.6×10^7 cm^2/s. The array of the ABS with current meters and the 5200 m isobath are shown. (See symbol 26 in Figure 4.1.1.)

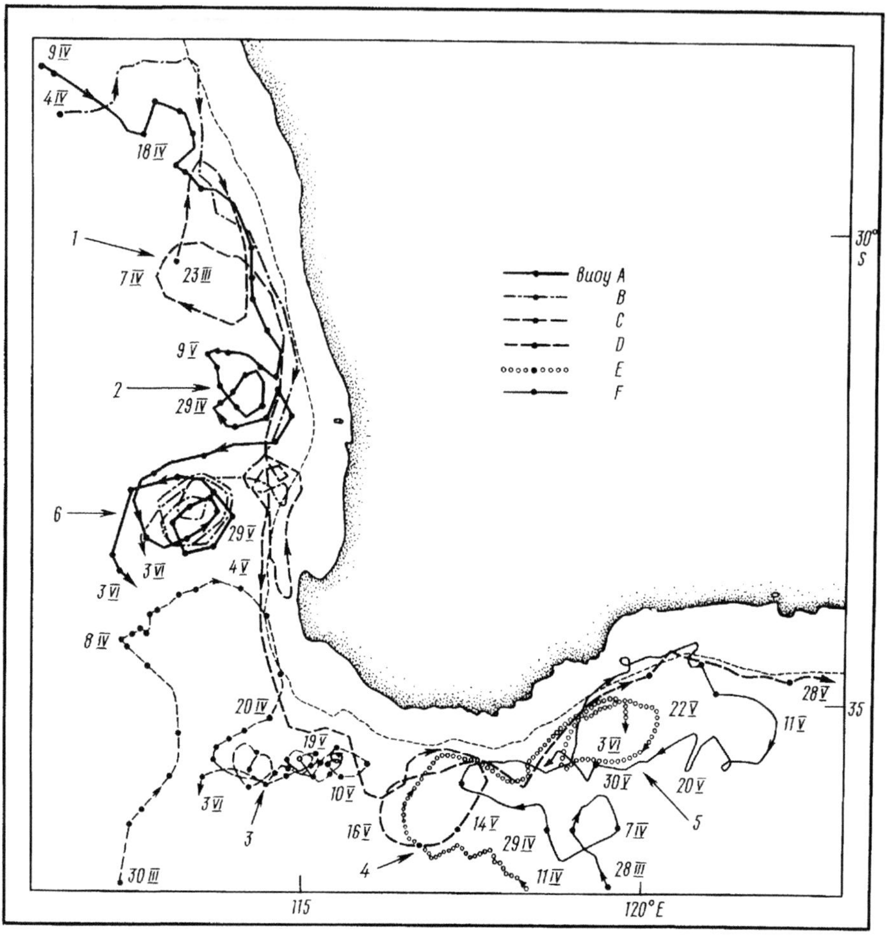

Fig. 5.3.15. Trajectories of motion of six satellite surface buoys in an ocean region adjoining south-west Australia in March–June 1976 (after Cresswell and Golding, 1980). The numbers with the arrows indicate the parts of the trajectories corresponding to cyclonic (1–5) and anticyclonic (6) motion of the buoys.

there existed above the seamount a local balance between the advection of relative vorticity in the mesoscale current field and the stretching of the vorticity lines along the vertical owing to the flow over the seamount. All the above-mentioned properties of the current field allowed the authors (Owens and Hogg, 1980) to interpret the observed anticyclonic eddy as a stratified Taylor column whose energy seemed to change depending on the velocity and structure of the large-scale flow incident on the seamount.

The hydrographic surveys of the ocean and surface current measurements by satellite buoys carried out by Australian oceanographers in the 1970s (Andrews, 1977; Cresswell and Golding, 1980) revealed the permanent existence of a geostrophic jet current off the south-west coast of Australia which first flows to the south and then to the east along the continental slope (Figure 5.3.15). The current is particularly clearly marked in the autumn–winter period when its velocity at the south-west extremity of Australia can attain 1.7 m/s. Figure 5.3.15 clearly indicates the formation of cyclonic circulations of water with horizontal scales of the order of 100 km on the ocean side of this current (parts 1–5 of the float trajectories in Figure 5.3.15). As is seen from Figure 5.3.15, on tracing one or two loops in circular motion the floats returned to the main current. It is remarkable that in this return the velocities of the floats increased from 0.1–0.4 m/s to 0.5–1.4 m/s and the water temperature they registered increased by 0.5–3 °C. At the same time the observations showed that at least formations 3 and 4 in Figure 5.3.15 had a tendency to move westward with a velocity of 6–7 km/day. One possible explanation for the combination of these westward motion of circulating centers and the intensive water exchange between them and the jet current can lie in the fact that this motion was primarily of a wave nature. An analysis of the results of a series of hydrographic surveys of the ocean region adjoining the south-west coast of Australia on its western side performed during the summer periods in 1972 and 1973 also led to the assumption that a system of baroclinic Rossby waves with predominantly meridional crests propagating in the westward direction from the Australian continent was formed in this part of the ocean (Andrews, 1977).

The character of the anticyclonic eddy, marked by the symbol 6 in Figure 5.3.15, was completely different in comparison with the above-mentioned cyclonic formations. It was undoubtedly formed as a result of the separation of an anticyclonic meander of a jet current. One of the two floats drawn in this eddy made four circulations and the other made ten before leaving the limits of the eddy (Figure 5.3.15). It is remarkable that the radii of rotation of the two floats increased with time and the rotation velocities increased approximately proportionally to the radii (from 20–50 cm/s to 90–120 cm/s) while rotation periods of 4–7 days were retained (Cresswell, 1977). The measurements showed that the eddy did not move for 30 days and then began to move north-north-east with a velocity of 5 cm/s.

In this section, so far, we have discussed only eddies in the open ocean. However, observations also indicate the existence of vortical disturbances of the large-scale circulation of waters in some inland seas, e.g. in the Mediterranean Sea (Cheney, 1978a; Burkov *et al.*, 1979; Garzoli *et al.*, 1982) and in the Black Sea (Blatov and Ivanov, 1979). As a rule, the total horizontal dimensions of these disturbances are smaller than in the open ocean and are about 100 km. It seems that most of these formations have a quasistationary character and are associated with mesoscale disturbances of the bottom relief and the continental slope line of inland seas.

4. EDDIES AT HIGH LATITUDES

In contrast to the Gulf Stream itself, the North Atlantic Current (NAC) flowing north-eastward from the southern extremity of the Newfoundland Bank to the Faroe–Shetland channel has a more smoothed and multijet character. Multiple CTD and XBT surveys showed that the NAC is unstable and that a great number of meanders and eddies develop in its field. Particularly sharp temperature field disturbances with a horizontal scale ('wavelength') of 300–350 km and an isothermal surface height difference attaining 400 m in the 150–700 m layer were found in June 1978 in two XBT sections across the NAC in the region of its intersection with the North Atlantic Ridge (symbol 27 in Figure 4.1.1) (Dickson and Gurbutt, 1979). A hydrographic survey in July 1965 performed in the region 52–54°N, 18–21°W adjoining the NAC axis on the south-east (symbol 30 in Figure 4.1.1) revealed a cyclonic eddy with a transverse dimension of 120 km and an isothermal surface height difference up to 400 m (Stommel *et al.*, 1977). The upper part of the eddy (above 200 m) was characterized by an absence of similarity between the temperature and salinity distributions, which indicates the advective nature of eddy motion in the upper ocean and can be regarded as an indirect indication of heat and salt exchange across the NAC.

In September 1979–April 1980 French, British, and Italian oceanographers carried out the 'Tourbillon' experiment in an Atlantic region (symbol 29 in Figure 4.1.1) with depth somewhat greater than 4000 m adjoining the Bay of Biscay on the west. The experiment covered an ocean region with a transverse dimension of 150 km and was specially intended to study synoptic ocean eddies (Le Groupe Tourbillon, 1983). The basic observations consisted of an eight-month period of current and water-temperature measurements mainly at depths of 700, 1500, 3000 and 4000 m at ten autonomous buoy stations (ABS); they also included four consecutive CTD surveys of the region (Figure 5.4.1) and current measurements by surface and subsurface free-drifting buoys in September–October 1979. The basic result was the discovery of a clearly marked anticyclonic eddy with a diameter about 100 km (Figure 5.4.1) which extended to the ocean bottom and passed across the observation region primarily in the westward direction with a velocity of about 2 cm/s. The maximum velocity of orbital motion of the water, attained at 500 m (above the main thermocline), was 30 cm/s. In Figure 5.4.1 the concentrated character of the eddy and the increase in the depth of isopycnic surfaces in the period between September 5 and 20, 1979 attract attention. These factors indicated a downward water motion in the eddy core. The eddy core consisted mainly of water of local origin. However, at depths of about 850 m it was observed that the eddy captured water of Mediterranean origin which was drawn in anticyclonic motion around the eddy center. The results of ABS measurements in March–April 1980 allow to assume that another eddy, which this time was cyclonic, travelled westward across the observation region. The average level of the current KE turned out to be approximately the same as in the MODE region (see Section 1), namely 57 cm^2/s^2 at a depth of 700 m, 8 cm^2/s^2 at 1500 m, and 3 cm^2/s^2 at 3 and 4 km. The ABS measurements revealed a northward time- and depth-averaged heat flux of 2.5×10^3 W/m^2, which amounts to 10% of the average density of the northward heat flux across the 47°N parallel in the whole ocean and the whole atmosphere (Vonder Haar and Oort, 1973).

During the period from February 1976 to November 1977 surface current measurements by 16 satellite-tracked free-drifting buoys were carried out (De Verdiere, 1983) in a region with center at 47°N, 11°W which is located east of the 'Tourbillon' experiment region and adjoins the Armorican Seamount on the north-east whose total horizontal dimension and elevation over the surrounding

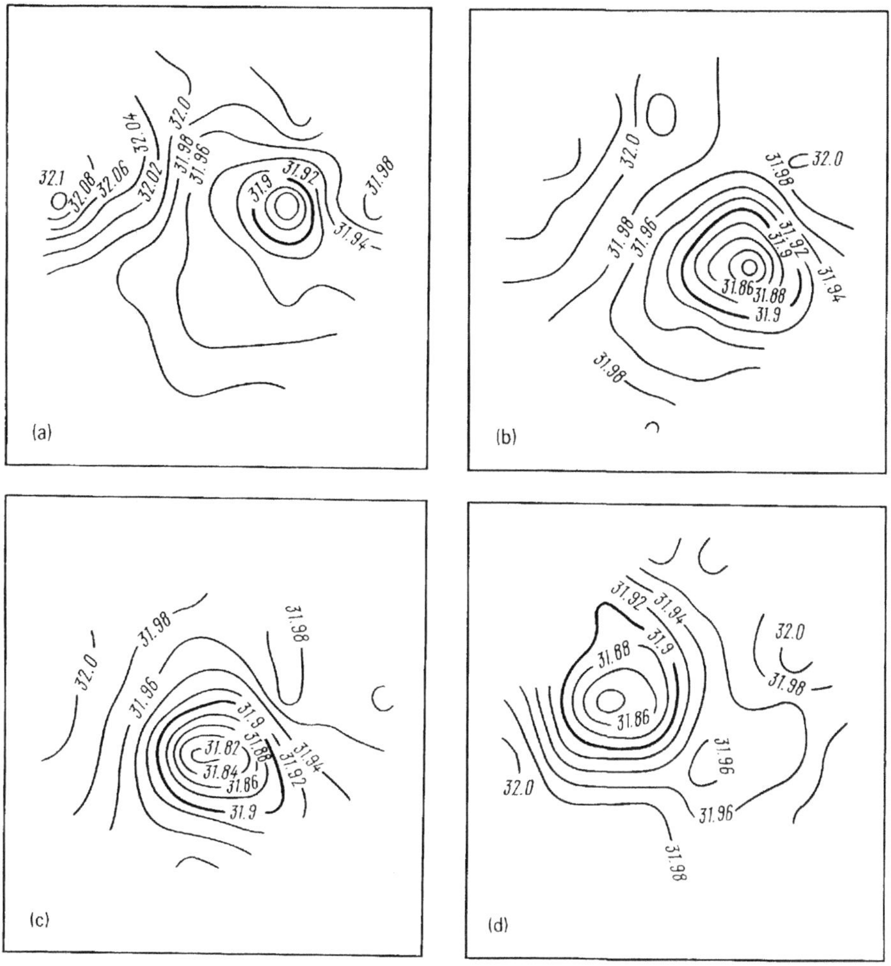

Fig. 5.4.1. Distribution of water density (the density anomaly *in situ*) in the 850 dbar isobaric surface on (a) September 3–8 and (b) September 20–24 and (c) October 1–5 and (d) October 13–17, 1979 according to the CTD survey data of the expedition 'Tourbillon' in the North-Eastern Atlantic (Le Groupe Tourbillon, 1983). The center of the region is at the point 47°00′N, 14°50′W. (See symbol 29 in Figure 4.1.1.)

ocean bottom are 80 km and 1000 m, respectively. The duration of measurements by individual buoys varied from 36 to 200 days. The measurements clearly indicated the almost permanent presence of an anticyclonic eddy over the north-east seamount slope with horizontal dimensions commensurable with those of the seamount, which was probably analogous to the 'Taylor columns' observed in some other Atlantic and Pacific regions (see Section 3). The typical velocity of the corresponding circular motion of the buoys was about 15 cm/s. The average dispersion rate of the buoys was not high and corresponded to estimates of the coefficients of turbulent diffusion in the zonal and meridional directions of 2.3×10^7 and 1.7×10^7 cm^2/s.

The Polar Front separating the warmer and somewhat more saline water of Atlantic origin from the colder and less saline subarctic water is located on the sea surface 100–200 km north-north-east of the Iceland–Faroe Ridge. The frontal surface is inclined and touches the ocean bottom along the crest of the Iceland–Faroe Ridge (depths of 400–500 m) or not very far to the south of it. Observations indicate quasigeostrophic currents with a general east-south-east direction along the frontal surface. Hydrographic and XBT surveys and sections of the region (symbol 30 in Figure 4.1.1) clearly indicate an intensive formation of meanders and eddies across the whole depth of the frontal surface down to the ocean bottom (Hansen and Meincke, 1979). The total transverse dimensions of meanders and eddies are 40–60 km, which is accounted for by the small local Rossby scale value (about 15 km). The character of the process is clearly demonstrated by Figure 5.4.2 where traces of two cyclonic eddies or meanders of the front over the crest of the Iceland–Faroe Ridge are clearly seen. Current measurements by subsurface buoys in this ocean region showed that disturbances of the type demonstrated in Figure 5.4.2 are in fact characterized by *cyclonic* vorticity of the current field with the absolute velocity value increasing from the bottom to the sea surface (Hansen and Meincke, 1979). This means that the appearance of synoptic disturbances is primarily caused not by the relief of the ocean bottom but by the baroclinic instability of a large-scale frontal current. The results of the 1975–1976 annual series of near-bottom measurements (17 m from the bottom) of temperature and velocity fluctuations over the crest of the Iceland–Faroe Ridge do not contradict the above conclusion (Willebrand and Meincke, 1980). Synoptic current fluctuations were mainly concentrated in the period range of 2–11 days and the root-mean-square fluctuation velocity was very high (30 cm/s). The notable correlation between the velocity and temperature oscillations indicated a northward heat flux of 1.2×10^5 W/m^2.

In July–September 1969 current and temperature measurements were carried out for one-and-a-half months in the southern part of the Norwegian Current in a

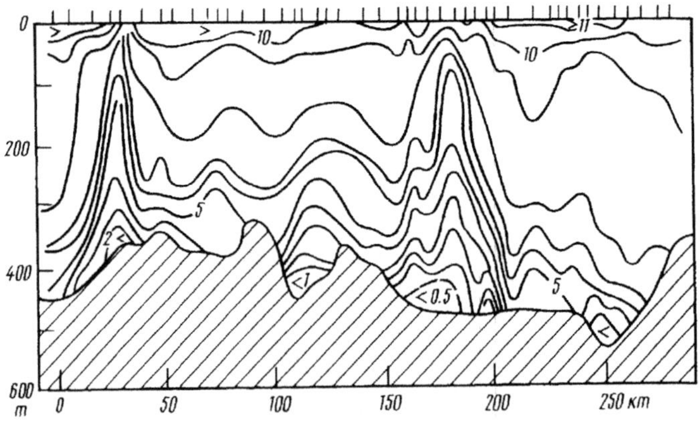

Fig. 5.4.2. Distribution of temperature (°C) in a section along the crest of the Iceland–Faroe Ridge (the WNW edge of the crest is on the left) on September 2–3, 1973 (after Hansen and Meincke, 1979). The marks at the top indicate the places of XBT soundings of the ocean. (See symbol 30 in Figure 4.1.1.)

layer between 100 and 500 m at six buoy stations with distances between neighboring stations ranging from 3 to 8 miles (Schott and Bock, 1980). The spectrum analysis of the current measurement data revealed the predominance of waves with wavelength from 30 to 45 km, periods of 2–3 days, and directions of propagation (azimuths) from 0 to 35°. The indicated parameters (with the exception of the direction of propagation) were in good agreement with the model of baroclinic instability of the Norwegian Current (Mysak and Schott, 1977). The results of these observations were used as a basis to estimate some of the components of the energy balance of synoptic disturbances. The estimates revealed the growth of this energy owing to the transformation of the APE of the large-scale current at a rate of 8–10 $\times$ 10^{-5} cm^2/s^3 at depths 100–200 m and also the reverse flux from the KE of fluctuations to the KE of the large-scale motion (the 'negative viscosity' effect) at a rate of 10–30 $\times$ 10^{-5} cm^2/s^3.

In July–September 1975, two-month current measurements at seven levels from 50 to 1000 m were carried out in the south-east part of the Norwegian Sea at an array of seven buoy stations emplaced along the perimeter and at the center of an equilateral triangle with sides equal to 60 miles (Figure 5.4.3) (Kort *et al.*, 1977). The experiment also involved three consecutive hydrographic surveys of a region with transverse dimension about 300 km, including the buoy station array. It can be concluded from Figure 5.4.3 that, on the whole, the energies of the mean (large-scale) and synoptic currents in the experiment region were of the same order, with a slight predominance of mean currents in the eastern part of the region (the axis of the Norwegian Current) and synoptic disturbances in its central and north-west parts. The hodographs of current directions at points B, C, and D (Figure 5.4.3) indicate the presence of velocity oscillations with periods of 20–25 days in the direction transverse to the mean north-north-east current. The oscillations were characterized by a high degree of depth coherence with a simultaneous sharp decrease of the absolute value of velocity with increasing depth. At the same time, as is seen from Figure 5.4.3, the coherence between the current fluctuations at neighboring points of the experiment region separated by a distance of only 17–30 miles was very low, particularly in the direction (WNW–ESE) transverse to the Norwegian Current. No direct correlation between the variations of the wind and current directions was found in the experiment region. The hydrographic surveys of the region confirmed the presence of synoptic meanders and eddies in the field of the Norwegian Current (Kort and Tarasenko, 1977).

The above results are in agreement with those of current and water temperature measurements at depths of 100, 200, and 400 m performed in August–September 1978 at a buoy station anchored in the southern part of the Norwegian Sea at the point 66°N, 2°E (Alekseev and Ivanov, 1983). The measurements indicated clearly marked current oscillations with a period of about 12 days in the direction transverse to the mean north-north-east current which seemed to be a kind of baroclinic Rossby wave. For the velocity of the mean current of 20–25 cm/s at a depth of 200 m and about 10 cm/s at 400 m, the amplitude of the transverse oscillations also attained 25 cm/s at 200 m and 10 cm/s at 400 m. The measurements also revealed a phase lag in the oscillations at 200 m relative to the oscillations at 400 m, which probably indicated the generation of oscillations owing to the

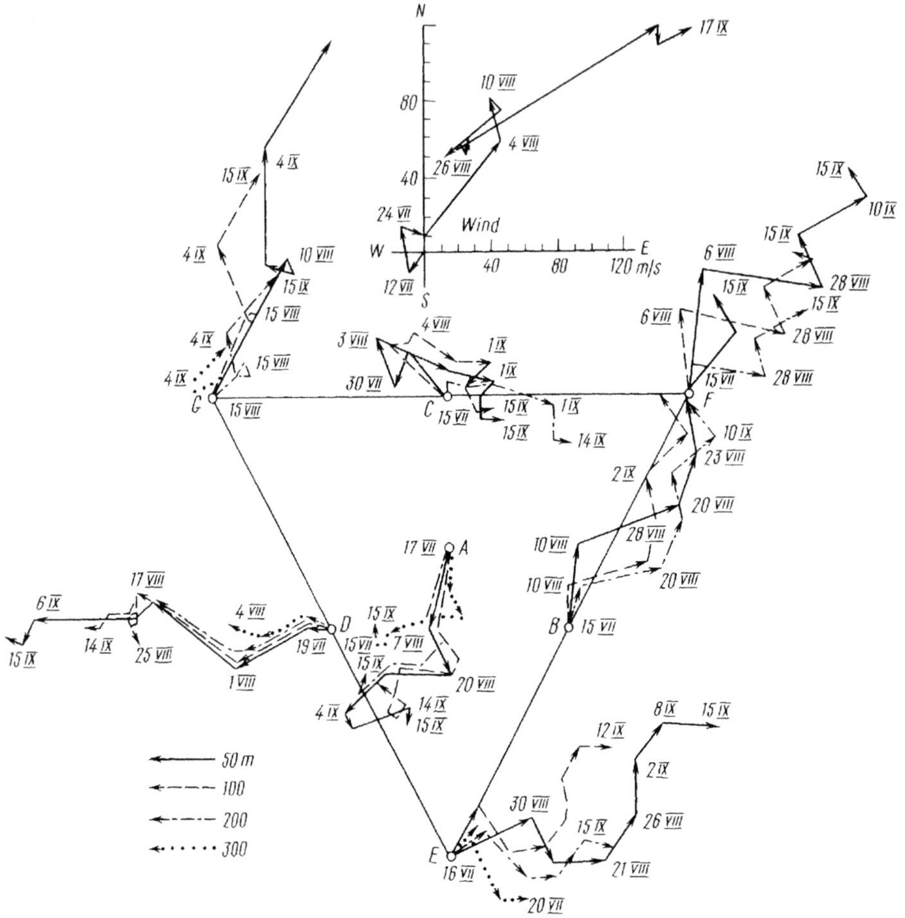

Fig. 5.4.3. Time variation of directions of synoptic currents at depths of 50, 100, 200, and 300 m at seven points in an experiment region in the Norwegian Sea in July–September 1975 (after Kort *et al.*, 1977). The side of the triangle is 60 miles long. Shown at the top is the direction of the wind. (See symbol 31 in Figure 4.1.1.)

baroclinic instability of the large-scale current. The many large-scale surveys of the entire Norwegian Sea or its southern part performed from 1976 to 1979 confirmed the presence of well-developed warm and cold synoptic disturbances in the circulation (Romantsov and Smirnov, 1983), some of which were quasistationary formations probably associated with large-scale and mesoscale irregularities of the bottom relief of the ocean.

The hydrographic surveys of the ocean, the satellite infrared surveys of the sea surface, and the surface current measurements by satellite-tracked buoys (Mysak, 1977; Kirwan *et al.*, 1978; Tabata, 1982) carried out in the 1950s, 1960s, and 1970s showed that the separation region of the California and Alaska Currents, the region of the Alaska Current, and the entire region of the Pacific west and south of the Alaska Current (Figures 5.4.4 and 5.4.5) are virtually saturated with non-

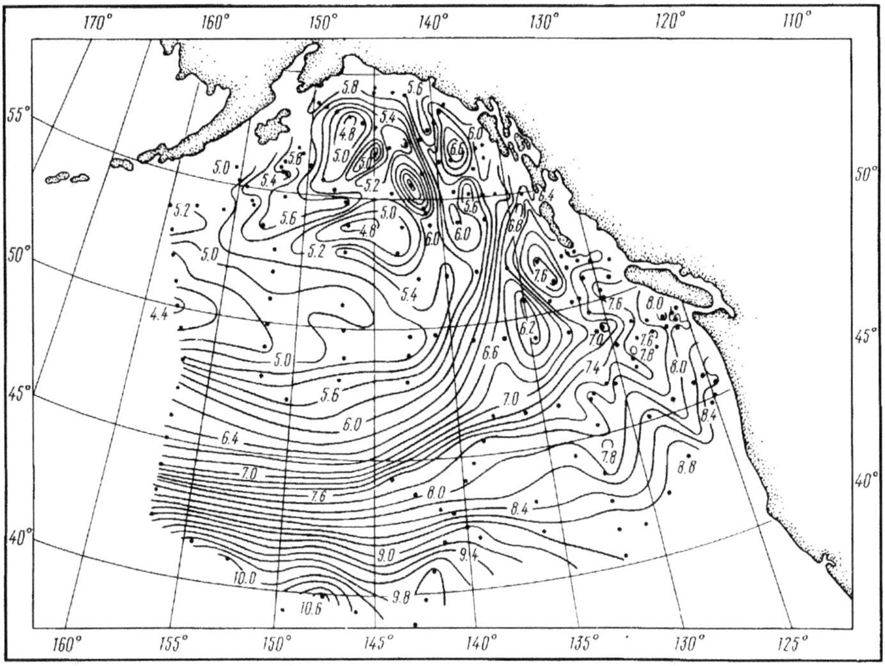

Fig. 5.4.4. Distribution of temperature (°C) in the isopycnic surface σ_t = 26.0 (depths from 40 to 160 m) in the North-Eastern Pacific in May–July 1961 (after Tabata, 1982). The points indicate the places of the hydrographic stations.

stationary cyclonic and anticyclonic disturbances of water circulation with horizontal dimensions from 100 to 300 km and dominant time scales of 10–20 days. Figure 5.4.4 clearly demonstrates the advective character of motion of the most powerful of these disturbances and allows us to assume that they play a very important part in the large-scale mixing of waters of different origin in the ocean region under study. Theoretical analysis (Mysak, 1977) allows us to assume that the main factor producing eddies in this ocean region is the baroclinic instability of large-scale currents. The big quasistationary anticyclonic eddy in the region 57°N, 138°W west of Baranof Island (Figures 5.4.4 and 5.4.5) was subjected to particularly thorough investigations (Tabata, 1982). One of the factors of the permanent presence of the eddy in this region is probably the mesoscale disturbances of the continental slope line in this part of the ocean (Figure 5.4.5). The dimension and intensity of the eddy and the position of its center varied periodically (probably with a period of about one year). On average, the eddy had a total transverse dimension of about 200 km; it extended at least to a depth of 1000 m and was characterized by an isopycnic surface height difference in the main pycnocline of approximately 100 m. The orbital velocity of water motion in the eddy measured by satellite-tracked buoys (Figure 5.4.5) attained 110 cm/s at a distance of 70 km from the eddy center and the corresponding water transport in a layer above 1000 m was, on average, equal to 6.5×10^6 m^3/s, which is comparable with the transport of the axial jet of the Alaska Current flowing along the continental slope (Tabata, 1982).

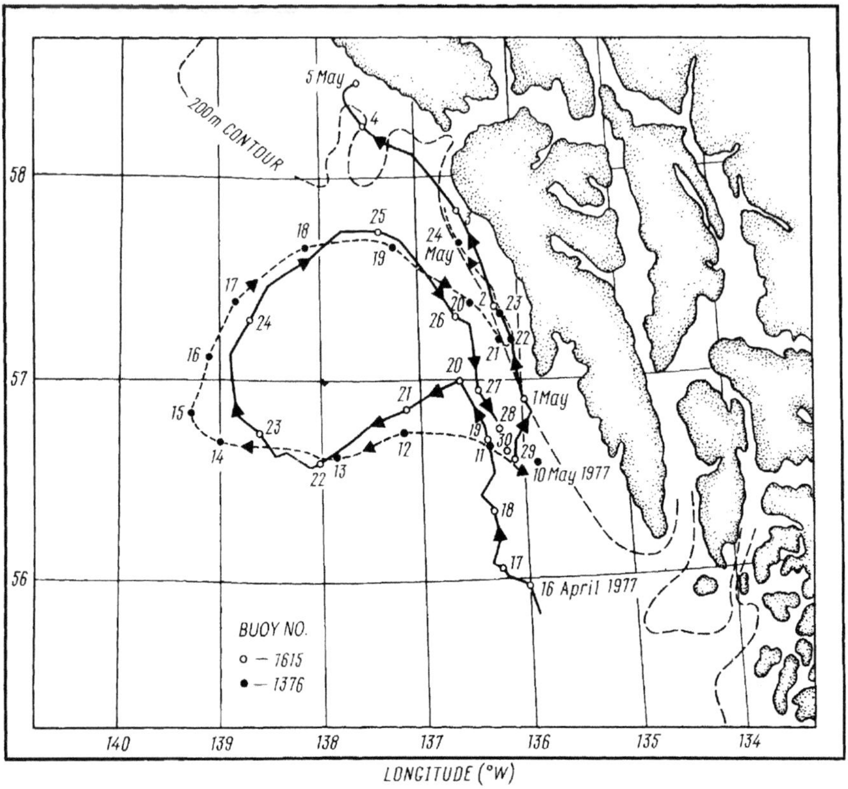

Fig. 5.4.5. Trajectories to two satellite-tracked surface floats in a region in the Pacific west of the Baranof Island in April–May 1977 (after Tabata, 1982).

The T–S–O_2 analysis showed that the eddy core in the upper ocean consisted of shelf water located permanently east of the eddy (Figure 5.4.5). In combination with the peculiar features of buoy trajectories in Figure 5.4.5, what has been said means that the disturbance under consideration is a quasistationary anticyclonic eddy-meander of the Alaska Current; it renews its water mass in the period of its greater 'openness' in the eastward direction.

The results of the hydrographic surveys of the ocean and the current measurements by ABS and satellite-tracked buoys carried out in the spring–summer periods of 1973, 1974, and 1977 in the south-east part of the Bering Sea in the region 55–56°N, 169–170°W (symbol 16 in Figure 4.2.1) were used to reveal the quasistationary eddy-meander of a large-scale current flowing to the north-west along the continental slope separating the north-east shallow-water part of the sea from its south-west deep-water part (Kinder and Coachman, 1977; Kinder *et al.*, 1980). The eddy turned out to be associated with the mesoscale disturbance of the continental slope line. The satellite surveys of the Sea of Okhotsk performed in February–April 1977, 1978, and 1979 (symbol 17 in Figure 4.2.1) indicated the presence of numerous essentially nonstationary cyclonic and anticyclonic eddy disturbances of currents with dimensions 100–150 km in the southern deep-water part of the sea; these disturbances manifest themselves in 'spirals' of floating ice. Lappo (1985) interprets these disturbances as a result of the instability of shelf waves propagating to the south-west along the continental slope of the Sea of Okhotsk.

As far back as the time of the drift of the Soviet stations 'North Pole-1' in 1937 (Shirshov, 1944) and 'North Pole-2' in 1950–1951 (Belyakov and Volkov, 1981), observations revealed episodic nonstationary currents under the ice in the Arctic Ocean. They were not related to the ice drift in a visible manner and were characterized by a maximum velocity in an intermediate ocean layer. According to the observations in 1937, the current velocity attained 30 cm/s at a depth of 100 m, and according to the observations in 1950–1951 it attained 60 cm/s at 150 m. The many current measurements and the many temperature and salinity measurements from drifting ice at all the subsequent 'North Pole' stations over the last 30 years provided continuous oceanographic and meteorological observations in the Central Arctic, have confirmed that these currents are typical of the Arctic Ocean (Belyakov and Volkov, 1980, 1981), and have allowed us to conclude that, as a rule, they are related to single eddies concentrated in an intermediate ocean layer approximately from 30 to 350 m, which corresponds to the clearly marked main pycnocline in the Arctic Ocean. Along with the single Sargasso Sea eddies occupying intermediate depths (see Sections 2 and 3 in the present chapter), almost all the observed eddies of the Arctic Ocean had well-mixed cores (Figure 5.4.6) and therefore since they were quasigeostrophic, they were characterized by the anticyclonic sign of rotation of water (Figure 5.4.7). In the cases demonstrated by Figures 5.4.6 and 5.4.7, along with the majority of the other cases where eddies have been observed in the Arctic Ocean, the drift velocity of the ice from which the

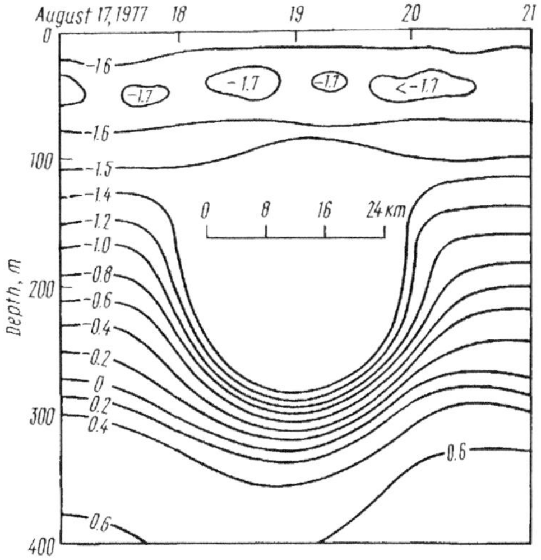

Fig. 5.4.6. Distribution of water temperature (°C) in a section across an intermediate anticyclonic eddy according to the data of sea soundings at six-hourly intervals at the ice station 'North Pole-23' in the Arctic Ocean (after Belyakov and Volkov, 1981). The figures at the top indicate the dates in August, 1977.

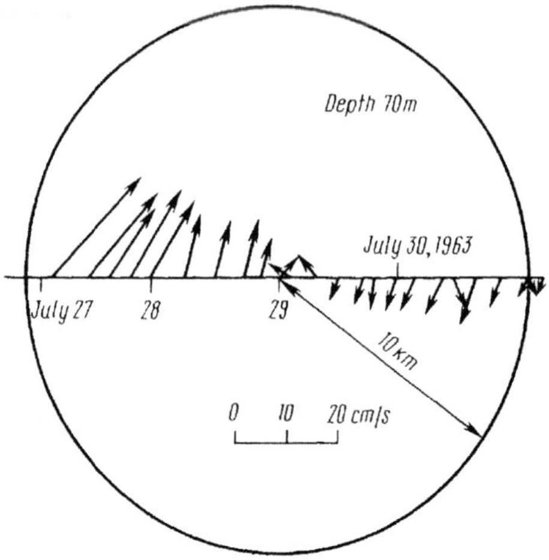

Fig. 5.4.7. Distribution of velocity vectors of currents (the north is at the top) at a depth of 70 m in a section across an anticyclonic eddy according to the data of current measurements at the ice station 'North Pole-12' in July 1963 (after Belyakov and Volkov, 1981). The circle shows the hypothetical total dimension of the eddy.

measurements were performed was much greater than the velocity of the eddy-carrying large-scale current in the intermediate ocean layer. Therefore, the distributions shown in Figures 5.4.6 and 5.4.7 can be approximately regarded as spatial and not temporal. As far as we can conclude from the observation data, the diameters of the eddies were 20–40 km for a typical Rossby scale value for the Arctic Ocean of 10 km. The current velocities in the eddies at depths of 100–250 m were, on average, 10–20 cm/s and sometimes attained 50–60 cm/s. In some cases intensive jets were registered in the eddy field; their width was only of the order of 1 km for a velocity difference between the jet axis and the surrounding water of 10–15 cm/s. The average amounts of APE and KE stored in a typical eddy of the Arctic Ocean were estimated as 32×10^{10} and 1.6×10^{10} J (Belyakov and Volkov, 1981).

Among the 1100 vertical soundings of current velocity performed in different years at the 'North Pole' stations in the Chukotsk–Alaska sector of the Arctic (Figure 5.4.8) analyzed by Belyakov and Volkov (1980), 350 soundings registered local currents with increased velocities at intermediate depths in the ocean. In many of these cases it was possible to identify, to a certain degree of reliability, the eddy structure of the currents (Figure 5.4.8) and to draw the preliminary conclusion that eddies occupy as much as 20–25% of the ocean area in the indicated Arctic region. In a sense, the results of temperature and salinity measurements in eddies are in agreement with this conclusion. They lead to the assumption that the cores of many of the anticyclonic eddies are formed from Pacific water which flows spontaneously into the Arctic Ocean through the Bering Straits (Belyakov and Volkov, 1981). Figure 5.4.8 shows that the ocean is more strongly saturated with eddies in regions of relatively high values of average vertical water density gradient in the upper part of the main pycnocline.

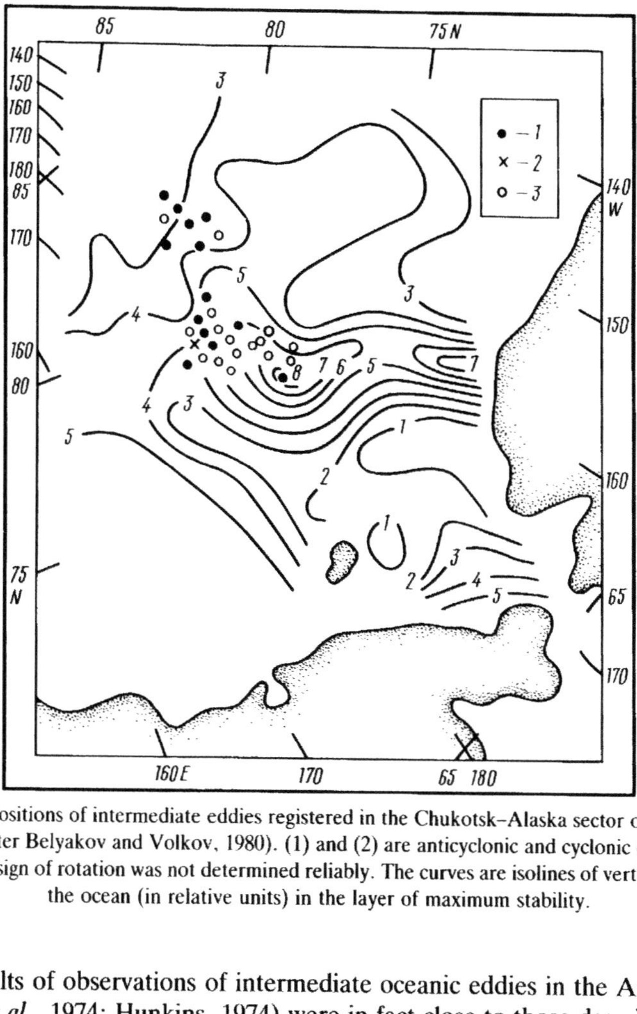

Fig. 5.4.8. Positions of intermediate eddies registered in the Chukotsk–Alaska sector of the Arctic in 1950–1964 (after Belyakov and Volkov, 1980). (1) and (2) are anticyclonic and cyclonic eddies; (3) are eddies whose sign of rotation was not determined reliably. The curves are isolines of vertical stability of the ocean (in relative units) in the layer of maximum stability.

The results of observations of intermediate oceanic eddies in the Arctic Ocean (Newton *et al.*, 1974; Hunkins, 1974) were in fact close to those described above. These observations were performed by American oceanographers at three drifting ice stations in the spring period of 1972 in the Beaufort Sea (75–76°N, 148–155°W). The current, water temperature, and salinity measurements registered three anticyclonic eddies and one cyclonic eddy with transverse dimensions of 15–35 km concentrated in the 30–350 m layer which were characterized by orbital velocities of water motion up to 40 cm/s at a depth of 150 m. During the annual drift in the Arctic of four drifting ice stations in 1975–1976, Manley and Hunkins (1978) found 34 features similar to those described above. The velocity maximum was observed at depths of 50–200 m with the average density of KE at its maximum equal to 5.9 J/m^3, whereas the average density of KE of large-scale currents attained its maximum at the sea surface and was equal to 0.2 J/m^3.

On August 5, 1973 an air photography survey of a region of the East Siberia Sea revealed simultaneously more than 50(!) nonstationary cyclonic and anticyclonic eddies which manifested themselves in the photographs of the sea surface in the form of 'spirals' of scattered fine-broken ice with transverse dimensions from 2 to 13 km (Gorbunov and Losev, 1978). The total area of the region occupied by the eddies exceeded 20 000 km^2. Both single eddies and a series of 2–6 eddies of one sign were observed. It is remarkable that during an air photography survey performed only two days earlier, no ice spirals were found. Gorbunov and Losev believe that the appearance of these eddies can be related to the reconstruction of the pattern of sea currents (the sea depth in the observation region was only 10–12 m) begin forced by variable wind.

In 1979–1981 Soviet oceanographers carried out purposeful investigations of synoptic disturbances in the central part of the Baltic Sea with depths from 80 to 150 m (Aitsam *et al.*, 1981). CTD surveys of the sea and ABS current measurements were both performed. Besides some other disturbances in the circulation of sea water, clearly marked synoptic eddies were found. They belonged to the following types: (a) cyclones with diameters of the order of 40 km (the estimation of the Rossby scale for the observation region yielded 10 km), in which the relative elevation of the main pycnocline reached 20–25 m and maximum velocities at the sea surface attained 15–20 cm/s; and (b) anticyclones with diameters about 20 km, an elevation of isopycnic surfaces in the upper part of the pycnocline and a sinking in its lower part, and maximum velocities in the intermediate ocean layer. Both types of eddies had a tendency to move along the isobaths of the bottom relief with velocities of 1–2 km/day. The T, S analysis showed that the strongest eddies transported water in their cores. The estimation of the values of the APE density averaged over the eddy volume yielded 1000 cm^2/s^2 for eddies of the first type and 200 cm^2/s^2 for eddies of the second type. An analysis of the observation data made it possible to conclude that the basic mechanism of generation of eddies was the baroclinic instability of mean ('permanent') sea currents.

The investigations of synoptic eddies at high latitudes of the Southern Hemisphere were primarily associated with the Antarctic Circumpolar Current (ACC) which is the most powerful current of the ocean in its total transport (150–200 $\times$ 10^6 m^3/s; Sarukhanyan, 1980b). The drift trajectories of surface buoys in Figure 5.4.9 give a good indication of the ocean zone where the ACC flows, if we exclude from consideration the two extreme northern trajectories west of Chile and also the trajectory south-east of Africa which corresponds to the Agulhas Return Current. The ACC axis coincides with the Antarctic Front (AF) separating in the upper 1000 m layer the subantarctic water in the north, with a typical temperature of 3–6 °C, from the colder (and somewhat less saline) antarctic water in the south which is characterized in the summer period by the presence at depths of 50–500 m of an intermediate low-temperature water layer with a minimum temperature slightly higher or lower than 0 °C. The AF is located near 60°S in the Pacific sector of the Southern Ocean and approximately at 50°S in the Indian and Atlantic sectors. As a rule, the ACC velocity in the axis in the upper ocean is $\times$10 cm/s, but in some regions, e.g. in the Drake Passage, it can attain 1 m/s. At depths greater than 2–2.5 km the average current velocity is of the order of 1 cm/s.

It is advisable to begin the review of synoptic eddies in the Southern Ocean with the Australian sector. As early as October–November 1967 the Soviet satellite 'Kosmos-184' photographed eddy-shaped cyclonic and anticyclonic spirals of float-

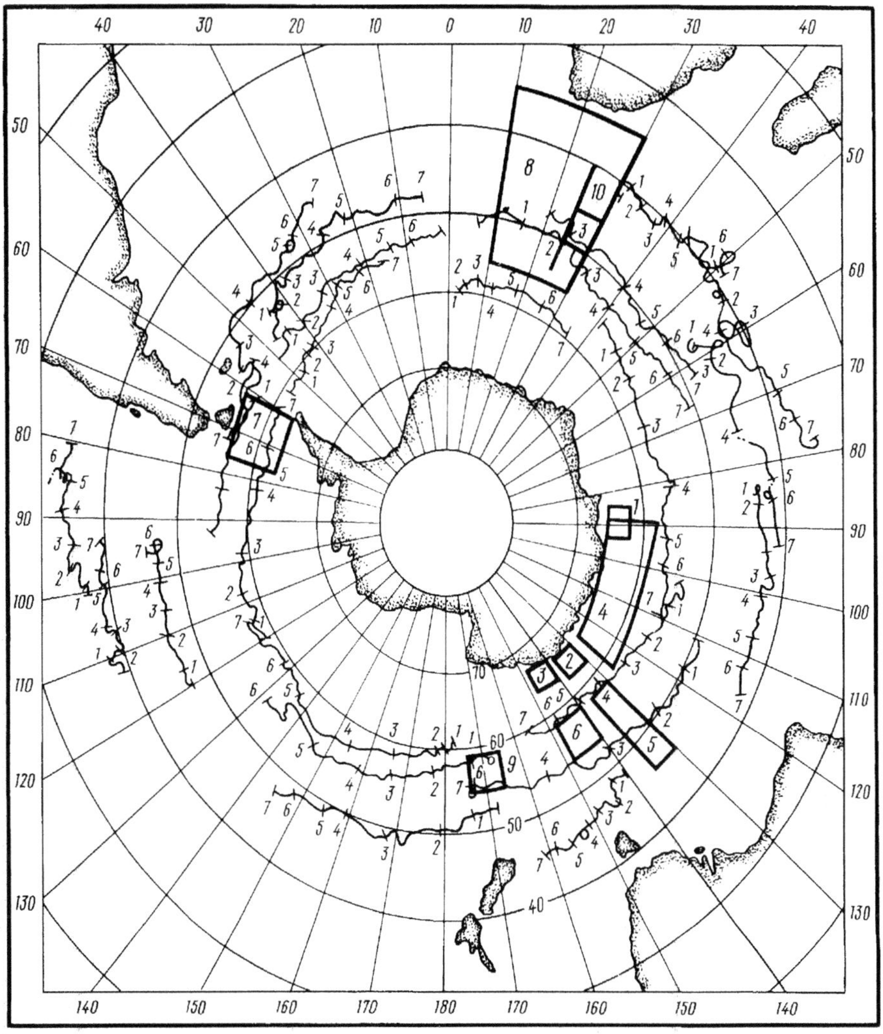

Fig. 5.4.9. Trajectories of satellite-tracked surface drifting buoys in the Southern Ocean in January–July 1979 (after Kort, 1981). The figures at the trajectories indicate the months of 1979. The individual symbols with figures mark the places of the experiments mentioned in the text.

ing ice at the Antarctic coast in the regions 90–95, 130–137, and 140–142°E (Preobrazhenskaya, 1971; symbols 1–3 in Figure 5.4.9). At 130–137°E a chain of three interrelated spiral disturbances was observed, in which the structure was similar to that of eddies observed in the Bering Sea (Section 3 of Chapter 4). The transverse dimensions of the eddies varied from 30 to 100 km. Geographically, these results are analogous to those found for the drift of two antarctic icebergs in the sector 90–120°E (symbol 4 in Figure 5.4.9) from observations carried out between January 1976 and April 1977 by location of the icebergs from satellite Nimbus F (Tchernia and Jeannin, 1980). At the beginning of the observations the two icebergs were located at 64–65°S. During the first 5–6 months they drifted in the general direction of the west and then west-north-west in the system of the near-continent Eastern Wind Current until they reached the region 62–63°S, 92–96°E where they were captured by the southern branch of the ACC, owing to which the westward drift of the icebergs was replaced by an eastward drift that was observed in the zone 60–63°S up to the end of the experiment. During the whole drift the motion of the two icebergs underwent clearly marked synoptic-scale disturbances, and both the absolute ('instantaneous') velocity of the drift and the dimensions and periods of the synoptic meanders and loops of the trajectories of the drift were much greater than in the field of the Eastern Wind Current. The amplitude of the meanders in the ACC field attained 150–180 km for a typical drift velocity of an iceberg in a meander equal to 10–20 km/day. The observations also revealed loop-shaped parts of trajectories with a typical loop diameter of 20–30 km, which indicated the existence of synoptic-scale eddies in the ACC field. The spectrum analysis of the observation data on the drift of the icebergs showed the predominance of current oscillations with periods of 15–20 days.

The first detailed observations of an individual synoptic eddy in the axial part of the ACC were carried out by a Soviet oceanographic expedition in January–February 1977 in the region 50–52°S, 131–133°E (symbol 5 in Figure 5.4.9) (Savchenko *et al.*, 1978). The bathometer observations, the XBT soundings, and the 16-day current measurements at two buoy stations to a depth of 3000 m registered a cyclonic eddy with a transverse dimension of 180 km and an isothermal surface depth difference in the 200–1000 m layer attaining 600 m(!), to which there corresponded a temperature drop about 3 °C. The current velocity in the eddy field attained 50 cm/s at depths of 400–500 m and 15 cm/s at a depth of 2000 m. The current measurements at 3000 m indicated supposedly the existence of a weak counter (anticyclonic) circulation of water in the deep layer. The T, S relation for the water in the central part of the eddy showed that its formation had resulted from the separation of a northward cyclonic meander of the AF; at the instant of observation the front was 3–4° south of the eddy's location. The eddy moved to the north-north-east with an average velocity of 3 cm/s. In September 1976 a satellite survey of the sea surface temperature performed in the region 54–58°S, 140–150°E (symbol 6 in Figure 5.4.9) revealed four warm anticyclonic eddies that were probably formed by the separation of anticyclonic meanders of the AF (Legeckis, 1978). The biggest of the eddies had a transverse dimension of about 150 km and was contoured by a temperature front with a temperature drop of 2 °C at a distance of 2 km.

In December 1982–January 1983 a Soviet expedition discovered and investigated another powerful cold cyclonic eddy in a sector of the Southern Ocean south of New Zealand (Figure 5.4.10; symbol 9 in Figure 5.4.9) (Ivanov *et al.*, 1984). The eddy had an elongated form with a total transverse dimension of about 200 km and was very clearly marked in the current and temperature fields. For example, the elevation of isothermal surfaces at the center of the eddy was about 1 km(!) and the surface current velocity attained 1.5 m/s in some parts of the eddy. Salinity in the upper ocean increased from 34.0‰ at the center of the eddy to 34.4‰ on its periphery. The core of the eddy was formed by antarctic water located stationarily

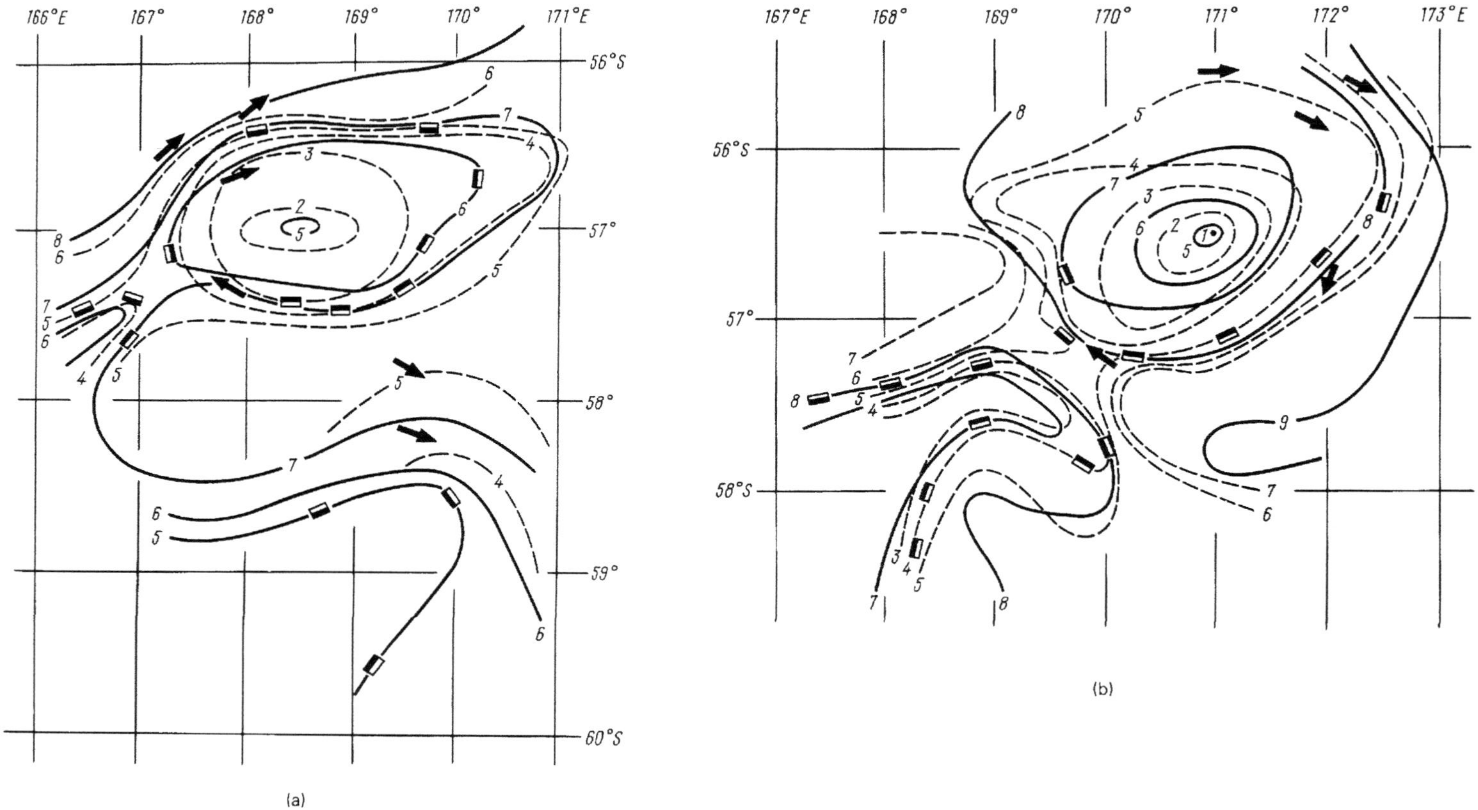

Fig. 5.4.10. Distribution of temperature (°C) at the sea surface (solid lines) and at a depth of 200 m (dashed lines) on (a) January 1–8 and (b) January 16–22, 1983 in a region of the Southern Ocean south of New Zealand (after Ivanov *et al.*, 1984). The arrows show the direction of drift of the ship for the cases when the drift velocity exceeded 1.5 knots. The half-blackened rectangles show the positions of sharp temperature fronts at the sea surface. (See symbol 9 in Figure 5.4.9.)

in this ocean region south of 59°S and differing sharply in its physico-chemical characteristics from the subantarctic water surrounding the eddy. As can be concluded from Figure 5.4.10, during the observation period the eddy was not entirely separated from a cyclonic AF meander located south-west of it, and moved to the east-north-east together with the meander with a velocity of about 10 km/day. It is also interesting that during a two-week period the dimension of the eddy considerably increased as it drew in the cyclonic motion of the subantarctic water surrounding the eddy. This is clearly seen from a comparison of the position of the 8 °C isotherm at the sea surface in (a) and (b) of Figure 5.4.10.

The first direct current measurements in the Drake Passage (symbol 7 in Figure 5.4.9) were performed by Soviet oceanographers at four buoy stations in December 1974–February 1975 (Bagryantsev *et al.*, 1976). They revealed strong synoptic oscillations of currents with a dominant period of about 15 days. The kinetic energies of the synoptic and mean currents in the 0–1000 m layer turned out to be of the same order (250–500 cm^2/s^2); at depths of 3000 and 3680 m, the energy of the synoptic currents (30–70 cm^2/s^2) substantially exceeded that of the mean currents (3–13 cm^2/s^2).

In 1975–1977 American oceanographers carried out a long-term program of current and water temperature measurements in the Drake Passage with the aid of three consecutive annual buoy station systems, each of which consisted of five or six stations and involved from one to five measurement levels at each of the stations. The statistical processing of these observation data (Pillsbury *et al.*, 1979; Sciremammano *et al.*, 1980) showed that at depths of 300–650 m in the central part of the Drake Passage the KE of the mean (large-scale) currents was approximately twice as great as the average energy of synoptic fluctuations of the currents (250–350 and 100–200 cm^2/s^2, respectively). These energies were approximately equal at depths of 2000–3000 m and varied at different stations from 10 to 40 cm^2/s^2. Throughout the entire ocean layer under study (from 300 to 3500 m), synoptic oscillations of currents were concentrated predominantly within the range of periods of 5–50 days, with a high depth coherence of oscillations. For a Rossby scale of 20 km in the central part of the Drake Passage, the transverse horizontal scale of the synoptic component of the velocity field (the correlation radius in the direction transverse to the direction of the current) was 30–40 km, the longitudinal scale was 60–80 km, and the horizontal scale of the synoptic component of the temperature field was 80 km.

A synoptic analysis of the results of the above-mentioned current and temperature measurements, and simultaneous hydrographic ship observations, showed that the synoptic oscillations of the current and temperature fields in the Drake Passage can be due to the following three factors (Sciremammano *et al.*, 1980): (1) the meridional displacements of the ACC axis located along the AF; (2) the meandering of the ACC axis; and (3) the separation of meanders of the ACC axis and further propagation of the cold cyclonic and warm anticyclonic eddies thus formed. The formation and further propagation of two cyclonic eddies of this origin were registered in the central part of the Drake Passage by CTD and XBT sections and surveys in March–April 1976 (Joyce *et al.*, 1981) and in January–February 1979 (Peterson *et al.*, 1982). In both cases the interval from the beginning of the formation of a meander to the complete separation of an eddy from the AF was 3–4 weeks. The separation process of the second of the above-mentioned eddies is very clearly seen in Figure 5.4.11 where, to the core of the eddy, there corresponds a region of minimum temperature in the northern part of the section. The two eddies were characterized by a total transverse dimension of 100–120 km; the current velocity of the upper ocean in the eddy of 1976 attained 50 cm/s, and attained 90 cm/s in the eddy of 1979. During the separation process from the AF, and further drift, the first eddy moved to the north-north-east with a velocity of 5 –10 cm/s and the other eddy moved to the north-north-west with approximately the same velocity. The volume-integrated APE (see Section 2) and KE of the first eddy were, respectively, 5.1×10^{14} and 3.4×10^{14} J. The *T*, *S* analysis reliably confirmed that the cores of these eddies in the Drake Passage were formed by antarctic water. The authors (Joyce *et al.*, 1981; Peterson *et al.*, 1982) introduced an 'isopycnic' temperature anomaly for the water masses of the eddies. This is defined as the difference between the water temperature at a given point inside the eddy and the water temperature outside the eddy corresponding to the same value of the density of the water for which the temperature inside the eddy is taken. The isopycnic salinity anomaly is introduced in like manner, after which it becomes possible to estimate the volume-integrated isopycnic anomalies of the contents of heat and salts which, obviously, are measures of the transport of heat and salts across the AF associated with the eddy formation process at the front. The calculations give the following values of the indicated integrated anomalies: -1.2×10^{19} J and -2.5×10^{11} kg for the eddy of 1976 and -0.8×10^{19} J and -2.0×10^{11} kg for the eddy of 1979. The annual average rate of transfer of heat from the ocean to the atmosphere in the entire region of the Southern Ocean south of the AF is estimated as 4×10^{14} W (Gordon and Taylor, 1975). If we assume that

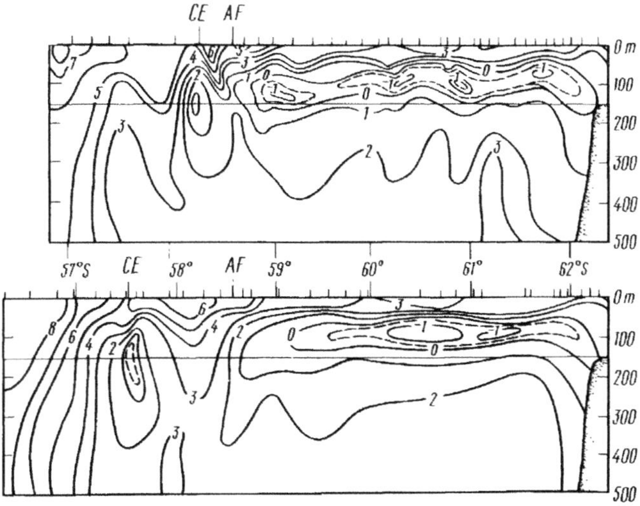

Fig. 5.4.11. Distributions of temperature (°C) in a section across the Drake Passage (symbol 7 in Figure 5.4.9) on January 14–18 and January 23–24, 1979 (after Peterson *et al.*, 1982). The symbols AF and CE show the positions of the Antarctic Front and the center of the cyclonic eddy; the marks on the sea surface line indicate the places of thermal soundings of the ocean.

the eddies formed at the AF do not come into repeated contact with the front (which is hardly a realistic assumption), then it follows that about 1400 cold cyclonic eddies, whose power is equal to that of the above-mentioned eddies and the same number of warm anticyclones formed from AF anticyclonic meanders, are required to compensate for the annual loss of heat of the indicated region of the Southern Ocean. Such a large number of eddies can hardly occur. However, the indicated number of eddies seems to be overestimated several times because, as is shown by experimental data (e.g. see Ivanov *et al.*, 1984; Bubnov *et al.*, 1984), AF eddies in the Drake Passage have substantially smaller dimensions than analogous eddies in other regions of the Southern Ocean. Peterson *et al.*, (1982) believe that the penetration of cold AF cyclones into the subantarctic ocean zone, and the mixing of their waters with the surrounding subantarctic water, plays an important part in the formation of the intermediate antarctic water which, as is known, propagates in the intermediate ocean layer from the ACC zone to the north in all three oceans.

Among the twelve XBT sections across the Drake Passage performed from November 1976 to March 1977 (Sievers and Emery, 1978), four sections registered very clearly marked local temperature minima north of the AF. They were analogous to those presented in Figure 5.4.11 and indicated the presence of cut-off or nearly cut-off cyclonic eddies. In two cases in which the eddies were particularly strong and specified the existence of notable temperature minima not only in the intermediate ocean layer but also on its surface, local disturbances in the temperature of the near-water air layer and in the direction and power of the wind were registered in the corresponding ocean regions. In September 1976 a satellite survey of the sea surface temperature in the Drake Passage registered the formation of a warm eddy from an anticyclonic meander of the AF (Legeckis, 1978).

The results of the above-mentioned three-year series of current and temperature measurements at buoy stations in the Drake Passage in 1975–1977 were also used to estimate the mean meridional transfer of heat produced by eddies in this ocean region (Bryden, 1979; Sciremammano, 1980). The coefficient of correlation between oscillations of temperature and the meridional component of velocity in the synoptic range (periods between 10 and 100 days) at depths of 1500–3000 m in the central part of the Drake Passage was estimated to be in the order of 0.30–0.55. The average meridional heat flux was estimated as 1.7 W/cm^2 with a spread from 0.9 to 2.8 W/cm^2 for different series of observation data. About 75–90% of these resulting heat fluxes for all the series, without exception, was realized by the passage of only 3–5 eddies a year and the total time of the passage was 15–25% of the observation period. The last fact is very important for the correct planning of observations of eddies in the Southern Ocean. To correctly estimate the resultant meridional transport of heat in the antarctic zone of the ocean it is also necessary to know whether the process of reverse absorption of eddies by the AF generating them (similar to the process in the Gulf Stream region) is typical of this zone. It is clear that this requires special observations.

As the density of water depends on its temperature, the meridional transport of heat in the Antarctic zone of the ocean produced by eddies must be accompanied by the transformation of APE in the large-scale ACC into the APE and KE of eddies. Bryden (1979) used the values of the parameters of the ACC and the estimates of the above-mentioned heat flux he had found (from the data of the above-mentioned current and temperature measurements in the Drake Passage) to calculate the typical time of renewal of the resultant energy of eddies. He obtained a time scale lying within the limits of 15–55 days, which turned out to be close to the typical time of transfer of KE to the ACC from the wind. The baroclinic instability of the ACC is a natural factor transforming the APE of the ACC into eddy energy. The vertical phase shift of oscillations of current velocity discovered during current measurements in the Drake Passage in 1975 (Bryden, 1979) can be regarded as a confirmation of this assumption.

In December 1978–February 1979 a Soviet oceanographic expedition performed a hydrographic survey of a vast ocean region south of Africa (Figure 4.3.6; symbol 8 in Figure 5.4.9) and also carried out two-month current measurements in a layer from 200 to 4000 m at four buoy stations along the 20°E meridian from 47 to 52°S (Sarukhanyan, 1980a). The results of the survey were described in Section 3 of Chapter 4. Here we note additionally that the survey clearly revealed the strongest part of the ACC in the zone 46–52°S and the presence of synoptic-scale meanders and eddies in the ACC field (Figure 4.3.6). The current measurements showed that the synoptic disturbances in water circulation penetrated to the deepest parts of the ocean.

In December 1982–February 1983 Soviet oceanographers again carried out an expedition in the African sector of the Southern Ocean (Bubnov *et al.*, 1984; Koshlyakov *et al.*, 1985; symbol 10 in Figure 5.4.9). First, a CTD section along 25°E was performed. This revealed traces of a powerful cyclonic eddy in the region 47–48.5°S which first manifested itself by elevating isotherms more than 1 km(!) in the deep ocean layer (Figure 5.4.12). In this section the ACC occupied a zone from 44°S to the southern edge of the section, and the AF was located at 49–49.5°S. Further CTD surveys showed that the section intersected the western periphery of the eddy and that its center was approximately at 47°20′S, 26°E. Therefore, to survey the eddy a buoy system was emplaced; from 5 to 15 January 1983 it consisted of six ABS with current meters at depths 75, 300, 600, 1000, 1700, and 2600 m, and from January 15 to February 5 it consisted of four ABS with meters at 75, 300, 600, and 1700 m (Figure 5.4.13). This was the first direct long-term spatial survey of currents in the field of a synoptic eddy performed in the Southern Ocean. The velocity vectors in Figure 5.4.13 were found from the measurement data after inertial and tidal currents had been filtered out, and hence they are sums of the velocity vectors of large-scale and synoptic currents. As can be concluded from Figure 5.4.13, the eddy had an elliptic form with an axes ratio of the order of 1.5 and a minor semi-axis (the distance from the center of the eddy to the point of maximum velocity) about 50 km. During the observation period the major axis of the eddy turned clockwise through 80°. The center of the eddy moved with a velocity of the order of 1–1.5 km/day; it first travelled a distance of about 15 km to the south-east and then a distance of about 35 km to the north-west. The current velocity in the northern (north-eastern) part of the eddy, where the synoptic component of the current added to the northern jet of the ACC, attained 65 cm/s at a depth of 300 m (Figure 5.4.13) and 45 cm/s at a depth of 1700 m. Hence, the barotropic component of the current was very clearly marked.

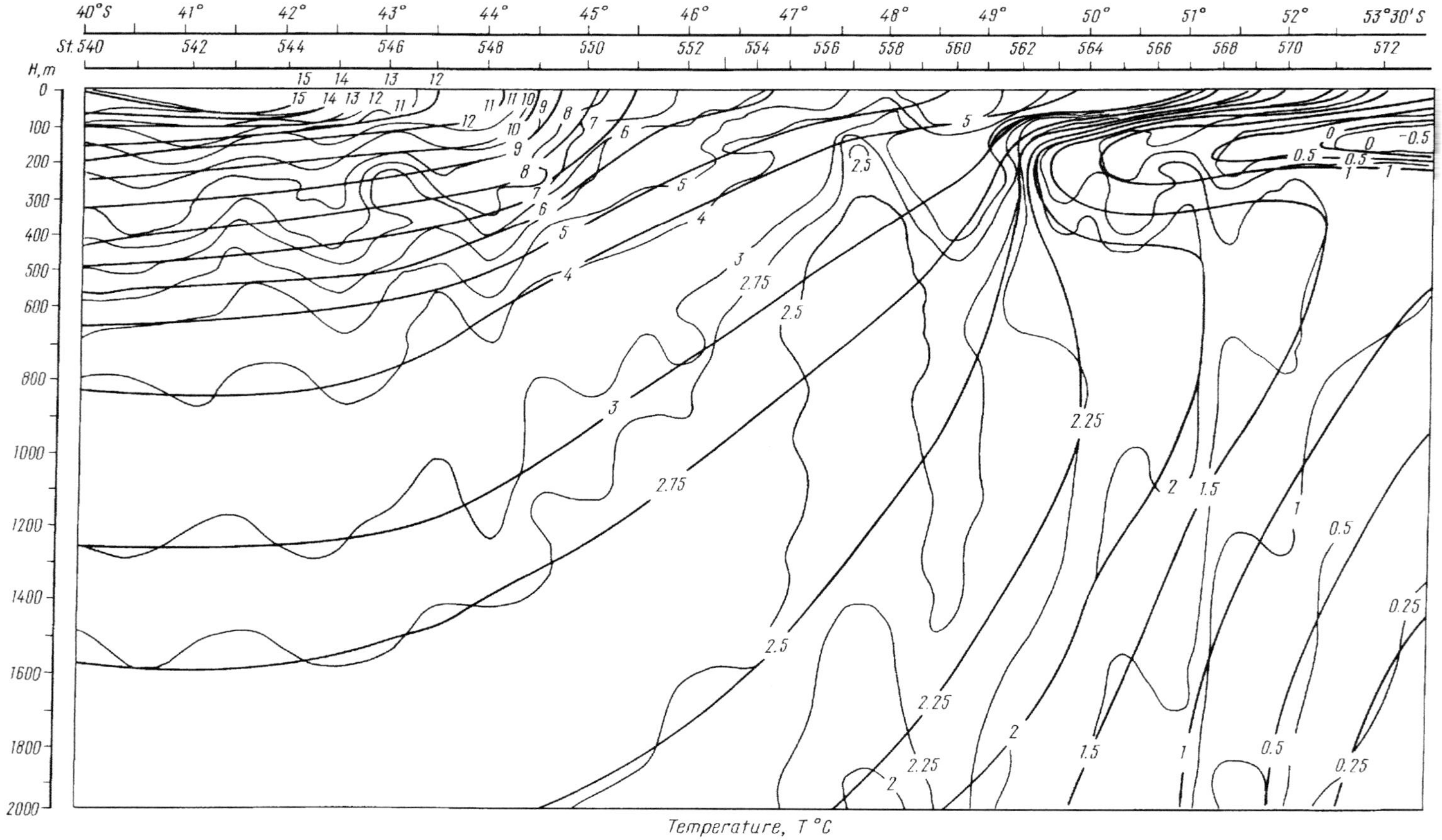

Fig. 5.4.12. Observed and smoothed (large-scale) distributions of water temperature (°C) in a section across the Antarctic Circumpolar Current along 25°E on December 23–31, 1982 (after Koshlyakov *et al.*, 1985). The positions and numbers of CTD stations are shown at the top. (See symbol 10 in Figure 5.4.9.)

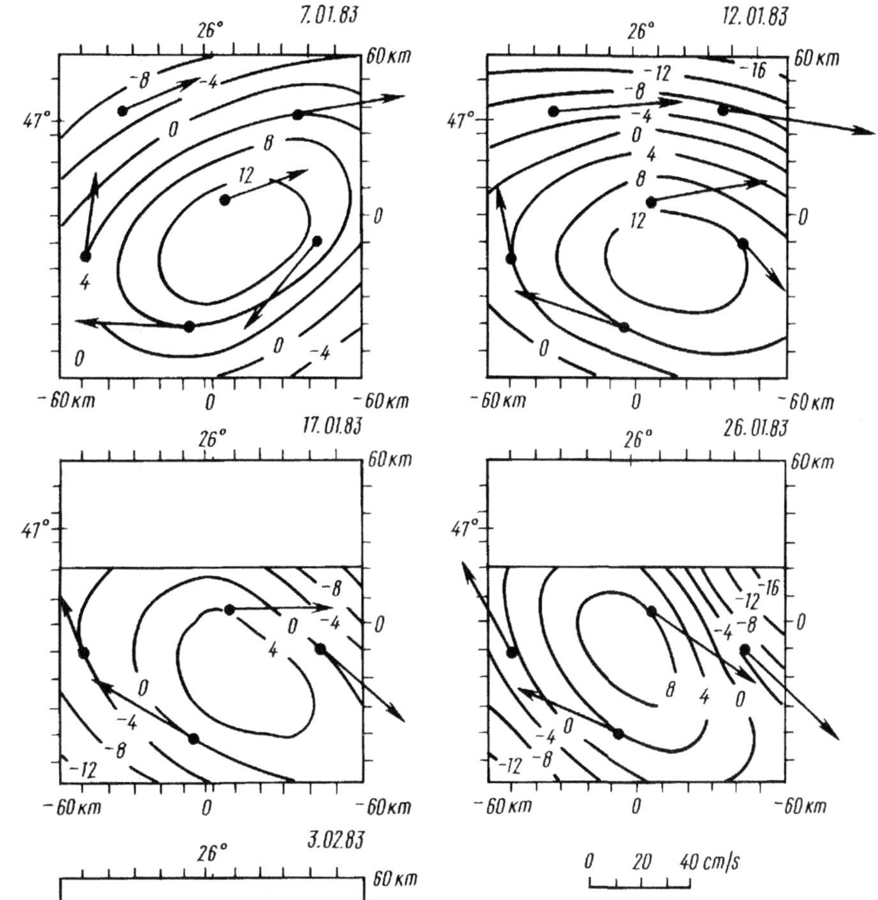

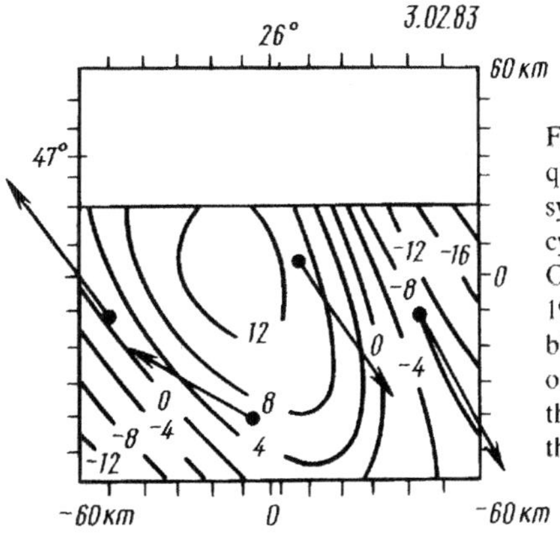

Fig. 5.4.13. Velocity vectors and streamlines of the quasistationary current (the large-scale current plus the synoptic current) at a depth of 300 m in the region of a cyclonic eddy in the African sector of the Southern Ocean for different dates in 1983 (after Bubnov *et al.*, 1984). The stream function is given in 10^7 cm^2/s. The black dots show the positions of the ABS. The marks on the sides of the squares indicate the step (10 km) of the grid at whose points the optimal interpolation of the velocity field was performed. (See symbol 10 in Figure 5.4.9.)

Figure 5.4.14 together with Figure 5.4.12 show very clearly that the core of the cyclonic eddy under consideration was formed of antarctic water permanently located immediately south of the AF zone. At the same time, the contrast between the water mass of the eddy and the background subantarctic water surrounding the eddy (curves 1 and 4 in Figure 5.4.14) was very sharp. This result, together with the facts that the eddy remained in the observation region during the entire obser-

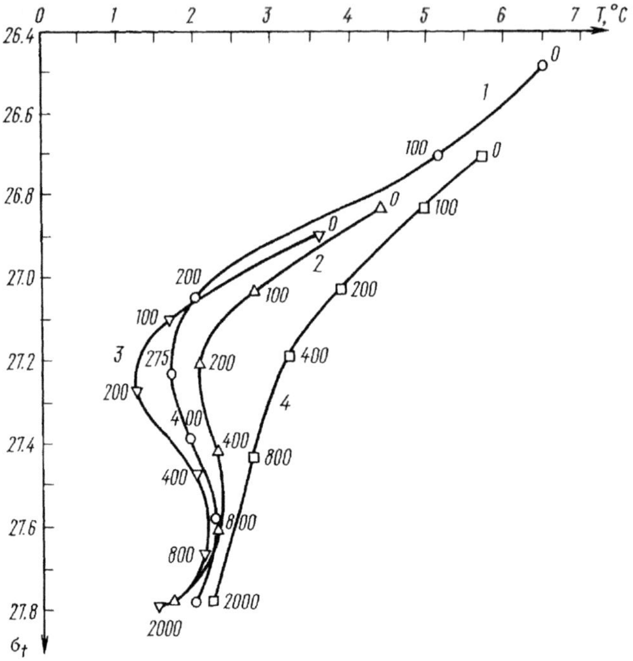

Fig. 5.4.14. $T - \sigma_t$ curves for (1) the center of a cyclonic eddy (at 47°20′S, 25°50′E; Figure 5.4.13) according to the data of the CTD survey of the eddy on February 7–12, 1983 and for (2) 49°30′, (3) 50°, and (4) 47°20′S, respectively, according to the smoothed distributions of temperature (Figure 5.4.12) and water density at a section along 25°E (after Koshlyakov *et al.*, 1985). The figures at the curves indicate the depth in meters.

vation period (Figure 5.4.13) and that there was a completely analogous eddy in the same ocean region in the dynamic charts constructed from the data of the 1978–1979 expedition (Figure 4.3.6), leads to the conclusion that the disturbance under study is a quasistationary eddy-meander of the axial ACC jet flowing along the AF. From time to time it opens on its southern side and thus renews the water of antarctic origin in its core. While the eddy is in the state of separation from the AF, as occurred during the observation period, there is an exchange of water between the eddy and the surrounding ocean owing to the mesoscale mixing inside the isopycnic water layers constituting the fine structure of the oceanic pycnocline, and a great number of such layers were found during the expedition. The permanent presence of 'eddy-meander' in the ocean region under study is probably related to the large-scale relief of the ocean bottom and, in particular, to the Africa–Antarctica Ridge located south-east of the observation region.

To estimate the transfer of heat across the AF owing to the above-mentioned eddy, the distributions of the isopycnic and horizontal synoptic anomalies of water temperature in the body of the eddy were calculated; the smoothed temperature distribution at the section along 25°E (Figure 5.4.12) in the zone of the eddy was taken as the background distribution in these calculations. The relative position of the curves of the isopycnic and horizontal anomalies of heat content in the eddy water

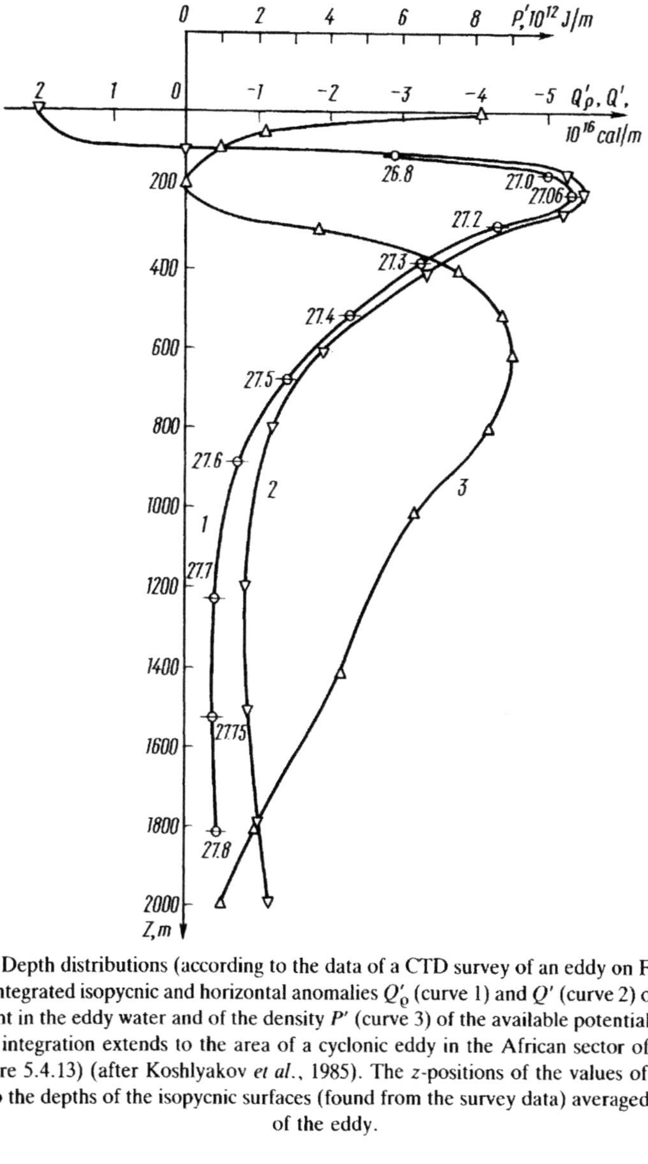

Fig. 5.4.15. Depth distributions (according to the data of a CTD survey of an eddy on February 7–12, 1983) of the integrated isopycnic and horizontal anomalies Q'_ϱ (curve 1) and Q' (curve 2) of the densities of heat content in the eddy water and of the density P' (curve 3) of the available potential energy of the currents (the integration extends to the area of a cyclonic eddy in the African sector of the Southern Ocean) (Figure 5.4.13) (after Koshlyakov *et al.*, 1985). The z-positions of the values of σ_t on curve 1 correspond to the depths of the isopycnic surfaces (found from the survey data) averaged over the area of the eddy.

(Figure 5.4.15) is remarkable. It shows that the upper part of the eddy, approximately above a depth of 700 m, was formed almost entirely by antarctic water, whereas with increasing depths the effect of drawing the local subantarctic water in the cyclonic rotation became stronger. If we assume that Q'_ϱ (Figure 5.4.15) remained constant in the entire ocean layer from a depth of 1800 m to the bottom (at a depth of 5400 m), then the isopycnic anomaly of heat content integrated over the whole volume of the eddy is equal to -4.2×10^{19} cal, which is 16(!) and 24(!) times as great as the analogous anomalies in the two above-mentioned cyclonic eddies in the Drake Passage. Taking into account the above estimate of the annual

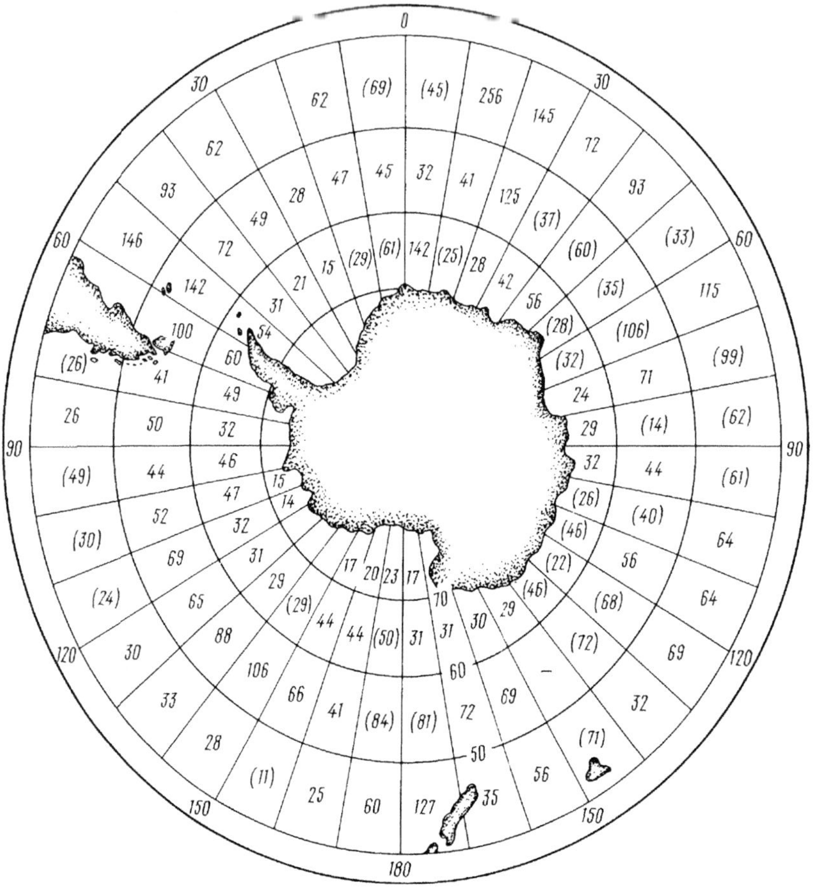

Fig. 5.4.16. Distributions of the root-mean-square values of the synoptic disturbances of the dynamic height (in 10^{-2} m^2/s^2) of the sea surface relative to the 1000 dbar reference surface in the Southern Ocean (after Lutjeharms and Baker, 1980). Given in parentheses are the values calculated from less than ten hydrographic stations. The blank squares symbolize that there were less than two stations.

average rate of heat transfer from the ocean to the atmosphere south of the AF (4×10^{14} W), we find that in this case only about 70 pairs of cyclones and anticyclones as powerful as the eddy described above are required to compensate for this transfer of heat. This number of eddies is quite realistic and allows us to assume that AF eddies play a notable role in the formation of climatic meridional heat fluxes in the Southern Ocean. If we neglect the APE of the eddy stored in the layer below 2000 m (Figure 5.4.15), then the total APE of the eddy is 9.4×10^{15} J.

The chart in Figure 5.4.16 was constructed from all the available hydrographic and bathythermograph measurement data. The synoptic anomalies of the dynamic height of the sea surface were calculated relative to the average distribution of this quantity in the Southern Ocean. In agreement with the results mentioned in Section 3 of Chapter 4, the chart in Figure 5.4.16 demonstrates a particularly high level of synoptic disturbances in the regions of the Agulhas Return Current

(10–30°E) and the Falkland Current (50–60°W). Increased synoptic activity is also observed along the whole ACC axis and, particularly, in those places where it narrows and thus intensifies (60–70°W, 130–150°W, 170°E–170°W, and 60–80°E).

In 1979, within the framework of the international meteorological program GARP, more than 300 free-drifting surface floats launched in the Southern Ocean were periodically tracked by the satellite 'Tiros-N' (Kort, 1981). Figure 5.4.9 demonstrates the trajectories of some of the floats for the first half of 1979. It can be concluded from the figure that the total track of a surface float around the Antarctic lasted about 40 months, which corresponds to an average velocity of eastward drift of 18 cm/s. Synoptic disturbances of this large-scale transfer manifested themselves mainly in wave-shaped bendings of float trajectories with a wavelength of 200–400 km, which probably indicated the development of Rossby waves transverse to the ACC. Besides these wave-shaped bendings, the trajectories shown in Figure 5.4.9 involve altogether 20 loop-shaped parts which probably indicate the presence of cut-off synoptic eddies. The major part of these loops was also concentrated in regions of particularly strong large-scale currents, namely the Agulhas Return Current and the beginning of the Falkland Current in the region of the Scotia Sea. In the other parts of the antarctic circle the loops in the trajectories were obviously single. On average, the diameter of the loops was 140 km and the average velocity of the floats in the loop-shaped parts of the trajectories was 25 cm/s.

5. SYNOPTIC EDDIES IN THE WORLD OCEAN

The first attempt to generalize quantitatively the properties of synoptic eddies on the scale of the World ocean was made by Wyrtki *et al.* (1976) who constructed charts of the kinetic energy of large-scale (mean) currents (KELC) and the average kinetic energy of synoptic currents (KESC) in the surface layer of the world ocean (Figure 5.5.1). The charts were calculated from the ship drift data obtained from 1900 to 1972 (altogether about four million measurements). The average multiyear drift in a given 5° square was interpreted as the action of a large-scale current, and any deviation of an individual diurnal drift from the average multiyear drift was interpreted as the action of a synoptic current. First, the general similarity between the KESC and KELC distribution in Figure 5.5.1 attracts attention. The two distributions attain their maximum values in the regions of western boundary currents (the Gulf Stream, the Kuroshio, and the Somali, Guiana, Agulhas, East Australian, and Brazil Currents), in the equatorial zone of the ocean, and near the ACC zone[2]; in these regions the value of the KESC is approximately 2–4 times as great as that of the KELC. The minimum values of the KELC and KESC are observed in central parts of subtropical gyres; here the KESC value is 15–30 times as great as the KELC value. The average values of the KELC and KESC over the

[2] The large values of KESC in the equatorial zone of the ocean and in the region of the Somali Current are in fact specified by seasonal disturbances in the ocean circulation rather than by synoptic disturbances.

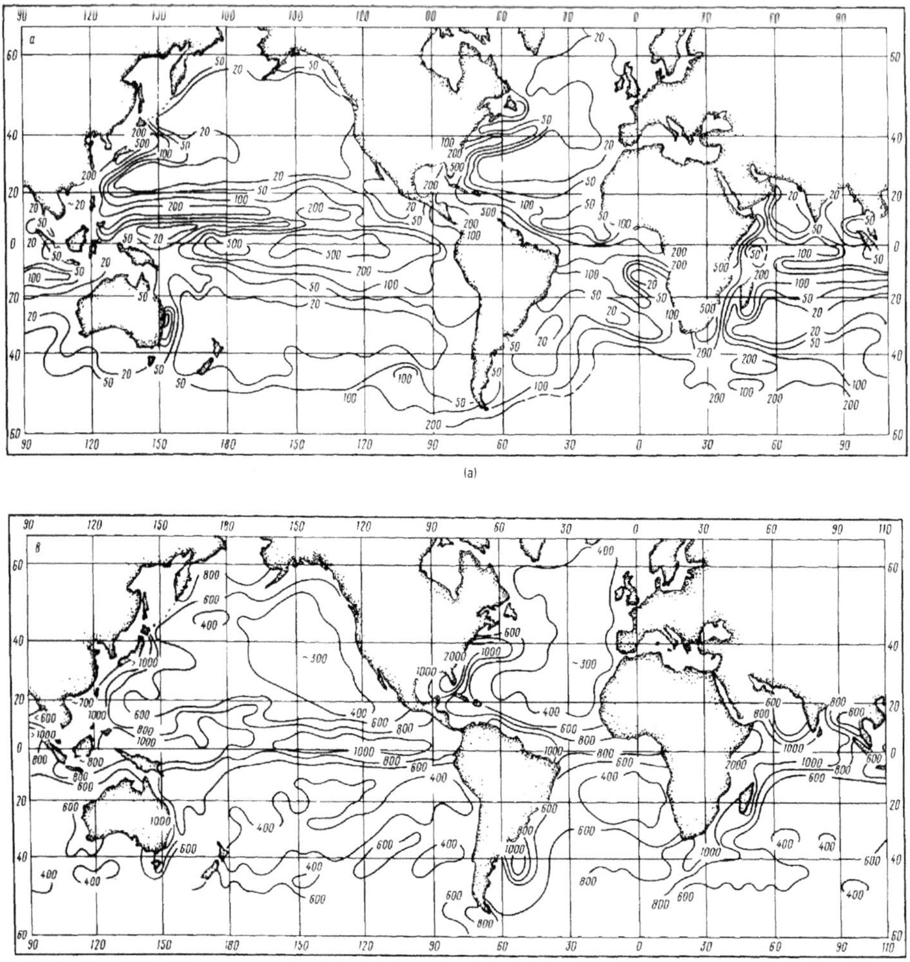

(b)

Fig. 5.5.1. Distribution of (a) the kinetic energy of mean currents and (b) the average kinetic energy of synoptic currents in the surface layer of the world ocean (after Wyrtki *et al.*, 1976). The energy is measured in cm^2/s^2.

whole ocean turned out to be equal to 100 and 500 cm^2/s^2, respectively. All the above-mentioned specific features of the KELC and KESC distributions do not contradict the idea that the main mechanism of generation of synoptic eddies in the world ocean is the baroclinic instability of large-scale currents.

Richardson (1983) calculated the average distribution of kinetic energy of synoptic-scale surface currents in the subtropical and temperate zones of the North Atlantic from the drift data of 110 satellite surface buoys with a total duration of 55 buoy-years. Although the resulting energy distribution turned out to be qualitatively similar to the one presented in Figure 5.5.1(b), the values of energy density in the Gulf Stream region were several times as great as those obtained by Wyrtki *et al.* (1976), the maximum energy (which exceeded 3000 cm^2/s^2) being attained in the axial part of the Gulf Stream between Cape Hatteras and the Grand Newfoundland Banks. In relation to these details, preference should be given to buoy data since the diurnal averaging of ship drift data undoubtedly led to considerable underestimation of the energy of synoptic currents (Richardson, 1983). The buoy data relating to the region of the North Atlantic Current indicated rather reliably the existence of seasonal variations in the energy level of synoptic current disturbances which follow the seasonal variations of the wind stress field approximately with a three-month lag (Figure 5.5.2). This result is in good agreement with the above-mentioned fact of an annual variability in the energy of synoptic eddies in the POLYMODE region (Figure 5.2.7) with maximum energy in the same period of the year (April–May); it allows us to assume that there exists a similar variability in the eddy field throughout the entire northern half of the Atlantic.

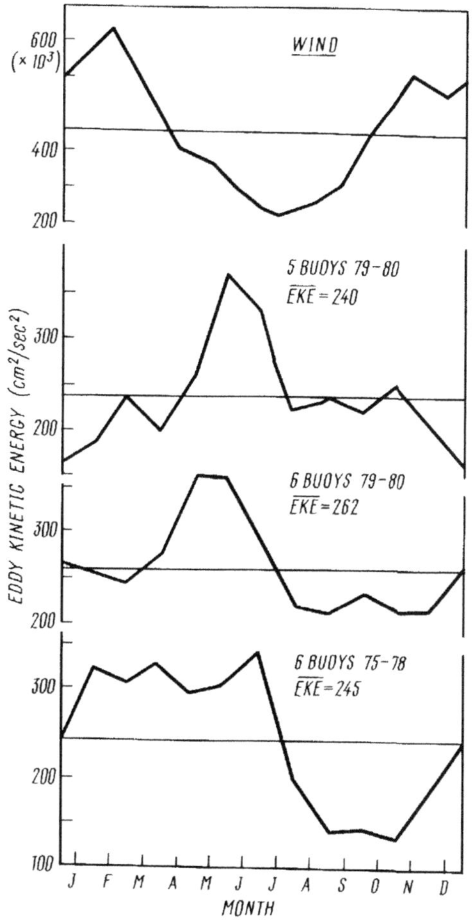

Fig. 5.5.2. Climatological mean annual variations of the kinetic energy of the near-water wind averaged over the area of the region 40–54°N, 10–38°W and the average annual variations of the kinetic energy of synoptic-scale currents according to the drift data of three groups of satellite-tracked buoys in different years for the same ocean region (after Richardson, 1983).

The above-mentioned results are completed by results of measurements of synoptic disturbances of the sea surface level performed with the aid of satellite altimeters. The first successful measurements of this kind, by satellite GEOS 3 from April 1975 to November 1976, confirmed the high level of development of synoptic disturbances of water circulation in the regions of the East Australian and Falkland Currents and in the Agulhas Current–Agulhas Return Current region (see Section 3 of Chapter 4) and gave maximum values (over the ocean area) of the root-mean-square synoptic deviation of the sea surface level from its average position equal to 80, 60, and 60 cm, respectively, for the three above-mentioned ocean regions (Gordon *et al.*, 1983).

The authors of the chart shown in Figure 5.5.3 (Cheney *et al.*, 1983) believe that, owing to the procedures of smoothing and interpolating the satellite altimeter data which preceded the construction of the chart, the root-mean-square values of synoptic disturbances of the sea surface level in different ocean regions given by the chart are two or three times as small as the actual values. Nevertheless, the chart clearly indicates a particularly high level of development of these disturbances in three regions of the world ocean; namely, the region of the Gulf Stream, the region of the convergence of the Falkland and Brazil Currents, and the Agulhas Current–Agulhas Return Current region. Next in intensity of their synoptic disturbances are the region of the Kuroshio west of the Emperor Seamounts, the region of the East Australian Current, and a number of parts of the Antarctic Circumpolar Current: at 50–75°E, at 8–100°E, in the Drake Passage, and in the regions where the current intersects seamount ridges (at 150–175°E, 130–145°W, and 25–40°E). Finally, an increased disturbance level is observed in the zone of the Polar Front between Greenland, Iceland, and Scotland, in the southern part of the Bering Sea, in the regions of the Somali, Madagascar, and West Australian Currents, and also in the region of the Eastern Wind Current off the coast of the Antarctic Continent. Even the enumeration of these regions implies good agreement between the chart in Figure 5.5.3 and the results of the regional investigations of ocean eddies described earlier in this book. Moreover, the measurement accuracy turned out to be sufficient for revealing a somewhat higher level of synoptic disturbances in the zones of the North Equatorial Current in the Pacific, the Equatorial Counter Current and the Guiana Current in the Atlantic, and in the southern part of the Indian Ocean in comparison, for instance, with the South Atlantic or the eastern parts of the subtropical gyres in the northern and southern halves of the Pacific (Figure 5.5.3). According to the chart in Figure 5.5.3, 70% of the area of the world ocean is characterized by small (less than 5 cm) values of the root-mean-square synoptic disturbance of the sea surface level; this percentage is perhaps somewhat overestimated since the chart does not include the Arctic Ocean, the extreme southern part of the Southern Ocean, and the inland seas where the synoptic activity of the ocean is rather high. However, on the whole, the chart in Figure 5.5.3 indicates that satellite altimeter measurements are highly promising for studying synoptic variability of the ocean.

The data of satellite altimeter measurements on which the chart in Figure 5.5.3 is based were also used to calculate spatial spectra of synoptic disturbances of the sea surface level (Fu, 1983) (Figure 5.5.4). The one-dimensional spectra $E_1(k)$ in

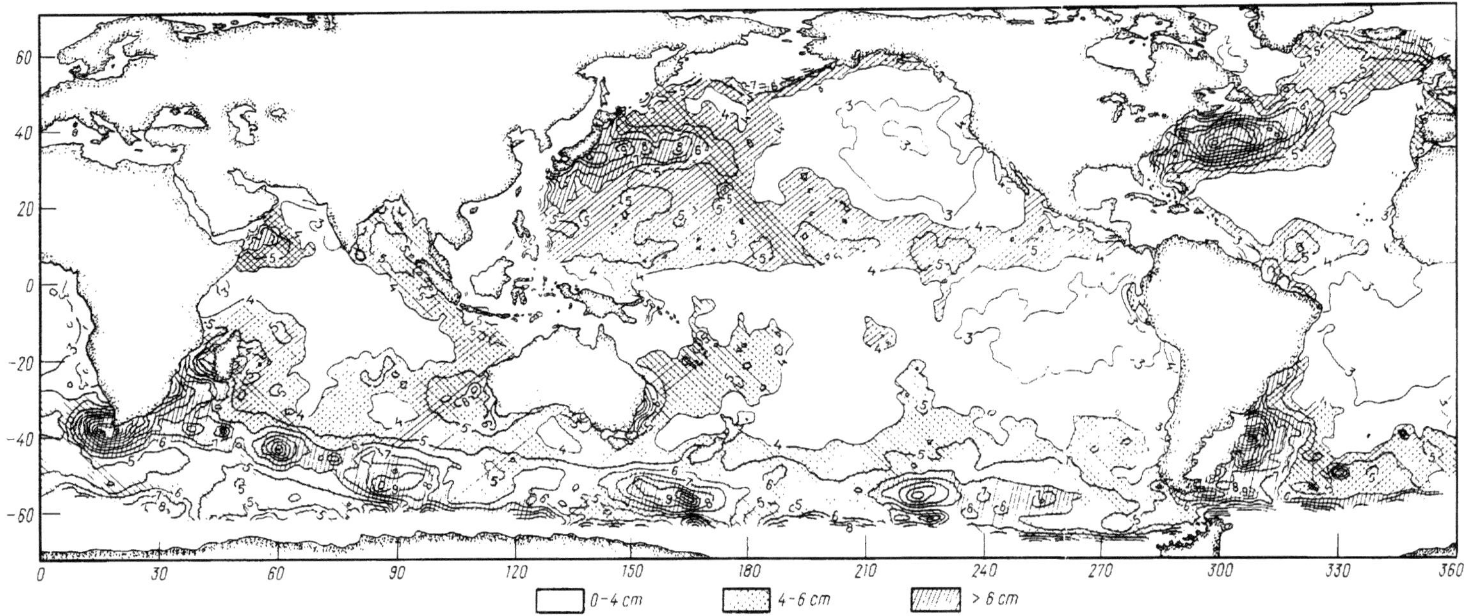

Fig. 5.5.3. Distribution of the root-mean-square values of synoptic disturbances of the sea surface level (cm) in the world ocean according to the data of altimeter measurements from the satellite SEASAT from September 15 to October 10, 1978 (after Cheney *et al.*, 1983).

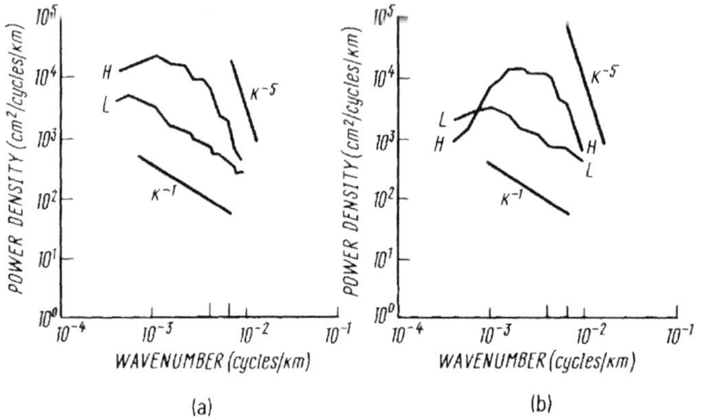

Fig. 5.5.4. (a) Mean one-dimensional spatial spectra of synoptic disturbances of the sea surface level for ocean regions with a (H) high and (L) low disturbance energy, and (b) the corresponding spectra as functions of wavenumbers (the integrals of the two-dimensional spectra corresponding to (a) along all directions) according to the data of altimeter measurements from the satellite SEASAT from September 15 to October 10, 1978 (after Fu, 1983). The marks on the wavenumber axis indicate wavelengths equal to 250 and 150 km.

Figure 5.5.4(a) were calculated directly from the altimeter data along individual satellite tracks under the assumption that the synoptic disturbances of the field of the sea surface level were isotropic; the wavenumber spectra $E(k)$ in Figure 5.5.4(b) are related to $E_1(k)$ by the expression $E(k) = -k[\mathrm{d}E_1(k)/\mathrm{d}k]$. The location of the maxima of the spectra corresponding to regions of high synoptic activity in the range of wavelengths of several hundred kilometers, and the fact that the slope of these spectra corresponds to k^{-5} in the range of large wavenumbers, are in agreement with the interpretation of the field of synoptic disturbances of the water circulation in these regions as quasi-two-dimensional large-scale turbulence (Rhines, 1977; Mirabel' and Monin, 1980). On the other hand, the predominance of disturbances with greater wavelengths and the smaller slope of the spectra for regions with a low synoptic activity (Figure 5.5.4) probably indicates that the disturbances are primarily wavelike. Here it should be borne in mind that, by virtue of some specific features of the data used in the calculations, the spectra presented in Figure 5.5.4 describe only those synoptic disturbances of the sea surface level that have a period of less than 24 days.

The average distributions of the intensity of synoptic disturbances in the water circulation in the main thermocline in the northern parts of the Atlantic Ocean (Dantzler, 1977; Emery, 1983; Burkov *et al.*, 1984) and Pacific Ocean (Emery, 1983) were studied using the numerous data of XBT (Dantzler, 1977; Emery, 1983) and hydrographical (Burkov *et al.*, 1984) observations accumulated by the middle of the 1970s and the beginning of the 1980s. The distribution of the APE of synoptic currents (APESC) shown in Figure 5.5.5 was calculated by the formula APESC $= \frac{1}{2}\,\overline{N^2}\,\overline{\zeta^2}$, where $\overline{N^2}$ is the average value of the square of the Brunt–Väisälä frequency over the main thermocline and $\overline{\zeta^2}$ is the variance of the deviations of the depth of the 15 °C isothermal surface (of the 13 °C isothermal surface in the regions east of the North Atlantic Ridge) from its average position in the

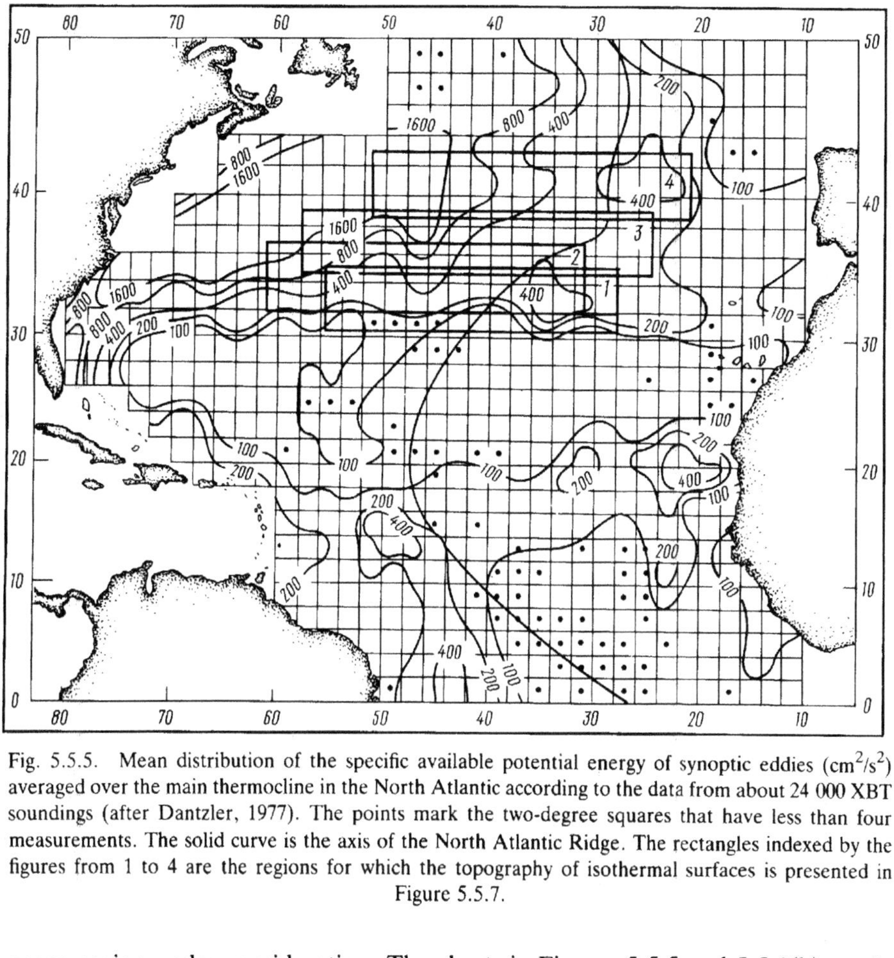

Fig. 5.5.5. Mean distribution of the specific available potential energy of synoptic eddies (cm^2/s^2) averaged over the main thermocline in the North Atlantic according to the data from about 24 000 XBT soundings (after Dantzler, 1977). The points mark the two-degree squares that have less than four measurements. The solid curve is the axis of the North Atlantic Ridge. The rectangles indexed by the figures from 1 to 4 are the regions for which the topography of isothermal surfaces is presented in Figure 5.5.7.

ocean region under consideration. The charts in Figures 5.5.5 and 5.5.1(b) are in good agreement not only qualitatively but also quantitatively, which indicates the predominant baroclinicity of synoptic currents and confirms the well-known theoretical conclusion (see Chapter 2) that their APE and KE are comparable. The very large values of the APESC in the north-western part of the region in Figure 5.5.5 are mainly due to Gulf Stream meanders and rings. The sharp decrease of the APESC to the south and east from the Gulf Stream coincides, in principle, with the POLYMODE results (see Section 2 of the present chapter). The local APESC maximum in the region south-west of the Azores and the area of increased APESC values in the zone 12–22°N in Figure 5.5.5 correspond to the south-east Gulf Stream jet and the North Equatorial Current ('Polygon-70'!) and hence they indicate local sources of synoptic eddies probably related to the baroclinic instability of the above-mentioned large-scale currents.

The average distribution of the specific APESC in the North Atlantic at a depth of 300 m calculated by Emery (1983) from the data of about 112 000 XBT soundings turned out to be close to the one presented in Figure 5.5.5, probably with the only major distinction that the APESC level in the Gulf Stream region was still higher than in Figure 5.5.5 and attained a maximum of 3000 cm^2/s^2 in the region 38°N, 60°W. A similar distribution constructed for the northern half of the Pacific from the data of 94 000 XBT soundings gave an APESC maximum in the Kuroshio region south of Japan, which attained 2500 cm^2/s^2 (Emery, 1983).

Along with the chart in Figure 5.5.5, the chart in Figure 5.5.6(b) demonstrates the comparative narrowness of the zone of increased values of synoptic disturbances of the temperature that is stretched along the Gulf Stream–North Atlantic Current axis (Figure 5.5.6(a)) and the sharp meridional gradients of the average amplitudes of these disturbances at the northern and southern boundaries of this zone, which is further confirmation of the fact that the majority of the Gulf Stream rings do not travel far from the current generating them. Remarkable features in Figure 5.5.6(b) are the absolute maximum of synoptic disturbances of temperature in the region south and south-east of Newfoundland, the relative maxima in the zones of the Equatorial and Guiana Currents and in the Gulf of Mexico, and also the wide belt of increased values of disturbances in the Polar Front zone between Baffin Land, Greenland, Iceland, Scotland and Norway. An analysis of the variations of the root-mean-square synoptic disturbances of temperature with increasing depth (Burkov *et al.*, 1984) showed that the surface of maximum values of these disturbances in the tropical and subtropical zones of the western part of the North Atlantic almost coincides with the 10 °C isothermal surface, which approximately corresponds to the middle of the main thermocline. For example, this surface lies approximately at a depth of 500 m (the root-mean-square temperature disturbances are about 4 °C) in the axial part of the Gulf Stream, at a depth of 1000 m in the central part of the Sargasso Sea (where the disturbances are somewhat smaller than 1 °C), and at a depth of about 400 m in the axial part of the North Equatorial Current (where the disturbances are somewhat higher than 1 °C). This behavior is in complete agreement with the concept that eddies are generated in the main thermocline owing to the baroclinic instability of large-scale geostrophic currents. The average intensity of synoptic disturbances of temperature at a depth of 200 m in the Gulf Stream–North Atlantic Current region in the winter period of January, February, and March turned out to be somewhat higher than in the summer period of July, August, and September (Burkov, *et al.*, 1984), which is in complete agreement with the analogous seasonal variations of eddy energy in the POLYMODE region (Figure 5.2.7).

The charts of the isothermal surface topography shown in Figure 5.5.7 were constructed from the data of XBT surveys, each of which was performed by five ships on parallel courses. The distances between the XBT soundings in each of the ship tracks was 30 km and the distance along the meridian between the tracks varied from 50 to 75 km. First we note the general agreement between Figure 5.5.7 and Figures 5.5.5 and 5.5.6(b) in relation to the geographical distribution of the energy of synoptic disturbances. At the same time, Figure 5.5.7 demonstrates one possible character of synoptic formations producing distributions of the intensity of synoptic disturbances in the corresponding parts of Figures 5.5.5 and 5.5.6(b). A clearly marked Gulf Stream cyclone (ring) is seen in the western part of swath 2 and three more rings with smaller diameters are seen in the western part of swath 3. At

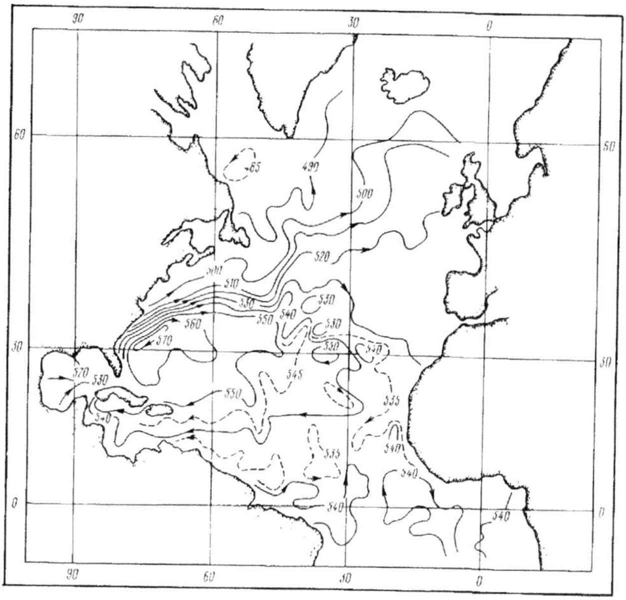

(a)

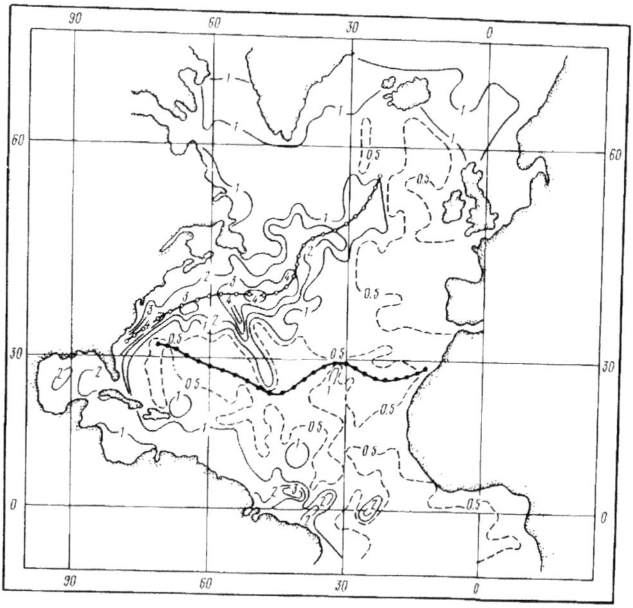

(b)

Fig. 5.5.6. (a) Climatological mean dynamic topography (in dynamic centimeters) of the 200 dbar isobaric surface relative to the 1500 dbar reference surface, and (b) the distribution of the root-mean-square synoptic disturbances of water temperature (°C) at a depth of 200 m in the North Atlantic according to the data of 60 000 hydrographic soundings of the ocean (after Burkov *et al.*, 1984). The curves with light and dark circles in Figure 5.2.6(b) are, respectively, the lines of maximum and minimum values of the root-mean-square deviations of temperature.

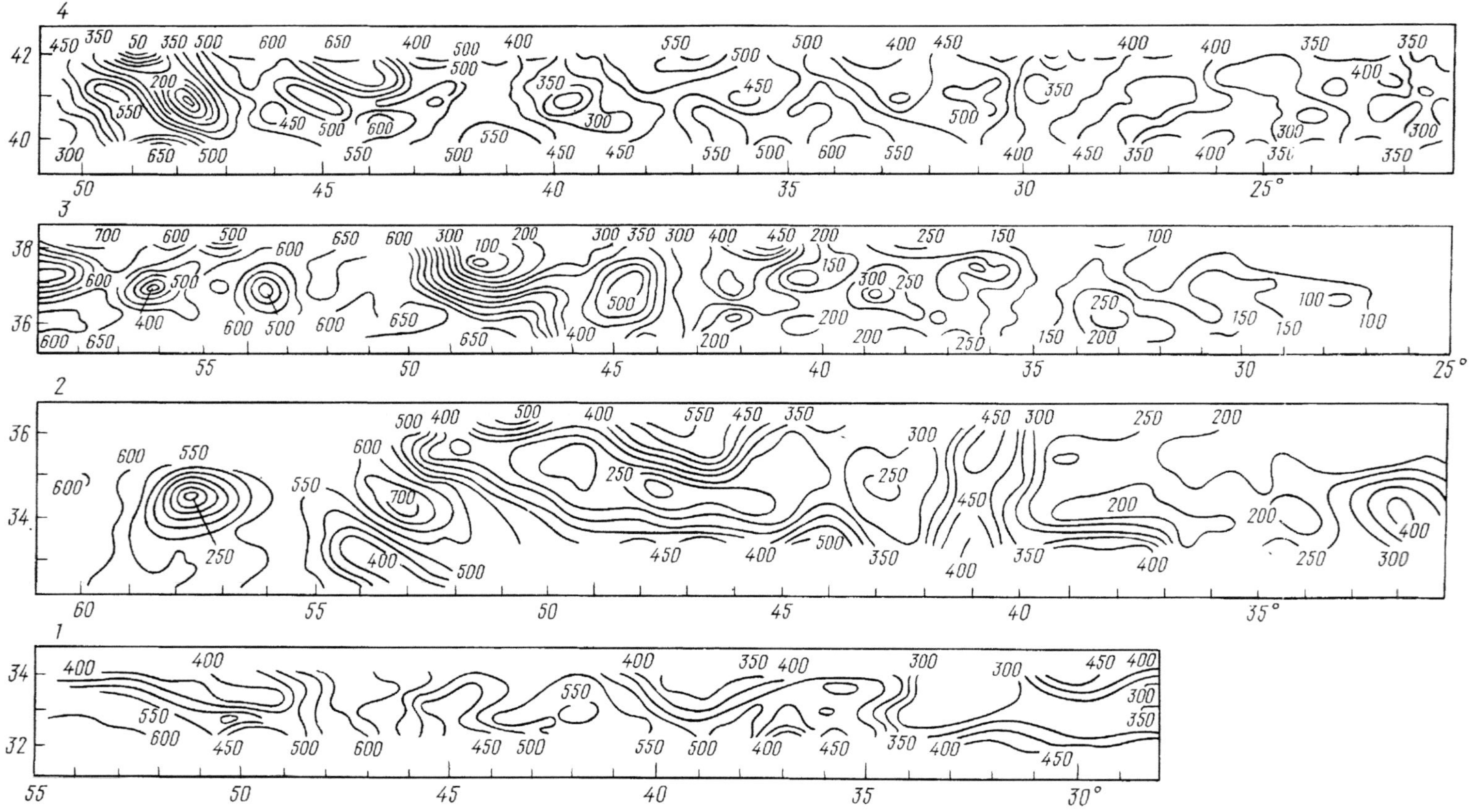

Fig. 5.5.7. Topography (depth in meters) of (1) 14 °C, (2 and 3) 15 °C, and (4) 12 °C isothermal surfaces according to the data of XBT surveys on (1) November 30–December 5, 1976, (2) October 14–23, 1977, (3) April 30–May 10, 1977, and (4) July 5–13, 1976 (after Emery *et al.*, 1980). The positions of the survey regions are shown in Figure 5.5.5.

least two of these rings were surrounded on all sides by regions in which synoptic currents were almost absent; this fact obviously contradicts the widespread belief that Gulf Stream cyclones radiate Rossby waves. Another remarkable structural element of the field of synoptic disturbances in the swaths of XBT surveys in Figure 5.5.7 is the presence of narrow jets of mainly zonal direction which are particularly clearly marked in swaths 1 and 2. The depth difference of the 15 °C isothermal surface across the quasizonal jet in the middle part of swath 2 was equal to 250–300 m, which is only twice as small as that across the main Gulf Stream jet. The appearance of such quasizonal baroclinic jets of synoptic currents in the open ocean seems to be a consequence of the dispersive properties of a field of nonlinear baroclinic Rossby waves (Rhines, 1977). The jet in the region 46–50°W in swath 3 in Figure 5.5.7, with a 15 °C isotherm depth difference of 500 m(!), was probably the south-east branch of the Gulf Stream, which has already been mentioned in Section 1 of Chapter 4 and Section 3 of this chapter. Near the above-mentioned concentrated jets in the central parts of swaths 2 and 3 and in the western part of swath 4 (Figure 5.5.7) there were cyclonic and anticyclonic eddies; generally this 'coexistence' of jets and eddies (probably strongly interacting) closely resembled the observations in the POLYMODE experiment region (Figures 5.2.4–5.2.6). The strong cyclone at 47–49°W in the western part of swath 4 was also very clearly marked in the temperature distribution at the sea surface. The salinity distribution in its field showed that it had been formed not long before the measurements from a cyclonic meander of the beginning of the North Atlantic Current. Among the other eddies and jets, almost all the quasizonal jets in swath 1, all the jets *east* of 41°W in swath 2, the very strong jet at 47–50°W in swath 3 which was mentioned above, and all the disturbances *west* of 41°W in swath 4 produced notable temperature signals at the sea surface. The temperature disturbances in the eastern parts of swaths 3 and 4 were considerably weaker than the disturbances in their western parts, and they resembled a quasigeostrophic field of baroclinic Rossby waves. On the whole, the patterns of synoptic disturbances in Figure 5.5.7 are very complicated and versatile and give the impression of a complex synthesis of interacting large-scale turbulence and quasigeostrophic planetary waves.

At the present time the scope of the observation data on synoptic ocean eddies is insufficient for an exhaustive generalization of all the properties of eddies on the scale of the world ocean; however, the results presented in this and the foregoing chapters make it possible to draw the following preliminary conclusions:

1. The most universal mechanism of generation of synoptic eddies on the scale of the world ocean is probably the baroclinic instability of large-scale currents (BILC) (see Chapter 3). As a consequence, Rossby-scale baroclinic eddies dominate in the ocean. In some cases ('Polygon-70' and POLYMODE) the phase shift of synoptic oscillations of the current velocity with increasing depth, which is characteristic of the BILC phenomenon, was revealed experimentally and an estimate was made of the corresponding rate of transformation of the APE of the large-scale current (LSC) into eddy energy. In accordance with the dominant mechanism of eddy generation, the geographical distribution of their energy in the world ocean is similar to the APE distribution of large-scale currents.

2. The properties of eddies are mainly determined by the character of the LSC

that generated them. The eddies (rings) of narrow and strong frontal currents of the type of the Gulf Stream, the Kuroshio, the Agulhas Current, and the axial part of the Antarctic Circumpolar Current are probably a consequence of the baroclinic–barotropic instability of these currents (see Chapter 3) and are generated owing to the meandering of the currents as a whole and the further separation of the meanders. This process of eddy formation specified their quasisingle character, their high concentration of energy, and the location of the resulting cyclones and anticyclones on different sides of frontal currents. The rings carry away their water mass in their motion and are mixed very slowly with the surrounding water. As a rule, the lifetime of an individual ring is determined by the process of its repeated interactions with the current that generated it. This process must also exert a determining effect on the formation of the effective heat flux across the frontal current related to the rings.

3. At present the details of the generation process of synoptic eddies owing to the baroclinic instability of broad and comparatively slow large-scale currents in the open ocean are not quite clear. It can be assumed that this process must first manifest itself in the main pycnocline in the formation of current velocity oscillations transverse to the LSC direction which belong to the type of baroclinic Rossby waves. Further, as can be judged from individual observations in, for instance, the regions of California or the Norwegian Current, there sometimes occurs spontaneous local intensification of some of the meandering LSC jets accompanied by the further sharpening of the meanders and separation of eddies. The eddies thus formed are not concentrated as strongly as the rings of frontal currents and can probably serve as generators of baroclinic Rossby waves. The whole process under consideration (which is in many respects hypothetical) results in the formation of a complicated 'two-component' system of relatively strong eddies and disturbances of the ocean circulation, mainly of a wave nature, surrounding (and interacting with) these eddies (POLYMODE!). It seems that the higher the APE level of the LSC, the greater is the fraction of energy of synoptic disturbances concentrated in the intensive eddies. The intensive eddies transport their water in the upper ocean. However, owing to active interactions between individual eddies and between the eddies and the wave background, the lifetime of each of the eddies as an individual formation is not long, which specifies the behavior of the entire field of synoptic disturbances similar to that of large-scale quasigeostrophic turbulence. The contribution of eddies in the open ocean to the average horizontal heat fluxes in the ocean do not seem to be very substantial.

4. The generation of eddies in the open ocean in the main pycnocline caused by BILC is accompanied by a transfer of energy of synoptic disturbances to the ocean depth, most likely owing to the work of forces of pressure; as a consequence, this leads to the formation of synoptic disturbances mainly of a wave nature in the deep ocean layer. When the eddy generation process is temporarily stopped or sharply weakened, a total or almost total barotropization of the eddy field can take place, which is accompanied by an expansion of the horizontal dimensions of the eddies. As a result, the system of eddy disturbances acquires the character of a field of nonlinear quasibarotropic Rossby waves. Many ocean regions are likely to be

characterized by cyclicity (possibly with a year-long period) of the process of interaction of LSC and eddies, including the reconstruction of the baroclinic LSC, some time after the barotropization of the eddies; however, the mechanism of this reconstruction is not quite clear. At the present time the scope of available observation data is insufficient to estimate the importance of the part played in this reconstruction by the possible reverse transfer of energy from eddies to the LSC (the so-called 'negative viscosity' effect) for the world ocean as a whole.

5. The bottom relief of the ocean is an important factor which, in many cases, determines a very specific character of the eddy generation process caused by the instability of large-scale oceanic currents. In these cases the behavior of the eddies is in many respects determined by the character and parameters of the irregularities of the relief and, primarily, by their horizontal scale, height, and the degree to which they are isolated. For example, when a current flows over seamounts whose horizontal dimensions are comparable with the local value of the Rossby scale and whose height is small in comparison with the ocean depth, the bottom relief can generate baroclinic eddies which are entirely concentrated in deep ocean layers, the velocity of orbital water motion in their field increasing towards the ocean bottom. On the other hand, when a current flows over large-scale seamounts whose tops essentially penetrate into the main pycnocline or when it intersects clearly marked large-scale forms of bottom relief (subwater ridges, seamount chains, etc.), as a rule, the bottom relief leads to the formation of quasistationary synoptic-scale baroclinic disturbances in the regions of these irregularities (eddies, eddy-meanders, and groups of eddies). These disturbances have an ordinary vertical structure with orbital velocity increasing towards the sea surface. The predominantly anticyclonic baroclinic eddies that were observed in the regions of the oceanic upwelling can serve as an example of synoptic disturbances formed under the joint action of the wind-driven effect, the slope of the ocean bottom, and the instability of coastal currents appearing in the upwelling.

6. Particularly remarkable are mesoscale and synoptic anticyclonic single eddies with well-mixed cores which are entirely concentrated in intermediate ocean layers and are transported by large-scale oceanic currents thousands of kilometers away from the regions of their birth. The degree to which these eddies are widespread in the world ocean is likely to be much higher than is implied by the observation data accumulated at the present time (the Sargasso Sea and the Arctic Ocean). At present the processes by which these eddies are generated is not quite clear.

When the seasonal pycnocline is well developed and, particularly, when it is also separated from the main pycnocline (as, for instance, is the case in the Sargasso Sea), small baroclinic eddies can be formed in the upper ocean, their horizontal dimensions being determined by the local value of Rossby scale typical of the seasonal pycnocline. As a rule, these eddies result from the instability of frontal currents in the upper ocean layer, and hence their generation is related to the disturbing effect exerted on the density field in this layer by motions with a greater scale and, primarily, by the main synoptic eddies.

7. As can be assumed at the present time, the generation of ocean eddies in the world ocean as a whole is the basic factor of energy sinks for large-scale

currents. The eddies in turn expend their energy mainly by their interaction with other processes with smaller scales primarily, on the generation of mesoscale quasihorizontal turbulence which produces the mixing of ocean waters within the limits of isopycnic layers forming the fine structure of the ocean, and also on work against buoyancy forces in the vertical reconstruction of the structure.

CHAPTER 6

Applied Problems

1. SYNOPTIC EDDIES AND FORMATION OF WEATHER AND CLIMATE

The most important properties of synoptic eddies relating to their influence on the weather are the intensity and the extent of temperature disturbances they produced. The intensity of a disturbance can be defined as the difference δT_s between the temperature of the surface water layer in the eddy and the temperature of the surrounding water. Temperature disturbances result in disturbances in the fields of energy fluxes between the sea surface and the air, among which the major role is played by the fluxes of latent (LE) and sensible (H) heat. These fluxes depend on wind velocity, vertical gradients of temperature T, and specific humidity q in the near-water air layer and on its stratification. If the air transformation is ignored, the changes of fluxes H and LE in the presence of a disturbance are equal to $H\ \delta T_S/(T_S - T_Z)$ and $LE\ \delta q_S/(q_S - q_Z)$. Here the gradients are replaced by the finite differences between the values of the quantities at the sea surface and at the measurement level. The fluxes of heat, both latent and sensible, can increase several times during the air motion above a warm Gulf Stream ring and they decrease and can change sign above a cold ring. This distinction between the effects of warm and cold rings increases still further owing to the fact that warm rings are formed nearer to the continent, and in winter (i.e. in the season of maximum heat transfer) the air above them is cooler and drier whereas cold rings are formed farther from the continent and the air moving above them has time to be heated and humidified. There are also some other distinctions between warm and cold rings which are important in relation to their effect on the atmospheric air. For example, in both types of Gulf Stream rings the main temperature contrast corresponds to the interface between the so-called slope waters and Gulf Stream waters, and a secondary front is formed between the Gulf Stream and Sargasso Sea waters. Here the circulation in both types of rings is such that it contributes to the diffusion of secondary fronts and the preservation of main fronts (Fedorov *et al.*, 1979). Therefore, if the initial meanders are of the same size, it seems that the temperature disturbances corresponding to anticyclonic rings must occupy greater areas than the temperature disturbances of cyclonic rings. Further, the waters in the central parts of warm rings are cooled from above, whereas in cold rings they are heated. Therefore, the stratification of the upper water layer in warm rings becomes unstable and heat exchange with the atmosphere involves a thicker water layer than in cold rings where stability increases and heat exchange with the deeper

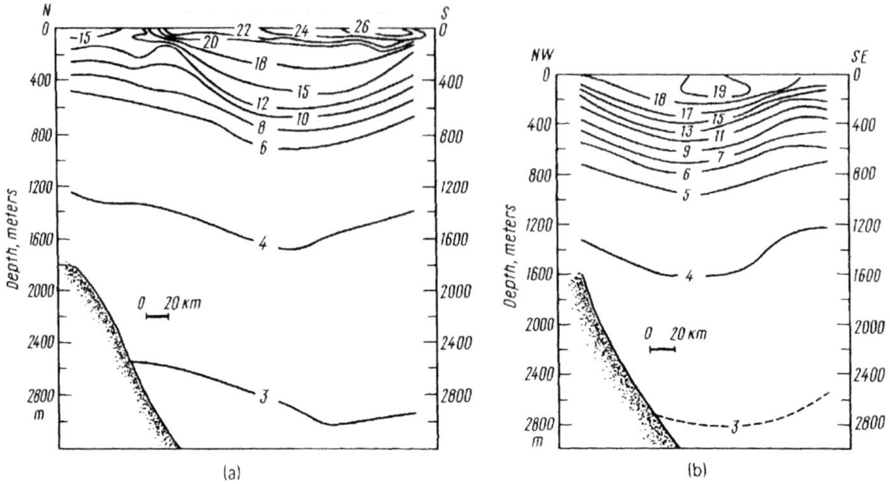

Fig. 6.1.1. Temperature section across an anticyclonic ring according to the data of CTD–XBT surveys in (a) September 28–October 1, 1977 and (b) November 30–December 12, 1977 (after Fornshell and Criess, 1979).

layers is hampered. Therefore, cold rings seem to be 'disguised' in the sea surface more frequently and faster.

Thus, warm anticyclonic rings must exert a more notable effect on the atmosphere than cold cyclonic rings.

We consider a specific example of the effect exerted by a warm ring on the atmosphere. On September 1, 1977 infrared satellite photographs detected a newly formed anticyclonic ring (Fornshell and Criess, 1979) which was permanently observed until January 24, 1978. In approximately five months the eddy travelled not less than 360 miles with an average velocity of 4.5 km/day. At the end of September and at the beginning of December hydrographic surveys of the eddy were carried out (Figure 6.1.1). The size of the eddy (determined from the intersection of the 15 °C isothermal surface with the 200 m level) decreased from 185 to 148 km and the surface water temperature decreased from 26–27 to 19 °C during the two-month interval between the surveys. The difference between the sea surface temperature in the eddy and in the surrounding water decreased from 11 to 3–4 °C. The depth of the mixed layer increased from 50 to 100 m. The upper layer, 200 m thick, was cooled by 2 °C over 60 days and by 1 °C over a week when there were two storms. This yields cooling rates of 339 and 1357 W/m^2, respectively. For comparison we note that the solar constant is 1360 W/m^2. According to the data of Timofeev (1970), in December and November the heat flux to the atmosphere in the region where the trajectory of the anticyclonic eddy passed was, on average, 190 and 270 W/m^2. Hence, the heat transfer from the eddy to the atmosphere was more intensive than that from the ocean in ordinary conditions in this region.

Let us compare values that have been obtained for energy flux from the sea surface in a warm ring with the fluxes required for the maintenance and development of an atmospheric cyclone. For the kinetic energy (KE) of a cyclone to be conserved it is necessary that there be an influx of available potential energy (APE) and its transformation into KE whose amount compensates for the dissipation of energy owing to friction. The dissipation rate for the entire atmosphere is estimated as 5 W/m^2. A typical value for the generation of KE in a developing cyclone, without including friction, amounts to 18.2 W/m^2 (Palmen and Newton, 1969). Thus, a developing cyclone must have energy sources with specific power of 23.2 W/m^2. Here the fluxes of latent and sensible heat from the sea surface make a significant contribution. For example, this is confirmed by the empirical fact that cyclones intensify when emerging on the warm sea surface.

If the amount of KE in a young cyclone with diameter of 1000 km is to be maintained at a constant level, there must exist an energy source with a power of 4.2×10^{12} W. It should be taken into account that an increase in the amount of KE in a developing cyclone must be accompanied by a simultaneous increase in the amount of APE. According to Lorenz (1955), the KE in the atmosphere amounts to 10% of the APE, and the data presented by Vulis and Monin (1975) imply that this fraction is approximately equal to 24%. Using these estimates we see that, with the inclusion of friction, the development of the cyclone under consideration requires an energy influx at a rate of 16.12×10^{13} or 7.78×10^{13} W, which corresponds to the specific power of 205 and 93 W/m^2 (cf. 339 and 1347 W/m^2).

An anticyclonic ring whose area amounts approximately to one-thirtieth of the area of the cyclone under consideration can provide, on average, 0.91×10^{13} W, and during intrusions of cold air it can provide 3.6×10^{13} W. These values exceed the fluxes required for maintaining the cyclone and are commensurable with fluxes characteristic of a developing cyclone. However, for a real cyclone not only the value of the energy flux is important but also the area from which the energy is transferred and the part of the cyclone where the heat is transferred to its system. On average, the influx of heat in cold air must decrease the APE of the cyclone, whereas, by contrast, its influx in the warm sector must contribute to the development of the cyclone. Moreover, of importance is the duration of the energy transfer to the given volume of air which depends on the velocity of air motion. It should also be taken into consideration that warm rings, in contrast to cold ones, do not travel far from the Gulf Stream and therefore the identification of their effect in a pure form is almost impossible. Therefore, the answer to the problem of the effect of the heat flux from the surface of an anticyclonic ring on a synoptic atmospheric disturbance is ambiguous. We can only estimate this effect qualitatively as potentially important.

As was mentioned above, cyclonic rings produce a weaker effect on atmospheric processes since the heat transfer from their surface decreases or stops, which contributes to the frequent appearance of stable stratification in the near-water air layer.

Richardson (1980) gives the following description of a cyclonic ring that was encountered in April 1977:

> "Immediately on entering the central part we noticed a seaweed smell, similar to that of a seashore on a summer day. The water was obviously green and turbid as compared to the clear and deep blue of the nearby Gulf Stream and Sargasso Sea. The temperature of the surface water in the center was 15 °C,

nearly 10 °C colder than the surrounding water; this difference was reflected in the air temperature."

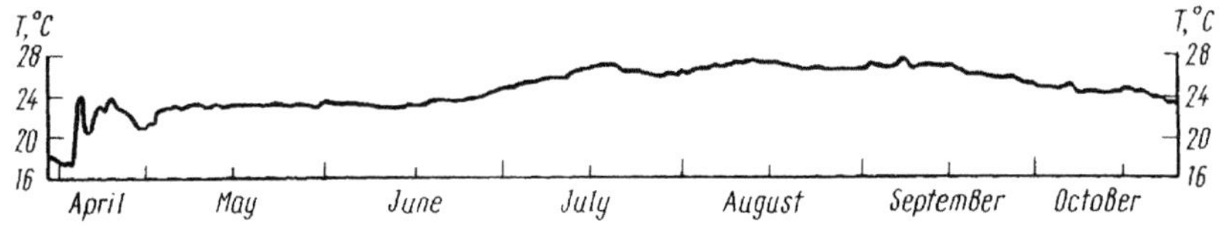

Fig. 6.1.2. Variations of the surface water temperature in a cold ring in March–October 1977 according to the data of Richardson (1980).

Figure 6.1.2 demonstrates the variation of the sea surface temperature measured by a satellite-tracked free-drifting buoy. We see that the most significant temperature variations occurred in the initial period of life of the ring. Some sharp changes (e.g. on April 20) related to the interaction between the ring and the Gulf Stream and took place on the background of the general increase of temperature caused by processes of energy transfer across the sea surface. The surface water temperature in the central part of the ring attained a quasistationary level in two weeks. In June–August, when the eddy was in a stable south-west motion, gradual heating of the surface waters took place.

Warm rings are formed in the Gulf Stream system north of the axis of the current and cold rings are formed south of it, whereas in the region of Kuroshio–Oyashio Current where the hydrological situation is more complicated and is characterized by at least two fronts not only warm anticyclonic Kuroshio rings but also cold cyclonic rings of the Oyashio Current can be observed between the axes of the two currents, i.e. north of the Kuroshio (Cheney, 1977; Burkov and Pavlova, 1979). This complicates the structure of the temperature field and, consequently, of the heat-exchange and moisture-exchange fields.

The eddies in the opean ocean also exert an effect on the local weather, but to a smaller degree in comparison with the rings. This is accounted for by the fact that they are much less clearly marked in the temperature field at the sea surface and, as a rule, the intensity δT_S of the corresponding disturbances does not exceed 1–4 °C.

As was shown by observations in the POLYMODE region in the Sargasso Sea, the existence of close-packed eddies in the open ocean leads to disturbances in the zonal temperature distribution and the formation of local hydrological surface fronts. These fronts manifest themselves in the local weather, which is confirmed by direct observations. For instance, using the data of complex studies of an eddy in the region 27°N, 70°W in December 7, 1977–January 8, 1978, Sizov *et al.* (1978) revealed the presence of a surface temperature front. It was found that the regime of heat exchange above the eddy substantially differed from the regime of the surrounding regions. This manifested itself outwardly in the concentration of lower level clouds above the southern part of the eddy and their absence above the northern part. Using the data of another expedition, Voskresenskaya and Kosnyrev (1979) reduced the observation data on air temperature and water vapor pressure to one diurnal cycle phase and obtained fields of these elements where disturbances with dimensions typical of oceanic synoptic eddies were clearly seen.

There are observations which seem to show that when a tropical cyclone remains in one and the same area for a sufficiently long time a ring-shaped circulation develops in the ocean and the upwelling of cold water appears. This produces a closure of the chain of negative feedback: cooling of surface water – a decrease in the intensity of evaporation – a decrease in the amount of heat liberated in the condensation of vapor in the core of a tropical cyclone. This kind of development can be observed rather frequently in nature. For example, during the expedition 'Typhoon-78' in the subtropical zone of the North-West Pacific in June 23–28 a spot of anomalously cold water ($\delta T_S = 3$ °C was found (Pudov and Bezzabotnov, 1980). It was bounded on the south-west side by a clearly pronounced temperature front and the water motion had a cyclonic character. Assumptions were made that this spot had appeared as a result of the action of a typhoon. An analysis of the cloud distribution in the July–September period, during which the charting of the regions not covered by clouds in the area 20–40°N, 140–160°E was carried out, showed that the only region where there were no clouds during the whole period was a region with dimensions of 150–300 km, approximately coinciding with the cold spot.

Up to this point we have discussed the effect of synoptic eddies on the local weather and on synoptic processes in the atmosphere. We can also speak about their effect on the climate in a broad sense, meaning the contribution of eddies to the meridional heat transfer in the ocean. This transfer, in the scale of the Hemispheres or in the scale of individual oceans, was estimated in a number of studies. The work by Vonder Haar and Oort (1973) is widely known. They used satellite observation data on the radiation heat flux transferred from the underlying surface–atmosphere system and also the existing and reliable estimates of the meridional heat flux in the atmosphere at different latitudes in the Northern Hemisphere. The difference between these quantities gives the heat flux in the ocean across the corresponding parallels. It turned out to be comparable with the meridional heat flux in the atmosphere; north of 30°N the major contribution comes from atmospheric transfer, and south of this latitude it comes from transfer in the ocean. The oceanic transfer attains maximum approximately at 20°N where it is equal to 3.4×10^{15} W (the actual value of the transfer can lie within the limits from 2.0 to 4.2×10^{15} W). Trenberth (1979) performed analogous calculations for the Southern Hemisphere.

Direct calculations of heat transfer in the ocean yield values approximately twice as small as those following from the Vonder Haar–Oort calculations. For example, the calculations for 24°N in the Atlantic (e.g. see Bryden and Hall, 1980) gave 1.1×10^{15} W, which is in agreement with the calculation of the heat transfer from the Atlantic to the atmosphere north of 24°N. The discrepancies between the results of global and direct calculations are rather large. Nevertheless, they do not contradict each other since the resulting estimates lie within the limits of the errors of the methods owing to the insufficient amount of data existing at present. Taking into consideration the confidence intervals established by Vonder Haar and Oort we assume the estimate of Bryden and Hall for the Atlantic (24°N) to be the most probable.

As is shown by observations, frontal zone rings are formations characterized by the predominance of the transport form of water motion. In other words, a water mass moves as a whole, undergoing transformation in the course of time. Therefore the formation of anticyclonic and cyclonic rings turns out to be a very effective mechanism of heat transfer in the ocean across hydrological fronts. Anticyclonic rings formed north of the fronts exchange heat with the surrounding cold water and cyclonic rings south of the fronts obtain heat from the subtropical water surrounding them. This results in a northward heat transfer. Estimating the average dimension of a ring as 200 km and the initial temperature difference between the ring core and the environment as 10 °C, and considering only the upper 1 km layer, Mintz (1979) found that each pair of rings (a cold ring and a warm ring) absorbed by the environment provides a transfrontal transfer of approximately 1.26×10^{21} J. If we assume that every year six or seven pairs of rings are formed and absorbed by the environment, the resulting transfrontal heat transfer is approximately 8.4×10^{21} J/yr or 2.7×10^{14} W. This amounts to almost one quarter of the entire transfer inside the Atlantic at 24°N.

The situation is quite the opposite in the estimation of the contribution of eddies in the open ocean. In most cases the relationship between the spatial and temporal scales of eddies (e.g. determined from the data of the 'Polygon-70' and MODE experiments) are in good agreement with the dispersion relations for the lowest baroclinic modes of Rossby waves. If this is the case, synoptic eddies in the open ocean cannot make a significant contribution to the meridional heat transfer in the ocean.

Gill (1980) tried to estimate the contribution of mid-ocean eddies without confining himself to any assumptions concerning their nature. Taking the value of the heat transfer across 24°N in the Atlantic (10^{15} W) obtained by Bryden and Hall (1980), and assuming that the whole flux is due to eddies and is uniformly distributed over the whole ocean width of 6000 km and depth of 1000 m, he obtained an average heat flux density of 17 J/cm^2·s. An eddy providing this heat flux density must have an average transport velocity and a temperature disturbance such that their product is equal to 4 °C·cm/s. The existing measurement data for northern tropical and subtropical latitudes in the Atlantic yield substantially smaller values and lead to the conclusion that eddies do not make a significant contribution to the meridional heat flux in the North Atlantic in the zone between 14 and 35°N, i.e. where this flux attains maximum. At the same time, their role can be significant in some other ocean regions, e.g. south of the Gulf Stream. For instance, according to the observations at 37°30′N, 55°W, the above-mentioned quantity attained about 10 °C cm/s at 600 m, about 4 °C·cm/s at 1000 m and about 1 °C·cm/s at 1500 m (Gill, 1980). We stress that in this case the direction of the heat transfer by eddies was opposite to that of the resultant transfer, i.e. it was directed towards the equator.

Using the data of Trenberth (1979), Gill found an analogous criterion of eddy efficiency in the Antarctic Circumpolar Current at 60°S. According to Trenberth, the meridional heat transfer across this parallel along the whole circle of latitude is about 10^{15} W. Under the assumption that here the eddy energy propagates to the whole ocean depth, the criterion turned out to be equal to 0.3 °C·cm/s. The observations in the Drake Passage and south of New Zealand show that within the framework of this quantity the eddy contribution to the meridional heat transfer in the antarctic waters can be substantial.

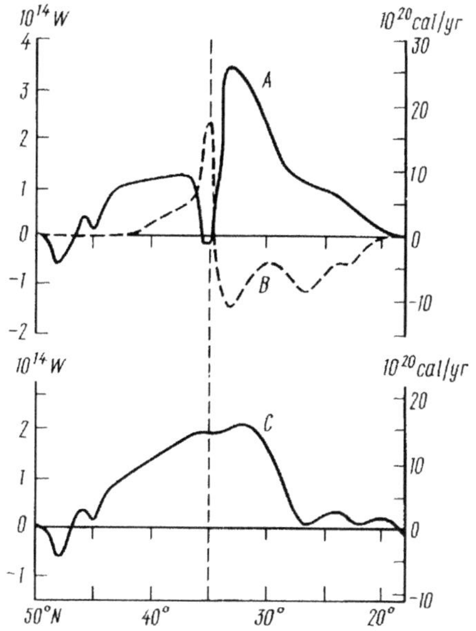

Fig. 6.1.3. Meridional heat transfer in a model ocean (the northward transfer is assumed to be positive) (after Mintz, 1979). A: transfer by mean circulation; B: transfer by nonstationary eddies; C; total transfer.

Of interest are the results obtained in some eddy-resolving models. Figure 6.1.3 shows the meridional heat flux in the ocean for one of these models (Mintz, 1979).

Curve B in the figure demonstrates the meridional heat transfer $\int\int \overline{v'_\varphi T'}\,\mathrm{d}z\,\mathrm{d}\lambda$ by eddy currents. According to Mintz, the major (if not the entire) contribution to this transfer comes from nonstationary synoptic eddies. The figure implies that south and north of the frontal zone the heat transfer to the pole is carried out by mean currents, whereas the transfer in the frontal zone itself is entirely due to synoptic eddies. It is also interesting to note that south of the frontal zone the heat transfer by eddies is directed towards the equator and constitutes in this region approximately half the amount of heat transferred northward by mean currents. Mintz believes that this equatorward transfer is also due to synoptic eddies owing to which the APE of the zonal motion is transformed into APE of eddies in a broad zone south of 33°N. In this zone, below the surface layer, the density increases and the temperature decreases towards the equator, owing to which the zonal flow is baroclinically unstable.

Thus, at present the following preliminary conclusions can be drawn from existing data:

(1) frontal rings and eddies in the open ocean exert an effect on the local weather;
(2) frontal rings possess a sufficient amount of APE and, in principle, can affect synoptic processes in the atmosphere;
(3) rings make a significant contribution to the transfrontal heat transfer and hence they are an important mechanism of the meridional heat transfer in the ocean;

(4) eddies in the open ocean do not seem to play any significant role in meridional heat transfer and do not affect the climate in the literal sense.

2. SYNOPTIC VARIABILITY OF HYDROCHEMICAL AND HYDROBIOLOGICAL CHARACTERISTICS

Investigations of synoptic eddies in the world ocean revealed a substantial synoptic-scale variability of hydrochemical parameters. It turned out that it was possible to use the distribution of hydrochemical characteristics to detect an eddy formation, to determine its dimensions, structure, direction and velocity of its propagation, and inclination of the axis of the eddy, and also to study the evolution of its structure. One of the important aspects of studying the dynamics of synoptic eddies is the problem of water mass transport by eddy formations. The solution of this problem can be facilitated by an investigation of the hydrochemical structure of an eddy field and its biological parameters; this can be used as an indicator of water masses. However, it should be noted that at the present time the existing data on synoptic disturbances of hydrochemical and hydrobiological characteristics are insufficient. The most complete information about these characteristics has been obtained for Gulf Stream rings and Sargasso Sea eddies.

Gulf Stream rings are formed on both sides of the current owing to the separation of meanders from the main flow; every year from five to eight rings are formed on each side (Fuglister, 1972). The eddies formed south or east of the Gulf Stream are rings of Gulf Stream water inside which the cold slope water is contained. These cyclonic eddies with cold cores exert a strong effect on the variability of hydrochemical and biological parameters of the north-western part of the Sargasso Sea (Boyd *et al.*, 1978; Wiebe, 1976, 1982; Wiebe and Boyd, 1978; Wiebe *et al.*, 1976). The rings transport the slope water in their cores to the Sargasso Sea; this water has a comparatively low temperature and salinity and a high content of oxygen and nutrients and also contains organisms of subarctic origin. The minimum oxygen concentration in the slope water is usually observed at depths of 100–200 m; the layer of minimum concentration in rings lies, depending on their age, at depths of 300–600 m (in the Sargasso Sea it is usually located at depths of 800–1000 m (Figure 6.2.1)). The Sargasso Sea contains organisms of subtropical and tropical origin.

When a cold cyclonic ring is formed, the water of its core differs greatly from both the water of the Gulf Stream itself and the Sargasso Sea water and it is very close in its properties to the slope water. The observed temperature drop between the centers of young cyclones and the surrounding Sargasso Sea water can attain more than 10°; salinity in the ring is usually 1‰ lower, and oxygen content is approximately 2 ml/l higher, than in the surrounding water; the ring water has a high concentration of nutrients; rings are characterized by rather high values of plant biomass and primary production (with a concentration of chlorophyll 'a' exceeding 3 μg/l). In the course of time, owing to both the heat from the sea surface and the turbulent exchange with the surrounding Sargasso Sea waters, and also because of multiple interaction with Gulf Stream waters (Richardson, 1980),

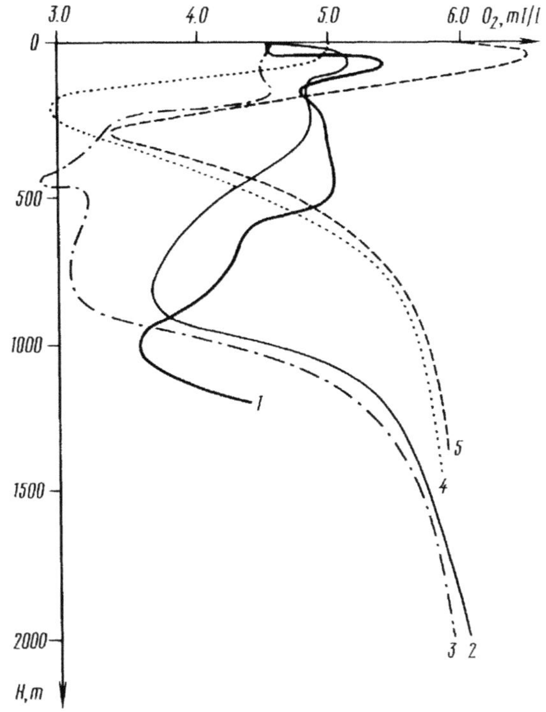

Fig. 6.2.1. Vertical distributions of oxygen (ml/l). (1) In a cyclonic formation in the Sargasso Sea (station 2229; the 25th cruise of the research vessel *Akademik Kurchatov*, July 1977); (2) in an anticyclonic formation in the Sargasso Sea (station 2338; the 25th cruise of the research vessel *Akademik Kurchatov*, October 1977); (3) in an intensive cyclonic eddy in the Sargasso Sea (station 2885; the 33rd cruise of the research vessel *Mikhail Lomonosov*, January 1978); (4) in a 10-month-old cyclonic Gulf Stream ring (station 11; the 38th cruise of *Knorr*, April 1974); (5) in slope waters (station 658; the 8th cruise of the research vessel *Mikhail Lomonosov*).

destruction of the rings takes place which manifests itself in an increase of temperature and salinity, in a decrease of the oxygen and nutrient content, and in the sinking of the layer of minimum concentrations of oxygen. No eddies older than three months were marked in the field of oceanographical characteristics on the sea surface; the upper layer water was analogous to the surrounding water in its properties; eddy disturbances were observed in the water below 200–300 m.

In the process of transformation of cyclonic rings the biological community of their waters undergoes substantial changes: namely, in the course of time the biomass and the dimensions of the organisms decrease; and the rate of change of phytoplankton exceeds that of zooplankton and fishes. For example, the chlorophyll concentration and, hence, the phytoplankton biomass at the center of ring Bob decreased eight times three months after its formation (Wiebe, 1982). The changes of zooplankton were considered (Boyd *et al.*, 1978) for the case of one of the euphausiid species (*Nematoscelis megalops*). These organisms are typical representatives of the slope water and their propagation is limited by the Gulf Stream. They live in some definite temperature and salinity conditions (~10° and

34–36‰) and cannot endure sharp variations of these parameters (not more than 4° and 0.5‰, respectively). The observations showed that, as the rings grew older, the amount of *Nematoscelis megalops* decreased, and this euphasiid species was not found in the 17-month-old ring. Simultaneously with the decrease of the amount of *Nematoscelis megalops*, their vertical distribution also changed. In the slope water these organisms are found in the upper 350 m layer whereas in rings their distribution in a water column is determined by the location of the 10° isotherm. As the rings grew older, *Nematoscelis megalops* went down, following the sinking temperature and salinity isosurfaces. The major part of the population in the six-month-old ring lived deeper than 350 m, and some individual specimens were found even at a depth of 800 m.

Owing to the changes of the physico-chemical conditions in the ring, the physiological state of the population also changed, which manifested itself in a reduction of the rate of increase of the length of individual specimens, their weight, and respiration rate (the oxygen consumption per 1 g of dry weight of animals in the slope water was ten times as great as that of animals in the nine-month-old ring). This also manifested itself in a change of the chemical constitution of the specimens; namely, the amounts of lipids, nitrogen, and carbon were reduced. Ultimately, the degeneration and extinction of the population took place; no acclimatization was observed; and subarctic zooplankton species were replaced by subtropical and tropical species of the Sargasso Sea.

The data on the fish biomass for various rings are more limited but, generally, its distribution coincided with that of zooplankton and the same trend was observed, namely, as the rings grew older, the fish biomass approached its level in the Sargasso Sea.

Thus, slope water and Sargasso Sea organisms play the role of indicators of the original biotic communities of the rings and their subsequent evolution. Cyclonic Gulf Stream rings which preserve themselves as individual physical formations can obviously remain regions of increased productivity for a long time. It is believed that the process of formation and propagation of rings with cold cores is a large-scale invasion of the plankton community of the Sargasso Sea by slope water organisms and can be regarded as a natural ecological experiment during which changes of the environment result in a change of the community (Wiebe *et al.*, 1976).

Anticyclonic rings with warm cores are formed on the northern side of the Gulf Stream and move to the south-west between the North America shelf and the Gulf Stream in the slope water region. When these rings are formed, their cores consist of Sargasso Sea water surrounded by a ring of Gulf Stream water; their diameters are 100–200 km. During 1981–1982 scientists from a number of institutes in the U.S.A. carried out an extensive program of investigation of rings with warm cores; at present the data obtained are being analyzed (Joyce and Wiebe, 1983). During the course of the program it was found that the average lifetime of a ring with a warm core is about 4.5 months, after which the ring merges with the Gulf Stream. However, it turned out that the lifetimes of individual rings are substantially different and depend on the region of their formation; namely, eddies appearing east of the New England Ridge exist for less than three months and west of it they exist for six to nine months. The lifetime of rings and the properties of the water in their cores are greatly affected by the interaction of the rings with the Gulf Stream. For example, it was found that in a ten-day period of intensive interaction between a ring and the Gulf Stream, the parameters of the ring's core water underwent changes that would have taken six months to complete in its individual existence.

Detailed studies of a warm ring which lasted 7.5 months showed that when the ring was formed (the end of February), its core consisted of Sargasso Sea water (Joyce and Wiebe, 1983). A well-developed homogeneous layer (approximately above 500 m) with a temperature of about 18° was observed; the vertical disturbances of the 10° isotherm in the main thermocline exceeded 300 m. At the end of the winter period the water temperature in the homogeneous layer decreased to 15.7°. Further, owing to the spring–summer heating the surface layer temperature gradually increased and the seasonal thermocline was formed. Simultaneously, there was a decrease in the thickness of the homogeneous layer, the salinity of its water, the 10° isotherm depth, and the diameter of the ring. During the entire lifetime of the ring its chemical and biological characteristics also changed. In March and April the total plankton biomass in the upper 1000 m layer was much smaller at the center of the ring in comparison with the slope water. The spring development of phytoplankton, which began at the end of April, led to an increase of biomass both in the ring core (by 50%) and in the surrounding slope water (× 3). Further, the plankton biomass in the ring sharply increased, but decreased in the slope waters. As a result, in June the observed plankton biomass was higher in the ring than in the slope water. By August, approximately 200 days after the formation of the ring, the plankton biomasses in the ring core and in the slope water were level, whereas the biomass in the core was substantially greater than in the Sargasso Sea and the Gulf Stream.

Joyce and Wiebe (1983) compared the development of zooplankton in rings with warm and cold cores and drew the conclusion that in warm rings the biomass increases and attains its level in slope water faster than it decreases in cold rings down to the biomass level in the Sargasso Sea.

Some interesting results were obtained in studying anticyclonic eddies in the south-western part of the Tasman Sea. These eddies were identified as regions with increased primary production (Tranter *et al.*, 1980). They are generated by the East Australia Current transporting southward warm waters with low concentrations of nutrients. It could have been expected that the warm eddies would be low-productive. However, in reality the phytoplankton production at the centers of anticyclones was much higher than in the surrounding waters and in the current itself. Australian scientists believe that this phenomenon is caused by the transport of nutrients to the surface owing to the winter cooling and the appearance of convection. However, since the water at the centers of anticyclonic rings is warmer than the surrounding water, it takes more time to attain thermal equilibrium; therefore, the cooling of surface waters goes on until spring and is accompanied by their sinking to the depth of the main thermocline. As a consequence, in the winter period the water at anticyclone centers is enriched with nutrients. The data on the seasonal variations of the ring 'F' (35–37°S, 152–154°E) are presented in Table 6.2.1.

TABLE 6.2.1

Seasonal variations of water parameters (after Tranter *et al.*, 1980)

Observation time in 1978	*T* at depth of 250 m	*T* at the surface	*H* (the mixed layer, m)	NO_3, μmol/l	Chlorophyll 'a', TU*
September (winter):					
Eddy center	17	17.7	215–320	2.9–3.2	100–200
Surrounding waters	15	15.3–20.2	37–88	1.1–3.3	200
November (summer):					
Eddy center	17	18.0–18.2	60–65	0.3–0.8	200–600
Surrounding waters	15	17.5–18.2	60	0.6	200–400

*TU: fluorescence units.

According to Table 6.2.1, phytoplankton production increased five times from the end of winter to the beginning of summer; the nitrate content decreased from 3 to 0.3–0.8 μM/l. The time the eddy can exist as an area of increased productivity is limited by the amount of nutrients stored in the winter isothermal layer. Phytoplankton production will decrease every year owing to the loss of some of the nutrients together with detritus, the gradual cooling of the water, and the mixing in the ring becoming weaker.

In December 1982–February 1983 Soviet oceanographers carried out a study of synoptic variability in the region of the Antarctic Circumpolar Current (ACC) south of New Zealand within the framework of the program 'the Southern Ocean'. In this region the ACC is divided into two branches by the Macquarie Ridge; the northern branch comes close to the continental slope of the Campbell Plateau and the southern branch goes above the South Pacific Rise, and between them a vast frontal zone is located. The ACC undergoes strong meandering, which creates conditions suitable for the generation of synoptic eddies when ACC meanders are separated. It is obvious that the low-salinity cold water cores discovered by New Zealand scientists when investigating this region during the IGY were these cut-off ACC meanders (Burling, 1961).

The synoptic-scale surveys carried out in an experiment area located in the zone of the South Polar Front (56–58°S, 167–172°E) made it possible to find a cyclonic eddy of a frontal type and observe its evolution. The first and second surveys were performed in the periods from January 1 to 6 and from January 17 to 29, 1983, respectively, and observations extended to a depth of 1500 m. From the first to the second survey the horizontal dimensions of the eddy varied from 90 to 80 miles along the major axis and from 60 to 50 miles along the minor axis.

The entire central part of the eddy above 1500 m was occupied by waters greatly differing in their properties from the surrounding waters. For example, the surface water at the eddy center was 2° colder than the surrounding water. The strongest temperature field disturbances were observed at a depth of 200 m where the temperature differences between the center and the periphery exceeded 5° and decreased with increasing depth. The water of the central part of the eddy had a subsurface temperature minimum, which is a usual feature of antarctic waters in the summer period, whereas the subantarctic surrounding waters did not have this

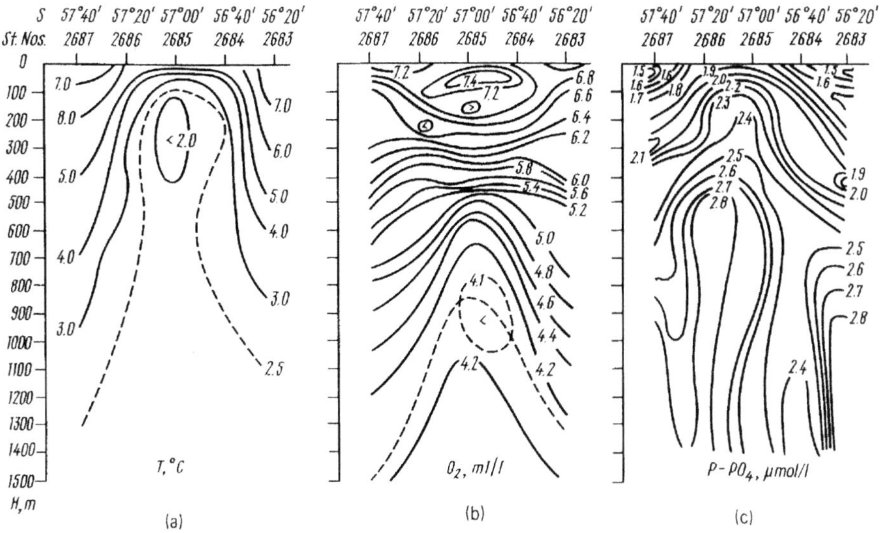

Fig. 6.2.2. Distributions in the section along 168°E across a frontal cyclonic eddy of the Antarctic Circumpolar Current (January 1–8, 1983; the 30th cruise of the research vessel *Dmitriĭ Mendeleev*) of: (a) water temperature, ° C; (b) oxygen, ml/l; (c) phosphates, μmol/l. (After Chernyakova and Borodkin, 1983.)

minimum (Ivanov *et al.*, 1983). The isohaline and isoxygen deflections have opposite signs in the seasonal and main thermoclines, which is a specific feature of the vertical salinity and oxygen distributions (Figure 6.2.2). For example, in the seasonal thermocline the isolines of these characteristics lying above 300 m deflected downward and the water was low-saline (< 34‰) and had a high oxygen content (> 7.4 ml/l) here. Below 500 m, under a layer of large gradients, a reconstruction of the salinity and oxygen fields took place; at the center of the cyclone the upwelling of more saline deep waters with low oxygen content was observed. The nutrient concentrations had increased values in the entire water column (Figure 6.2.2(b)). The analysis of (T, S) and (T, O_2) relations showed that the water at the center of the cyclone had characteristics analogous to the parameters of the water in the southern branch of the ACC south of 60°S and substantially differed from the background waters.

According to the observation data, the cyclonic eddy moved to the north-east with an average velocity of 11 cm/s. The distribution of hydrochemical characteristics obtained during the second survey for a depth of 200 m is presented in Figure 6.2.3. There were two cyclonic formations (one at the center of the experiment area and the other on its south-west periphery) which were characterized by low temperatures and high concentrations of oxygen and phosphates. The water characteristics of the first of these cyclonic eddies underwent substantial changes during the period between the surveys; namely, the temperature and salinity of the surface layer water increased and the content of oxygen and nutrients decreased. The silicate concentration in the surface layer decreased five to seven times in comparison with the first survey and at depths of 20–30 m its lowest concentration

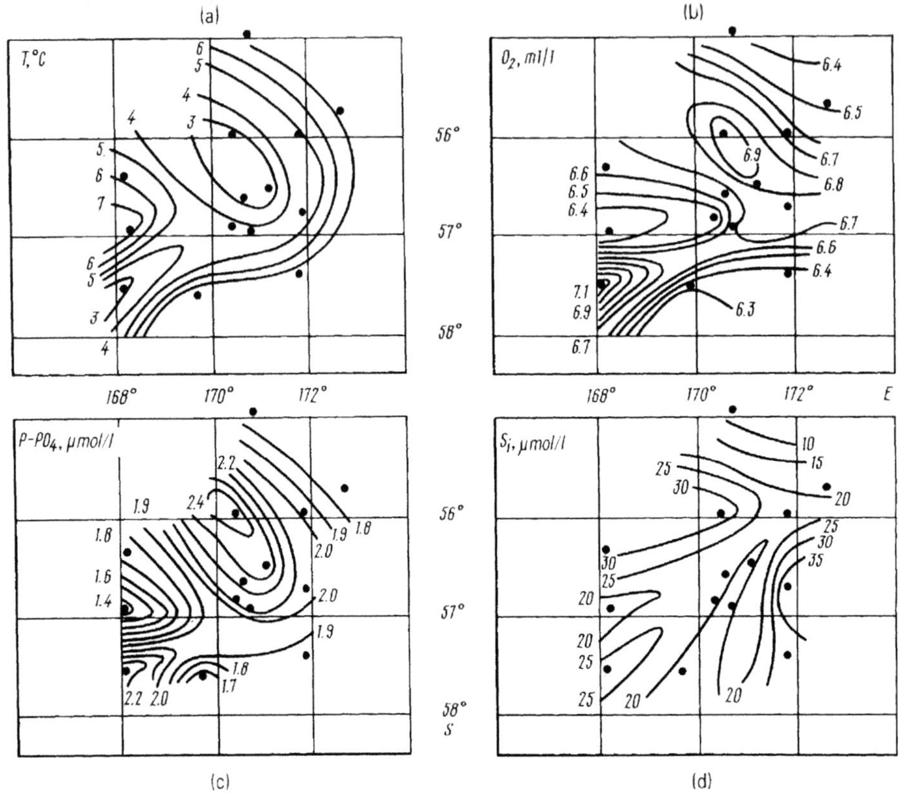

Fig. 6.2.3. Distributions at a depth of 200 m in a region of the Antarctic Circumpolar Current south of New Zealand (January 17–29, 1983, the 30th cruise of the research vessel *Dmitriĭ Mendeleev*; 56–58°S, 167–172°E) of: (a) water temperature, °C; (b) oxygen, ml/l; (c) phosphates, μmol/l; (d) silicates, μmol/l. (After Chernyakova and Borodkin, 1983.)

in this region was observed (2 μmol/l). The same depths contained a subsurface oxygen maximum with a saturation value up to 103–104%; the content of oxygen in the surrounding water was either close to 100% or 3–4% lower than the saturation value. The sharp decrease of the silicate concentration and the appearance of the subsurface oxygen maximum are related to the diatom bloom in the ring waters. According to the corresponding data (Ponomareva, 1983), the plankton biomass increased three to four times during this period. The ring waters also differed from the surrounding waters in the species composition of zooplankton; it contained species typical of waters of the southern branch of the ACC; in the surrounding waters either there were no such species or they were not numerous.

Hence, it is clear that the frontal cyclone under consideration was formed of waters of the southern branch of the ACC owing to the separation of a meander from this branch and then moved under the action of the mean current in the north-east direction transporting the antarctic water mass in its core. It is obvious that the ACC meandering, which is primarily related to the bottom topography of the region under consideration, is a permanently acting eddy-formation source and a mechanism of the northward transport of antarctic waters. This can be confirmed by the cyclone in the south-west corner of the experiment area which was found during the second survey; it continued the chain of eddies appearing in the region of the quasistationary meander at 59°S.

The distributions of hydrochemical characteristics in the section along 168°E (Figure 6.2.4; stations 2685 and 2705 have the same coordinates) underwent substantial changes between the first and second surveys. During the first survey (station 2685) a cyclonic eddy with cold water of the southern branch of the ACC was located here; 12 days later, during the second survey, the entire ocean layer under study was occupied by warm water with a low nutrient concentration and the behavior of the isolines had anticyclonic character. As in the case of the cyclonic eddy, the vertical salinity and oxygen distributions had a two-layer structure. However, in the warm eddy the salinity was increased (>34.5‰) and the oxygen content was decreased in the 0–300 m layer; their isolines in the seasonal thermocline were convex upward. Below the intermediate layer, approximately beginning with a depth of 500 m, both the salinity and oxygen fields underwent reconstruction and their isolines were convex downward along with the isolines of other parameters, which indicated the downwelling of waters in the anticyclone. This water in its characteristics corresponds to the water of the northern branch of the ACC but lack of data does not make it possible to reliably describe the mechanism of its appearance in this region (Chernyakova and Borodkin, 1983).

The synoptic variability of hydrochemical and hydrobiological characteristics relating to the eddy dynamics in the open ocean is considered below in view of the observation data obtained during the POLYMODE program in 1977–1978 in the

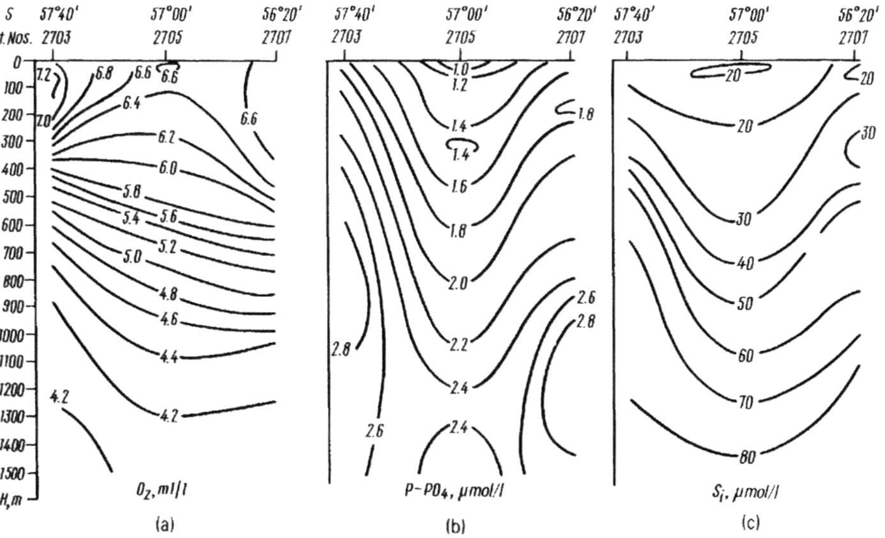

Fig. 6.2.4. Distributions in the section along 168°E in a region of the Antarctic Circumpolar Current south of New Zealand (January 17–29, 1983; the 30th cruise of the research vessel *Dmitrii Mendeleev*) of: (a) oxygen, ml/l; (b) phosphates, μmol/l; (c) silicates, μmol/l. (After Chernyakova and Borodkin, 1983.)

western part of the Sargasso Sea in the experiment region with center at 29°N, 70°W. The hydrochemical conditions in the region under study are determined by its position in the subtropical convergence area. High oxygen concentrations, large pH values, and a very low nutrient content were observed in a layer from the sea surface to a depth of 400–500 m. Extreme concentrations of chemical elements were observed at depths of 800–1000 m. The most typical feature of the vertical thermal structure in the summer period is the presence of a double thermocline split by the '18-degree' water lying at depths of 200–500 m; in the layer occupied by this water an intermediate oxygen maximum (~5.2 ml/l) was located and, simultaneously, the pH values increased.

During the observation period about 20 of cyclonic and anticyclonic eddy formations travelled across the experiment region. Along with the 'close-packed' eddies of different sign, a very strong single eddy was observed. All the eddy formations manifested themselves in their hydrochemical characteristics most clearly in the 200–1200 m layer. The general features of the behavior of the vertical distribution of chemical parameters typical of the Sargasso Sea were retained in both the cyclonic and anticyclonic eddies. The passage of 'close-packed' eddies produced disturbances in the fields of characteristics and led to the downwelling of waters in anticyclones and the upwelling in cyclones; the greatest displacement of the profiles relative to one another took place in the region of the main thermocline and attained 200 m (Figure 6.2.5). The data on the oxygen content are presented in Table 6.2.2.

The variability of thickness of the '18-degree' water layer, depending on the sign of an eddy and its energy, can be judged from the oxygen distributions in the cyclonic and anticyclonic 'close-packed' eddies and in the strong single cyclone (Figure 6.2.6(a), (b), and (c)). The '18-degree' water layer had its maximum thickness at the anticyclone center (stations 2337 and 2338) and extended from 200 to 500 m; its thickness in the cyclonic eddy was much smaller and was located between 100 and 250 m; there was almost no '18-degree' water in the single cyclone. The oxygen content also varied in all the eddies under study; at a depth of 400 m it was 5.1, 4.7, and < 3 ml/l, respectively; the anticyclone contained a clearly marked intermediate oxygen maximum and the single cyclone contained a minimum oxygen content.

The distributions of hydrochemical characteristics in the POLYMODE area varied in accordance with the development and motion of synoptic eddies. In the survey during May 1–8, 1978 a system of strong anticyclonic formations was marked in the central part of the area in the form of a region with increased oxygen concentrations and decreased nutrient content (Figure 6.2.7(a)). The repeated survey of May 31–June 7 showed that the system of warm eddies was displaced to the west, weakly marked cyclonic formations were located at the center of the area, and a new powerful anticyclone was approaching the area from the east. The corresponding reconstruction took place in the fields of hydrochemical characteristics, namely the nitrate, phosphate, and silicate concentrations in the experiment area increased by 2–3, 0.2–0.3, and 3–4 μmol/l, respectively, and the oxygen content decreased by 0.5 ml/l (Figure 6.2.7(b)). The distribution of hydrochemical characteristics clearly reflected the structure of the eddy field in the experiment

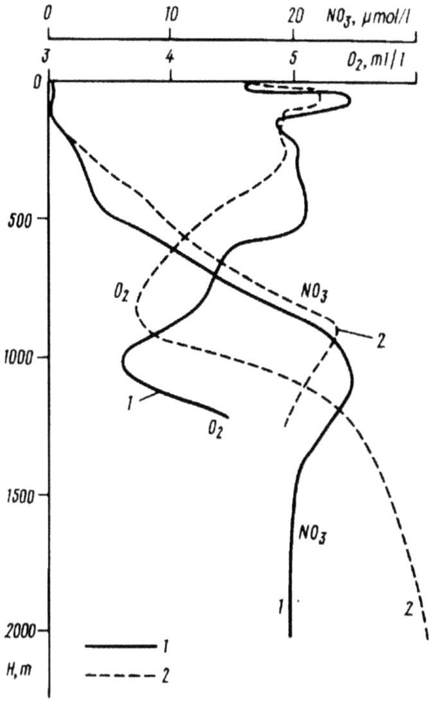

Fig. 6.2.5. Vertical distributions of oxygen (ml/l) and nitrate (μmol/l) in the POLYMODE region: (1) in an anticyclonic formation (station 2338; the 25th cruise of the research vessel *Akademik Kurchatov*, October 1977); (2) in a cyclonic formation (station 2229; the 25th cruise of the research vessel *Akademik Kurchatov*, July 1977. (After Chernyakova, 1978.)

area. For comparison we present a chart of the 15° isotherm depth whose average position in the region under study corresponds to a depth of 600 m (Figure 6.2.7) (Chernyakova and Romanov, 1980).

An analysis of relations T, O_2 and σ_t, O_2 found from the data of summer observations in 1977–1978 revealed that they were in close agreement with the so-called 'standard curve' constructed for the Sargasso Sea from data of long-term observations (Richards and Redfield, 1955). It was shown that the 'close-packed' eddies in the POLYMODE region contained waters of the Sargasso Sea itself (Figure 6.2.8) (Chernyakova, 1978; Chernyakova and Romanov, 1980).

The Sargasso Sea is one of the least productive regions of the ocean. Downwelling in the convergence region, and the development of the seasonal thermocline, lead to a low content of nutrients in waters of the euphotic layer and to their oligotrophy; the primary production in the summer season amounts to $0.1-0.2\ gC/m^2 \cdot$ day (Ryther and Menzel, 1960). Biological studies carried out in the experiment area, which included both cyclonic and anticyclonic eddies, showed the presence of species that are characteristic of Sargasso Sea waters. No species of shelf origin were found; the biological productivity of plankton was within the limits typical of the Sargasso Sea waters (Radchenko, 1978; Roukhiyaïnen and Yunev, 1978). In summer there is a closed circle of nutrients above the seasonal

TABLE 6.2.2*

Variability of the oxygen concentration in the region of the POLYMODE mooring array (ml/l)

H, m	$\bar{O}_2$	σO_2	O_2 minimum	O_2 maximum	ΔO_2
0	4.56	0.07	4.36	4.72	0.36
10	4.62	0.10	4.33	4.99	0.66
20	4.68	0.16	4.43	5.10	0.67
30	4.83	0.24	4.46	5.22	0.76
50	5.19	0.16	4.69	5.46	0.80
75	5.18	0.18	4.88	5.58	0.70
100	5.04	0.16	4.66	5.30	0.64
125	4.93	0.18	4.67	5.13	0.46
150	4.86	0.18	4.56	5.05	0.56
200	4.82	0.14	4.42	5.08	0.66
250	4.86	0.14	4.47	5.04	0.57
300	4.84	0.16	4.37	5.09	0.72
400	4.67	0.23	4.17	5.10	0.93
500	4.35	0.18	4.04	5.09	1.05
600	4.14	0.12	3.81	4.49	0.68
800	3.74	0.17	3.38	4.24	0.68
1000	4.08	0.32	3.44	4.66	1.22
1200	5.04	0.32	4.49	5.62	1.13
1500	5.74	0.18	5.19	6.00	0.81
2000	5.98	0.09	5.75	6.07	0.32

*The table is compiled according to the data of 66 stations of the 25th cruise of the research vessel *Akademik Kurchatov* in July–September 1977 (after Chernyakova, 1978).

thermocline. Synoptic eddy formations redistribute them in the photosynthesis layer and produce spots with increased nutrient content and thus exert an effect on the biological productivity of the waters. For example, according to the data of Roukhiyaĭnen and Yunev (1978), the propagation of eddies produces favorable conditions for an increase in the photosynthetic activity of phytoplankton, which results in an increase in the plant biomass and chlorophyll concentration. It turned out that the increase in photosynthetic activity of phytoplankton occurs 12–24 h after the nutrient inflow (Burlakova and Kholodov, 1979). This time lag can be one of the factors causing the absence of any direct correlation between the nutrients' distribution and productivity characteristics obtained simultaneously. According to the summer observation in 1977, no correlation existed between the oxygen content and the phytoplankton standing crop (the correlation coefficient was 0.26) and also between the number of plant cells and the amount of phosphates (the correlation coefficient was 0.08) (Roukhiyaĭnen and Yunev, 1978). In relation to productivity, the most promising areas were the zones of eddy interaction and the least promising areas were the anticyclone centers. Investigations into the composition of the dominant fouling species in the experiment region in winter and summer 1978 showed that the fouling species attained their maximum sizes in the water upwelling zones which were associated with synoptic front regions and also with cyclonic formations.

In the winter of 1977–1978 a strong single cyclonic eddy was found and investi-

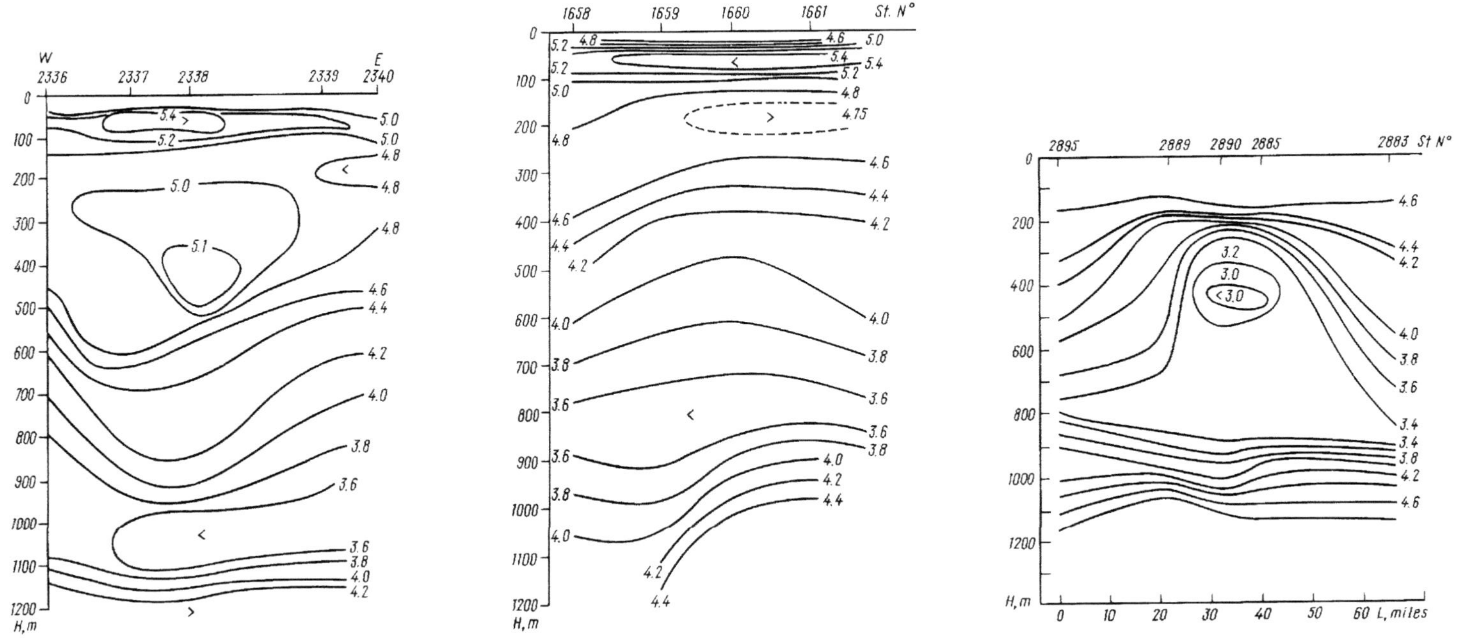

(a) (b) (c)

Fig. 6.2.6. Vertical distributions of oxygen (ml/l) in the POLYMODE region in the section across: (a) an anticyclonic eddy (October 1–2, 1977; the 25th cruise of the research vessel *Akademik Kurchatov*); (b) a cyclonic eddy (July 20, 1978; the 16th cruise of the research vessel *Akademik Vernadskiĭ*); (c) an intensive cyclonic eddy (January 6–7, 1978; the 33rd cruise of the research vessel *Mikhail Lomonosov*).

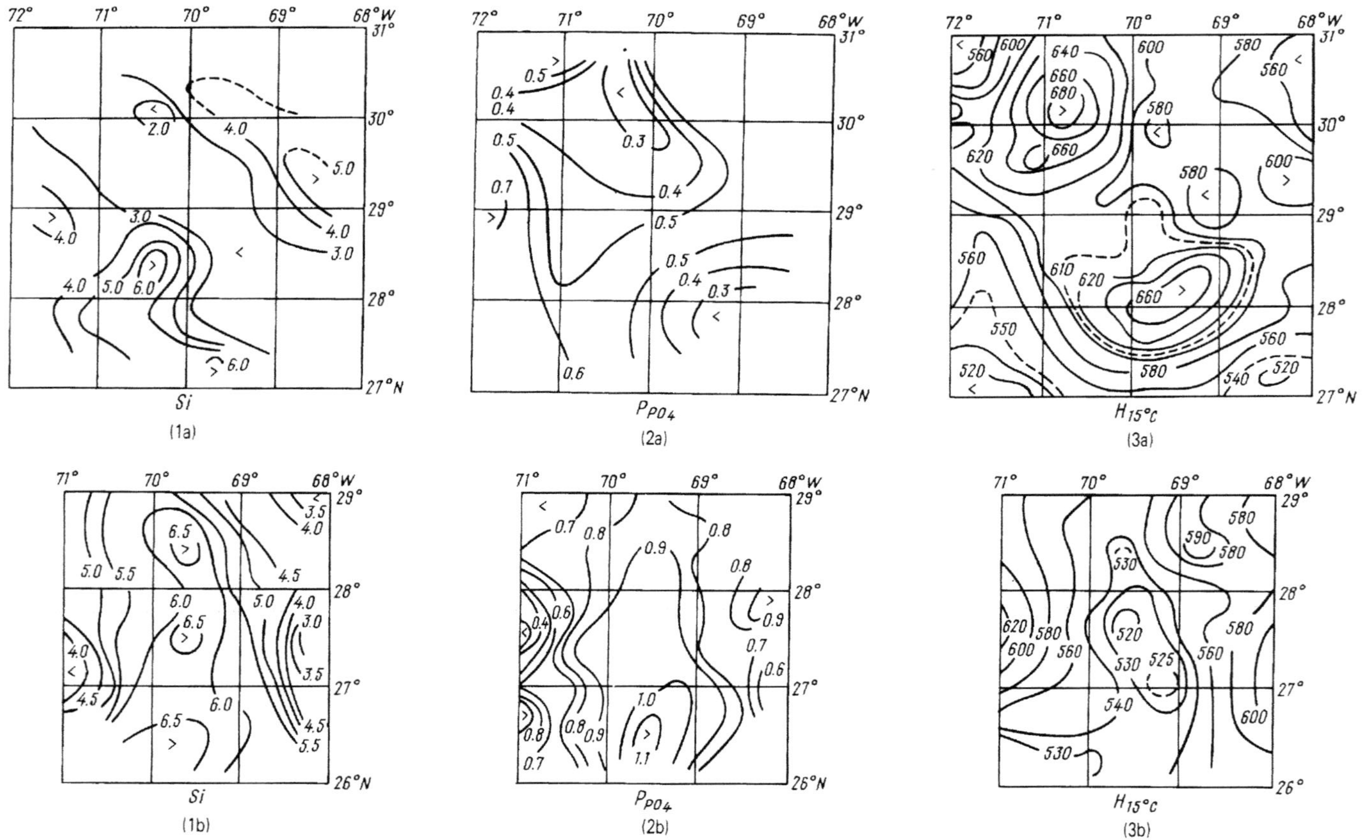

Fig. 6.2.7. Distributions of silicates and phosphates (in μmol/l) at a depth of 600 m and of 15 °C isotherm depths (in meters) in the POLYMODE region: (1a, 2a, 3a) May 1–8; (1b, 2b, 3b) June 1–7, 1978 (the 34th cruise of the research vessel *Mikhail Lomonosov*). (After Chernyakova and Romanov, 1980.)

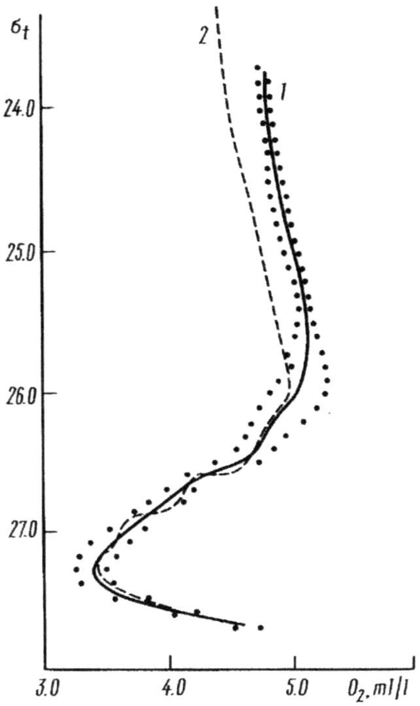

Fig. 6.2.8. σ_t–O_2 dependence according to POLYMODE data. (1) The mean curve (May–June 1978; the 34th cruise of the research vessel *Mikhail Lomonosov*) (after Chernyakova and Romanov, 1980); (2) the standard curve for Sargasso Sea waters (after Richards and Redfield, 1955).

gated in the POLYMODE region. Its cupole-shaped structure was clearly observed in the 200–1000 m layer; the elevation of the 15° isotherm exceeded 200 m. Its hydrochemical and hydrophysical parameters were substantially different from those of both Gulf Stream rings and Sargasso Sea synoptic eddies. It contained no '18-degree' water layer and hence no intermediate oxygen maximum. The maximum phosphate concentrations were observed at a depth of 600 m, their position being at 1000–1100 m in this region in the absence of eddies. The layer of oxygen minimum had an anomalous structure since it consisted of two parts, an upper part at 300–400 m and a lower part at 700–800 m, which were separated by waters of a higher oxygen content at depths of 500–600 m (Figure 6.2.6(c)). A comparison of the data of two surveys (carried out on January 5–7 and February 21, 1978) made it possible to find changes in the distribution of the oxygen content, namely the oxygen concentration in the upper 300–400 m layer increased by more than 0.5 ml/l over a period of 1.5 months; the thickness of the layer decreased and its depth increased. The existence of lenses of high-salinity waters (37.0–37.2‰) at depths of 150–200 m allowed Paramonov, *et al.* (1978) to assume that the eddy was formed in the south-eastern part of the subtropical gyre. Comparing the oxygen distribution in this eddy with the averaged data for the North Atlantic we can find that such low concentrations (<3 ml/l) that were encountered in the cyclonic eddy at depths of 300–400 m are not typical of the western part of the Sargasso Sea and are usually observed in the south-eastern part of the subtropical gyre.

Investigators have recently concentrated on studying specific features of hydrodynamical and biological conditions in regions of seamounts in the open ocean. The data obtained show that these regions have a very high biological productivity and in many regions of the Emperor Seamount Chain and the Hawaiian Ridge fish resources are already intensively used. High productivity of waters is also observed above the seamounts of the South Pacific Ridge, above the Norfolk and Lord Howe Ridges, and above many others; in some cases the productivity of seamount regions is comparable with that of high-productive shelf regions (Darnitskiĭ, 1979a, b, 1980a,b; Darnitskiĭ and Boldyrev, 1977; Kozlov and Darnitskiĭ, 1981; Bradford *et al.*, 1982).

It was shown that eddies of various scales are permanently formed in the zones of interaction of mean oceanic currents with seamounts. As a consequence of the Proudman–Taylor effect, an anomalous distribution of characteristics is observed in eddy fields above seamount tops. This distribution has the form of isolated columns with maximum concentrations of nutrients at their centers. The vertical disturbances of temperature and salinity isosurfaces can attain 200–400 m and those of oxygen and nutrients can attain 500 m. As example, Figure 6.2.9 presents the distributions of nutrients above some seamount tops. Owing to the two-dimensional eddy effect, the nutrients going upward to subsurface layers are kept at the center of the water column and can cause an outbreak of bioproductivity. However, topographic eddies seem to have a short lifetime, and when the main current strengthens, they are cut off and carried away by the current. They form a trace of eddy disturbances downstream behind the seamount, sometimes extending to 200–500 km. In particular, behind the Milluoky Seamount of the Emperor Seamount Chain five such sign-alternating eddies with diameter about 150 km were observed (Darnitskiĭ, 1979b). Eddy motion exerts a strong effect on a biological community by creating favorable or unfavorable food environments for fish. To a certain extent this very factor accounts for the variations of the biological productivity in seamount regions. For example, in the region of the Wanganella Bank, which is the top of the Norfolk Ridge located in the Tasman Sea, the maximum plankton biomass in 1976 was 3990 mg/m^3 and corresponded to good conditions for fisheries; in 1977 the plankton biomass did not exceed 115–150 mg/m^3 and the conditions for fisheries were unfavorable. In the latter case, currents strengthened and eddies separated and were driven to the north. For almost identical thermal conditions the concentration of nutrients in 1977 was twice as low as that in 1976. Hence, the variation of the biological productivity of seamount regions is related to the varying intensity of ocean currents and also, consequently, to eddy formation. Fish shoals can obviously follow eddy systems and choose optimal conditions for life in other ocean regions (Darnitskiĭ, 1979a). Eddy formations similar to Taylor columns are not always clearly identifiable since the columns of quasihomogeneous water do not always reach the sea surface (Figure 6.2.9).

3. ACOUSTIC APPLICATIONS

Synoptic eddies – the description of whose dynamics and physical properties is the subject of this book – are of particular interest from the viewpoint of acoustics. On

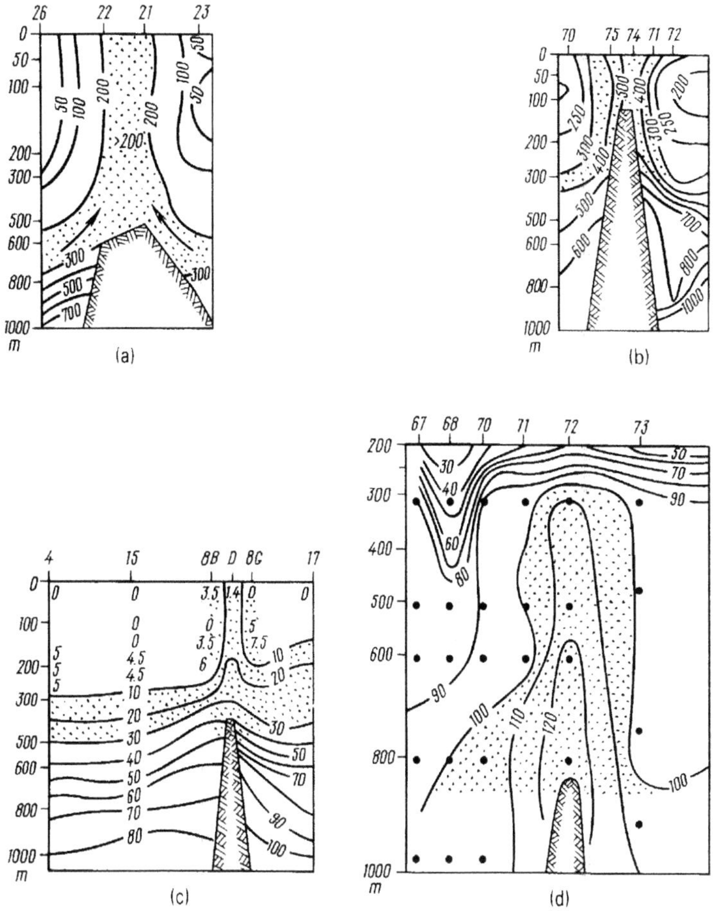

Fig. 6.2.9. Distribution of silicates (μg/l): (a) above the Pulkov Seamount (the Pacific, the Hawaiian Ridge, July 1978); (b) above the Wanganella Bank (the Pacific, 32°S, 167°E, July 1976). Distribution of phosphates (μg/l): (c) in the region of the Ecuador Seamount (the Pacific, 29°N, 179°W, December 1972); (d) in the region of the Sala and Gomes Ridge (the South-East Pacific, November 1975). (After Kozlov and Darnitskiĭ, 1981.)

the one hand, these eddies create additional difficulties in processing and interpreting sound signals and, on the other hand, they constitute an independent object admitting of the application of acoustic methods of investigation. Here we briefly discuss the effect of the synoptic variability of the ocean on sound fields.

Hamilton (1974) was probably the first to encounter a manifestation of this variability during the experiment on long-range propagation of sound along two paths (Antigua–Eleuthera and Antigua–Bermuda) in the Sargasso Sea. Under the study were fluctuations of the time of propagation of a sound signal along the two paths and the mutual correlation of these fluctuations. The observations lasted for 27 months. Fluctuations of 500 ms were registered. The root-mean-square fluctuations at the observation points were 200 ms and the correlation interval was several months. It was noted that no correlation between the fluctuations measured in

Eleuthera and Bermuda was observed. This can be accounted for by the fact that the sound paths intersected various synoptic-scale inhomogeneities of the oceanic structure.

Later Clark and Kronengold (1974) carried out detailed experiments to study the fluctuations of amplitude A and phase θ of a CW acoustic signal at short and long distances (across the Florida Strait and along the Eleuthera–Bermuda route, respectively). The sound pressure at the reception point had the form $p = A \cos (\theta - \sigma t)$, where σ is the frequency and A and θ are slowly varying functions of time. In both cases the sound paths intersected regions with clearly marked dynamic ocean activity. In contrast to Hamilton's experiment, the signals in this case propagated along several paths, added together, and interfered at the observation point (multiray reception). The results of processing the time series detected not only fluctuations produced by internal waves but also slow phase fluctuations which were related to the synoptic variability of the sound velocity field.

The direct measurement data demonstrate the complicated behavior of a real sound velocity field in both the vertical and horizontal planes. As a rule, the distinctions from the mean climatic (time-average) distributions are great, particularly in the upper ocean. As an example, Figure 6.3.1 shows a horizontal slice of the sound velocity field at a depth of 800 m in the Sargasso Sea (Beckerle and La Casce, 1973). The figures at the isolines indicate the sound velocity (m/s). A clearly marked cellular (eddy) structure of the field can be seen.

It is believed that the mesoscale variability of the ocean can be interpreted in the form of intensive isolated eddies with a warm or cold core which are formed owing to the instability of oceanic currents on the background of a less clearly marked eddy field of, mainly, barotropic and first-mode baroclinic Rossby waves (Korotaev *et al.*, 1979).

The geographical spread of eddies, and the substantial distortions of mean climatic sound velocity distributions and stationary currents that they produced, stimulated special experiments and model calculations. Their purpose is to estimate the fluctuations of sound signal parameters caused by eddies and to elaborate methods of forecasting these fluctuations in real conditions.

The acoustic effect of eddies is specified by their internal structure. We consider the so-called 'cold eddies', i.e. isolated eddies with cold cores. They include, for instance, the cyclonic Gulf Stream eddies ('rings'). These eddies are studied most thoroughly from the viewpoint of oceanology and acoustics. Figure 6.3.2 demonstrates a vertical section of the sound velocity field in one of the Gulf Stream eddies (Shank *et al.*, 1975). We can see that sound velocity variations during the passage of an eddy can attain 35 m/s at some depths, which must necessarily manifest itself in the form of fluctuations of the phase of the sound signal and its pressure level $L = 20 \log (A/A_0)$. Here A_0 is a fixed value of the amplitude and the sound pressure level is measured in decibels (dB).[1] In one of the experiments (Shank *et al.*, 1975) explosive sound sources were fired so that the sound paths intersected a Gulf Stream eddy in different directions. It turned out that the sound pressure level of

[1] The decibel scale is suitable for describing quantities ranging within broad limits. To a K-fold increase of amplitude A there corresponds an increment of the sound pressure level L by $20 \log K$ dB.

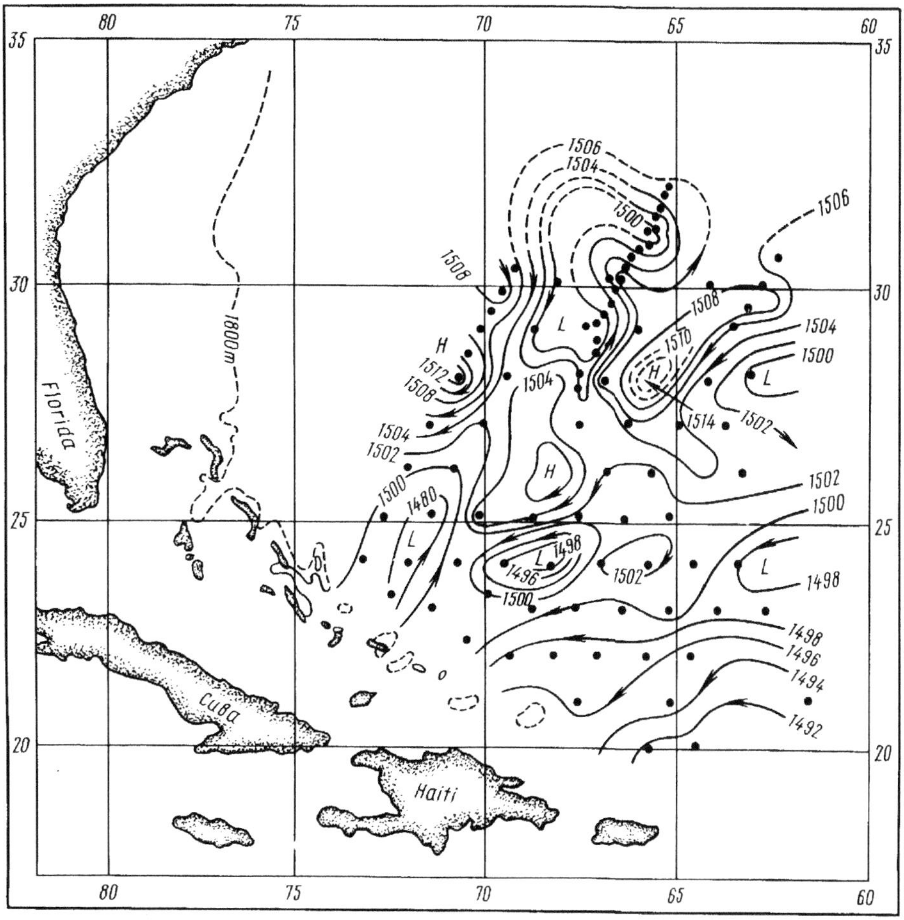

Fig. 6.3.1. Horizontal section of a sound velocity field (marked on the axes are the geographical coordinates in degrees).

the received signals substantially depended on the direction of the source. For example, the sound pressure level of the signal passing through the eddy center exceeded that of the signal arriving along the tangent to the eddy by 23 dB.

In another experiment (Beckerle *et al.*, 1980) sound sources were set on neutrally buoyant floats drifting in a region south-east of the Gulf Stream. The receiver located at Bermuda registered fluctuations of the sound pressure level of 12–18 dB caused by the passage of a Gulf Stream ring across the sound paths.

The vertical section of the sound velocity field shown in Figure 6.3.2, and some other similar sections, served as a basis for numerical calculations making it possible to model the effect of a cold eddy on the propagation of sound in various radiation and reception conditions. In particular, the calculation performed by Weinberg and Zabalgogeazcoa (1977) demonstrated the dynamics of fluctuations of the sound pressure level and the phase of a CW sound signal with a frequency of 200 Hz in the case of intersection of a sound path by the eddy shown in

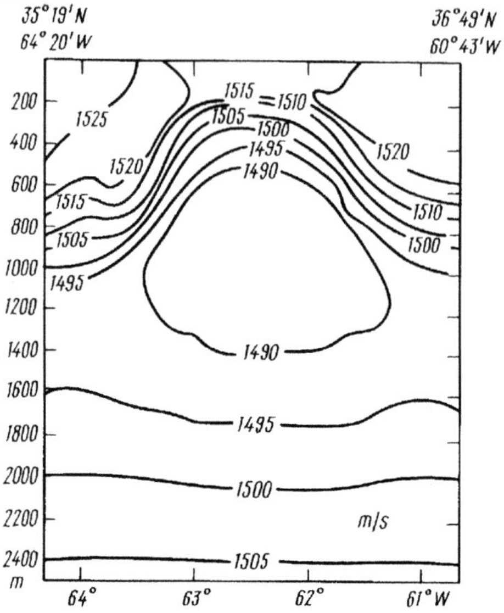

Fig. 6.3.2. Sound velocity isolines in a Gulf Stream eddy (February 4–7, 1973).

Figure 6.3.2. The depths of the source and the receiver were 200 m and 3000 m, respectively, and the distance between them was 445 km. The calculation results are presented in Figure 6.3.3. Shown at the top and bottom are the sound pressure level in dB and the phase in cycles, respectively (one cycle corresponds to a phase increment of 2π). The axis of abscissas represents time; the reference time in the plots corresponds to the moment when the eddy was located exactly between the source and the receiver. The time it took the eddy to entirely intersect the sound path was 80 days. We see that fluctuations of the sound pressure level attain 20 dB and the change of phase over a period of 40 days is equal to 180 cycles (this corresponds to 0.9 s in terms of the time of propagation of the signal). The amplitude fluctuations turned out to be commensurable with the corresponding fluctuations caused by internal waves (Steinberg *et al.*, 1972).

The effect of an eddy manifests itself not only in local changes of amplitude and phase of a sound signal but also in displacements of the boundaries of convergence zones and acoustic shadow zones as a whole. (The convergence zones correspond to the areas of substantial concentration of sound energy related to the arrival of one or several sound rays at each point of such a zone; in the shadow zones there are no rays or the intensities of the signals arriving along them are small.) Vastano and Owens (1973) applied the ray calculation method to study the structure of a sound field in the presence of a cold eddy. The axis of the main sound channel, indicating the position of deep minimum in the vertical sound–velocity profile $c(z)$, was elevated up to 400 m as it approached the core of the eddy. A sound source with frequency of 100 Hz was located in this model at a depth of 200 m above the edge of the eddy. The diameter of the eddy was about 120 km. The sound path passed through the eddy center. The dependences of the sound field's pressure level on the distance to the source in the range 0–180 km (horizontal cuts) for two

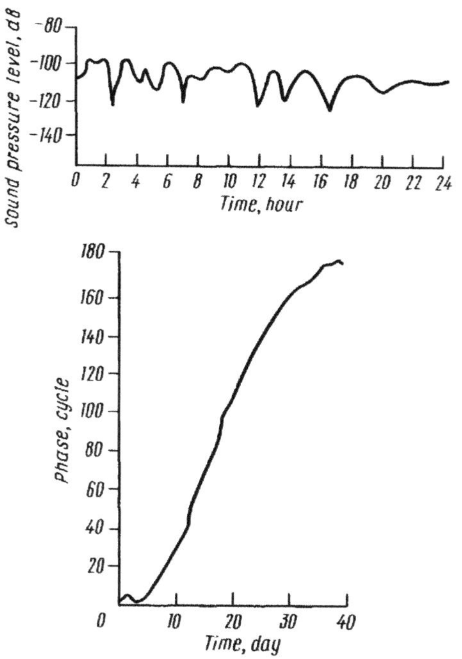

Fig. 6.3.3. Fluctuations of the sound pressure level and the phase of a CW sound signal propagating across a Gulf Stream eddy.

depths of 300 and 1000 m were calculated. These cuts were compared with the corresponding cuts in the absence of the eddy. It turned out that the boundary of the convergence zones was displaced towards the source. At a depth of 300 m the first and second convergence zones were displaced approximately by 5 km and became much narrower. At a depth of 1000 m the first and second regions shifted approximately by 20 and 30 km, respectively, towards the source. The sound intensity distributions inside these regions also changed substantially.

Since the effect of eddies on the conditions of sound propagation is very strong, and the information about the deep structure of the eddies is usually scarce, an attempt was made to parametrize a Gulf Stream ring and thus the distortions of the sound velocity field introduced by it on the basis of global parameters of the ring such as its radius, the depth of its influence, its direction of rotation, its maximum speed, and its transitional velocity (Henrick *et al.*, 1977). It was possible to partly measure the above-mentioned parameters by remote methods, e.g. from Earth's artificial satellites, and thus to control, at least crudely, dynamic and thermodynamic characteristics of the ocean. Despite the fact that it was impossible to reconstruct the internal structure of a real eddy with an accuracy required in acoustics, this few-parameter model turned out to be suitable for numerical experiments to analyze the effect of various eddy parameters on the propagation of sound.

Baer (1980) investigated the effect of the cold Gulf Stream eddy parametrized by Henrick *et al.* (1977) on the shape of the illuminated and shadow zones and also on the angular sound intensity spectrum. The maximum deviation of the sound velocity from the mean climatic value in the model was about 12 m/s, which corresponds to an eddy which is 'energetically' weak. The calculation was performed for frequencies of 25 and 100 Hz. At the beginning, the source at 600 m depth was located in the eddy center. The effect of the eddy manifested itself even at a distance of 20 km from the source where a shadow zone was formed in surface ocean layers. Owing to the eddy the configuration of the shadow zone changed. Its lower boundary was elevated approximately up to 600 m. The convergence zones in the surface layers came nearer to the source. For the third convergence zone and farther, these displacements were 25 km and remained invariable as the distance increased. In the bottom ocean layer, deeper than 3 km, there was almost no change in the pressure level of the sound field. The analysis of the angular sound intensity spectrum showed that for a frequency of 100 Hz the acoustic energy flux concentrated in an angle of 5° to the horizon was ten times higher in the presence of the eddy than without it (3.4 times higher for a frequency of 25 Hz). This indicated a substantial reorganization of the ray structure of the sound field. The numerical analysis of the effect of eddy motion along the sound path, i.e. from the source to the receiver, for a fixed distance between them equal to 250 km was also performed. The variations of the sound pressure level attained 20–30 dB. They were close to those obtained in the paper by Weinberg and Zabalgogeazcoa (1977).

Any eddy is a disturbance of not only the sound velocity field but also of currents. These currents can either accelerate or decelerate the propagation of a sound signal. Moreover, the length of the sound path and its orientation relative to the eddy are important for the propagation of sound. In the examples considered above the current was not taken into account and the orientation of the sound route was fixed. In some papers (Henrick, 1980; Itzikowitz *et al.*, 1982, 1983) attempts were made to consider all these effects jointly.

As a model, the eddy considered in the paper by Henrick *et al.* (1977) was again taken. The sound frequency was 400 Hz. A number of simplifications made it possible to find an analytical expression for propagation times (or phases) of sound signals arriving along different rays and also for the sound pressure level of the resultant multiray signal. These quantities depend on the radius and maximum current speed in the eddy, which were varied and also on the positions of the source and the receiver relative to the eddy center. The calculations were performed for short (0–50 km) and long (1000 km) distances.

It was found that the fluctuations of signal propagation times are maximum when the sound path passes through the eddy center. In this case the fluctuations increase with increasing path length in the eddy, the eddy diameter, and the maximum current speed in the eddy. A change of the position and the orientation of the sound route relative to the eddy can result in fluctuations of propagation times and sound pressure level up to 15 ms and 25 dB, respectively. It was also found that currents related to the eddy make relatively large contributions to the variability of the times of arrival (7–15%) and the sound pressure level (up to 10 dB).

One of the manifestations of oceanic eddies is the horizontal (lateral) refraction of sound rays produced by them. This refraction and its related effects were studied in the paper by Munk (1980). He used a plane eddy model. For simplicity it was assumed that the eddy was circular, had a radius r, and that the source D and receiver F were equidistant from the eddy center E (Figure 6.3.4). The sound velocity $c(r)$ inside the eddy depended only on the distance to its center: $c(r) = c_0 [1 + \nu(1 - r^2/r_0^2)]^{-1}$ and coincided with the sound velocity c_0 in the surrounding medium at the boundary $r = r_0$. The parameter ν denotes the maximum relative fluctuation of the velocity of sound. The case $\nu > 0$ corresponds to a cold eddy and $\nu < 0$ corresponds to a warm eddy. The parameter ν attains a value of 0.03 for intensive Gulf Stream eddies, but usually $|\nu| \lesssim 0.01$.

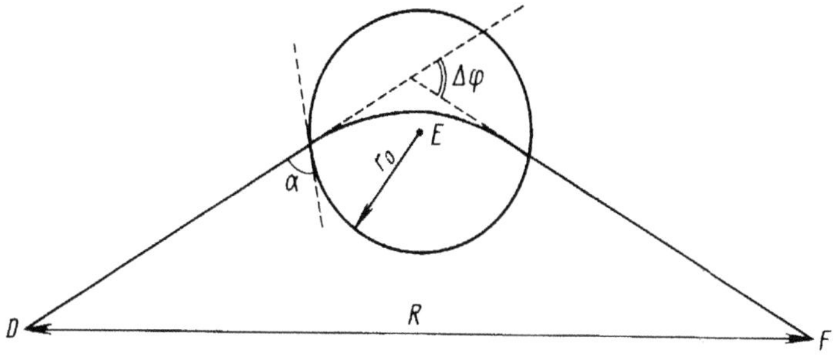

Fig. 6.3.4. Horizontal refraction in a plane eddy.

The fluctuations of the time of propagation of the sound signal between the source and the receiver are given by the expression

$$\tau = 2\left(\frac{\nu r_0}{c_0}\right) \sin^2 \alpha \left[\tfrac{2}{3} \sin \alpha + \chi \cos^2 \alpha\right],$$

where τ is the difference between the times of propagation of the signal in the medium without and with an eddy; $\chi = \nu R/r_0$, in which R is the distance between the source and the receiver, and α is the angle between the ray and the tangent to the eddy boundary (see Figure 6.3.4). For $r_0 = 100$ km, $\nu = 0.01$, $c_0 = 1.5$ km/s, and $\alpha = \pi/2$ we obtain $\tau \approx 1$ s. This value is commensurable with the one obtained in Hamilton's (1974) experiment, which indicates that the eddy model is chosen reasonably.

Simple analytical calculations performed by Munk showed that the angular deviation of the ray from the initial direction is equal to $\Delta\varphi = 2\nu \sin(2\alpha)$ rad. The maximum angular deviation is observed for rays passing at a certain distance from the eddy center. This depends on ν and lies within the limits 1.0–3.5°.

The horizontal refraction of rays caused by an eddy can sometimes lead to a multiray situation when several rays arrive at the observation point. The appearance of the multiray structure is determined by the quantity χ. For $|\chi| \ll 1$, which is usually the case, and on condition that the eddy is warm, the arrival of two rays is in fact possible. In this case one of the rays arrives from the exterior of the eddy in the vicinity of its outer edge and the other is refracted inside the eddy. The interference of signals arriving along different rays leads to additional variability of the sound field. Moreover, it becomes impossible to identify the arriving rays with those in a medium without an eddy.

The sound pressure level also fluctuates owing to the horizontal refraction of rays. This is due to the focusing or defocusing properties of an eddy in relation to rays intersecting it along different trajectories. For $|\chi| \ll 1$ a cold eddy should be expected to produce weak focusing of rays passing near the eddy center and strong defocusing of rays in directions tangent to the eddy. The situation is opposite for a warm eddy ($\nu < 0$): the rays deviate in the direction from the eddy center, time fluctuations τ are negative, weak defocusing of rays passing through the eddy center is observed, and strong focusing is possible in directions close to the tangential direction.

Clark and Weinberg (1980) calculated another plane model of a cold eddy. Its distinction from Munk's model was that the sound velocity distribution in a real eddy (see Figure 6.3.2) at a depth of 500 m was chosen as $c(r)$. The maximum lateral deviation turned out to be close to 1° for this model.

It should be noted that plane eddy models are a rather crude approximation to the reality. For instance, within the framework of these models it is impossible to forecast the horizontal refraction of rays, the major part of whose trajectories lies in deep ocean layers outside the range of the eddy action. Therefore these estimates should be regarded as estimates 'from above'.

Baer (1981) managed to obtain a more realistic estimate of the magnitude of the horizontal refraction in a cold eddy. A CW source with a frequency of 25 Hz was located in the model at a depth of 600 m. The sound path, 850 km long, intersected the eddy at an angle γ (the angle $\gamma = 0$ corresponds to the direction from the source to the eddy center). The eddy had a radius 125 km and a thickness of 2 km and corresponded to the model of Henrick *et al.* (1977). The eddy core center was located at a depth of 800 m and the maximum fluctuations of the sound velocity were 25 m/s. The calculation of the sound field was carried out by the parabolic equation method generalized to three-dimensional problems. The work of a 7.7 km horizontal antenna located at a depth of 600 m and oriented along the normal to the sound path was modelled. At each step along the distance the 'apparent' direction to the source was determined from the data of the antenna processing. For $\gamma = 5°$ the deviation of the actual direction from the apparent direction in the transmission through the eddy attained 0.5°.

Polyanskaya (1982) performed another calculation of the horizontal refraction of sound rays passing through a three-dimensional Gulf Stream eddy. The eddy whose vertical section is presented in Figure 6.3.2 was assumed to be cylindrical symmetric. The sound source was located near the eddy edge at a depth of 200 m. The construction of the rays revealed that the maximum lateral refraction did not exceed 0.22°. Deviation of the rays from the corresponding vertical planes of their emergence at distances up to 280 km from the source did not exceed several tens of meters.

As could be expected, the resulting estimates of the magnitude of the horizontal refraction for three-dimensional eddies are smaller than those predicted by Munk, although it is quite possible that the angular deviations can be greater for some other conditions of the problem (for some other source and receiver depths, mutual position of the source and the eddy, etc.).

As was already mentioned, there are qualitative distinctions between the effects of warm and cold eddies. These are not only distinctions in the character of the horizontal refraction. The effects of warm and cold eddies on the structure of the sound velocity field and refraction in the vertical plane are also substantially different. As an example, we consider a warm anticyclonic eddy believed to be formed owing to the baroclinic instability of the East Australian Current (Nilsson *et al.*, 1977). This was observed in March 1975 south-east of Sydney between 34 and 36°S. The warm eddy core was located at a depth of 200 m and the radius of the eddy was 60 km. The effect of the eddy on the vertical profile of the sound velocity manifested itself in the appearance of an additional 'eddy' underwater sound channel with thickness of 175 m between the deep sound channel at a depth of 1200 m and the subsurface channel. To study the effect of such an eddy on the propagation of sound a special experiment was performed (Nysen *et al.*, 1978). A ship with a receiver was located near the eddy center, explosive sound sources were used, and the frequency band 25–2000 Hz was employed. It was important to discover the contribution of the eddy channel to the illumination of the shadow zone nearest to the source, and to determine how the sound energy was redistributed between the subsurface, deep, and eddy channels. It turned out that the first illuminated zone was moved away from the source by a distance about 8 km in comparison with the ordinary conditions in this region. The investigations also showed that there exists an optimal frequency for the propagation of sound in the eddy channel which is equal to 100 Hz in the case under consideration. Above this frequency there appeared additional losses owing to scatter in the eddy channel and absorption in the medium. Below 100 Hz sound energy losses were caused by an energy leakage from the eddy channel to the deep channel. As the thickness of the eddy channel decreases towards the periphery of the eddy, the energy leakage from the channel rapidly increases. Moreover, additional and not very clearly marked convergence zones were observed at distances of 17, 30, and 43 km, which corresponds to a 13 km ray cycle in the eddy channel. It was also found that energy was transferred from the subsurface channel to the eddy channel. The signal in the subsurface channel was audible at distances up to 7 km, whereas at the upper boundary of the eddy channel it was audible at distances up to 30 km. For greater distances it was lost in the general noise of the area. Thus the existence of an eddy channel opens up additional possibilities of receiving sound in a region which would be entirely shadowed in the absence of a warm eddy.

As already mentioned, along with intensive isolated eddies of this or that nature, Rossby waves also make a contribution to the formation of the synoptic variability of the ocean. Baroclinic Rossby waves move fluid particles hundreds of meters along the vertical, which results in displacements of the axis of the underwater sound channel and in sound velocity fluctuations up to 10 m/s at some depths. The ray calculation of fluctuations of phase θ of a harmonic sound signal with a frequency of 406 Hz under the action of a first-mode baroclinic Rossby wave in conditions simulating those in the Eleuthera–Bermuda path was first carried out by Baer and Jacobson (1975). It was shown that phase fluctuations vary in time with a Rossby wave frequency. The magnitudes of the fluctuations are determined by the orientation of the sound path relative to the direction of propagation of the Rossby wave and also by the mutual position of the source and the observation point.

Maximum fluctuations should be expected when the source and receiver lie in the vicinity of the axis of the deep sound channel, and the direction of propagation of the Rossby wave is orthogonal to that of the sound path. According to the calculations, at a distance of 1250 km phase fluctuations can attain 100 and more cycles, which exceeds the values found experimentally (Clark and Kronengold, 1974).

Similar estimations of the disturbing effect of Rossby waves on the sound velocity field were performed for a region in the North-West Pacific, which is substantially different from the Sargasso Sea in its climatic conditions (Kurtepov, 1983). The effect of a first-mode baroclinic Rossby wave of western direction with a wavelength of 120 km was analyzed. The data on the mean climatic vertical temperature $T(z)$ and salinity $S(z)$ distributions were taken from atlases (Moiseev *et al.*, 1980; Panfilova *et al.*, 1981). The amplitude of the maximum vertical displacements (150 m) in the wave was chosen so that the temperature fluctuations caused by this wave (at a depth of 1 km unattainable for the seasonal variability) should be close to the observed fluctuations registered by Moiseev *et al.* (1980). The sound velocity field, with inclusion of the Rossby wave effect, were calculated using special programs.

Shown on the left in Figure 6.3.5 is the mean climatic sound velocity profile (solid curve) and the limits of its variability owing to Rossby waves (dashed line). Shown on the right are the isolines of the sound velocity field at a certain time (a vertical section in the east-to-west direction). They would be horizontal in the absence of Rossby waves. The maximum sound velocity fluctuations are 7.3 m/s.

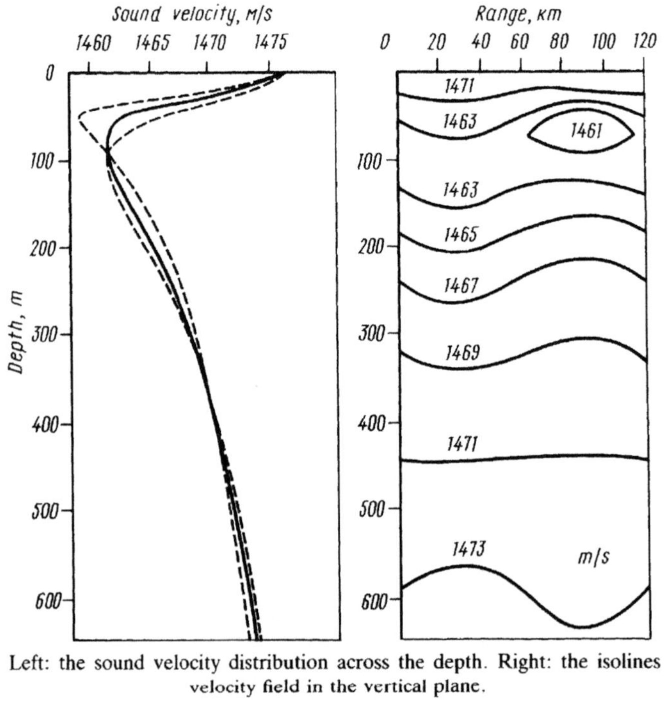

Fig. 6.3.5. Left: the sound velocity distribution across the depth. Right: the isolines of the sound velocity field in the vertical plane.

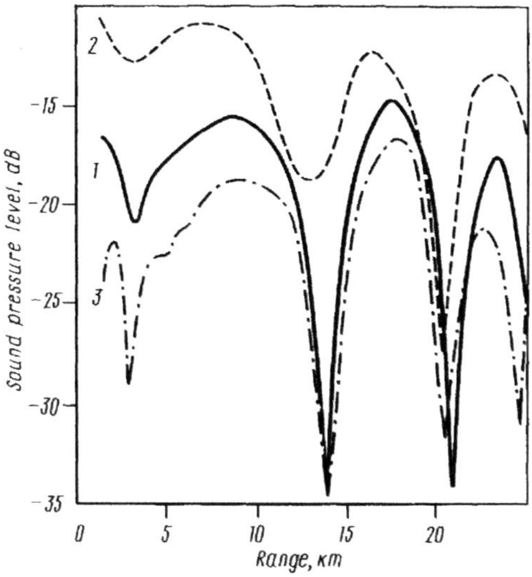

Fig. 6.3.6 Variability of the pressure levels of a sound field under the action of a Rossby wave.

Figure 6.3.6 demonstrates the results of the wave calculation of the pressure level of a sound field in a medium with a Rossby wave for a sound source with a frequency of 50 Hz. The depths of the source and the receiver are 50 m and the distance between them is 1.5–25 km. Curve 1 in the figure corresponds to the mean climatic sound velocity profile. Curve 2 represents a similar section in the direction orthogonal to that of the propagation of the Rossby wave for the extremal case when, owing to Rossby wave dynamics or a change of the position of the source, this section lies in the region of maximum elevation of fluid particles. Curve 3 demonstrates an analogous horizontal cut for another extremal case when the downward displacement of fluid particles is at maximum. As is seen from Figure 6.3.6, the pressure level of the sound field is extremely sensitive to the action of Rossby waves and the orientation of the sound path. Fluctuations of the sound pressure level can attain 10 dB and more. For other orientations of the sound route with respect to the direction of propagation of Rossby waves, the fluctuations of the sound pressure level relative to the 'undisturbed' values (curve 1) are also large. We note that, as in the case of an isolated eddy, the effect of Rossby waves can manifest itself even at short distances from the source. This imposes certain requirements on the hydrological provision of acoustic experiments. Namely, the vertical sections of the sound velocity field along the sound path must be performed with a sufficiently short spatial step of the order of 10 km to reveal the possible synoptic inhomogeneities. Otherwise it cannot be guaranteed that such 'fine' acoustic effects as the detailed structure of interference, the fluctuations of times of arrival of sound signals, sound attenuation, etc., are measured and correctly interpreted.

If for some reason the sound velocity field along the sound path is unknown or is measured with too long a step, then acoustic calculations have to be performed using available data, e.g. a single sound velocity profile $c(z)$ at the source site or a profile averaged over several rare hydrographic stations. So how great is the error if we try to forecast the sound field in the real ocean with synoptic inhomogeneities by calculating it from a single $c(z)$ profile? The paper by Emery *et al.* (1979) gives a partial answer to this question. In this paper the data of measurements performed in a region in the North Pacific with coordinates 36–38°N, 155°E–175°W (north of Hawaii) were employed to reveal a sound velocity field with clearly marked synoptic inhomogeneities. Further, a numerical experiment on the propagation of sound was carried out. The depth of the source was about 300 m and the frequency was 50 Hz. The horizontal coordinate was varied with respect to the longitude so that the source did not leave the limits of the region under study. The calculation of the pressure level of the sound field was carried out for distances 0–250 km and at all depths for each of the chosen positions of the source. Altogether 100 such calculations (realizations) were performed; they differed in the positions of the source or, which is the same, in the behavior of the sound velocity field along the path; the results were subjected to statistical processing. The processing reduced to the derivation of the statistic–average pressure level of the sound field for all depths and distances, and also the standard deviations of the sound pressure level from the average values (the root-mean-square level fluctuations).

It turned out that along the first 20 km the statistic–average sound pressure level differed very slightly from individual realizations (the standard deviations were small). At these distances synoptic inhomogeneities did not manifest themselves notably. Further, the difference increased with distance so that the depth-averaged standard deviation of the sound pressure level increased from 4 to 5.5 dB at distances from 25 to 150 km and then remained almost constant, attaining 6 dB at 250 km. The standard deviations of the depth-averaged sound pressure level of 5–6 dB indicate substantial ambiguity in forecasting a sound field from the statistic-average value. If it is taken into account that in deep layers the standard deviation is obviously smaller than the depth-averaged value (e.g. see Baer, 1980) and, consequently, in surface layers it is greater, the reliability of the forecast is still lower, particularly in the upper 1.5–2 km ocean layer.

Summing up the results of the above-mentioned experiments and theoretical investigations we enumerate the basic acoustic effects caused by the synoptic variability of the ocean and give, when possible, their numerical estimates:

- sound velocity field fluctuations (up to 35 m/s);
- the deformation of the boundaries of convergence and shadow zones (up to 25–30 km along the horizontal and up to 1 km along the vertical);
- fluctuations in the time of propagation of a sound signal (up to 0.5 s);
- fluctuations in sound pressure level (up to 30 dB);
- the deformation of ray structures in the vertical plane;
- the horizontal (lateral) refraction (0.2–0.5°);
- sound pressure level fluctuations due to different orientations of the sound path relative to synoptic inhomogeneities (5–10 dB).

These numerical estimates are not universal but they seem to demonstrate the modern state of these areas of research.

Thus, according to the existing data, the effect of the synoptic variability on sound fields is strong. However, fluctuations of acoustic parameters of sound signals are not always an undesirable phenomenon. On the contrary, in principle, the same fluctuations can be used as the initial data to reconstruct the internal structure of the ocean; namely, sound velocity, temperature and salinity fields as well as currents, and wave motion spectra. Some of the methods of such 'acoustic tomography' of the ocean and the estimation of their possibilities in relation to the study of synoptic-scale phenomena are presented in the papers by Munk and Wunsch (1979, 1982a, b).

These authors put forward a scheme of an acoustic experiment to verify the possibility of a continuous long-term observation of the synoptic variability of the ocean in a 1000 × 1000 km square. To this end a system of acoustic sources and receivers must be emplaced along the perimeter of this area (at depths near the axis of the deep sound channel). It is intended to use the information about the fluctuations of the times of arrival of sound signals along all the possible ray trajectories. Besides the global character, the advantages of this acoustic monitoring include the prospect of a long-term simultaneous scanning of not only the subsurface layers but also the entire ocean depth. The suggested method is applicable to subpolar regions as well. It is also significant that in this method integrated characteristics of the ocean are measured and small-scale factors which strongly distort the data of local experiments are filtered out.

The method is based on the assumption (which is not always justifiable) that deformations of ray trajectories caused by the synoptic variability of the ocean are sufficiently small. This makes it possible to relate the fluctuations of time of arrival of a sound signal along a ray to the fluctuations of the sound velocity field along the corresponding 'undisturbed' ray in a medium with mean climatic sound velocity field. The parameters of the 'undisturbed' rays are calculated on a computer.

The following scheme is a reconstruction of the sound velocity field. The region under study is split into three-dimensional cells whose sizes are small in comparison with those of the expected inhomogeneities. The sound velocity in the nth cell is represented in the form $c_n = c_n^0 + \Delta c_n(t)$, where t is time, c_n^0 is the mean climatic sound velocity distribution, and $\Delta c_n(t)$ are synoptic fluctuations which are assumed to be constant within the limits of the cells. For $|\Delta c_n| \ll c_n^0$ the fluctuations of the time of propagation of the signal along the mth ray is approximately expressed in the form $\Delta T_m = \Sigma_{n=1}^{N} E_{mn}\, \Delta c_n$ $(m = 1, \ldots, M)$. Here N is the number of cells and M is the number of rays. The constant coefficients E_{mn} are nonzero only in the cells intersected by the undisturbed ray. They are easily calculated if the trajectory of the undisturbed ray and the configuration of the cells it intersects are known. The resulting linear relations between ΔT_m and Δc_n are used to set the inverse problem. The aim of the inversion is to reconstruct the fluctuations $\Delta c_n(t)$ from the measured values of $\Delta T_m(t)$.

The solution of these problems encounters traditional difficulties. As a rule, the number N of unknowns in the system of algebraic linear equations under study exceeds the number M of equations. The problem is indeterminate, which results in an ambiguity of the solution. Another difficulty is the high sensitivity of solutions to the accuracy of the measurement of ΔT_m which is determined in the case under consideration not only by technical possibilities of the instruments but also by errors in the form of acoustic noise and internal waves. To identify a single solution close to the real sound velocity field among the multitude of solutions, we have to use *a priori* information about the field. At present there are several methods of such regularization of inverse problems. The quality of some of these methods was estimated by means of numerical experiments in the above-mentioned papers by Munk and Wunsh.

Along with the purely mathematical difficulties, there are some other difficulties of a fundamental character. For instance, currents which make a significant contribution to fluctuations in the times of wave arrivals (Itzikowitz *et al.*, 1983) are not taken into account. This can lead to large errors in the reconstruction of sound velocity fields. Moreover, calculations by Mercer and Booker (1983) have shown that it is not always possible to identify the rays arriving at the reception point by comparing them with undisturbed rays. Intensive eddies can distort ray trajectories very strongly. For example, the turning points of rays can be displaced by hundreds of meters along the vertical. The disappearance of some of the rays and the splitting of others are also possible. It is important to emphasize that ray trajectories change not only in the region occupied by inhomogeneities but also outside it. Therefore, generally speaking, the fluctuations ΔT_m are due not only to the fluctuations Δc_n produced by the inhomogeneities themselves but are also due to the propagation of the signal across a region in the medium without inhomogeneities along a new trajectory different from the undisturbed trajectory. In other words, the calculation of fluctuations in the times of arrival of sound signals ΔT_m by integrating over undisturbed trajectories is incorrect, at least for some of the rays. Moreover, as is shown by calculations, the relationship between ΔT_{ml} and the vector $(\Delta c_1, \ldots, \Delta c_N)$ can be nonlinear and hence the linearization with respect to the parameter $\Delta c_n/c_n^0$ performed in the derivation of the basic system of inversion equation does not apply to all the rays.

What has been said implies the important conclusion that, as a reference sound velocity field c_n^0, we should take not the mean climatic distribution of the field but its accurately measured state at the time of the beginning of the tomographic scanning of the ocean. The inversion should be performed sufficiently frequently to have time to observe the variation of the sound velocity field and the deformation of ray trajectories. It is advisable to choose a sound velocity field reproduced by inversion and the rays calculated for it as reference values at the next inversion step, etc.

Along with the numerical modelling of acoustic tomography problems, preliminary experiments were simultaneously carried out to answer the following questions.

Given the existing sound sources and the means of signal processing, is it possible to identify a sufficiently large number of rays in the conditions of a real ocean with synoptic inhomogeneities, and at what distances from the source can this be done?

Is it possible to measure the fluctuations in arrival times along the observed rays with an accuracy sufficient for tomographic purposes?

Spindel and Spiesberger (1981) managed to separate out and identify six sound rays on the background of noise in an experiment carried out over a period of 19 days in the Sargasso Sea north-west of Bermuda along a path 300 km long. The measured fluctuations of times of arrival of signals along these rays correlated with the passage of a Gulf Stream meander across the path. The recognition of rays was performed by comparing the times of arrival of signals along these rays with the times of arrival calculated in advance for the undisturbed rays (time selection).

In another experiment (Spiesberger and Worcester, 1981) carried out over 48 days in a region west of Bermuda at a distance of 900 km from the source, 16 rays were reliably identified. The fluctuations in the times of arrival of acoustic signals were measured to within an accuracy of 10 ms, which is in complete agreement with the requirements put forward by Munk and Wunsch (1979). It was also noted that, for such long distances, importance should be given to additional fluctuations in the sound signal propagation times related to the different conditions of sound propagation along or against tidal currents.

As was already mentioned, the problem of identifying sound rays plays a key role in tomography. It is not always possible to solve this problem by the time selection method. It sometimes happens that signals arrive at the reception point from above and from below simultaneously, or with a small interval. In these conditions it is advisable to use the angular selection of rays; that is, the application of acoustic antennas. A four-element antenna of this kind was successfully tested at a distance of 215 km in an experiment (Worcester, 1981) for 22 days in the Pacific south-west of San Diego.

In February–June 1981 a test experiment on the acoustic tomography of the ocean was carried out (Berhinger *et al.*, 1982). The experiment covered an area of 300 × 300 km with a center at the point 26°N, 70°W. Four sources and five receivers (which included two directional receivers) were emplaced. They were all set on subsurface moorings near the axis of the sound channel. The displacements of the sources and the receivers in the course of the experiment caused by the buoy drift were registered with a high accuracy with the aid of special bottom transponders. The timekeeping was guaranteed by the use of high-precision clocks (a rubidium atomic frequency standard) at each of the sources and receivers. The experiment was performed in three stages. At the beginning instrumental measurements of the sound velocity field were carried out in the region under study for 19 days. The data of these measurements were introduced into the corresponding program for calculating the geometry of undisturbed rays and times of arrival along them. Further, all the fluctuations of the times of arrival along the individual rays averaged over several days were attributed to the effect of the synoptic variability of the ocean and the values of these fluctuations were used to solve the inverse problem of reconstructing the sound velocity field in the experiment area. The instrumental survey was repeated after a period of two months. At this stage it was intended to compare the tomographic chart of the sound velocity field with the real one. Unfortunately, three receivers failed, which made it impossible to perform the comparison. An intermediate comparison (30 days after the beginning of the tomographic reconstruction) of the sound velocity field at a depth of 350 m with the AXBT measurement data indicated a certain kind of similarity between them.

Although the experiment was not quite successful the members of the tomographic group are optimistic about the possibilities and prospects of the acoustic tomography. In particular, it is advisable to try to reconstruct the temperature, salinity, and density fields using *T, S* curves and the equations of state of the sea water. We can estimate the component of the current velocity along the direction of a sound path from the difference between the times of propagation of signals in the opposite directions between the two corresponding points. There are projects to cover vast areas of the world ocean by a tomographic network, which will make it possible to approach the problem of oceanic effects on the Earth's weather and climate from a new point of view.

If at least some of these plans are realized, acoustic tomography will be among the most powerful (and inexpensive) methods of investigating the ocean.

Bibliography

1. Agafonova, E.G., Galerkin, L.I., and Monin, A.S.: 1972, *Okeanologiya* **12**, No. 6, 950–955 (*in Russian*).
2. Aitsam, A., Elken, J., Pavelson, J., and Talpsepp, L.: 1981, in *The Investigation and Modelling of Processes in the Baltic Sea. Part I*, Tallin, pp. 70–98.
3. Alekseev, G.V. and Ivanov, N.I.: 1983, *Trudy Arkticheskogo i Antarkticheskogo nauchn.-issled. instituta* **382**, 76–79 (*in Russian*).
4. Andrews, J.C.: 1977, *Deep-Sea Res.* **24**, No. 12, 1133–1148.
5. Andrews, J.C. and Scully-Power, P.: 1976, *J. Phys. Oceanogr.* **6**, No. 5, 756–765.
6. Antipov, S.V., Nezlin, M.V., Rodionov, V.K., Snezhkin, E.N., and Trubnikov, A.S. 1983. *Zh. Eksper. Teor. Fiz.* **84**, No. 4, 1357–1372 (*in Russian*).
7. Antipov, S.V., Nezlin, M.V., Snezhkin, E.N., and Trubnikov, A.S.: 1982, *Zh. Eksper. Teor. Fiz.* **82**, No. 1, 145–160 (*in Russian*).
8. Antonova, R.A., Zhvaniya, B.P., Lominadze, D.G., Nanobashvili, D.I., and Petviashvili, V.I.: 1983, *Pis'ma v ZhETF* **37**, No. 11, 545–548 (*in Russian*).
9. Aretinskiĭ, G.Yu., Dykman, V.Z., Efremov, O.I., and Kiseleva, O.A.: 1980, in *Sinopticheskie vikhri v okeane* (*Synoptic Eddies in the Ocean*), Naukova Dumka, Kiev pp. 167–177 (*in Russian*).
10. *Background Papers and Supporting Data on the Practical Salinity Scale 1978. UNESCO Technical Papers in Marine Science 37, UNESCO/ICES/SCOR/IAPSO Joint Panel on Oceanographic Tables and Standards* (UNESCO, 1981).
11. Baer, R.N.: 1980, *JASA* **67**, No. 4, 1180–1185.
12. Baer, R.N.: 1981, *JASA* **69**, No. 1, 70–75.
13. Baer, R.N. and Jacobson, M.J.: 1975, *JASA* **57**, No. 3, 569–576.
14. Bagryantsev, N.V., Borisov, B.G., Sarukhanyan, E.I., and Smirnov, N.P.: 1976, *Trudy AANII* **344**, 115–128 (*in Russian*).
15. Baranov, E.I.: 1971, *Okeanologicheskie Issled.*, No. 22, 220–278 (*in Russian*).
16. Baranov, E.I.: 1979, *Trudy GOIN*, No. 150, 62–70 (*in Russian*).
17. Barrett, J.R.: 1971, *Deep-Sea Res.* **18**, No. 12, 1221–1231.
18. Batchelor, G.K.: 1969, *The Physics of Fluids Suppl. II* **12**, No. 12, 233–239.
19. Beckerle, J.C., Baxter, L., Porter, R.P., and Spindel, R.C.: 1980, *JASA* **68**, No. 6, 1750–1767.
20. Beckerle, J.C. and La Casce, E.O.: 1973, *Deep-Sea Res.* **20**, No. 7, 673–675.
21. Belyakov, L.N. and Volkov, V.A.: 1980, *Doklady Akad. Nauk SSR* **254**, No. 3, 752–754 (*in Russian*).
22. Belyakov, L.N. and Volkov, V.A.: 1981, in *Fondy Arkticheskogo i Antarkticheskogo instituta (Materials of Arctic and Antarctic Research Institute)* (*in Russian*).
23. Benney, D.J.: 1966, *J. Math. Phys.* **45**, No. 1, 52–63.
24. Berestov, A.L.: 1979, *Izv. Akad. Nauk SSSR Fiz. Atm. Okeana* **15**, No. 6, 648–654 (*in Russian*).
25. Berestov, A.L.: 1981a, *Izv. Akad. Nauk SSSR Fiz. Atm. Okeana* **17**, No. 1, 82–87 (*in Russian*).
26. Berestov, A.L.: 1981b, *Izv. Akad. Nauk SSSR Fiz. Atm. Okeana* **17**, No. 6, 653–656 (*in Russian*).
27. Berestov, A.L.: 1985, *Izv. Akad. Nauk SSSR Fiz. Atm. Okeana* **21**, No. 3, 332–334 (*in Russian*).
28. Berestov, A.L. and Monin, A.S.: 1980, *Uspekhi Mekhaniki (Advances in Mechanics)* **3**, No. 3, 3–34 (*in Russian*).
29. Berhinger, D., Birdsall, T., Brown, M., Cornuelle, B., Heinmiller, R., Knox, R., Metzger, K.,

Munk, W., Spiesberger, J., Spindel, R., Webb, D., Worcester, P., and Wunsch, C.: 1982, *Nature* **299**, No. 5879, 121–125.
30. Bernstein, R.L., Breaker, L., Whritner, R.: 1977, *Science* **195**, No. 4276, 353–359.
31. Bernstein, R.L. and White, W.B.: 1974, *J. Phys. Oceanogr.* **4**, No. 4, 613–624.
32. Bernstein, R.L. and White, W.B.: 1977, *J. Phys. Oceanogr.* **7**, No. 1, 123–126.
33. Bernstein, R.L. and White, W.B.: 1981, *J. Phys. Oceanogr.* **11**, No. 5, 692–704.
34. Bernstein, R.L. and White, W.B.: 1982, *J. Phys. Oceanogr.* **12**, No. 2, 154–159.
35. Blatov, A.S. and Ivanov, V.A.: 1979, in *Kompleksnye issledovaniya Chernogo morya* (*Complex Studies of the Black Sea*), MGI, Sebastopol, pp. 43–51 (*in Russian*).
36. Boland, F.M. and Church, J.A.: 1981, *Deep-Sea Res.* **28A**, No. 9, 937–959.
37. Boland, F.M. and Hamon, B.V.: 1970, *Deep-Sea Res.* **17**, No. 4, 777–794.
38. Boyd, S.H., Wiebe, P.H., and Cox, I.L.: 1978, *J. Mar. Res.* **36**, No. 1, 143–159.
39. Bradford, J.M., Heath, R.A., Chang, F.H., and Hay, C.H.: 1982, *Deep-Sea Res.* **29**, No. 12A, 1501–1516.
40. Brekhovskikh, L.M.: 1971, in *Proc. Joint Oceanogr. Assembly*, Tokyo, pp. 201–202.
41. Brekhovskikh, L.M., Fedorov, K.N., Fomin, L.M., Koshlyakov, M.N., and Yampol'skiĭ, A.D.: 1971a, *Deep-Sea Res.* **18**, No. 12, 1189–1206.
42. Brekhovskikh, L.M., Koshlyakov, M.N., Fedorov, K.N., Fomin, L.M., and Yampol'skiĭ, A.D.: 1971b, *Doklady Akad. Nauk SSSR* **198**, No. 6, 1434–1437 (*in Russian*).
43. Bruce, J.: 1979, *J. Geophys. Res.* **84**, No. 12, 7742–7748.
44. Bryden, H.L.: 1976, *Deep-Sea Res.* **23**, No. 12, 1165–1174.
45. Bryden, H.L.: 1979, *J. Mar. Res.* **37**, No. 1, 1–21.
46. Bryden, H.L.: 1982, *J. Mar. Res.* **40**, No. 4, 1048–1067.
47. Bryden, H.L. and Hall, M.M.: 1980, *Science* **207**, No. 4433, 884–886.
48. Bubnov, V.A., Burkov, V.A., Grachev, Yu. M., and Koshlyakov, M.N.: 1984, *Doklady Akad. Nauk SSSR* **275**, No. 3, 737–740 (*in Russian*).
49. Bubnov, V.A. and Fedorov, K.N.: 1980, *Okeanologicheskie Issled.*, No. 32, 1417 (*in Russian*).
50. Buchwald, V.T.: 1973, *Geophys. Fluid Dynamics* **5**, No. 4, 359–367.
51. Bulatov, N.V.: 1980, *Trudy DVNIGMI* (*Far-East Hydrometeorological Research Institute*), No. 80, 34–45 (*in Russian*).
52. Bulgakov, N.P., Dzhiganshin, G.F., and Belous, L.M.: 1977, *Morskie Gidrofiz. Issled.* **4**, 30–39 (*in Russian*).
53. Bulgakov, N.P., Glushchuk, B.A., Kozlov, V.F., Novozhilov, V.N., and Starodubtsev, E.I.: 1972, *Subarkticheskiĭ front severo-zapadnoĭ chasti Tikhogo okeana* (*Subarctic Front of the North-west Pacific*), Pacific Institute of Oceanology, DVNTs AN SSSR, (*in Russian*).
54. Bulgakov, N.P. and Paranichev, L.G.: 1980, in *Sinopticheskie vikhri v okeane* (*Synoptic Eddies in the Ocean*), 'Naukova Dumka', Kiev, pp. 85–89 (*in Russian*).
55. Burkov, V.A., Krivosheya, V.G., Ovchinnikov, I.M., and Savin, M.T.: 1979, *Okeanologiya* **19**, No. 1, 19–26 (*in Russian*).
56. Burkov, V.A., Kut'ko, V.P., and Kharlamov, A.I.: 1984, *Trudy VNIIGMI MTsD* (*All-Union Research Institute of Hydrometeorological Information*; *the World Data Center*), No. 102, 3–22 (*in Russian*).
57. Burkov, V.A. and Ovchinnikov, I.M.: 1960, *Trudy IOAN* **40**, 121–134 (*in Russian*).
58. Burkov, V.A. and Pavlova, Yu.V.: 1979, *Okeanologiya* **19**, No. 4, 584–592 (*in Russian*).
59. Burkov, V.A. and Pavlova, Yu.V.: 1980, *Okeanologiya* **20**, No. 3, 415–424 (*in Russian*).
60. Burkov, V.A. and Sizova, Yu.V.: 1979, *Okeanologiya* **19**, No. 3, 373–380 (*in Russian*).
61. Burlakova, Z.P.: 1979, in *Morskie Gidrofiz. Issled.* **85**, No. 2, 102–111 (*in Russian*).
62. Burling, R.W.: 1961, *N.Z. Oceanogr. Inst. Mem.* **10**, No. 61, 44–46.
63. Byshev, V.I., Grachev, Yu.M., and Ivanov, Yu.A: 1976, *Okeanologiya* **16**, No. 2, 216–221 (*in Russian*).
64. Charney, J.G.: 1971, *J. Atm. Sci.* **28**, No. 6, 1087–1095.
65. Charney, J.G.: 1973, in *Dynamic Meteorology* (ed. P. Morel), Reidel, Dordrecht, pp. 97–351.
66. Charney, J.G. and Flierl, G.R.: 1981, in *Evolution of Physical Oceanography. Scientific Surveys in Honor of Henry Stommel* (eds B. A. Warren and C. Wunsch), MIT Press, Cambridge, Mass., pp. 504–549.

67. Charney, J.G. and Phillips, N.A.: 1953, *J. Meteorol.* **10**, No. 2, 71–99.
68. Chefranov, S.G.: 1979 *Spektral'nyĭ analiz kvasidvumernoĭ atmosfernoĭ turbulentnosti i problemy predskazuemosti* (*Spectral Analysis of Quasi-Two-Dimensional Atmospheric Turbulence and Predictability Problems*), Moscow, Preprint IFA AN SSSR (*in Russian*).
69. Cheney, R.: 1977, *J. Geophys. Res.* **82**, No. 34, 5459–5468.
70. Cheney, R.E.: 1978a, *J. Geophys. Res.* **83**, No. C9, 4593–4597.
71. Cheney, R.E.: 1978b, *Tracking a Kuroshio Cold Ring with Free-drifting Surface Buoy*, Preprint, NAVOCEANO Technical Note 3700–78, Washington, D.C., p. 17.
72. Cheney, R.E., Gemill, W., Shank, M.K., and Richardson, P.L.: 1976, *J. Phys. Oceanogr.* **6**, No. 5, 741–749.
73. Cheney, R.E., Marsh, J.G., and Beckley, B.D.: 1983, *J. Geophys. Res.* **88**, No. C7, 4343–4354.
74. Cheney, R.E. and Richardson, P.L.: 1976, *Deep-Sea Res.* **23**, No. 2, 143–155.
75. Chernyakova, A.M.: 1978, *Okeanologiya* **18**, No. 5, 827–832, (*in Russian*).
76. Chernyakova, A.M. and Borodkin, S.O.: 1983, in *Otchet o rabotakh v 30-m reĭse NIS 'Dmitriĭ Mendeleev'*, No. 7, T.II, Ch. 1 (*Account of the 30th Voyage of the Research Vessel 'Dmitriĭ Mendeleev'*, Vol. II, Part 1, No. 7, pp. 300–316 (*in Russian*).
77. Chernyakova, A.M. and Romanov, A.S.: 1980, *Okeanologiya* **20**, No. 1, 64–71 (*in Russian*).
78. Chernyakova, A.M. and Yakushev, E.V.: 1981, *Okeanologicheskie Issled.*, No. 34, 102–113 (*in Russian*).
79. Clark, J.C. and Weinberg, N.L.: 1980, *JASA* **68**, No. 2, 703–706.
80. Clark, J.G. and Kronengold, M.: 1974, *JASA* **56**, No. 4, 1071–1081.
81. Clarke, R.A., Hill, H.W., Reiniger, R.F., and Warren, B.A.: 1980, *J. Phys. Oceanogr.* **10**, No. 1, 25–65.
82. Cochrane, J.D.: 1969, *J. Mar. Res.* **27**, No. 3, 327–334.
83. Cochrane, J.D., Kelly, F.J., and Olling, C.R.: 1979, *J. Phys. Oceanogr.* **9**, No. 7, 724–738.
84. Cresswell, G.R.: 1977, *Deep-Sea Res.*, **24**, No. 12, 1203–1209.
85. Cresswell, G.R. and Golding, T.J.: 1980, *Deep-Sea Res.* **27A**, No. 6A, 449–466.
86. Csanady, G.T.: 1979, *J. Geophys. Res.* **84**, No. C2, 777–780.
87. Dantzler, H.L.: 1977, *J. Phys. Oceanogr.* **7**, No. 4, 512–519.
88. Darnitskiĭ, V.B.: 1979a, in *Issledovanya po biologii ryb i promyslovoĭ okeanografii*, vyp. 10 (*Studies in Fish Biology and Production Oceanography*, No. 10), Vladivostok, pp. 14–25 (*in Russian*).
89. Darnitskiĭ, V.B.: 1979b, in *XIV Tikhookeanskiĭ kongress. Komitet G* (*14th Pacific Ocean Congress. Committee G*), Moscow, pp. 88–89 (*in Russian*).
90. Darnitskiĭ, V.B.: 1980a, *Trudy DVNII*, No. 86, 51–62 (*in Russian*).
91. Darnitskiĭ, V.D.: 1980b, *Trudy DVNII*, No. 86, 63–70 (*in Russian*).
92. Darnitskiĭ, V.B. and Boldyrev, V.Z.: 1977, in *Voprosy promyslovoĭ okeanologii Mirovogo okeana* (*Topics in Production Oceanology of the World Ocean*), pp. 73–75 (*in Russian*).
93. Davis, R.E. and Acrivos, A.: 1967, *J. Fluid. Mech.* **29**, Part 3, 593–608.
94. De Verdiere, A.C.: 1983, *J. Mar. Res.* **41**, No. 3, 375–398.
95. Dickson, R. and Gurbutt, P.: 1979, *POLYMODE News*, No. 64 (unpublished manuscript).
96. Dikiĭ, L.A.: 1976, *Gidrodinamicheskaya ustoĭchivost' i dinamica atmosfery* (*Hydrodynamic Stability and Dynamics of the Atmosphere*), Gidrometeoizdat, Leningrad (*in Russian*).
97. Drozdov, V.N., Monin, A.S., and Yushina, I.G.: 1981, *Doklady Akad. Nauk SSSR* **258**, No. 2, 331–334 (*in Russian*).
98. Duncan, C.P.: 1968, *J. Geophys. Res.* **73**, No. 2, 531–534.
99. Ebbesmeyer, C.C. and Taft, B.A.: 1979, *J. Phys. Oceanogr.* **9**, No. 6, 1073–1089.
100. Ebbesmeyer, C., Taft, B., Cox, J., McWilliams, J., and Owens, B.: 1978, *POLYMODE News*, No. 54 (unpublished manuscript).
101. Elliott, B.A.: 1982, *J. Phys. Oceanogr.* **12**, No. 11, 1292–1309.
102. Emery, W.J.: 1983, *J. Phys. Oceanogr.* **13**, No. 2, 269–291.
103. Emery, W.J., Ebbesmeyer, C.C., and Dugan, J.P.: 1980, *J. Phys. Oceanogr.* **10**, No. 6, 885–899.
104. Emery, W.J. and Magaard, L.: 1976, *J. Mar. Res.* **34**, No. 3, 365–385.
105. Emery, W.J., Reid, T.J., De Santo, J.A., Baer, R.N., and Dugan, J.: 1979, *JASA* **66**, No. 3, 831–841.

106. Enikeev, V.Kh. and Koshlyakov, M.N.: 1973, *Okeanologiya* **13**, No. 6, 947–962 (*in Russian*).
107. Enikeev, V.Kh., Kozubskaya, G.I., Koshlyakov, M.N., and Yaremchuk, M.I.: 1982, *Doklady Akad. Nauk SSSR* **262**, No. 3, 4573–4577 (*in Russian*).
108. Ertel, H.: 1942, *Meteorol. Z.* **59**, No. 9, 277–281 (*in German*).
109. Fedorov, K.N. and Belkin, I.M.: 1984, *Okeanologiya* **24**, No. 2, 204–216 (*in Russian*).
110. Fedorov, K.N., Bubnov, V.A., Ginzburg, A.I., and Paka, V.T.: 1979a, *Doklady Akad. Nauk SSSR* **246**, No. 5, 1227–1231 (*in Russian*).
111. Fedorov, K.N., Ginzburg, A.I., and Zatsepin, A.G.: 1980, *Okeanologicheskie Issled.*, No. 32, 29–37 (*in Russian*).
112. Fedorov, K.N., Ginzburg, A.I., Zatsepin, A.G., and Shapovalov, S.M.: 1979b, *Okeanologicheskie Issled.*, No. 30, 51–62 (*in Russian*).
113. Fedorov, K.N., Krasnopevtsev, A.Yu., and Varfolomeev, A.A.: 1978, in *Issledovanie izmenchivosti fizicheskikh protsessov v okeane* (*Investigation of the Variability of Physical Processes in the Ocean*), IOAN, Moscow, pp. 35–40 (*in Russian*).
114. Fedorov, K.N., Plakhin, E.A., Prokhorov, V.I., and Sedov, V.G.: 1974, in *Atlanticheskiĭ gidrofizicheskiĭ poligon-70* (*Atlantic Hydrophysical Polygon-70*), Nauka, Moscow, pp. 236–282 (*in Russian*).
115. Fjortoft, R.: 1953, *Tellus* **5**, No. 3, 225–230.
116. Flierl, G.R.: 1978, *Dyn. Atm. Oceans* **2**, No. 4, 341–381.
117. Flierl, G.R.: 1979, *Dyn. Atm. Oceans* **3**, No. 1, 15–38.
118. Flierl, G.R., Larichev, V.D., McWilliams, J.C., and Reznik, G.M.: 1980, *Dyn. Atm. Oceans* **5**, No. 1, 1–41.
119. Flierl, G.R., Stern, M.E., and Whitehead, J.A. Jr: 1983, *Dyn. Atm. Oceans* **7**, No. 4, 233–264.
120. Fomin, L.M. and Yampol'skiĭ, A.D.: 1978, *Okeanologiya* **18**, No. 4, 593–601 (*in Russian*).
121. Fornshell, J.A. and Criess, W.A.: 1979, *J. Phys. Oceanogr.* **9**, No. 5, 992–1000.
122. Fox, D.G. and Orszag, S.A.: 1973a, *J. Comp. Phys.* **11**, No. 4, 612–619.
123. Fox, D.G. and Orszag, S.A.: 1973b, *Phys. of Fluids* **16**, No. 2, 169–171.
124. Frankignoul, C. and Müller, P.: 1979a, *J. Phys. Oceanogr.* **9**, No. 1, 104–127.
125. Frankignoul, C. and Müller, P.: 1979b, *J. Phys. Oceanogr.* **9**, No. 6, 1207–1213.
126. Freeland, H.J. and Gould, J.: 1976, *Deep-Sea Res.* **23**, No. 10, 915–924.
127. Freeland, H.J., Rhines, P.B., and Rossby, T.: 1975, *J. Mar. Res.* **33**, No. 3, 383–404.
128. Frye, H.W. and Pugh, J.D.: 1971, *J. Acoust. Soc. Amer.* **50**, No. 1, Part 2, 384–386.
129. Fu, Lee-Lueng: 1983, *J. Geophys. Res.* **88**, No. C7, 4331–4341.
130. Fu, L.L., Keffer, T., Niiler, P.P., and Wunsch, C.: 1982, *J. Mar. Res.* **40**, No. 3, 809–848.
131. Fuglister, F.C.: 1963, *Progress in Oceanography* **1**, 265–385.
132. Fuglister, F.C.: 1972, in *Studies in Physical Oceanography* (ed. A. Gordon) Vol. 1 Gordon and Breach, New York, London, Paris, pp. 137–168.
133. Fuglister, F.C. and Worthington, L.V.: 1951, *Tellus* **3**, No. 1, 1–14.
134. Galerkin, L.I.: 1976, *Doklady Akad. Nauk SSSR* **229**, No. 6, 1458–1461 (*in Russian*).
135. Galerkin, L.I. and Gritsenko, A.M.: 1980, *Doklady Akad. Nauk SSSR* **251**, No. 5, 1254–1257 (*in Russian*).
136. Galin, M.B.: 1974, *Izv. Akad. Nauk SSSR Fiz. Atm. Okeana* **10**, No. 5, 455–466 (*in Russian*).
137. Gandin, L.S.: 1963, *Ob"ektivnyĭ analiz meteorologicheskikh poleĭ* (*Objective Analysis of Meteorological Fields*), Gidrometeoizdat, Leningrad (*in Russian*).
138. Gandin, L.S.: 1964, *Trudy Gl. Geofiz. Observ.* **165**, 47–59 (*in Russian*).
139. Gavrilin, B.L. and Mirabel', A.P.: 1972, *Trudy GMTs SSSR* **100**, 117–127 (*in Russian*).
140. Gavrilin, B.L., Mirabel', A.P., and Monin, A.S.: 1972, *Izv. Akad. Nauk SSSR Fiz. Atm. Okeana* **8**, No. 5, 483–493 (*in Russian*).
141. Garzoli, S., Parisi, V., and Paschini, E.: 1982, *Deep-Sea Res.* **29**, No. 1A, 77–86.
142. Gill, A.E.: 1974, *Geophys. Fluid Dynamics* **6**, No. 1, 29–47.
143. Gill, A.E.: 1975, *Deep-Sea Res.* **22**, No. 10, 647–652.
144. Gill, A.E.: 1980, in *SCOR WG 34 Publication on Ocean Eddies*, Ch. 17.
145. Gill, A.E., Green, J.S.A., and Simmons, A.J.: 1974, *Deep-Sea Res.* **21**, No. 7, 499–528.
146. Godfrey, J.S., Cresswell, G.R., Golding, T.J., Pearce, A.F., and Boyd, R.: 1980, *J. Phys. Oceanogr.* **10**, No. 3, 430–440.

147. Gorbunov, Yu.A. and Losev, S.M.: 1978, *Trudy AANII* **354**, 52–57 (*in Russian*).
148. Gordon, A.L., Ki-iti Horai, and Donn, M.: 1983, *J. Geophys. Res.* **88**, No. C1, 755–762.
149. Gordon, A.L. and Taylor, H.W.: 1975, in *Numerical Models of Ocean Circulation*, National Academy of Sciences, Washington, D.C., pp. 54–56.
150. Gould, W.J., Schmitz, W.J., and Wunsch, C.: 1974, *Deep-Sea Res.* **21**, No. 11, 911–932.
151. Grachev, Yu.M., Enikeev, V.Kh., and Koshlyakov, M.N.: 1982a, *Izv. POLIMODE*, No. 8, 13–31 (*in Russian*).
152. Grachev, Yu.M., Enikeev, V.Kh., Kozubskaya, G.I., Koshlyakov, M.N., and Mikhaĭlichenko, Yu.G.: 1982b, *Izv. POLIMODE*, No. 8, 3–12 (*in Russian*).
153. Grachev, Yu.M., Koshlyakov, M.N., Nechaev, D.A., Sazhina, T.G., and Yaremchuk, M.I.: 1984, *Okeanologiya* **24**, No. 4, 549–557 (*in Russian*).
154. Grachev, Yu.M., Koshlyakov, M.N., Neelov, I.A., and Chalikov, D.V.: 1984, *Doklady Akad. Nauk. SSSR* **275**, No. 3, 740–744 (*in Russian*).
155. Grishin, G.A., Dotsenko, S.F., and Korotaev, G.K.: 1979, *Morsk. Gidrofiz. Issled.* **85**, No. 2, 134–141 (*in Russian*).
156. Gründlingh, M.L.: 1977, *Deep-Sea Res.* **24**, No. 10, 903–913.
157. Gründlingh, M.L.: 1978, *Deep-Sea Res.* **25**, No. 12, 1209–1224.
158. Gunn, J.T. and Watts, D.R.: 1982, *J. Mar. Res.* **40**, No. 1, 1–18.
159. Hagan, D.E., Olson, D.B., Schmitz, J.E., and Vastano, A.C.: 1978, *J. Phys. Oceanogr.* **8**, No. 6, 997–1008.
160. Haidvogel, D.B.: 1979, *Dyn. Atm. Oceans* **3**, Nos 2–4, 181–190.
161. Haidvogel, D.B. and Holland, W.R.: 1978, *J. Phys. Oceanogr.* **8**, No. 3, 393–413.
162. Hamilton, G.R.: 1974, in *International Workshop on Low-frequency Propagation and Noise, Woods Hole, Massachusetts, Oct. 14–19*, pp. 7–30.
163. Hamon, B.V.: 1965, *Deep-Sea Res.* **12**, No. 6, 899–921.
164. Hansen, B. and Meincke, J.: 1979, *Deep-Sea Res.* **26A**, No. 9A, 1067–1082.
165. Hansen, D.V.: 1970, *Deep-Sea Res.* **17**, No. 3, 495–511.
166. Harris, T.F.W., Legeckis, R., and Van Forest, D.: 1978, *Deep-Sea Res.* **25**, No. 6, 543–548.
167. Harris, T.F.W. and Van Forest, D.: 1978, *Deep-Sea Res.* **25**, No. 6, 549–561.
168. Harrison, D.E.: 1979, *Rev. Geophys. Space Phys.* **17**, No. 5, 969–979.
169. Harrison, D.E.: 1980, *J. Phys. Oceanogr.* **10**, No. 6, 900–905.
170. Harrison, D.E. and Heinmiller, R.H.: 1983, *J. Phys. Oceanogr.* **13**, No. 5, 859–872.
171. Hart, J.E.: 1979, in: *Ann. Rev. Fluid Mech.* **11**, Palo Alto, California 147–172.
172. Hasegawa, A., MacLennan, C.G., and Kodama, Y.: 1979, *Phys. Fluids* **22**, Part 11, 2122–2129.
173. Hasselmann, K.: 1967, *J. Fluid. Mech.* **30**, Part 4, 737–739.
174. Henrick, R.F.: 1980, *JASA* **67**, No. 1, 121–134.
175. Henrick, R.F., Siegman, W.L., and Jacobson, M.J.: 1977, *JASA* **62**, No. 4, 860–870.
176. Herring, J.R.: 1975, *J. Atm. Sci.* **32**, No. 12, 2254–2271.
177. Herring, J.R., Orszag, S.A., Kraichnan, R.H., and Fox, D.G.: 1974, *J. Fluid Mech.* **66**, Part 3, 417–444.
178. Herring, J.R., Riley, J.J., Patterson, G.S., and Kraichnan, R.H.: 1973, *J. Atm. Sci.* **30**, No. 6, 997–1006.
179. Holland, W.R.: 1977, in *The Sea*, Vol. 6 (ed. E. D. Goldberg), Wiley, New York pp. 3–45.
180. Holland, W.R.: 1978, *J. Phys. Oceanogr.* **8**, No. 3, 363–392.
181. Holland, W.R. and Lin, L.: 1975, *J. Phys. Oceanogr.* **5**, No. 4, 658–669.
182. Holland, W.R. and Rhines, P.B.: 1980, *J. Phys. Oceanogr.* **10**, No. 7, 1010–1031.
183. Holloway, G. and Hendershott, M.C.: 1977, *J. Fluid Mech.* **82**, Part 4, 747–765.
184. Horn, L.H. and Bryson, R.A.: 1963, *J. Geophys. Res.* **68**, No. 4, 1059–1064.
185. Hunkins, K.: 1974, *Deep-Sea Res.* **21**, No. 12, 1017–1033.
186. Huppert, H.E. and Bryan, K.: 1976, *Deep-Sea Res.* **23**, No. 8, 655–679.
187. Ichiye, T.: 1955, *Records of Oceanogr. Works Japan* **2**, No. 2, 82–89.
188. Ichiye, T.: 1956, *Oceanogr. Mag.* **7**, No. 2, 115–132.
189. Ichiye, T. and Ichiye, C.: 1956, *Oceanogr. Mag.* **8**, No. 2, 123–141.
190. Il'in, A.M.: 1972, *Mat. Sbornik* **87** (129), No. 4, 529–553 (*in Russian*).
191. Iselin, C. O'D. and Fuglister, F.C.: 1984, *J. Mar. Res.* **7**, No. 3, 317–329.

192. Itzikowitz, S., Jacobson, M.J., and Siegman, W.L.: 1982, *JASA* **71**, No. 5, 1131–1142.

193. Itzikowitz, S., Jacobson, M.J., and Siegman, W.L.: 1983, *JASA* **73**, No. 5, 1556–1566.

194. Ivanenkov, V.N.: 1979, in *Khimiya okeana*, T.1. *Khimiya vod okeana*. §7.3 (*Chemistry of the Ocean*, Vol. 1. *Chemistry of Ocean Water*, §7.3), Nauka, Moscow, pp. 188–229 (*in Russian*).

195. Ivanov, A.F. and Paramonov, A.N.: 1980, in *Sinopticheskie vikhri v okeane* (*Sypnotic Eddies in the Ocean*), Naukova Dumka, Kiev, pp. 107–114 (*in Russian*).

196. Ivanov, Yu.A. and Byshev, V.I.: 1972, *Izv. Akad, Nauk SSSR Fiz. Atm. Okeana* **8**, No. 3, 344–348 (*in Russian*).

197. Ivanov, Yu.A., Krasnopevtsev, A. Yu., Morozov, E.G., and Shadrin, I.F.: 1983, in *Otchet o rabotakh v 30-m reĭse NIS 'Dmitriĭ Mendeleev'*, T.2, Ch. I (*Account of the 30th Voyage of the Research Vessel 'Dmitriĭ Mendeleev'*, Vol. 2, Part I), pp. 91–103 (*in Russian*).

198. Ivanov, Yu.A., Krasnopevtsev, A.Yu., Morozov, E.G., Shadrin, I.F., and Shcherbinin, A.D.: 1984, *Doklady Akad Nauk. SSSR* **275**, No. 4, 1012–1015 (*in Russian*).

199. Ivanov, Yu.A. and Novitskiĭ, A.G.: 1975, *Okeanologiya* **15**, No. 5, 786–789 (*in Russian*).

200. Joyce, T.M. and Wiebe, P.H.: 1983, *Oceanus*, **26**, No. 2., 34–44.

201. Joyce, T.M., Patterson, S.L., and Millard, R.C. Jr: 1981, *Deep-Sea Res.* **28A**, No. 11, 1265–1287.

202. Julian, P.R., Washington, W.M., Hembree, L., and Ridley, C.: 1970, *J. Atm. Sci.* **27**, No. 3, 376–387.

203. Kamenkovich, V.M.: 1973, *Osnovy dinamiki okeana* (*Fundamentals of Ocean Dynamics*), Gidrometeoizdat, Leningrad, (*in Russian*).

204. Kamenkovich, V.M.: 1977, *Fundamentals of Ocean Dynamics* (English Transl.), Elsevier, Amsterdam, Oxford, New York.

205. Kamenkovich, V.M., Belotserkovskiĭ, S.O., and Panteleev, M.S.: 1985a, *Izv. POLIMODE*, No. 15, 3–23 (*in Russian*).

206. Kamenkovich, V.M., Larichev, V.D., and Kharkov, B.V.: 1981, *Okeanologiya* **21**, No. 6, 949–959 (*in Russian*).

207. Kamenkovich, V.M., Larichev, V.D., and Kharkov, B.V.: 1985b, *Izv. POLIMODE*, No. 14, 3–11 (*in Russian*).

208. Kamenkovich, V.M. and Monin, A.S.: 1978, in *Fizika okeana*, T.1. *Gidrofizika okeana* (*Physics of the Ocean*, Vol. 1. *Hydrophysics of the Ocean*), Nauka, Moscow, pp. 85–112 (*in Russian*).

209. Kamenkovich, V.M. and Reznik, G.M.: 1978, in *Fizika okeana*. T.2. *Gidrodinamika okeana* (*Physics of the Ocean*, Vol. 2. *Hydrodynamics of the Ocean*), Nauka, Moscow, pp. 300–358 (*in Russian*).

210. Kao, S.K. and Wendell, L.L.: 1970, *J. Atm. Sci.* **27**, No. 3, 359–375.

211. Karpman, V.I.: 1973, *Nélineĭnye volny v dispergiruyushchikh sredakh* (*Nonlinear Waves in Dispersive Media*), Nauka, Moscow (*in Russian*).

212. Kawai, H.: 1972, in *Kuroshio, Physical Aspects of the Japan Current* (eds H. Stommel and K. Yoshida), Univ. Washington Press, Seattle, pp. 235–352.

213. Kawai, H.: 1980, in *The Kuroshio IV*, Saikon, Tokyo, pp. 250–273.

214. Kaz'min, A.C. and Sklyarov, V.E.: 1981, *Issledovanie Zemli iz Kosmosa*, No. 6, 48–57 (*in Russian*).

215. Kenyon, K.: 1964, in *Notes on the* 1964 *Summer Study Program in Geophys. Fluid Dynamics at the Woods Hole Oceanogr. Inst., Student Lectures*, Vol. II, pp. 69–83.

216. Kenyon, K.: 1967, *Proc. R. Soc.* **A299**, No. 1456, 141–144.

217. Kibel', I.A.: 1940, *Izv. Akad. Nauk SSSR Ser. Geograf. Geofiz.*, No. 5, 627–638 (*in Russian*).

218. Killworth, P.D.: 1980, *Dyn. Atm. Oceans* **4**, No. 3, 143–184.

219. Kinder, T.H. and Coachman, L.K.: 1977, *J. Phys. Oceanogr.* **7**, No. 6, 946–952.

220. Kinder, T.H., Schumacher, J.D., and Hansen, D.V.: 1980, *J. Phys. Oceanogr.* **10**, No. 8, 1228–1245.

221. Kirwan, A.D., McNally, G., and Coehlo, J.: 1976, *J. Phys. Oceanogr.* **6**, No. 5, 750–755.

222. Kirwan, A.D., McNally, G.J., Reyna, E., and Merrel, W.J.: 1978, *J. Phys. Oceanogr.* **8**, No. 6, 937–945.

223. Kitano, K.: 1975, *J. Phys. Oceanogr.* **5**, No. 2, 245–252.

224. Kizner, Z.I.: 1984a, in *Tonkaya struktura i sinopticheskaya izmenchivost' moreĭ i okeanov. Tezisy dokladov II Vsesoyuzn. simposiuma*, Chast' II, *Tallin 1984)* (*Fine Structure and Synoptic Variability of Seas and Oceans. Abstracts of the 2nd All-Union Symposium*, Part 2, *Tallin, 1984)*, Tallin, pp. 22–24 (*in Russian*).

225. Kizner, Z.I.: 1984b, in *Tonkaya struktura i sinopticheskaya izmenchivost' moreĭ i okeanov. Tezisy dokladov II Vsesoyuzn. simposiuma,* Chast' II, *Tallin, 1984*) (*Fine Structure and Synoptic Variability of Seas and Oceans. Abstracts of the 2nd All-Union Symposium*, Part 2, *Tallin, 1984*), Tallin, pp. 19–21 (*in Russian*).
226. Kizner, Z.I.: 1984c, *Doklady Akad. Nauk SSSR* **275**, No. 6, 1495–1498 (*in Russian*).
227. Kizner, Z.I.: 1985, *Izv. Akad. Nauk SSSR Fiz. Atm. Okeana* **21**, No. 3, 330–332 (*in Russian*).
228. Kobaladze, Z.L., Patariya, A.D., and Khantadze, A.G.: 1981, *Izv. Akad. Nauk SSSR Fiz. Atm. Okeana* **17**, No. 6, 649–653 (*in Russian*).
229. Kogelman, S. and Keller, J.B.: 1973, *SIAM J. Appl. Math.* **24**, No. 3, 352–361.
230. Konyaev, K.V. and Merinova, G.I.: 1983, *Izv. Akad. Nauk SSSR Ser. Fiz. Atm. Okeana* **19**, No. 10, 1042–1052 (*in Russian*).
231. Konyaev, K.V. and Sabinin, K.D.: 1980, *Doklady Akad. Nauk SSSR* **253**, No. 4, 970–973 (*in Russian*).
232. Konyaev, K.V. and Sabinin, K.D.: 1981, *Izv. Akad. Nauk SSSR Ser. Fiz. Atm. Okeana* **17**, No. 11, 1191–1200 (*in Russian*).
233. Kool', L.V., Romanov, Yu.A., Samoĭlenko, V.S., and Shishkov, Yu.A.: 1974, in *Atlanticheskiĭ gidrofizicheskiĭ poligon-70* (*Atlantic Hydrophysical Polygon-70*), Nauka, Moscow, pp. 20–45 (*in Russian*).
234. Korotaev, G.K.: 1980, *Struktura, dinamika i energetika sinopticheskoĭ izmenchivosti okeana* (*Structure, Dynamics, and Energetics of the Sypnotic Variability of the Ocean*), Preprint No. 7, MGI AN Ukr. SSR, Sebastopol (*in Russian*).
235. Korotaev, G.K. and Kosnyrev, V.K.: 1980, in *Sinopticheskie vikhri v okeane* (*Sypnotic Eddies in the Ocean*), Naukova Dumka, Kiev, pp. 121–125 (*in Russian*).
236. Korotaev, G.K., Kosnyrev, V.K., and Shevchenko, E.A.: 1979, *Morskie Gidrofiz. Issled.* **85**, No. 4, 167–174 (*in Russian*).
237. Korotaev, G.K. and Shapiro, N.V.: 1980, in *Sinopticheskie vikhri v okeane* (*Sypnotic Eddies in the Ocean*), Naukova Dumka, Kiev, pp. 183–187 (*in Russian*).
238. Kort, V.G.: 1981, *Okeanologiya* **21**, No. 3, 405–415 (*in Russian*).
239. Kort, V.G.: 1982, *Doklady Akad. Nauk SSSR* **263**, No. 1, 190–193 (*in Russian*).
240. Kort, V.G., Byshev, V.I., and Tarasenko, V.M.: 1974, in *Atlanticheskiĭ gidrofizicheskiĭ poligon-70* (*Atlantic Hydrophysical Polygon-70*), Nauka, Moscow, pp. 181–188 (*in Russian*).
241. Kort, V.G., Krasnopevtsev, A.Yu., and Shadrin, I.F.: 1983, *Okeanologiya* **23**, No. 1, 5–12 (*in Russian*).
242. Kort, V.G., Pavlova, Yu.V., and Fomin, L.M.: 1980, *Izv. POLIMODE*, No. 2, 5–13 (*in Russian*).
243. Kort, V.G. and Tarasenko, V.M.: 1977, *Okeanologiya* **17**, No. 4, 578–583 (*in Russian*).
244. Kort, V.G., Titov, V.B., and Osadchiĭ, A.S.: 1977, *Okeanologiya* **17**, No. 5, 769–773 (*in Russian*).
245. Koshlyakov, M.N.: 1961, *Trudy Instituta Okeanologii AN SSSR* **38**, 31–55 (*in Russian*).
246. Koshlyakov, M.N.: 1973, *Okeanologiya* **13**, No. 5, 760–767 (*in Russian*).
247. Koshlyakov, M.N.: 1974, *Problemy diagnosticheskikh i eksperimental'nykh issledovaniĭ okeanskoĭ tsirkulyatsii* (*Problems of Diagnostic and Experimental Investigations of Ocean Circulation*) (Doctoral thesis, IOAN, SSSR, (*in Russian*).
248. Koshlyakov, M.N., Borisenko, Yu.D., Brekhovskikh, A.L., Vinogradov, A.S., Drozdov, V.N., Moroshkin, K.V., Pavel'son, Yu.Kh., and Rzheplinskiĭ, D.G.: 1980a, *Okeanologicheskie Issled.*, No. 31, 42–55 (*in Russian*).
249. Koshlyakov, M.N. and Enikeev, V.Kh.: 1977, *POLYMODE News*, No. 23.
250. Koshlyakov, M.N., Galerkin, L.I., and Chyong Din' Khien: 1970, *Okeanologiya* **10**, No. 5, 805–814 (*in Russian*).
251. Koshlyakov, M.N. and Grachev, Yu.M.: 1973, *Deep-Sea Res.* **20**, No. 6, 507–526.
252. Koshlyakov, M.N. and Grachev, Yu.M.: 1974, in *Atlanticheskiĭ gidrofizicheskiĭ poligon-70* (*Atlantic Hydrophysical Polygon-70*), Nauka, Moscow, pp. 163–180 (*in Russian*).
253. Koshlyakov, M.N., Grachev, Yu.M., and Chyong Din' Khien: 1972, *Okeanologiya* **12**, No. 4, 728–734 (*in Russian*).
254. Koshlyakov, M.N., Grachev, Yu.M., and Enikeev, V.Kh.: 1980b, *Doklady Akad. Nauk SSSR* **252**, No. 3, 573–577 (*in Russian*).

255. Koshlyakov, M.N., Grachev, Yu.M., Mikhaĭlichenko, Yu.G., Sazhina, T.G., and Yaremchuk, M.I.: 1984a, *Okeanologiya* **24**, No. 1, 5–14 (*in Russian*).
256. Koshlyakov, M.N., Grachev, Yu.M., Nechaev, D.A., Sazhina, T.G., and Yaremchuk, M.I.: 1984b, *Doklady Akad. Nauk SSSR* **276**, No. 2, (*in Russian*).
257. Koshlyakov, M.N., Grachev, Yu.M., Sazhina, T.G., and Yaremchuk, M.I.: 1985, *Okeanologiya* **25**, No. 7 (*in Russian*).
258. Koshlyakov, M.N. and Monin, A.S.: 1978, *Ann. Rev. Earth Planet. Sci.* **6**, 495–523.
259. Koshlyakov, M.N. and Yaremchuk, M.I.: 1984, *Izv. Akad Nauk SSSR Ser Fiz. Atm. Okeana* **20**, No. 8, 749–753 (*in Russian*).
260. Koshlyakov, M.N. and Yenikeyev, V.Kh.: 1977, *POLYMODE News*, No. 23.
261. Kozlov, V.F.: 1983, *Modeli topograficheskikh vikhreĭ v okeane* (*Models of Topographic Eddies in the Ocean*), Nauka, Moscow. (*in Russian*).
262. Kozlov, V.F., Darnitskiĭ, V.D., and Ermakov, M.I.: 1982, *Trudy DVNII*, No. 96, 3–25 (*in Russian*).
263. Kraichnan, R.H.: 1967, *Phys. Fluids* **10**, No. 7, 1417–1423.
264. Kraichnan, R.H.: 1971, *J. Fluid Mech.* **47**, Part 3, 525–535.
265. Kraichnan, R.H.: 1974, *J. Fluid Mech.* **64**, Part 4, 737–762.
266. Kraichnan, R.H.: 1975, *J. Fluid Mech.* **67**, Part 1, 155–175.
267. Kreiss, H. and Oliger, J.: 1973, *Methods for the Approximate Solution of Time-dependent Problems* (GARP Publications Series, No. 10, WMO, ICSU).
268. Kurtepov, V.M.: 1983, *Akusticheskiĭ Zhurnal* **29**, No. 1, 80–83 (*in Russian*).
269. Kushnir, V.M.: 1980, in *Sinopticheskie vikhri v okeane* (*Synoptic Eddies in the Ocean*), Naukova Dumka, Kiev, pp. 156–163 (*in Russian*).
270. Lai, D.Y. and Richardson, P.L.: 1977, *J. Phys. Oceanogr.* **7**, No. 5, 670–683.
271. Lamb, H.: 1932, *Hydrodynamics* (6th edn), Cambridge Univ. Press.
272. Lambert, R.B.: 1974, *Deep-Sea Res.* **21**, No. 7, 529–546.
273. Lappo, S.S.: 1985, in *Issledovaniya techeniĭ okeana* (*Investigations of Ocean Currents*), Nauka, Moscow (*in Russian*) (in the press).
274. Larichev, V.D.: 1974, *Izv. Akad. Nauk SSSR Fiz. Atm. Okeana* **10**, No. 4, 763–770 (*in Russian*).
275. Larichev, V.D. and Reznik, G.M.: 1976a, *Okeanologiya* **16**, No. 2, 200–206 (*in Russian*).
276. Larichev, V.D. and Reznik, G.M.: 1976b, *Doklady Akad. Nauk SSSR* **231**, No. 5, 1077–1079 (*in Russian*).
277. Larichev, V.D. and Reznik, G.M.: 1976c, *Okeanologiya* **16**, No. 3, 381–388 (*in Russian*).
278. Larichev, V.D. and Reznik, G.M.: 1976d, *Okeanologiya* **16**, No. 6, 961–967 (*in Russian*).
279. Larichev, V.D. and Reznik, G.M.: 1982, *Doklady Akad. Nauk SSSR* **264**, No. 1, 229–233 (*in Russian*).
280. Larichev, V.D. and Reznik, G.M.: 1983, *Okeanologiya*, **23**, No. 5, 725–734 (*in Russian*).
281. Larsen, L.H.: 1965, *J. Atm. Sci.* **22**, No. 2, 222–224.
282. Legeckis, R.: 1973, *Deep-Sea Res.* **20**, No. 6, 507–526.
283. Legeckis, R.: 1978, *J. Geophys. Res.* **83**, No. C9, 4501–4522.
284. Legeckis, R. and Gordon, A.L.: 1982, *Deep-Sea Res.* **29**, No. 3A, 375–401.
285. Le Groupe Tourbillon: 1983, *Deep-Sea Res.* **30**, No. 5A, 475–511.
286. Leith, C.E.: 1968, *Phys. Fluids* **11**, No. 3, 671–672.
287. Leith, C.E.: 1971, *J. Atm. Sci.* **28**, No. 2, 145–161.
288. Leith, C.E. and Kraichnan, R.H.: 1972, *J. Atm. Sci.* **29**, No. 6, 1041–1058.
289. Lewis, E.L. and Perkin, R.G.: 1981, *Deep-Sea Res.* **28**, No. 4A, 307–328.
290. Lighthill, M.J.: 1969, *Philos. Trans. R. Soc. London* **A265**, No. 1159, 45–92.
291. Lilly, D.K.: 1969, *The Physics of Fluids Suppl. II* **12**, No. 12, 240–249.
292. Lilly, D.K.: 1971, *J. Fluid Mech.* **45**, Part 2, 395–415.
293. Lilly, D.K.: 1972, *Geophys. Fluid Dyn.* **4**, No. 1, 1–28.
294. Long, R.R.: 1964, *J. Atm. Sci.* **21**, No. 1, 197–200.
295. Longuet-Higgins, M.S. and Gill, A.E.: 1967, *Proc. R. Soc. Ser.* **A299**, No. 1456, 120–140.
296. Lorenz, E.N.: 1955, *Tellus* **7**, No. 2, 157–167.
297. Lorenz, E.N.: 1972, *J. Atm. Sci.* **29**, No. 2, 258–264.
298. Lutjeharms, J.R.E.: 1981, *Deep-Sea Res.* **28A**, No. 11, 1289–1302.

299. Lutjeharms, J.R.E. and Baker D.J.: 1980, *Deep-Sea Res.* **27A**, No. 2A, 145–159.
300. Lutjeharms, J.R.E., Bang, N.D., and Duncan, C.P.: 1981, *Deep-Sea Res.* **28A**, No. 9, 879–899.
301. Magaard, L.: 1977, *J. Phys. Oceanogr.* **7**, No. 3, 359–364.
302. Makino, M., Kamimyra, T., and Taniuti, T.: 1981, *J. Phys. Soc. Japan* **50**, No. 3, 980–989.
303. Manabe, S., Smagorinsky, J., Holloway, J.L., and Stone, H.M.: 1970, *Monthly Weather Rev.* **98**, No. 3, 175–212.
304. Manley, T. and Hunkins, K.: 1978, *Trans. Amer. Geophys. Union* **59**, No. 12, 1099.
305. Mann, C.R.: 1967, *Deep-Sea Res.* **14**, No. 3, 337–359.
306. Masuzawa, J.: 1955, *Oceanogr. Mag.* **7**, No. 1, 29–47.
307. Masuzawa, J.: 1957, *Rec. Oceanogr., Wrs. Japan* **3**, No. 1, 1–7.
308. Mathäus, W.: 1974, *Beitr. Meereskunde*, Heft 33, 73–78 (*in German*).
309. Maxworthy, T. and Redecopp, L.G.: 1976, *Icarus* **29**, No. 2, 261–271.
310. Maxworthy, T., Redecopp, L.G., and Weidman, P.D.: 1978, *Icarus* **33**, No. 2, 388–409.
311. Mazeika, P.A.: 1973, *Deutsche Hydrogr. Z.*, Heft 2, 49–73.
312. Mazeika, P.A., Burns, D.A., and Kinder, T.H.: 1980, *J. Geophys. Res.* **85**, No. C5, 2743–2758.
313. McCartney, M.S., Schmitz, W.J., and Worthington, L.V.: 1978, *J. Geophys. Res.* **83**, No. C2, 901–914.
314. McDowell, S.E. and Rossby, H.T.: 1978, *Science* **202**, 1085–1087.
315. McWilliams, J.C.: 1976a, *J. Phys. Oceanogr.* **6**, No. 6, 810–827.
316. McWilliams, J.C.: 1976b, *J. Phys. Oceanogr.* **6**, No. 6, 828–861.
317. McWilliams, J.C.: 1983, *J. Mar. Research* **41**, No. 3, 427–460.
318. McWilliams, J.C. and Flierl, G.R.: 1976, *Deep-Sea Res.* **23**, No. 4, 285–300.
319. McWilliams, J.C., Flierl, G.R., Larichev, V.D., and Reznik, G.M.: 1981, *Dyn. Atm. Oceans* **5**, No. 4, 219–238.
320. McWilliams, J.C. and Robinson, A.R.: 1974, *Deep-Sea Res.* **21**, No. 5, 359–368.
321. McWilliams, J.C. and Zabusky, N.J.: 1982, *Geophys. Astrophys. Fluid Dynamics* **19**, Nos 3–4, 207–227.
322. Mercer, J.A. and Booker, J.B.: 1983, *J. Geophys. Res.* **88**, No. C1, 689–699.
323. Merilees, P.E. and Warn, H.: 1975, *J. Fluid Mech.* **69**, Part 4, 625–630.
324. Mikhaĭlova, E.N. and Shapiro, N.B.: 1980, *Izv. Akad. Nauk SSSR Fiz. Atm. Okeana* **16**, No. 8, 823–833 (*in Russian*).
325. Mintz, Y.: 1979, in *Rep. IOC Study Conference on Climate Models: Performance, Intercomparison, and Sensitivity Studies*, Vol. 11, *Wash., D.C., 3–7 April 1978*, GARP Publ. Series No. 22, WMO, Geneva, pp. 607–687.
326. Mirabel', A.P.: 1974, *Izv. Akad. Nauk SSSR Fiz. Atm. Okeana* **10**, No. 2, 211–216 (*in Russian*).
327. Mirabel', A.P. and Monin, A.S.: 1979, *Uspekhi Mekhaniki (Advances in Mechanics)* **2**, No. 3, 47–95 (*in Russian*).
328. Mirabel', A.P. and Monin, A.S.: 1980, *Izv. Akad. Nauk SSSR Fiz. Atm. Okeana* **16**, No. 10, 1011–1023 (*in Russian*).
329. Moiseev, L.K., Galerkin, L.I., Kut'ko, V.P., Oleĭnikov, S.A., Panfilova, S.G., Stepanov, V.N., and Shcherbinin, A.D.: 1980, in *Pole temperatury vody severnoĭ chasti Tikhogo okeana (Water Temperature Field in the North Pacific)* Gidrometeoizdat, Moscow, p. 22 (*in Russian*).
330. Monin, A.S.: 1969, *Prognoz pogody kak zadacha fiziki (Weather Forecast as a Problem of Physics)*, Nauka, Moscow (*in Russian*).
331. Monin, A.S.: 1970, *Dokl. Akad. Nauk SSSR* **193**, No. 5, 1038–1041 (*in Russian*).
332. Monin, A.S.: 1973, *Izv. Akad. Nauk SSSR Fiz. Atm. Okeana* **9**, No. 10, 1063–1068 (*in Russian*).
333. Monin, A.S. and Fedorov, K.N.: 1983, *Doklady Akad. Nauk SSSR* **270**, No. 5, 1213–1216 (*in Russian*).
334. Monin, A.S., Kamenkovich, V.M., and Kort, V.G.: 1974, *Izmenchivost' Mirovogo okeana (Variability of the Ocean)*, Gidrometeoizdat, Leningrad (*in Russian*).
335. Monin, A.S., Kamenkovich, V.M., and Kort, V.G.: 1977, *Variability of the Ocean*, (English Transl.), Wiley, New York.
336. Monin, A.S. and Seidov, D.G.: 1982, *Doklady Akad. Nauk SSSR* **268**, No. 2, 454–457 (*in Russian*).
337. Monin, A.S. and Seidov, D.G.: 1983, *Priroda*, No. 1, 34–42 (*in Russian*).

338. Mountain, D.C. and Shuhy, J.L.: 1980, *J. Mar. Res.* **38**, No. 2, 203–213.
339. Müller, P.: 1982, *The Dynamics of Synoptic-Scale Motions in the Ocean*, Mitteilungen des Instituts für Meereskunde der Universität Hamburg, No. 25.
340. Munk, W.H.: 1980, *J. Phys. Oceanogr.* **10**, No. 4, 596–664.
341. Munk, W.H. and Wunsch, C.: 1979, *Deep-Sea Res.* **26**, No. 2A, 123–161.
342. Munk, W.H. and Wunsch, C.: 1982a, *Deep-Sea Res.* **29**, No. 12A, 1415–1436.
343. Munk, W.H. and Wunsch, C.: 1982b, *Phil. Trans. R. Soc. Ser.* **A307**, No. 1499, 439–464.
344. Muromtsev, A.M.: 1972, *Trudy Gos. Okeanogr. Instituta*, No. 106, 160–174 (*in Russian*).
345. Mysak, L.A.: 1977, *J. Phys. Oceanogr.* **7**, No. 6, 904–917.
346. Mysak, L.A. and Schott, F.: 1977, *J. Geophys. Res.* **82**, No. 15, 2087–2095.
347. Neelov, I.A.: 1982, *Okeanologiya* **22**, No. 6, 875–884 (*in Russian*).
348. Neelov, I.A. and Chalikov, D.V.: 1981, *Okeanologiya* **21**, No. 1, 5–11 (*in Russian*).
349. Neiman, V.G.: 1970, *Doklady Akad. Nauk SSSR* **195**, No. 5, 948–952 (*in Russian*).
350. Newton, J., Agaard, K., and Coachman, L.: 1974, *Deep-Sea Res.* **21**, No. 9, 707–719.
351. Nezlin, M.V.: 1981, *Pis'ma v ZhETF* **34**, No. 2, 83–86 (*in Russian*).
352. Nikitin, O.P.: 1977, *Ocean Modelling*, No. 8, 4–8.
353. Nikitin, O.P.: 1985, *Okeanologiya* (*in Russian*) (in the press).
354. Nilsson, C.S., Andrews, J.C., and Scully-Power, P.: 1977, *J. Phys. Oceanogr.* **7**, No. 5, 659–669.
355. Nilsson, C.S. and Cresswell, G.R.: 1981, *Prog. Oceanogr.* **9**, 133–183.
356. Nishida, H. and White, W.B.: 1982, *J. Phys. Oceanogr.* **12**, No. 2, 160–170.
357. Nishiyama, K., Konaga, S., and Ishizaki, H.: 1980, *Oceanogr. Mag.* **31**, No. 1–2: 27–32.
358. Novikov, E.A.: 1978, *Izv. Akad. Nauk SSSR Fiz. Atm. Okeana* **14**, No. 6, 668–671 (*in Russian*).
359. Novitskiĭ, A.G.: 1977, *Okeanologiya* **17**, No. 1, 23–26 (*in Russian*).
360. Nowlin, W.D., Hubertz, J.M., and Reid, R.O.: 1968, *J. Mar. Res.* **26**, No. 2, 185–186.
361. Nysen, P.A., Scully-Power, P., and Browning, D.G.: 1978, *JASA* **63**, No. 5, 1381–1388.
362. Obukhov, A.M.: 1960, *Izv. Akad. Nauk SSSR Ser. Geophys.*, No. 3, 432–440 (*in Russian*).
363. Obukhov, A.M.: 1962, *Doklady Akad. Nauk SSSR* **145**, No. 6, 1239–1242 (*in Russian*).
364. Odulo, A.B. and Pelinovskiĭ, E.N.: 1978, *Okeanologiya* **18**, No. 5, 773–777 (*in Russian*).
365. Ogura, Y.: 1952, *J. Meteor. Soc. Japan Ser. II* **30**, No. 2, 53–58.
366. Ogura, Y.: 1958, *J. Meteor.* **15**, No. 4, 375–382.
367. Ogura, Y.: 1962, *Phys. Fluids* **5**, No. 4, 395–401.
368. Olson, D.B.: 1980, *J. Phys. Oceanogr.* **10**, No. 4, 514–528.
369. Onsager, L.: 1949, *Nuovo Cimento Suppl.* **6**, No. 2, 279–287.
370. Oort, A.H. and Vonder Haar, T.H.: 1976, *J. Phys. Oceanogr.* **6**, 781–800.
371. Orszag, S.A.: 1971, *Stud. Appl. Math.* **50**, No. 2, 293–327.
372. Orszag, S.A. and Patterson, G.S.: 1972, *Phys. Rev. Lett.* **28**, 76–79.
373. Owens, W.B. and Hogg, N.G.: 1980, *Deep-Sea Res.* **27A**, No. 12, 1029–1048.
374. Owens, W.B., Luyten, J.R., and Bryden, H.L.: 1982, *J. Mar. Res.* **40**, Supplement, 509–524.
375. Ozmidov, R.V.: 1962, *Trudy Instituta Oceanologii AN SSSR* **60**, 114–129 (*in Russian*).
376. Ozmidov, R.V.: 1965, *Izv. Akad. Nauk SSSR Fiz. Atm. Okeana* **1**, No. 4, 439–448 (*in Russian*).
377. Ozmidov, R.V. and Yampol'skiĭ, A.D.: 1965, *Izv. Akad. Nauk SSSR Ser. Fiz. Atm. Okeana* **1**, No. 6, 615–622 (*in Russian*).
378. Palmen, E. and Newton, C.W.: 1969, *Atmospheric Circulation System*, Academic Press, New York, London.
379. Panfilova, S.G., Galerkin, L.I., Kut'ko, V.P., Moiseev, L.K., Oleĭnikov, V.N., and Shcherbinin, A.D.: 1981, in *Klimaticheskaya ismenchivost,' solenosti vody severnoĭ chasti Tikhogo okeana* (*Climatic Variability of Water Salinity in the North Pacific*), Gidrometeoizdat, Moscow, p. 22 (*in Russian*).
380. Paramonov, A.N., Ivanov, A.F., and Koveshnikov, L.A.: 1978, in *Eksperimental'nye issledovaniya po mezhdunarodnoĭ programme POLIMODE (resul'taty 17 reĭsa NIS 'Akademik Vernadskiĭ' i 33 reĭsa NIS 'Mikhail Lomonosov')* (*Experimental Investigations According to the International Program POLYMODE (the Results of the 17th Voyage of the Research Vessel 'Akademik Vernadskiĭ' and the 33rd Voyage of the Research Vessel 'Mikhail Lomonosov')*), Sebastopol, pp. 122–127 (*in Russian*).
381. Parker, C.E.: 1971, *Deep-Sea Res.* **18**, No. 11, 981–993.

382. Pavlychev, V.P.: 1975, *Izv. TINRO* (Pacific Institute of Fishery and Oceanography), No. 96, 3–18 (*in Russian*).
383. Pedlosky, J.: 1965, *J. Mar. Res.* **23**, No. 3, 207–209.
384. Pedlosky, J.: 1979, *Geophysical Fluid Dynamics*, Springer-Verlag, New York.
385. Pelinovskiĭ, E.N.: 1978, *Okeanologiya* **18**, No. 2, 192–195 (*in Russian*).
386. Peregudin, A.S.: 1982, *Trudy DVNII*, No. 96, 53–62 (*in Russian*).
387. Peterson, R.G., Nowlin, W.D. Jr, and Whitworth III, T.: 1982, *J. Phys. Oceanogr.* **12**, No. 7, 712–719.
388. Petviashvili, V.I.: 1980, *Pis'ma v ZhETF* **32**, No. 11, 632–635 (*in Russian*).
389. Petviashvili, V.I.: 1983, *Pis'ma v Astronomicheskiĭ Zh.* **9**, No. 4, 253–256 (*in Russian*).
390. Petviashvili, V.I. and Yan'kov, V.V.: 1982, *Doklady Akad. Nauk SSSR* **267**, No. 4, 825–828 (*in Russian*).
391. Phillips, N.A.: 1963, *Rev. Geophys.* **1**, No. 2, 123–176.
392. Phillips, N.A.: 1966, *J. Geophys. Res.* **71**, No. 16, 3833–3891.
393. Pillsbury, R.D., Whitworth, T., Nowlin, W.D., and Sciremammano, F.: 1979, *J. Phys. Oceanogr.* **9**, No. 3, 469–482.
394. Pokudov, V.V. and Tunegolovets, V.P.: 1981, in *Resul'taty ekspeditsii KISZ-80*, Chasti 1–2 (*Results of the Expedition KISZ-80*, Parts 1–2), Gidrometeoizdat, Leningrad (*in Russian*).
395. Pokudov, V.V. and Tunegolovets, V.P.: 1983, *Vikhri Filippinskogo morya. Resul'taty ekspeditsii KISZ-80*, Chast' 3 (*Eddies of the Philippine Sea. Results of the Expedition KISZ-80*, Part 3), Gidrometeoizdat, Leningrad (*in Russian*).
396. Pokudov, V.V. and Vel'yaots, K.O.: 1976, *Trudy DVNIGMI* (Far-East Hydrometeorological Research Institute), No. 62, 3–61 (*in Russian*).
397. Pokudov, V.V. and Vel'yaots, K.O.: 1978, *Trudy DVNIGMI*, No. 71, 139–148 (*in Russian*).
398. Pokudov, V.V. and Vel'yaots, K.O.: 1979, *Trudy DVNIGMI*, No. 77, 61–69 (*in Russian*).
399. Pokudov, V.V., Yarichin, V.G., and Volkov, Yu.N.: 1978, *Trudy* DVNIGMI, No. 71, 3–138 (*in Russian*).
400. Polyanskaya, V.A.: 1982, in *Voprosy Sydostroeniya. Ser. Akustika*, vyp. 15 (*Shipbuilding Problems. Ser. Acoustics*, No. 15), 98–108 (*in Russian*).
401. Ponomareva, L.A.: 1983, in *Otchet o rabotakh v 30-m reĭse NIS 'Dmitriĭ Mendeleev'*, T.1, Ch. II, (*Account of the 30th Voyage of the Research Vessel 'Dmitriĭ Mendeleev'*, Vol. 1, Part II), pp. 1–25 (*in Russian*).
402. Pouquet, A., Lesieur, M., Andre, J.C., and Basdevant, C.: 1975, *J. Fluid Mech.* **72**, Part 2, 305–319.
403. Preobrazhenskaya, T.I.: 1971, in *Antarctida. Doklady komissii, 1968 (Antarctic. Commission Reports, 1968)*, Nauka, Moscow, pp. 135–137 (*in Russian*).
404. Price, J.F. and Rossby, H.T.: 1982, *J. Mar. Res.* **40**, Supplement, 543–558.
405. Pudov, V.D. and Bezzabotnov, V.S.: 1980, in *Taifun-78 (Typhoon-78)*, Gidrometeoizdat, Leningrad, pp. 118–121 (*in Russian*).
406. Radchenko, L.A.: 1978, in *Eksperimental'nye issledovaniya po mezhdunarodnoĭ programme POLIMODE (resul'taty 16-go reĭsa NIS 'Akademic Vernadskiĭ') (Experimental Studies According to the International Program POLYMODE (the Results of the 16th Voyage of the Research Vessel 'Akademic Vernadskiĭ'))*, MGI, Sebastopol, pp. 134–141 (*in Russian*).
407. Redecopp, L.G.: 1977, *J. Fluid Mech.* **82**, Part 4, 725–745.
408. Redecopp, L.G. and Weidman, P.D.: 1978, *J. Atm. Sci.* **35**, No. 5, 790–804.
409. Resin, E.I.: 1980, *Okeanologiya* **20**, No. 4, 590–599 (*in Russian*).
410. Reznik, G.M.: 1976, *Okeanologiya* **16**, No. 4, 571–577 (*in Russian*).
411. Reznik, G.M.: 1983, *Okeanologiya* **23**, No. 3, 373–381 (*in Russian*).
412. Reznik, G.M.: 1984a, *Okeanologiya* **24**, No. 6, 869–873 (*in Russian*).
413. Reznik, G.M.: 1984b, *Int. J. Nonlinear Mech.* **19**, No. 2, 95–113.
414. Reznik, G.M. and Kozlov, O.V.: 1981, *Izv. Akad. Nauk SSSR Fiz. Atm. Okeana* **17**, No. 6, 632–638 (*in Russian*).
415. Reznik, G.M. and Soomere, T.E.: 1983a, *Izv. Akad. Nauk Est. SSR Fiz. Mat.* **19**, No. 3, 294–299 (*in Russian*).
416. Reznik, G.M. and Soomere, T.E.: 1983b, *Okeanologiya* **23**, No. 6, 923–927 (*in Russian*).

417. Reznik, G.M. and Soomere, T.E.: 1984a, *Okeanologiya* **24**, No. 3, 373–384 (*in Russian*).
418. Reznik, G.M. and Soomere, T.E.: 1984b, *Okeanologiya* **24**, No. 4, 558–565 (*in Russian*).
419. Rhines, P.B.: 1970, *Geophys. Fluid Dyn.* **1**, No. 3, 273–302.
420. Rhines, P.B.: 1971, *Deep-Sea Res.* **18**, No. 1, 21–26.
421. Rhines, P.B.: 1973, *Bound. Layer Meteor.* **4**, No. 3, 345–360.
422. Rhines, P.B.: 1975, *J. Fluid Mech.* **69**, Part 3, 417–443.
423. Rhines, P.B.: 1977, in *The Sea*, Vol 6 (ed. E. D. Goldberg), Wiley, New York, pp. 189–318.
424. Rhines, P. and Bretherton, F.: 1973, *J. Fluid Mech.* **61**, Part 3, 583–607.
425. Richards, F.A.: 1965, in *Chemical Oceanography*, Vol. 1, Academic Press, London, New York, pp. 611–645.
426. Richards, F.A. and Redfield, A.C.: 1955, *Deep-Sea Res.* **2**, 182–199.
427. Richardson, P.L.: 1980, *J. Phys. Oceanogr.* **10**, No. 1, 90–104.
428. Richardson, P.L.: 1981, *J. Phys. Oceanogr.* **11**, No. 7, 999–1010.
429. Richardson, P.L.: 1983, *J. Geophys. Res.* **88**, No. C7, 4355–4367.
430. Richardson, P.L., Cheney, R.E., and Mantini, L.A.: 1977, *J. Phys. Oceanogr.* **7**, No. 4, 580–590.
431. Richardson, P.L., Cheney, R.E., and Worthington, L.V.: 1978, *J. Geophys. Res.* **83**, No. C12, 6136–6144.
432. Richardson, P.L., Mailard, C., and Stanford, T.B.: 1979, *J. Geophys. Res.* **84**, No. C12, 7727–7741.
433. Richman, J.G., Wunsch C., and Hogg, N.G.: 1977, *Rev. Geophys. Space Phys.* **15**, No. 4, 385–420.
434. Riser, S.C. and Rossby, H.T.: 1983, *J. Mar. Res.* **41**, No. 1, 127–162.
435. Rizzoli, P.: 1980, *Dyn. Atm. Oceans* **4**, No. 4, 261–294.
436. Rizzoli, P. and Hendershott, M.C.: 1980, *Dyn. Atm. Oceans* **4**, No. 4, 247–260.
437. Roden, G.I., Taft, B.A., and Ebbesmeyer, C.C.: 1982, *J. Geophys. Res.* **87**, No. C12, 9537–9552.
438. Romanova, N.N. and Tseitlin, V.Yu.: 1984, *Izv. Akad. Nauk SSSR Fiz. Atm. Okeana* **20**, No. 2, 115–124 (*in Russian*).
439. Romantsov, V.A. and Smirnov, N. P.: 1983, *Trudy AANII* **382**, 84–101 (*in Russian*).
440. Rossby, H.T. and Freeland, H.: 1977, in *Atlas of the Mid-Ocean Dynamics Experiment (MODE)* (eds V. Lee and C. Wunsch), MIT, Cambridge, Mass, pp. 91–104.
441. Rossby, H.T., Riser, S.C., and McDowell, S.E.: 1982, *On the Origin and Structure of a Small-Scale Lens of Water Observed in the North Atlantic Thermocline* (Preprint), Univ. of Rhode Island.
442. Rossby, T., Voorhis, A.D., and Webb, D.: 1975, *J. Mar. Res.* **33**, No. 3, 355–382.
443. Roukhiyaĭnen, M.M. and Yunev, O.A.: 1978, in *Eksperimental'nye issledovaniya po mezhdunarodnoĭ programme POLIMODE (resul'taty 16-go reĭsa NIS 'Akademik Vernadskiĭ') (Experimental Investigations According to the International Program POLYMODE (the Results of the 16th Voyage of the Research Vessel 'Akademik Vernadskiĭ'))*, MGI AN Ukr. SSR, Sebastopol, pp. 125–133 (*in Russian*).
444. Ryther, J.H. and Mezel, D.W.: 1960, *Deep-Sea Res.* **7**, 235–238.
445. Salmon, R.: 1977, in *Issledovaniya sinopticheskoĭ izmenchivosti okeana. Materialy sov.-amer. in-ta POLIMODE, Yalta, 1976 (Investigations of the Synoptic Variability of the Ocean. Materials of the Soviet–American Institute POLYMODE, Yalta*, 1976), MGI AN Ukr. SSR, Yalta, pp. 180–195.
446. Salmon, R.: 1978, *Geophys. Astrophys. Fluid. Dyn.* **10**, No. 1, 25–51.
447. Salmon, R., Holloway, G., and Hendershott, M.C.: 1976, *J. Fluid Mech.* **75**, Part 4, 691–704.
448. Saltzman, B. and Fleisher, A.: 1962, *J. Atm. Sci.* **19**, No. 2, 195–204.
449. Sanford, T.: 1978, *POLYMODE News*, No. 51 (unpublished manuscript).
450. Sarukhanyan, E.I.: 1980a, *Doklady Akad. Nauk SSSR* **250**, No. 3, 738–741 (*in Russian*).
451. Sarukhanyan, E.I.: 1980b, *Struktura i izmenchivost' Antarkticheskogo Tsirkumpolyarnogo techeniya (Structure and Variability of the Antarctic Circumpolar Current)*, Gidrometeoizdat, Leningrad (*in Russian*).
452. Saunders, P.M.: 1971, *Deep-Sea Res.* **18**, No. 12, 1207–1219.
453. Saunders, P.M.: 1972, *J. Atm. Sci.* **29**, No. 1, 197–199.
454. Savchenko, V.G., Emery, W.J., and Vladimirov, O.A.: 1978, *J. Phys. Oceanogr.* **8**, No. 5, 825–837.

455. Sazhina, T.G.: 1982, *Izv. POLIMODE*, No. 8, 31–46 (*in Russian*).
456. Sazontov, A.G.: 1981, *K teorii dvumernoĭ turbulentnosti* (*On the Theory of Two-Dimensional Turbulence*), Institute of Applied Physics of the USSR Academy of Sciences, Gorkiĭ, Preprint No. 30 (*in Russian*).
457. Schmitz, W.J.: 1977, *J. Mar. Res.* **35**, No. 1, 21–28.
458. Schmitz, W.J.: 1978, *J. Mar. Res.* **36**, No. 2, 295–310.
459. Schmitz, W.J., Niiler, P.P., Bernstein, R.L., and Holland, W.R.: 1982, *J. Geophys. Res.* **87**, No. C12, 9425–9440.
460. Schmitz, W.J., Robinson, A.R., and Fuglister, F.C.: 1970, *Science* **170**, No. 3963, 1192–1194.
461. Schott, R. and Bock, M.: 1980, *J. Geophys. Res.* **85**, No. C7, 4007–4014.
462. Schroeder, E. and Stommel, H.: 1969, *Progress in Oceanography* **5**, 31–40.
463. Sciremammano, F.: 1980, *J. Phys. Oceanogr.* **10**, No. 6, 843–852.
464. Sciremammano, F., Pillsbury, R.D., Nowlin, W.D., and Whitworth, T.: 1980, *J. Geophys. Res.* **85**, No. C7, 4015–4028.
465. Scott, A.S., Chu, F.Y.F., and McLaughlin, D.W.: 1973, *Proc. IEEE* **61**, No. 10, 1443–1483.
466. Seckel, G.: 1975, *Deep-Sea Res.* **22**, No. 6, 379–401.
467. Seidov, D.G.: 1978, *Izv. Akad. Nauk SSSR Fiz. Atm. Okeana* **14**, No. 7, 757–767 (*in Russian*).
468. Seidov, D.G.: 1980, *Izv. Akad. Nauk SSSR Fiz. Atm. Okeana* **16**, No. 1, 73–87 (*in Russian*).
469. Seidov, D.G.: 1981, *Chislennoe issledovanie krupnomasshtabnoĭ tsirkulyatsii i techeniĭ sinopticheskogo masshtaba v okeane* (*Numerical Investigation of Large-Scale Circulation and Synoptic-Scale Currents in the Ocean*), P.P. Shirshov Institute of Oceanology, USSR Academy of Sciences; Doctoral thesis, (*in Russian*).
470. Seidov, D.G.: 1985, in *Issledovanie techeniĭ okeana* (*Investigation of Ocean Currents*), Nauka, Moscow, pp. 28–46 (*in Russian*).
471. Semtner, A.J. and Holland, W.R.: 1978, *J. Phys. Oceanogr.* **8**, No. 5, 735–754.
472. Seyler, C.E., Salu, Y., Montgomery, D., and Knorr, G.: 1975, *Phys. Fluids* **18**, No. 7, 803–813.
473. Shank, M., Matthews, W., Crumper, A., Cheney, R., and Cecelski, A.: 1975, *Nav. Res. Rev.* **28**, No. 6, 1–12.
474. Shirshov, P.P.: 1944, *Nauchnye resul'taty dreifa stantsiĭ 'Severnyĭ polyus'* (*Scientific Results of the Drift of the Station 'North Pole'*), Doklad na obshchem sobr. AN SSSR, Moscow; 14–17 fevralya 1944 g. (*in Russian*).
475. Shoji, D.: 1972, in *Kuroshio, Physical Aspects of the Japan Current* (eds H. Stommel and K. Yoshida), Univ. of Washington Press, Seattle, pp. 217–234.
476. Shtokman, V.B. and Ivanovskiĭ, I.I.: 1937, *Meteorol. Gidrol.*, Nos. 3–5, 154–160 (*in Russian*).
477. Shtokman, V.B., Koshlyakov, M.N., Ozmidov, R.V., Fomin, L.M., and Yampol'skiĭ, A.D.: 1969, *Doklady Akad. Nauk SSSR* **186**, No. 5, 1070–1073 (*in Russian*).
478. Sieverse, H.A. and Emery, W.J.: 1978, *J. Geophys. Res.* **83**, No. C6, 3010–3022.
479. Sizov, A.A., Liverdy, V.P., Nagornova, G.G., and Shenogina, Z.B., in: *Eksperimental'nye issledovaniya po mezhdunarodnoy programme "POLYMODE"*. (Experimental investigations by the international program "*POLYMODE*") (Sebastopol: Marine Hydrophysical Institute; 1978): 26-32 (*in Russian*).
480. Sklyarov, V.E. and Fedorov, K.N.: 1980, *Issledovanie Zemli iz kosmosa*, No. 3, 5–13 (*in Russian*).
481. Sokolov, V.A. and Tikhonov, V.A.: 1984, *Issledovaniya protsessov vzaimodeistvija okeana i atmosfery* (*Investigation of the Interaction Processes between the Ocean and the Atmosphere*), Gidrometeoizdat, Moscow, pp. 136–143.
482. Solomon, H. and Ahlnäs, K.: 1978, *Deep-Sea Res.* **25**, No. 4, 403–410.
483. Soomere, T.E.: 1983, *Ob evolyutsii spektra energii sistemy slabonelineino-vzaimodeistvuyushchikh voln Rossbi so sluchainymi amplitudami* (*On the Evolution of the Energy Spectrum of a System of Weakly-Nonlinear Interacting Rossby Waves with Random Amplitudes*), Academy of Sciences of the Estonian SSR, Tallin (Preprint), (*in Russian*).
484. Spain, D.L., O'Gara, R.M., and Rossby H.T.: 1980, *SOFAR Float Data Report of the POLYMODE Local Dynamics Experiment*, Technical Report No. 80–1, Univ. of Rhode Island.
485. Spiendel, R.C. and Spiesberger, J.L.: 1981, *JASA* **69**, No. 4, 982–988.
486. Spiesberger, J.L. and Worcester, P.F.: 1981, *JASA* **70**, No. 2, 565–576.
487. Starr, V.P.: 1968, *Physics of Negative Viscosity Phenomenon*, McGraw-Hill, New York.

488. Starr, V.P. and Long, R.R.: 1953, *USAF Geophys. Res. Paper* **24**, No. 2, 103–113.
489. Steinberg, J.G., Clark, J.G., De Ferrary, H.A., Kronengold, M., and Yacoub, K.: 1972, *JASA* **52**, No. 5, 1521–1535.
490. Stepanov, V.N.: 1969, *Okeanologiya* **9**, No. 3, 387–397 (*in Russian*).
491. Stepanov, V.N.: 1974, *Mirovoĭ okean. Dinamika i svoistva vod* (*World Ocean. Dynamics and Properties of the Waters*), Znanie, Moscow (*in Russian*).
492. Stepanov, V.N., Sarkisyan, A.S., Demin, Yu.L., Drozdov, V.N., Bulatov, R.P., Galerkin, L.I., and Shcherbinin, A.D.: 1977, *Morskie Gidrofiz. Issled.* No. 1 (76), 34–46 (*in Russian*).
493. Stern, M.E.: 1975, *J. Mar. Res.* **33**, No. 1, 1–13.
494. Stommel, H.: 1965, *The Gulf Stream. A Physical and Dynamical Description* (2nd edn), Univ. of California Press, Berkeley.
495. Stommel, H., Meinke, J., and Zenk, W.: 1977, *POLYMODE News*, No. 22 (unpublished manuscript).
496. Stommel, H., Niiler, P., and Anati, D.: 1978, *J. Mar. Res.* **36**, No. 3, 449–468.
497. Stumpf, H. and Legeckis, R.V.: 1977, *J. Phys. Oceanogr.* **7**, No. 5, 648–658.
498. Sugiura, J.: 1957, *Oceanogr. Mag.* **9**, No. 1, 121–131.
499. Swallow, J.C.: 1971, *Phil. Trans. R. Soc. Lond.* **A270**, No. 1206, 451–460.
500. Swallow, J.C. and Fieux, M.: 1982, *J. Mar. Res.* **40**, Supplement, 747–755.
501. Swallow, J.C. and Hamon, B.V.: 1960, *Deep-Sea Res.* **6**, No. 2, 155–168.
502. Tabata, S.: 1982, *J. Phys. Oceanogr.* **12**, No. 11, 1260–1282.
503. Taft. B., Ebbesmeyer, C., McWilliams, J., Sayles, M., and Shen. C.: 1978, *POLYMODE News*, No. 51 (unpublished manuscript).
504. Tchernia, P. and Jeannin, P.: 1980, *Deep-Sea Res.* **27A**, No. 6A, 467–474.
505. The MODE Group,: 1978, *Deep-Sea Res.* **25**, No. 10, 859–910.
506. Thompson, R.O.R.Y.: 1971, *Deep-Sea Res.* **18**, No. 1, 1–19.
507. Thompson, R.O.R.Y.: 1977, in *Progress in Oceanography*, Vol. 7, Pergamon Press, pp. 135–162.
508. Thompson, R.O.R.Y. and Luyten, J.R.: 1976, *Deep-Sea Res.* **23**, No. 7, 629–635.
509. Timofeev, N.A.: 1970, *Atlas teplovogo balansa okeanov* (*Atlas of the Heat Balance of the Oceans*), MGI, Sebastopol (*in Russian*).
510. Titov, V.B., Zubin, A.B., and Dzhiganshin, G.F.: 1979, *Okeanologiya* **19**, No. 1, 27–30 (*in Russian*).
511. Tranter, D.I., Parker, R.R., and Cresswell, G.R.: 1980, *Nature* **284**, 10 April, 540–542.
512. Trenberth, K.E.: 1979, *Dyn. Atm. Oceans* **4**, 57–64.
513. Tsybaneva, T.B.: 1984, in *Tonkaya struktura i sinopticheskaya izmenchivost' moreĭ i okeanov. Tezisy dokladov II Vsesoyuzn. simposiuma. Chast' II. Tallin-1984* (*Fine Structure and Synoptic Variability of Seas and Oceans. Abstracts of the 2nd All-Union Symposium. Part 2. Tallin, 1984*), Tallin, pp. 181–182 (*in Russian*).
514. Turpaeva, E.P. and Yampol'skiĭ, A.D.: 1979, *Okeanologiya* **19**, No. 6, 1116–1123 (*in Russian*).
515. Vasilenko, V.M. and Mirabel', A.P.: 1977, *Okeanologiya* **17**, No. 6, 962–966.
516. Vasilenko, V.M. and Mirabel', A.P.: 1980, *Fiz. Atm. Okeana* **16**, No. 12, 1326–1330 (*in Russian*).
517. Vasilenko, V.M., Mirabel', A.P., and Ozmidov, R.V.: 1976, *Okeanologiya* **16**, No. 1, 55–60 (*in Russian*).
518. Vastano, A.C. and Owens, G.E.: 1973, *J. Phys. Oceanogr.* **3**, No. 4, 470–478.
519. Volosov, V.M.: 1976a, *Okeanologiya* **16**, No. 3, 389–396 (*in Russian*).
520. Volosov, V.M.: 1976b, *Okeanologiya* **16**, No. 5, 741–749 (*in Russian*).
521. Volosov, V.M.: 1976c, *Okeanologiya* **16**, No. 6, 949–960 (*in Russian*).
522. Volosov, V.M.: 1982, *Okeanologiya* **22**, No. 2, 165–169 (*in Russian*).
523. Volosov, V.M. and Zhdanov, M.A.: 1980a, *Okeanologiya*, **20**, No. 1, 5–10 (*in Russian*).
524. Volosov, V.M. and Zhdanov, M.A.: 1980b, *Okeanologiya* **20**, No. 4, 581–589 (*in Russian*).
525. Volosov, V.M. and Zhdanov, M.A.: 1982, *Okeanologiya* **22**, No. 5, 698–704 (*in Russian*).
526. Volosov, V.M. and Zhdanov, M.A.: 1983, *Okeanologiya* **23**, No. 2, 204–209 (*in Russian*).
527. Vonder Haar, T.H. and Oort, A.H.: 1973, *J. Phys. Oceanogr.* **3**, No. 2, 169–172.
528. Voorhis, A.D.: 1969, *Deep-Sea Res.* **16**, Supplement, 331–337.
529. Voorhis, A.D., Schroeder, E.H., and Leetman, A.: 1976, *J. Phys. Oceanogr.* **6**, No. 6, 953–961.
530. Voskresenskaya, E.N. and Kosnyrev, V.K.: 1979, *Morskie Gidrofiz. Issled.* **2**, 26–36 (*in Russian*).

531. Vukovich, F.M. and Crissman, B.W.: 1978, *J. Phys. Oceanogr.* **8**, No. 5, 838–845.
532. Vukovich, F.M., Crissman, B.W., Bushnell, M., and King, W.J.: 1979, *J. Phys. Oceanogr.* **9**, No. 11, 1214–1222.
533. Vulis, I.L. and Monin, A.S.: 1975, *Doklady Akad. Nauk SSSR* **221**, No. 3, 597–600 (*in Russian*).
534. Wadhams, P. and Squire, V.A.: 1983, *J. Geophy. Res.* **88**, No. C5, 2770–2720.
535. Wadhams, P., Gill, A.E., and Linden, P.F.: 1979, *Deep-Sea Res.* **26A**, No. 12A, 1311–1327.
536. Ward, F.: 1966, *Astrophys. J.* **141**, No. 2, 534–547.
537. Warren, B.A.: 1967, *Deep-Sea Res.* **14**, No. 5, 505–524.
538. Weidman, P.D. and Maxworthy, T.: 1978, *J. Fluid Mech.* **85**, Part 3, 417–431.
539. Weinberg, N.L. and Zabalgogeazcoa, X.: 1977, *JASA* **62**, No. 4, 888–894.
540. Wiebe, P.H.: 1976, *Oceanus* **19**, No. 3, 69–76.
541. Wiebe, P.H.: 1982, *Scient. Amer.* **246**, No. 3, 40–60.
542. Wiebe, P.H. and Boyd, S.H.: 1978, *J. Mar. Res.* **56**, No. 1, 119–142.
543. Wiebe, P.H., Hulburt, E.M., Carpenter, E.J., Jahn, A.E., Knapp, G.P., Boyd, S.H., Orthner, P.B., and Cox, J.L.: 1976, *Deep-Sea Res.* **23**, 695–710.
544. Wiin-Nielsen, A.: 1967, *Tellus* **19**, No. 4, 540–559.
545. Willebrand, J.: 1978, *J. Phys. Oceanogr.* **8**, No. 6, 1080–1094.
546. Willebrand J. and Meincke, J.: 1980, *Deep-Sea Res.* **27**, No. 12A, 1047–1066.
547. Willebrand, J., Philander, S.G.H., and Pacanowski, R.C.: 1980, *J. Phys. Oceanogr.* **10**, No. 3, 411–429.
548. Williams, G.P.: 1978, *J. Atm. Sci.* **35**, No. 8, 1399–1426.
549. Williams, G.P.: 1979, *J. Atm. Sci.* **36**, No. 5, 932–968.
550. Wilson, W.S. and Dugan, J.P.: 1978, *J. Phys. Oceanogr.* **8**, No. 3, 537–540.
551. Woods, J.D.: 1980, *Nature* **288**, No. 5788, 219–224.
552. Worcester, P.F.: 1981, *JASA* **70**, No. 6, 1743–1747.
553. Worthington, L.W.: 1976, *The Johns Hopkins Oceanographic Studies* **6**.
554. Wunsch, C.: 1972, *Deep-Sea Res.* **19**, No. 8, 577–593.
555. Wyrtki, K.: 1967, *Deutsche Hydr. Z.* **20**, Heft 4, 176–186.
556. Wyrtki, K.: 1982, *J. Phys. Oceanogr.* **12**, No. 7, 746–749.
557. Wyrtki, K., Magaard, L., and Hager, J.: 1976, *J. Geophys. Res.* **81**, No. 15, 2641–2646.
558. Yaremchuk, M.I.: 1982, *Izv. POLIMODE*, No. 8, 59–67 (*in Russian*).
559. Yasui, M.: 1961, *Oceanogr. Mag.* **12**, 157.
560. Zantopp, R. and Leaman, K.: 1982, *J. Geophys. Res.* **87**, No. C3, 1927–1934.
561. Zhidkov, V.I. and Dzhiganshin, G.F.: 1978, in *Eksperimental'nye issledovaniya po mezhdunarodnoĭ programme POLIMODE* (*Experimental Studies According to the International Program POLYMODE*), MGI AN Ukr. SSR, Sebastopol, pp. 128–135 (*in Russian*).
562. Zyryanov, V.N.: 1981, *Izv. Akad. Nauk SSSR Fiz. Atm. Okeana* **17**, No. 10, 1072–1083 (*in Russian*).

Subject Index

GPSR Compliance

The European Union's (EU) General Product Safety Regulation (GPSR) is a set of rules that requires consumer products to be safe and our obligations to ensure this.

If you have any concerns about our products, you can contact us on ProductSafety@springernature.com

In case Publisher is established outside the EU, the EU authorized representative is:

Springer Nature Customer Service Center GmbH
Europaplatz 3
69115 Heidelberg, Germany

Batch number: 10755284

Printed by Printforce, the Netherlands